Gas Installation Technology

Gas Installation Technology

Third Edition

Andrew S. Burcham CertEd, MCIPHE, Eng Tech MIGEM
Colchester Institute
Colchester, UK

Stephen J. Denney CertEd, MCIPHE, M.W.M.Soc (Snr)
Colchester Institute
Colchester, UK

Roy D. Treloar

WILEY Blackwell

Edition History
John Wiley & Sons, Ltd (2e, 2010)
Blackwell Publishing Ltd (1e, 2005)

Registered Office(s)
John Wiley & Sons, Inc., 111 River Street, Hoboken, NJ 07030, USA
John Wiley & Sons Ltd, The Atrium, Southern Gate, Chichester, West Sussex, PO19 8SQ, UK

For details of our global editorial offices, customer services, and more information about Wiley products visit us at www.wiley.com.

Library of Congress Cataloging-in-Publication Data

Names: Burcham, Andrew S., author. | Denney, Stephen J., author. | Treloar,
    Roy D., author.
Title: Gas installation technology / Andrew S. Burcham, Colchester
    Institute, Colchester, UK, Stephen J. Denney, Colchester Institute,
    Colchester, UK, Roy D. Treloar.
Description: Third edition. | Hoboken, NJ, USA : Wiley-Blackwell, 2024. |
    Revised edition of: Gas installation technology / R.D. Treloar, 2nd ed.
    Chichester, West Sussex, U.K. ; Ames, Iowa : Wiley-Blackwell, 2010.
Identifiers: LCCN 2023047916 (print) | LCCN 2023047917 (ebook) | ISBN
    9781119908180 (PB) | ISBN 9781119908197 (ePDF) | ISBN 9781119908203
    (epub)
Subjects: LCSH: Gas-fitting. | Gas appliances–Installation.
Classification: LCC TH6810 .T74 2024 (print) | LCC TH6810 (ebook) | DDC
    696/.2–dc23/eng/20231208
LC record available at https://lccn.loc.gov/2023047916
LC ebook record available at https://lccn.loc.gov/2023047917

Cover Design: Wiley
Cover Image: © KTPhoto/Getty Images

Set in 9.5/12.5pt STIXTwoText by Straive, Chennai, India
Printed and bound by CPI Group (UK) Ltd, Croydon, CR0 4YY

C9781119908180_020524

Contents

Introduction

The highly respected second edition of Gas Installation Technology was completed in 2010, and as with any technical work, revisions are needed in light of new thinking, industry updates, advances and innovations. This third edition builds on the fine foundation of the work completed by Roy Treloar and continues to cover all areas of the gas industry that operatives are likely to encounter.

In addition to being fully updated to current standards, the layout has been revised and includes a number of additions. An appendix has been included along with a comprehensive index and section-numbered text for ease of referencing. Commercial Catering and Commercial Laundry have been given their own dedicated parts, and there are additional parts focusing on Educational Establishments and the ACS Assessment Process. Full-colour photographs are now displayed throughout.

The information contained within continues to purposely group together all the various aspects of gas work to include both natural gas and LPG within domestic and commercial installations. This overcomes the problem of repeating topics and allows all relevant information to be contained within one volume. There are variances to this, hence the need to sometimes identify specific installation types and procedures, and this has been done where necessary.

Although many recent industry standards and documents have adopted new style unit symbols in their specification text, we have continued to use the more familiar original notations throughout this book, as shown below and defined in the Appendix:

New style unit symbols	Original style as used within this book
$m\ s^{-1}$	m/s
$m^3\ h^{-1}$	m^3/h
$m^3\ s^{-1}\ m^2$	m^3/s/m^2

The book is not designed to be read from cover to cover, and the reader will invariably need to dip into it to retrieve information on a specific problem or interest. A particular topic of interest can be found in one of the following ways:

- First, by referring to the contents page which identifies the subject areas of the book.
- Second, by referring to the index and choosing a term related to the subject in question.

This book will be beneficial to both new entrants to the gas industry and highly experienced engineers alike. It will prove an invaluable on-site guide for all gas engineers, and we hope that it will also prove useful during training and in preparation for your ACS assessments.

It should be noted that Building Regulations differ depending on where you are working in the United Kingdom. As it has not been possible to list all regional differences, this publication primarily refers to Building Regulations (England). Therefore, operatives should ensure that they are working to the correct regulations by contacting their Local Authority Building Control.

With the completion of this revised edition, it is our sincere hope and expectation that *Gas Installation Technology* will continue to be widely used and respected by all sectors of the gas industry.

Stephen Denney/Andrew Burcham

Acknowledgements

We would like to acknowledge Roy Treloar for his work in the first two editions of Gas Installation Technology without which this edition would not exist. We also express our gratitude to Colchester Institute, Leslie Bennett, Carmine Sagnella, Jemma Hyde and Geoffrey Eaton who assisted us in various ways during the production of this publication. Also, to Ian Cook, Kai Sillery and Andy Lord for contributing personal photographs displayed in Parts 2 and 9.

Our thanks also go to the following companies and organisations for their kind permission to reproduce photographs which are new to this edition: Alde International (UK) Ltd; Gas Safe Register; Johnson and Starley Ltd; Lifestyle Appliances Ltd; Maywick Ltd; Powrmatic Ltd; Rinnai UK; Titan Products Ltd; Truma Ltd.

We generously acknowledge that data in many of the tables within this book include extracts from various publications produced by organisations such as the British Standards Institute, the Institution of Gas Engineers and Managers, and the Building Engineering Services Association. These source documents have been referenced at the start of each relevant section.

Stephen Denney/Andrew Burcham

1

The Gas Industry

The Gas Industry

1.1   The gas industry has gone through major changes in the past few decades. Prior to, and during, the early 1960s most gas installation work in the UK was undertaken by British Gas. In 1968, a 22-storey block of flats in Canning Town, East London, was devastated by a major gas explosion, which persuaded the industry that a body was needed to oversee this kind of work. As a result, in 1970 a voluntary gas body was formed, called the Confederation of Registered Gas Installers (CORGI).

During the early 1970s, plumbers and heating engineers began to take a greater interest in undertaking gas work, thanks to central heating systems becoming a requirement in the average home. In 1972 the first Gas Safety Regulations were introduced, which identified the legal responsibilities to which the installer had to adhere.

With the introduction of the Approved Code of Practice (ACOP) in 1990, gas installers started to take updated training and assessments in gas working practices. By 1991, anyone working in the gas industry for financial gain had to be registered with a Health and Safety Executive (HSE) approved body.

From 1991 through to 2009 CORGI (renamed the Council for Registered Gas Installers) held the register, which all gas engineers needed to be a member of. On 1 April 2009, under an agreement with the HSE, the Gas Safe Register was launched and replaced CORGI as the only gas safety registration body in Great Britain and Northern Ireland. The Gas Safe Register also operates in the Isle of Man, Jersey and Guernsey and works on behalf of the relevant Health and Safety Authorities in those regions.

Since 1991 all gas engineers have needed to be assessed as competent in the aspect of gas work that they wish to undertake; undertaking any work without this assessment would mean that they are in breach of the law and working illegally. The assessment that an individual undertakes is called the Nationally Accredited Certification Scheme for Individual Gas Fitting Operatives (ACS). There are many different assessments, and these are listed in Section 1.5.

Gas Installation Technology, Third Edition. Andrew S. Burcham, Stephen J. Denney and Roy D. Treloar.
© 2024 John Wiley & Sons Ltd. Published 2024 by John Wiley & Sons Ltd.

Working in the Gas Industry

1.2   Today, if you wish to work in the gas industry, you must belong to a Gas Safe registered company or become registered in your own right. Becoming registered is no easy task, and currently there are really only two options available:

Option 1

Gain a qualification such as a Level 3 Diploma in Gas Engineering or a Level 3 Plumbing and Domestic Heating diploma with a gas pathway.

These qualifications are usually linked to an apprenticeship and are vocational, meaning employment with a Gas Safe registered company is necessary. These qualifications require the completion of in-centre and work-based training. This provides the knowledge, skills and experience individuals need to apply for Gas Safe registration after completion of the knowledge and practical assessments associated with the course. In addition, you will also be required to build an 'on the job' portfolio of work-based evidence whilst working under the direct supervision of a registered gas engineer. Only after you have successfully completed your qualification can you apply for Gas Safe registration.

Option 2

Enrol in an industry-recognised training programme that meets the requirements of IGEM/IG/1 'Standards of training in gas work'.

This standard, first published on 1 April 2014 by the Institution of Gas Engineers and Managers (IGEM), provides the agreed scope and structure of training for new entrants to the gas industry and will enable you to gain experience in the areas of gas work that you wish to undertake. The course will be an extended Managed Learning Programme (MLP) and will consist of a minimum number of classroom-based guided learning hours.

You will also need to complete a minimum number of hours 'on the job' training whilst working under the direct supervision of a Gas Safe registered engineer who is willing to sponsor and train you, acting as your mentor. During this period, you will need to complete a portfolio with authenticated evidence of installations that you have completed, demonstrating the relevant knowledge, skills and experience you have gained whilst working under the guidance of the registered installer.

You will then complete a range of theoretical and practical assessments at the training centre to prove your knowledge, skills and understanding of matters relating to gas safety. Upon successful completion of both on and off-site requirements, you will be issued with a certificate of training.

With your certificate of training, you can now apply to undertake the written and practical ACS assessments relevant to your MLP training course. Only after completing your MLP and passing your ACS assessments can you obtain work with a Gas Safe registered business or apply to become registered in your own right.

The contact details for the HSE gas registration scheme are:
Gas Safe Register
Website: www.gassaferegister.co.uk
Email: enquiries@gassaferegister.co.uk
Phone: Engineers: 0800 408 5577; Consumers: 0800 408 5500

Regardless of which option you choose, if in the future you would like to extend the scope or range of gas work undertaken, you will first need to provide evidence to the training centre that you have Gas Safe registration. There must also be a minimum period of 12 months since completion of your original core gas safety qualification or ACS. You will then be required to undertake further classroom-based training on your chosen subject before attempting the relevant ACS assessments.

Where an engineer wishes to extend the scope or range of gas work within 12 months of completing their core qualification or ACS, it is possible to undertake an MLP – Bridge with a recognised training provider. The training duration and content will follow the specification set out in IGEM/IG/1 for the relevant extensions. Gas Safe registration will be required.

Gas Safe Register

1.3 The Gas Safe Register is the only registration body for businesses and individuals working within the gas industry in Great Britain, Northern Ireland, Isle of Man, Jersey and Guernsey. By law, all individuals and businesses working with gas in these regions must be on the Gas Safe Register. Each operative is issued with a personal licence number but may not necessarily have their own individual registration. Often it is the business that is registered, but registration will only be granted providing the business employs at least one qualified gas engineer. All gas engineers that a business employs must be listed on the Gas Safe Register, which is publicly available for consumers to verify online.

As further proof that an operative is qualified and on the register, Gas Safe issues an identification card annually to all those registered. The operative should carry this as proof of registration and is encouraged to show this to their customers. On the reverse of the card is a list of the work categories that the operative is allowed to perform, see Figure 1.1.

Gas Safe carryout inspections of the work performed by businesses on the register to ensure compliance with the law. Gas Safe will also investigate gas safety complaints and, where necessary, apply sanctions for unsafe work or where there have been breaches to the rules of registration.

(a)

Figure 1.1 (a) An example of the Gas Safe Register ID card (front). (b) An example of the Gas Safe Register ID card (rear). Credit: Gas Safe Register

Domestic	Gas	LPG	Non-Domestic	Gas	LPG
Pipework	31/03/23	31/03/23			
Cooker	31/03/23	31/03/23			
Fire	31/03/23	31/03/23			
Water Heater	31/03/23	31/03/23			
Gas Boiler	31/03/23	31/03/23			
Comb Analysis	31/03/23	31/03/23			
The cardholder is deemed competent only in the categories of work identified by a date.					**3456789**

(b)

Figure 1.1 (*Continued*)

Although registration has been a legal requirement since 1991, there are still regular instances of individuals working illegally and putting people's lives at risk by leaving unsafe gas installations. Gas Safe play a critical role in investigating this illegal work and will pass investigation reports of illegal gas work performed by un-registered individuals and businesses to the HSE for further investigation and possible prosecution.

Working Without Registration

1.4 By carefully examining the Gas Safety (Installation and Use) Regulations you will find that it is possible to undertake some gas work without the need for registration, but these circumstances are very limited. For example, Regulation 2 lists areas to which the Gas Regulations do not apply. These include certain work within factories, mines, quarries, agricultural premises etc. However, when working in such premises, competence in gas work is still required in order to comply with the requirements of the Health and Safety at Work Act. In addition, the Gas Regulations stipulate that whoever works on a gas fitting or installation must be competent to do so whether or not they are required to be an approved member of a class of persons (Gas Safe). This would also apply to do-it-yourself gas work or those performing favours for friends and family. The law requires that an individual who intends to carry out these types of gas work still needs to be competent. This can be proved by successfully completing an appropriate training course followed by an assessment of competence sufficient for, and relevant to, the type of work being undertaken.

Nationally Accredited Certification Scheme for Individual Gas Fitting Operatives (ACS)

1.5 In order for gas operatives to prove competence and work in a particular aspect of the profession, they need to have undertaken the appropriate ACS gas assessment. There are many different assessments that are applicable to domestic, commercial, natural gas and liquified petroleum gas (LPG) installations; there are also some specialist and service provider assessments. In addition to the list of assessments identified below, there is a range of changeover assessments, providing conversions between the various core assessments, e.g. domestic to commercial (CCN1 to COCN1). These are listed under *changeover assessments* in Section 1.6.

The following specific ACS Assessments are organised alphabetically under these categories:

- **Domestic**
- **Emergency Service Provider/Meter Installations**
- **LPG**
- **Non-Domestic**

Domestic

CCN1 – Core domestic natural gas safety
CENWAT – Domestic central heating boilers and instantaneous water heaters ($\leq$70 kW)
CKHB1 – Domestic gas range cooker/boilers
CKR1 – Domestic gas cooking appliances
CMDDA1 – Carbon monoxide/carbon dioxide atmosphere and appliance testing
CPA1 – Combustion performance analysis
DAH1 – Domestic gas ducted air heaters ($\leq$70 kW)
DFDA1 – Appliances with forced draught burner in domestic dwellings
EFJLP1 – Domestic/non-domestic electro-fusion jointing of PE pipework and fittings
HTR1 – Domestic gas fires and wall heaters
HWB1 – Gas fired swimming pool boilers
LAU1 – Domestic gas laundry appliances
LEI1 – domestic gas leisure/miscellaneous appliances (Barbecues, greenhouse heaters, refrigerators, gas pokers, gas lighting and patio heaters)
MET1 – Domestic gas meters up to 6 m^3/h
WAT1 – Domestic gas instantaneous water heaters

Emergency Service Provider/Meter Installations

CESP1 – Core domestic and non-domestic emergency service provider gas work
CMA1 – Core domestic and non-domestic gas metering work
CMA2 LS – Core domestic limited scope meter work
CMA3 – Gas meter installer domestic natural gas with a maximum capacity not exceeding 6 m^3/h
CMET1 – Natural gas low-pressure diaphragm and rotary displacement meter installations
CMET2 – Natural gas diaphragm, rotary displacement and turbine meters not exceeding 7 bar
CMIT1 – Emergency service provider limited scope gas work on meter instrumentation
MET3 LS – Domestic natural gas meter installations with a capacity not exceeding 6 m^3/h (limited scope)

MET4 –Non-domestic diaphragm gas meter installations $\leq 40\,\text{m}^3/\text{h}$

REGT1 – Domestic medium pressure meter regulator installation and commission

REGT2 – Non-domestic medium pressure regulators and controls installation and commission

LPG

CCLP1 – Core LPG gas safety

CCLP1 B – Core LPG gas safety: Boats, yachts and other vessels

CCLP1 EP – Core domestic LPG gas safety for external pipework – limited scope (Must be taken in conjunction with VESLP1)

CCLP1 EPC – Domestic LPG external pipework connections

CCLP1 LAV – Core LPG gas safety: Leisure accommodation vehicles

CCLP1 MC & CABLP1 – Limited scope core gas safety for LPG mobile cabinet heaters & commission service, repair and break down domestic butane gas-fired mobile cabinet heaters

CCLP1 PD – Core LPG gas safety: Permanent dwellings

CCLP1 RPH – Core LPG gas safety: Residential park homes

EFJLP1 – Domestic/Non-domestic electro-fusion jointing of PE pipework and fittings

HTRLP2 – Domestic LPG closed flue gas fires

HTRLP3 – Domestic LPG caravan space heaters

LEILP1 – LPG single bottle supply leisure equipment (barbecues, greenhouse heaters, gas lighting and patio heaters)

REFLP2 – Domestic LPG refrigerators in motorised and touring caravans, and boats

VESLP1 – LPG single gas storage vessels & service pipework

VESLP2 – LPG single & multi supply gas storage vessels and service pipework

WAHLP1 – Domestic LPG boat/caravan warm air heaters

WATLP2 – Domestic LPG caravan gas water heaters

Non-Domestic

BMP1 – Gas boosters operating pressure up to 0.5 bar pressure

CCCN1 – Natural gas commercial catering core safety

CCLNG1 – Natural gas commercial laundry core safety

CCP1 – Commissioning plant and equipment

CDGA1 – Direct-fired commercial heating appliances and equipment

CGFE1 – Gas fuelled engines

CGLP1 – LPG-fired generators

CIGA1 – Indirect commercial gas-fired heating appliances and equipment

CLE1 – Commercial laundry appliances

CMC1 – Commercial mobile catering

COCN1 – Natural gas commercial heating appliances core safety

COCNPI1 LS – Natural gas core commercial gas safety pipework installer/commissioner (limited scope)

COMCAT1 – Commercial catering appliances with atmospheric burners, e.g. boiling burners, open/solid top (Ranges)

COMCAT2 – Commercial catering appliances with atmospheric burners, e.g. pressure type water boilers (pressure steamers, pressure steaming ovens)

COMCAT3 – Commercial catering appliances deep fat and pressure fryers

COMCAT4 – Commercial catering appliances fish and chip ranges

COMCAT5 – Commercial catering appliance forced draught burner appliances

CORT1 – Commercial overhead radiant plaque and tube heaters

EFJLP1 – Domestic/Non-domestic electro-fusion jointing of PE pipework and fittings (Natural gas & LPG)

ICAE1 LS – Limited scope commercial first fix appliances and equipment

ICPN1 – First fix commercial pipework

ICPN1 LS – First fix commercial pipework limited scope

TPCP1 – Testing and purging commercial gas pipework $>1\,m^3$ with MOP ≤ 16 bar

TPCP1A – Testing and purging low-pressure commercial gas pipework $\leq 1\,m^3$ with MOP ≤ 40 mbar at outlet of the primary meter regulator

Changeover Assessments

1.6 In order to undertake an assessment in the particular work category that you require, e.g. CENWAT (domestic boilers/water heaters), you also need to hold the specific core assessment (CCN1). This is because the core assessment is a prerequisite to all appliance assessments.

There are several core assessment categories and many have similar assessment criteria. Therefore, it is not always necessary to undertake the complete core for a specific range of appliances as a changeover assessment can be obtained. These include:

CoCATA – Changeover core domestic to non-domestic catering appliances

CoCCLNG1 – Changeover core domestic to non-domestic core laundry

CoCDN1 – Changeover non-domestic to domestic natural gas

CoDC1 – Changeover core domestic to catering core

CoDNCO1 – Changeover core domestic gas to non-domestic core gas safety

CoDNESP1 – Changeover core domestic gas to emergency service provider &/or gas meter installer

CoLPNG1 – Changeover domestic LPG to natural gas

CoNGLP1 – Changeover domestic natural gas to LPG – generic

CoNGLP1 CMC – Changeover non-domestic natural gas to LPG mobile catering

CoNGLP1 B – Changeover domestic natural gas to LPG – boats

CoNGLP1 LAV – Changeover domestic natural gas to LPG – leisure accommodation vehicles

CoNGLP1 PD – Changeover domestic natural gas to LPG – permanent dwellings

CoNGLP1 RPH – Changeover domestic natural gas to LPG – residential park homes

Importance of Maintaining a Current Core Assessment

1.7 Because the core is a mandatory requirement for any further appliance assessment it must always be maintained as a valid and current assessment certificate. Assessments are valid for five years from the date of issue. Should the core certification run out, then the validation of all other certificates received after the date of the core, including changeover, ceases, even though they may have time to run. As soon as the core category has been re-assessed and the assessment passed, the other certificates become valid again.

Working Within Scope of ACS Work Categories

1.8 It is important that engineers recognise the limits of their competence and remain within the scope of their ACS work categories. Questions may arise about the qualifications needed to work on domestic appliances in a commercial environment. In this situation gas engineers will need to be particularly careful not to step outside the scope of their qualifications. Gas engineers should risk assess the situation and confirm the following before deciding to work on a domestic appliance in non-domestic premises:

- The appliance is designed for domestic use and the manufacturer allows the installation in this setting.
- The appliance can be isolated from the gas supply by its own appliance isolation valve.
- The pipework downstream of an isolation valve is within the scope of a domestic engineer (see Part 5: Section 5.1).
- The appliance can be used independently and does not incorporate other appliances, such as a modular boiler installation.
- The appliance is not within a commercial catering environment.

Where any of these considerations cannot be confirmed, a suitable commercial qualification will be required.

Consider the following examples.

1. An operative holding CCN1 and CENWAT can work on a domestic boiler in commercial premises, providing the following conditions are met: The meter is no larger than a U16/G10; the installation does not exceed 35 mm in diameter and 0.035 m³ in volume; and no other non-domestic appliances are installed. Alternatively, where the gas supply can be isolated by means of an isolation valve and the installation from the isolation valve does not exceed 35 mm in diameter and 0.035 m³ in volume, the engineer can work on the boiler downstream from this valve.
2. An operative holding a commercial core can work on a domestic boiler in commercial premises if they also hold CENWAT. In these circumstances, they will not be required to hold the domestic core. However, they cannot work in domestic premises unless they obtain the domestic core. Likewise, where a commercial catering operative wishes to work on a domestic type of cooker in a commercial catering setting, they would need to acquire CKR1. In addition to CKR1, the domestic core would be required if they wanted to work on the appliance if fitted within a domestic property.
3. Where a domestic gas cooker is installed in a commercial catering setting along with other catering appliances, the installation must comply with BS 6173. So, although the appliance is of a domestic type, the gas operative would need to hold a suitable commercial catering qualification as well as CKR1. This would also apply to domestic-type cookers installed in food technology classrooms within educational establishments (see Parts 12 and 13).

Required ACS Assessments and Flowcharts

Individual ACS Requirements

1.9 At first sight the vast list of ACS assessment criteria can look quite daunting and it is difficult to choose which assessment to undertake. However, this is can be approached by using the flowcharts shown in Figures 1.2–1.5.

Figure 1.2   Domestic natural gas and LPG ACS assessments

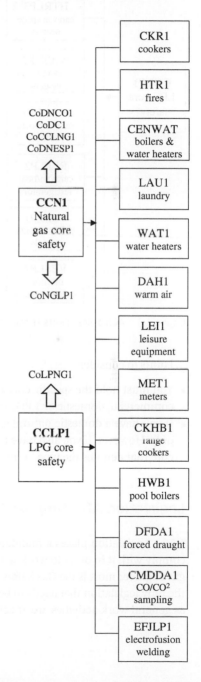

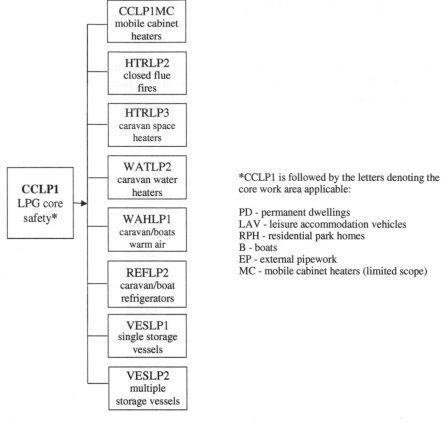

Figure 1.3   ACS assessments specific to LPG

Points to consider:

- You must hold the specific core for the area of gas work in which you wish to work, e.g., commercial, domestic or LPG, and have any prerequisite assessments.
- If you have a domestic natural gas core and CKR1, you can work on natural gas cookers only. However, if you also have the LPG core, you can also work on LPG cookers. It is the core that denotes the type of installation into which the cooker is installed.

Legislation Affecting Gas Operatives

1.10   Legislation places a mandatory/legal responsibility on the gas operative, who must comply with it in order to work within the industry. The piece of legislation that affects the gas engineer most is the Gas Safety (Installation and Use) Regulations. However, this is not the only legislation that needs to be observed. In fact, in this area, most actions undertaken as general work activities are affected by some piece of legislation.

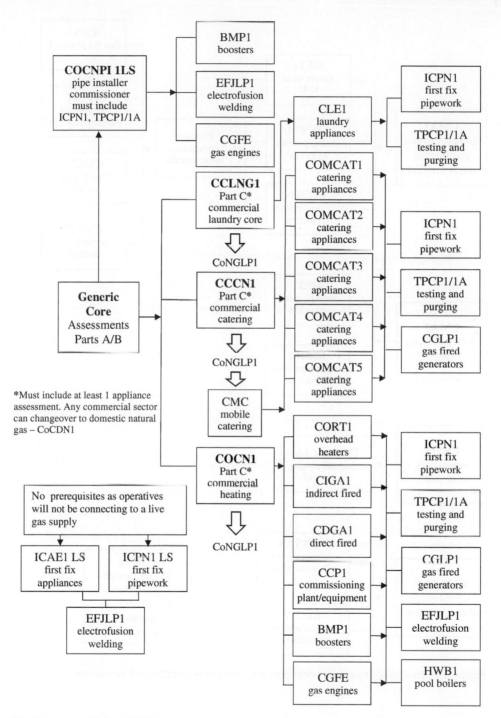

Figure 1.4 Commercial ACS assessments

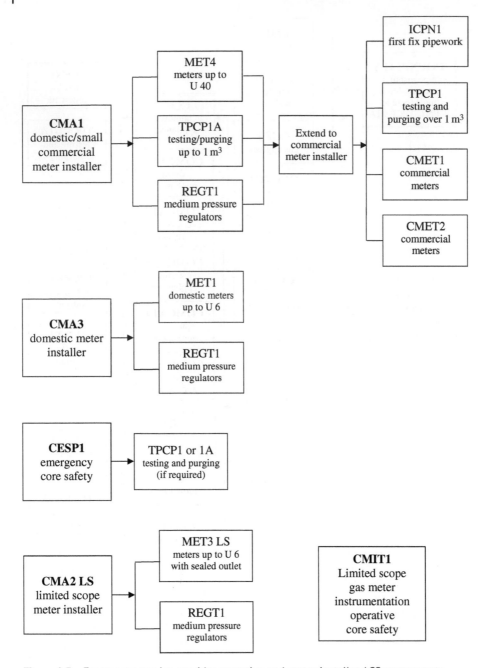

Figure 1.5 Emergency service provider operative and meter installer ACS assessments

Hierarchy of Legislation and Standards

1.11 By referring to Figure 1.6, which only represents a very small percentage of the **Legislative Documents** that must be observed, it will be seen that the law is divided into two parts, Acts of Parliament and Regulations. Acts of Parliament are the primary legislation in the United Kingdom whilst Regulations are secondary, or subordinate. Regulations are drawn up by government, often by authority given in an Act of Parliament, and are usually policed by an authority such as the HSE, local authority building control officers (BCOs) or water authorities.

Acts of Parliament and Regulations are updated as necessary and care needs to be taken to ensure that the latest version is referred to. As legislation often only states what can or cannot be done, without giving much detail, additional industry guidance documents are issued to help the installer practically comply with the law. For example, the Building Regulations have several supporting Approved Documents which give examples and solutions on how to comply with the enforceable legislation. The HSE also publish Approved Codes of Practice and guidance documents. Although these documents do not always have legal status, if the guidance in these documents is followed then there is usually the presumption that the law will be met. Where an individual chooses not to follow these guidance documents and they are prosecuted under health and safety law they will need to show that they have complied with the law in some other way or a court will find them at fault.

The next tier of documentation is referred to as **Normative Documents** and include gas industry standards such as those produced by the British Standards Institute (BSI), the Institution of Gas Engineers and Managers (IGEM) and Liquid Gas UK. These documents, along with manufacturer's instructions, are essential for gas operatives and are aimed at correct installation, commissioning and maintenance of gas appliances, pipework and associated equipment. Finally, organisations including Gas Safe publish technical bulletins (TBs) and IGEM's Gas Industry Unsafe Situations Procedure. These documents provide useful information for gas operatives as they carry out their daily activities; these are referred to as **Informative Documents**.

To help gas operatives understand and comply with current legislation and guidance, the Gas Safe Register publishes a quarterly list of all **legislative, normative** and **informative** documents. This list should be studied thoroughly as it is used, along with manufacturer's instructions, by Gas Safe Register inspectors to assess compliance of all work undertaken by those working in the industry. Further examples of these documents will be discussed in more depth later in the chapter.

It may be beneficial to view the legislative, normative and informative documents as a pyramid with legislation at the bottom and you, the gas operative, balancing at the top (see Figure 1.7). As illustrated, legislative Acts of Parliament hold up all other laws and guidelines. The gas operative sits at the top with the responsibility of personal compliance, this can be a delicate balancing act. But if the individual has a knowledge and understanding of the legislation relating to the gas industry, then he/she has a firm foundation to work from and is less likely to fall foul of the law. If an individual fails to comply it is possible that the courts will take enforcement action.

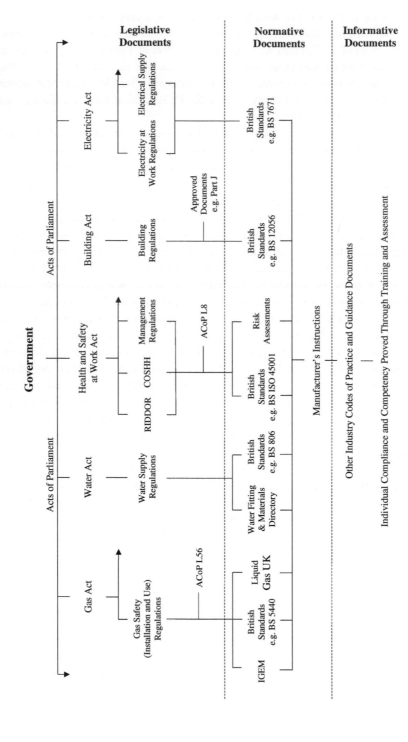

Figure 1.6 Examples of legislative, normative and informative documents

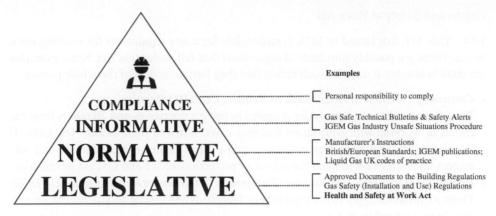

Examples

┌ Personal responsibility to comply

┌ Gas Safe Technical Bulletins & Safety Alerts
└ IGEM Gas Industry Unsafe Situations Procedure

┌ Manufacturer's Instructions
├ British/European Standards; IGEM publications;
└ Liquid Gas UK codes of practice

┌ Approved Documents to the Building Regulations
├ Gas Safety (Installation and Use) Regulations
└ **Health and Safety at Work Act**

Figure 1.7 Hierarchy of legislation and standards

Criminal and Civil Law

1.12 English law is divided into criminal law and civil law and different courts and procedures are followed for each.

Criminal law is penal law involving a crime against the State and is punishable by imprisonment and/or a fine. Action is taken by the police or by such bodies as the HSE or local authority.

Civil law pertains to the rights of private individuals and to the legal proceedings involving those rights. In a civil case the aim of a trial is to establish facts, based on evidence, and to determine liability. Civil law provides for compensation for the injured party, usually in the form of damages or an injunction demanding certain action to be taken.

Specific Legislation Affecting Gas Operatives

Gas Act

1.13 The Gas Act has seen many changes over the years and is responsible for putting in place the following Regulations, to name a few:

- *Gas Safety (Installation and Use) Regulations*
  These are the main regulations that are applicable to the gas operative, but it must be emphasised that they amount to only a few of many regulations that are applicable. These Gas Regulations are described in more detail at the end of this chapter.
- *Gas Safety (Rights of Entry) Regulations*
  Under these regulations the gas supplier may, with police assistance, if necessary, enter a property to make an installation safe.
- *Gas Safety (Management) Regulations*
  Under these regulations the gas supplier must submit a 'safety case' identifying their procedures. They must operate a full emergency gas service, to include a central telephone emergency number and must investigate any major gas incident, such as a poisoning or gas explosion.

Health and Safety at Work Act

1.14 This Act, first issued in 1974, is responsible for many regulations for working operatives. There are possibly hundreds of regulations that fall under this Act. Some examples are cited below, but it must be understood that they form only part of the whole picture.

- ***Control of Substances Hazardous to Health (COSHH) Regulations***
  These contain the statutory duties designed to protect operatives and all others from the effects of working with substances that may cause harm to their health. The COSHH Regulations make it a requirement to maintain a list of all hazardous materials and substances used, and have to hand the necessary protection advice and first aid information.
- ***Reporting of Injuries, Diseases and Dangerous Occurrences Regulations (RIDDOR)***
  These Regulations apply to the reporting of dangerous and unsafe situations and are covered in more detail in Part 8.
- ***Management of Health and Safety at Work (MHSW) Regulations and Construction (Design and Management) (CDM) Regulations***
  These Regulations place a wide range of duties on employers, contractors, designers, clients, etc., to ensure that health and safety are maintained throughout the construction process. This includes ensuring that adequate risk assessments are carried out.

Building Act

1.15 The Building Act 1984 consolidated previous legislation and allowed for the provision of the Building Regulations. There are differences in England, Wales, Northern Ireland and Scotland. The Isle of Man also has some variations. However, these regulations all identify the minimum requirements to be applied to building works as well as aspects affecting safety, energy conservation, etc. The Regulations themselves are quite a small document. However, alongside them sit a series of Approved Documents (or Scotland's Building Standards Technical Handbook), which set out detailed design requirements. Examples of those used in England and Wales are:

- ***Approved Document F:*** Ventilation;
- ***Approved Document G****: Sanitation, hot water safety and water efficiency;*
- ***Approved Document J:*** Combustion appliances and fuel storage systems;
- ***Approved Document L:*** Conservation of fuel and power.

Following the Grenfell Tower disaster on 14 June 2017, in which 72 people lost their lives when the 24-storey block of flats in North Kensington, West London was engulfed by a horrific fire, the Government commissioned an independent review of the Building Act and Regulations. As a consequence of the review there have been a number of amendments and wider ground-breaking reforms with the introduction of the **Building Safety Act 2022**.

Water Act

1.16 The earliest Water Acts made provision for local water byelaws. These were made to suit the various local water authority conditions. However, since 1999, all water installation work in England and Wales has had to comply with a set of mandatory ***Water Supply***

Regulations. As with building controls, Northern Ireland and Scotland have their own variations.

Electricity Act

1.17 The earliest Electrical Acts date back to the late 19th century when electricity was first produced. This Act is responsible for putting in place the following Regulations:

- *The Electricity Safety Quality and Continuity Regulations*
  This identifies the types of supply that may be used to serve a particular property, etc.
- *The Electricity at Work Regulations*
  These Regulations, amongst other things, identify specific tasks that must be undertaken when working on electrical supply systems. For example, the Regulations state the minimum tests to be carried out when checking an installation and also identify how the test equipment should be checked for correct operation prior to and after testing.

1.18 The effect of legislation does not stop here and many other laws are involved. For example, issued under the Road Traffic Act, the **Carriage of Goods Regulations** identifies the carriage of LPG cylinders in closed vans. Another piece of legislation to affect the operative is the **Environmental Protection Act**, which aims to protect the environment and requires an operative to have a licence to carry waste in their vehicle. It is difficult to keep abreast of new laws; however, ignorance is no defence in the eyes of the law.

Industry Documents and British Standards

1.19 As mentioned earlier, gas operatives need to be aware of an array of industry documents that must be followed in order to ensure that an installation conforms to the required standard.

Industry documents include manufacturer's instructions; British Standards; Institution of Gas Engineers and Managers (IGEM) procedures; Liquid Gas UK codes of practice, to name a few. As with legislation, these documents are continually being updated and reviewed so it is essential that the latest version of the document is used. Another problem is that one document often overlaps another and invariably they have different or conflicting views. The first requirement is to follow mandatory legislation; you then need to observe the specific manufacturer's instructions, followed by the British Standards, IGEM and Liquid Gas UK codes of practice. Finally, where assistance is still required, you should contact the Gas Safe Register technical helpline for further guidance.

Manufacturer's Instructions

1.20 If we look at the Gas Safety (Installation and Use) Regulations, we will find that the law requires an appliance to be installed in accordance with the manufacturer's instructions. The instructions themselves may well indicate that the appliance must be installed in accordance with a relevant British Standards or other industry document. Therefore, in essence, if you fail to install or test an appliance by the methods suggested in the British Standards, etc. this may well suggest that you are not complying with the manufacturer's instructions and, therefore the law.

The Gas Appliances (Enforcement) and Miscellaneous Amendments Regulations

(a)

(b)

Figure 1.8 (a) The UKCA (UK Conformity Assessed) marking is now applicable to most goods which previously required CE marking. (b) CE mark

1.21 These regulations are in place to protect consumers and end users from unsafe gas appliances or fittings. Under these regulations, manufacturers are required to demonstrate that their products for sale in the United Kingdom meet certain essential requirements. These requirements are there to ensure that the design and manufacture of gas-burning appliances, or fittings, ensure the health and safety of users, their domestic animals and property.

The regulations place the obligation on the manufacturer that, before placing a gas appliance or fitting on the Great Britain market, the manufacturer must have a relevant conformity assessment carried out on their product. Once complete the manufacturer must draw up a declaration of conformity and visibly affix a legible and indelible UKCA (UK Conformity Assessed) marking to the appliance or fitting or its data plate. Where this is not possible, due to the nature of the appliance or fitting, it must be affixed to the packaging or product label. This will help suppliers, consumers and end users determine that the gas appliance or fitting is fit for purpose, suitable and safe to use as the manufacturer intended. Installers may be more familiar with the European CE product marking which is still recognised on the Great Britain market (Figures 1.8a and 1.8b).

Please note that different rules apply to Northern Ireland.

British Standards

1.22 The following are a selection of useful British Standards that are applicable to gas work:

BS 1710 – Specification for identification of pipelines and services

BS 5440-1 – Flueing and ventilation for gas appliances of rated input not exceeding 70 kW net (1st, 2nd and 3rd family gases) – Specification for installation of gas appliances to chimneys and for maintenance of chimneys

BS 5440-2 – Flueing and ventilation for gas appliances of rated input not exceeding 70 kW net (1st, 2nd and 3rd family gases) – Specification for the installation and maintenance of ventilation provision for gas appliances

BS 5546 – Specification for installation and maintenance of gas-fired water-heating appliances of rated input not exceeding 70 kW net

BS 5864 – Installation and maintenance of gas-fired ducted air heaters of rated heat input not exceeding 70 kW net (2nd and 3rd family gases)

BS 5871-1 – Specification for the installation and maintenance of – Gas fires, convector heaters, fire/back boilers and heating stoves (2nd and 3rd family gases)

BS 5871-2 – Specification for the installation and maintenance of – Inset live fuel effect gas fires of heat input not exceeding 15 kW, and fire/back boilers (2nd and 3rd family gases)

BS 5871-3 – Specification for the installation and maintenance of – Decorative fuel effect gas appliances of heat input not exceeding 20 kW (2nd and 3rd family gases)

BS 5871-4 – Specification for the installation and maintenance of – Independent gas-fired flueless fires, convector heaters and heating stoves of nominal heat input not exceeding 6 kW (2nd and 3rd family gases)

BS 5854 – Code of practice for flues and flue structures in buildings

BS 5925 – Code of practice for ventilation principles and designing for natural ventilation

BS 6172 – Specification for installation, servicing and maintenance of domestic gas cooking appliances (2nd and 3rd family gases)

BS 6173 – Installation and maintenance of gas-fired catering appliances for use in all types of catering establishments (2nd and 3rd family gases)

BS 6230 – Specification for installation of gas-fired forced convection air heaters for commercial and industrial space heating (2nd and 3rd family gases)

BS 6400-1 – Specification for installation, exchange, relocation, maintenance and removal of gas meters with a maximum capacity not exceeding 6 m^3/h – Low pressure (2nd family gases)

BS 6400-2 – Specification for installation, exchange, relocation, maintenance and removal of gas meters with a maximum capacity not exceeding 6 m^3/h – Medium pressure (2nd family gases)

BS 6400-3 – Specification for installation, exchange, relocation and removal of gas meters with a maximum capacity not exceeding 6 m^3/h – Low and medium pressure (3rd family gases)

BS 6644 – Specification for the installation and maintenance of gas-fired hot water boilers of rated inputs between 70 kW (net) and 1.8 MW (net) (2nd and 3rd family gases)

BS 6798 – Specification for selection, installation, inspection, commissioning, servicing and maintenance of gas-fired boilers of rated input not exceeding 70 kW net

BS 6891 – Specification for the installation and maintenance of low-pressure gas installation pipework of up to 35 mm (*R*1 ¼) on premises

BS 6896 – Specification for installation and maintenance of gas-fired overhead radiant heaters for industrial and commercial heating (2nd and 3rd family gases)

BS 7624 – Installation and maintenance of domestic direct gas-fired tumble dryers of up to 6 kW heat input (2nd and 3rd family gases)

BS 7671 – Requirements for Electrical Installations. IET Wiring Regulations

BS 7967 – Guide for the use of electronic portable combustion gas analysers for the measurement of carbon monoxide in dwellings and the combustion performance of domestic gas-fired appliances

BS 7967-5 – Carbon monoxide in dwellings and other premises and the combustion performance of gas-fired appliances – Guide for using electronic portable combustion gas analysers in non-domestic premises for the measurement of carbon monoxide and carbon dioxide levels and the determination of combustion performance

BS 8446 – Installation and maintenance of open-flued, non-domestic gas-fired laundry appliances

BS EN 498 – Specification for dedicated liquefied petroleum gas appliances. Barbecues for outdoor use contact grills included

BS EN 721 – Leisure accommodation vehicles. Safety ventilation requirements

BS EN 1443 – Chimneys. general requirements

BS EN 1949 – Specification for the installation of LPG systems for habitation purposes in leisure accommodation vehicles and accommodation purposes in other vehicles

BS EN 50291 – Gas detectors. Electrical apparatus for the detection of carbon monoxide in domestic premises – Test methods and performance requirements

BS EN 50379-3 – Specification for portable electrical apparatus designed to measure combustion flue gas parameters of heating appliances – Performance requirements for apparatus used in non-statutory servicing of gas fired heating appliances

BS EN 13410 – Gas-fired overhead radiant heaters. Ventilation requirements for non-domestic premises

BS EN ISO 10239 – Small craft. Liquefied petroleum gas (LPG) systems

PD 54823 – Guidance for the design, commissioning and maintenance of LPG systems in small craft

Institution of Gas Engineers and Managers (IGEM) Publications

1.23 The following are a selection of useful IGEM Publications/Procedures (Utilisation Procedure [UP], General [G], Gas Measurement [GM], Industry Guidance [IG]):

IGE/UP/1 – Strength testing/tightness testing/direct purging of industrial and commercial gas installations

IGE/UP/1A – Strength/tightness testing/purging of small, low-pressure industrial & commercial installations

IGEM/UP/1B – Tightness testing/direct purging of small Liquefied Petroleum Gas/Air, NG/LPG installations

IGEM/UP/1C – Strength testing, tightness testing and direct purging of Natural Gas and LPG meter installations

IGEM/UP/2 – Installation pipework on industrial and commercial premises

IGEM/UP/6 – Application of compressors to Natural Gas fuel systems

IGE/UP/7 – Gas installations in timber-framed and light steel-framed buildings

IGE/UP/9 – Application of natural gas & fuel oil systems to gas turbines and supplementary and auxiliary-fired burners

IGEM/UP/10 – Installation of flued gas appliances in industrial and commercial premises

IGEM/UP/11 – Gas installations for educational establishments

IGEM/UP/12 – Application of burners and controls to gas-fired process plant

IGEM/UP/16 – Design for natural gas installations on industrial and commercial premises

IGEM/UP/17 – Shared chimney and flue systems for domestic gas appliances

IGEM/UP/18 – Gas installations for vehicle repair and body shops

IGEM/UP/19 – Design & application of interlock devices/associated systems in gas appliance installations in catering

IGEM/G/4 – Definitions for the gas industry

IGEM/G/5 – Gas in multi-occupancy buildings

IGEM/G/6 – Gas supplies to mobile dwellings

IGEM/G/11 – Gas industry unsafe situations procedure

IGEM/G/11 Supplement 1 – Responding to domestic CO alarm activations/reports of fumes after attendance by the emergency service provider or the Liquefied Petroleum Gas supplier

IGEM/GM/6 – Non-domestic meter installations. Standard designs

IGEM/IG/1 – Standards of training in gas work

IGEM/IG/1 Supplement 1 – Non-domestic training specification

IGEM/IG/1 Ed 2 Supplement 2 – Domestic training specification

Liquid Gas UK Codes of Practice

1.24   The following are a selection of Liquid Gas UK Codes of Practice (CoP):

CoP1 – Bulk LPG storage at fixed installations

CoP7 – Storage of full and empty LPG cylinders and cartridges

CoP17 – Purging LPG vessels and systems

CoP22 – Design, installation and testing of LPG piping systems

CoP24 – Use of LPG cylinders

CoP25 – LPG central storage and distribution systems for multiple consumers

CoP27 – The carriage of LPG cylinders by road & hazard information labelling requirements

CoP32 – LPG systems in leisure accommodation vehicles and road vehicles with habitation – Post-delivery inspection, commissioning and maintenance

GN2 – A guide for cabinet heater servicing

Industry Standards at Reduced Cost: the Gas Safe Register offer access to many of the IGEM, BSI and Liquid Gas UK documents listed in Sections 1.22–1.24. This is available to registered businesses via an optional subscription.

Gas Safety (Installation and Use) Regulations

1.25   The Gas Safety Regulations, which have been enacted under the Gas Act, are divided into a number of topics, with each set of regulations having a specific focus. Their subject matter is indicated in the brackets as follows:

- Gas Safety (Installation and Use) Regulations;
- Gas Safety (Rights of Entry) Regulations;
- Gas Safety (Management) Regulations.

Where this book makes reference simply to the Gas Regulations, this refers to the specific set of regulations dealing with Installation and Use.

The first set of the Gas Regulations was published in 1972, and it has since undergone many changes. The Regulations are available as a Statutory Instrument as laid before Parliament and within an approved code of practice (L56) as supplied by the HSE, see Section 1.34.

The Gas Regulations are divided into seven parts as follows.

Part A: General

1.26 This part begins by citing the dates on which the new Regulations came into effect. The bulk of this section deals with the general interpretation and application of the regulations and defines specific terms. For example, the term 'responsible person' refers to the occupier or owner of the property and not the gas operative, as is often mistakenly thought. This part also defines those buildings to which the regulations do not apply.

Part B: Gas Fittings – General Provisions

1.27 This part deals with the requirement for anyone carrying out gas work to be competent, even when working in areas to which the regulations do not apply. This part also makes reference to the safe working practices that must be complied with.

This section refers to employers and self-employed persons being classes of persons approved by the HSE. The regulations do not state that every operative needs to be approved. However, this comes within the rules of the Gas Registration scheme in that all businesses must provide the names of every competent and ACS-assessed operative working for them. This section also puts an onus on responsible persons to ensure that those working for them are adequately qualified. This part does more than define competency, it also goes on to emphasise the need to ensure that gas is not freely discharged from a pipe, and that a naked flame is not used to assist in finding a gas leak.

Within this section, there is a prohibition to make any alteration at a premises with a gas supply that could compromise safety or become a danger to a person in or about the property. This includes work done by homeowners or other trades. For example, the installation of cavity wall insulation that could block combustion air vents or the installation of an extractor fan that could affect the pull of an open flue. Any work that could potentially have an adverse effect on a gas installation must be carefully planned and applied to anyone, not just gas engineers.

Part C: Meters and Regulators

1.28 This section deals with the positioning and labelling of meters, both primary and secondary. It also deals with the supply regulator and, where applicable, any meter bypass.

Part D: Installation Pipework

1.29 This part of the regulations considers the safe use and location of pipework. It considers equipotential bonding of pipework, testing and purging, and ventilation of voids containing gas pipework. Where pipework is placed within commercial premises, it also identifies appropriate marking.

There is also the requirement for large consumers of gas with a gas supply of $\geq 50\,mm$ internal diameter to affix in a suitable position a permanent line diagram with details of the gas installation within the building.

Part E: Gas Appliances

1.30   This part of the regulations covers the necessary testing and checks that must be undertaken when working on an appliance and it also includes testing and checking the flue and ventilation. This is to ensure that every gas appliance is left in a safe condition to use and will not constitute a danger to any person. Included in this section is the prohibition to install non-room-sealed appliances in certain locations.

Under Part E, gas operatives have the legal duty to install gas appliances and fittings in accordance with the manufacturer's instructions. In addition, the manufacturer's instructions must be left with the consumer, and it is an offence for an operative to take them away on completion of a contract. This ensures that the appliance can be maintained to the specified standards as prescribed by the manufacturer.

Within this part is the duty that a responsible person must prevent the use of a gas appliance or installation if it is known, or there is reason to suspect, that it constitutes a danger to any person.

Part F: Maintenance

1.31   This part places specific duties on the owners of properties to ensure that any gas appliance, installation or flue within the property, or place of work, are maintained in a safe condition.

Landlords also have duties under this section and are required to ensure that any gas appliance and flue installed within their properties receive annual gas safety checks. All appropriate records of these checks need to be maintained.

Part G: Miscellaneous

1.32   This final part deals with miscellaneous issues, such as how to deal with gas escapes, exemption certificates, review of the regulations and cites regulations that have been revoked or amended.

What Is Not in the Gas Regulations

1.33   It should be remembered that the Regulations do not specify how work should be carried out. They simply identify what should and should not be done; the installer will need to refer to the various industry documents and manufacturer's instructions in order to comply with the law. Nor do the Gas Regulations identify what materials should be used. These may, however, be specified in other legislation, such as the Building Regulations, which also need to be complied with. Particular attention should be made to Approved Document J, which deals with the requirements for flues.

Approved Code of Practice (ACOP)

1.34   The HSE has published ACOP L56, entitled *Safety in the installation and use of gas systems and appliances*. This ACOP, and associated guidance, advises on how to meet the requirements of the Gas Regulations and is applicable to anyone who installs, services, maintains or repairs gas appliances and fittings. In addition, landlords will find this document a useful reference as they also have duties under the Gas Regulations.

The Future of Gas

1.35 Many will have heard of the climate crisis and the United Kingdom government's aim to achieve 'Net Zero' carbon dioxide (CO_2) emissions by the year 2050. Net zero targets have been set due to the effect CO_2 emissions from the burning of fossil fuels are having on the global climate, including those released from gas-fired appliances. It is the international scientific consensus that human-produced emissions of CO_2 need to fall to limit the effects of global warming. Global temperatures are said to be rising faster than predicted and governments around the world are working towards vastly reducing, or removing completely, fossil fuels from their energy mix.

In October 2021 The UK government released its Heat & Buildings Strategy, which encouraged the installation of low-carbon heating such as heat pumps. It proposed the gradual, but complete, removal of fossil fuels for heating. The Future Homes Standard, due to be released in 2025, is expected to ensure that all new homes will be fitted with low-carbon heating and phase out the installation of all natural gas boilers from 2035. This has left many gas engineers asking 'What does the future hold for the gas industry?' Whilst this legislation, if enacted, would bring to an end natural gas and LPG as we know it, it does not mean the immediate end of gas. With a gas boiler having a 10–15 year lifespan, gas engineers will be needed at least until 2050 for natural gas and LPG alone.

In August 2021, the UK government published its UK Hydrogen Strategy with the ambition to produce 5 GW of low-carbon hydrogen by 2030. In an update in December 2022, this ambition was doubled to 10 GW by 2030, with at least 50% of this being produced from electrolytic hydrogen. The use of hydrogen to heat buildings is not yet an established technology, and the government's strategic decisions on the role hydrogen could play in heating buildings in the future are not planned until 2026. However, small pilot schemes have already taken place with the potential for a town pilot by 2030.

At the time of writing, a consultation on improving boiler standards and efficiency is considering whether all domestic sized boilers sold on the UK market should be hydrogen-ready by 2026 with gas boiler/heat pump hybrid systems being considered as an option to help transition to a low carbon future. This is a dynamic and fast moving sector with exciting opportunities for all in the heating industry, but engineers must remain alert to any changes in legislation affecting the heating of homes and other buildings.

In preparation for the change to decarbonisation of heat in the UK, industry has agreed a new set of labelling to assist installers in identifying the suitability of appliances for the fuel provided:

- Hydrogen blend – capable of running on a 20% blend of hydrogen in the natural gas network.
- Hydrogen ready – able to run on natural gas and/or a 20% hydrogen blend, but can be converted to run on 100% hydrogen by a registered gas engineer.
- 100% hydrogen – manufactured ready to run on 100% hydrogen without the need for conversion.

Figure 1.9   Hydrogen appliance labels

The labelling that has been agreed by the membership of the Heating and Hotwater Industry Council is shown in Figure 1.9.

Hydrogen is just one of a number of suggested low-carbon alternatives that may be able to help the UK meet its net zero targets, but if it can, then the future of the gas industry may be secure for many decades to come.

Note: Net zero refers to the balance between the amount of greenhouse gases (carbon dioxide, methane, etc.) emitted into the atmosphere and the amount that is removed. Net zero is achieved when the amount of gases emitted is no more than the amount removed. It can be achieved through reducing emissions and/or removing emissions from the atmosphere.

2

Gas Utilisation

Gas: Its Origin

2.1   The word 'gas' is derived from the Greek word meaning chaos, possibly because its molecules are continually moving in all directions.

Natural gas is not a new discovery. The use of natural gas was mentioned in China around 900 BC, and in about AD 1000, wells were drilled and the gas was piped through bamboo tubes. Natural gas was discovered in Europe in England in 1659. Unfortunately, because of the difficulty of transporting the gas over long distances, it remained unused in an economy which was based on coal, oil and electricity. Later, a gas was manufactured, known as town gas. It was produced by a process of mixing coal gas and water gas. Coal gas was produced by heating coal to approximately 1000 °C, whereas water gas was produced by passing steam over red-hot coke. This early form of gas had a high proportion of hydrogen (around 45%) and included about 15% carbon monoxide (CO). This resulted in a gas that was both highly explosive and very toxic. However, this gas is no longer in use, a conversion took place in the UK between 1967 and 1977 to supply a large proportion of the UK with natural gas.

Natural gas is a gas that has, over millions of years, accumulated beneath the Earth's surface. It is the result of the breakdown of organic matter in the natural process of degradation. As the gas rises, it becomes trapped by the impervious layers of the Earth; this is shown in Figure 2.1. There are many types of natural gas, including methane, ethane, propane and butane. However, in general, the term 'natural gas' usually refers to gas that has a high proportion of methane.

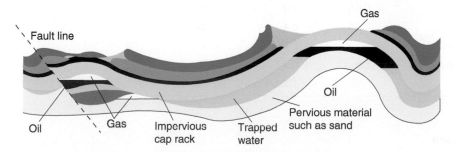

Figure 2.1   Section through Earth's strata showing natural gas traps

Gas Installation Technology, Third Edition. Andrew S. Burcham, Stephen J. Denney and Roy D. Treloar.
© 2024 John Wiley & Sons Ltd. Published 2024 by John Wiley & Sons Ltd.

Liquefied petroleum gas (LPG) is also found in the oil and gas fields and is obtained from crude oil or as a condensate product of natural gas. A variety of LPGs are produced by the many refineries and the composition varies with the process used. The term LPG applies to the group of hydrocarbon gases that can be liquefied by applying moderate pressures or temperature (see Section 2.8).

Substitute Natural Gas (SNG)

2.2 It is possible to manufacture natural gas, either for adding to an existing supply for peak loads or as a direct substitute. It is made from other petroleum products, such as LPG or light distillate grade liquid fuels.

Toxicity and Odour

2.3 Some gases are toxic or poisonous and inhaling them may result in death. Carbon monoxide (CO) is one such gas and was a constituent of town gas as supplied before conversion to natural gas. Fuel gases such as natural gas and propane, are non-toxic as they do not contain any CO, but they can produce it if they are not fully burnt. Natural gases such as methane, propane, etc., are odourless and an odorant, such as diethyl sulphide and ethyl butyl mercaptan, are added at the point of distribution to give the gas a recognisable smell.

Note: CO, when it is produced from natural gas, is an odourless gas.

Constituents of Gases

2.4 The natural gases are not commercially used as pure methane or propane but are mixed with other gases in order to improve their burning quality. Table 2.1 shows the constituents of the gases.

Table 2.1 Typical percentages by volume of the constituents of gases

Constituent	Symbol	Natural gas	LPG: Propane
Methane	CH_4	90	–
Ethane	C_2H_6	5.3	1.5
Propylene	C_3H_6	–	12.0
Propane	C_3H_8	1.0	85.9
Butane	C_4H_{10}	0.4	0.6
Nitrogen	N_2	2.7	–
Carbon dioxide	CO_2	0.6	–
		100.0%	100.0%

Hydrogen

2.5 Due to government targets to tackle carbon emissions, research and development is currently ongoing to further the evolution of the UK gas supply by converting it to hydrogen.

Initially, it is proposed to use a 20% mix that will be added to the natural gas networks with a view to converting to 100% hydrogen in the future. The 20% mix enables existing appliances to continue to be used but when 100% hydrogen is introduced, the appliances will need to be exchanged.

Characteristics and Properties of Gases

What Is Gas?

2.6   All substances are made up of tiny particles called molecules. In a solid, each molecule is strongly attracted to its neighbouring molecule by what is known as cohesion and there is very little space between each molecule to allow any movement. In liquids, cohesion still exists but because of a change in pressure or temperature, the molecules have added kinetic energy and move about more vigorously, giving the material its fluid properties. Whereas the molecules of a gas, have no cohesion, are free from adjoining molecules and can move in any direction.

The three prime gases used in the gas supply industry, namely natural gas and the two types of LPG, propane and butane, have different qualities. Table 2.2 contains their various characteristics and properties. These values are only a guide, as they will vary under different atmospheric conditions. The Quality/Unit column is explained in the following sections.

Chemical Formulae

2.7   A chemical formula is the shorthand abbreviation used to identify a material. As mentioned in Section 2.6, substances are made up of molecules, and each molecule in turn is made up of atoms. From Table 2.2, it can be seen that there is 1 carbon and 4 hydrogen atoms for every 1 methane gas molecule, hence its formula: CH_4. (See Table 2.1 for a more accurate list of the constituent parts of natural gas and LPG.)

Boiling Point

2.8   Gases such as methane (natural gas) are found only in liquid form at extreme temperatures or pressures, far above those in our atmosphere. You can see from the Table 2.2 that, at atmospheric pressure, the temperature would need to be above $-162\,°C$ before the liquid would change to a gas. This means that $-162\,°C$ is the boiling point of liquid methane. (Gas Vaporisation is explained further in Section 2.19, in the context of LPG.)

Specific Gravity (Also Called Relative Density)

2.9   Specific gravity (SG) is the term used to compare one substance with another. Liquids and solids are compared with water, which has an SG of 1. Those with an SG less than 1 will float on water, and those with an SG greater than 1 will sink. Gases, however, are compared with dry air under the same atmospheric conditions; air has an SG of 1. Therefore, any gas with an SG greater than 1, such as propane, will sink in air; those with an SG less than 1, such as methane, will rise.

Table 2.2 Physical characteristics and properties of gases (typical values)

Quality/unit	Natural methane	LPG propane	LPG butane
Chemical formula	CH_4	C_3H_8	C_4H_{10}
Boiling point	$-162\,°C$	$-40\,°C$	$-2\,°C$
Specific gravity of liquid fuel	–	0.5	0.58
Specific gravity of gas vapour	0.58	1.5	2.0
Gross calorific value	$38.5\ MJ/m^3$	$95\ MJ/m^3$	$121.5\ MJ/m^3$
Gas family	2nd	3rd	3rd
Flammability limits	5–15%	1.7–10.9%	1.4–9.3%
Air/gas ratio	9.81:1	23.8:1	30.9:1
Oxygen/gas ratio	2:1	5:1	6.5:1
Flame speed	0.36 m/s	0.46 m/s	0.38 m/s
Ignition temperature	$704\,°C$	$470\,°C$	$372\,°C$
Maximum flame temperature	$1000\,°C$	$1980\,°C$	$1996\,°C$
Nominal system operating pressure	21 ± 2 mbar	37 mbar	28 mbar
Liquid storage vapour pressure	–	6–7 bar	1.5–2 bar

Graham's Law of Diffusion (see Section 2.74) proved that a light gas will mix twice as fast as a gas that is four times its weight. You can experience this when purging a natural gas system, compared with butane or propane installations; you can see from Table 2.2 that they are heavier and therefore more difficult to disperse.

Calorific Value (CV) Gross and Net

2.10   Gas is sold based on its gross calorific value This figure varies at different times and in different locations due to the gas field supplying the fuel. The gross CV is the amount of heat energy produced by the complete consumption of a known quantity of fuel. The CV for various fuels will be found to differ and one can see that more heat can be obtained from burning 1 m³ of propane than from burning 1 m³ methane. When the fuel gas is consumed, water is formed as a by-product. When this water condenses from vapour to liquid state, latent heat is given up. This latent heat is not usually available for use as it passes out through the flue system, so it cannot be regarded as part of total heat energy available. Therefore, it is not counted when quoting energy efficiencies, and appliances are invariably specified as a net CV.

Gas Family

2.11   The amount of heat given off from a burner depends on several factors, which can be divided into two groups:

- Group 1 relates to the SG;
- Group 2 depends on the size of the injector and pressure of the gas within the system.

The factors in Group 2 are dependent on the design or adjustment of the appliance, whereas those in Group 1 are dependent on the properties of the gas being supplied. These two characteristics can be linked to give what is known as a 'Wobbe number', which, in turn, is used to define the international family of gases into which the fuel can be categorised. The Wobbe number is calculated as $(CV \div \sqrt{SG})$.

Example: Natural gas with a gross CV of 38.5 and SG of 0.58 would have a Wobbe number of: $38.5 \div \sqrt{0.58} = 50.55$.

Knowing the Wobbe number, we can see from Table 2.3 that this is natural gas.

Table 2.3   Families of gases

Family	Gas type	Approx. Wobbe no.
1st	Manufactured (town gas)	24–29
2nd	Natural	48–53
3rd	LPG	72–87

Flammability Limits (Explosive Limits)

2.12   If you open a matchbox full of gas in a room, wait a second, then try to ignite the gas, nothing will happen: there is too much air. Conversely, if you strike a match in a room that is completely full of gas, again no combustion will occur: there will not be enough air. The gas will only ignite if there is a certain percentage range of air/gas mixture – this is referred to as its 'flammability limits'. Table 2.2 gives the lower flammable limit (LFL) for natural gas as 5% and the upper flammable limit (UFL) as 15%, therefore it will only burn within this range (5–15%).

Air/Gas Ratio

2.13   As we have seen, air needs to be mixed with the gas for it to burn. The flammability limits defined for natural gas suggest a mixture of 5–15% gas. However, how much air is required for all the fuel to be consumed? Some air may be present for combustion to occur, but is it sufficient? If there is insufficient air, combustion will be incomplete, and CO will be produced (see Section 2.23). This is what is meant by the term 'air/gas ratio'. From Table 2.2 we see that natural gas needs 9.81 volumes of air for complete combustion of 1 volume of gas. (See Section 2.26, for an explanation of why more air is required for LPG.)

Oxygen/Gas Ratio

2.14   This is basically very similar to the air/gas ratio explained in Section 2.13. However, air is a mixture of gases: approximately, 80% nitrogen and 20% oxygen (i.e. one-fifth oxygen). Nitrogen plays no part in the combustion process and so can be ignored here.

Rounding up the numbers, natural gas has an air/gas ratio of 10 : 1. Only 1/5 of the gas in air is oxygen and we know that 1/5 of 10 is 2. Therefore, we can estimate that the required oxygen/gas ratio is 2 : 1. To put it simply, twice as much oxygen is required than natural gas to ensure complete combustion, see Figure 2.2.

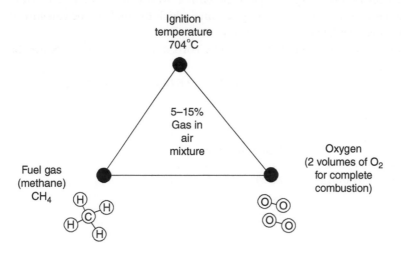

Ignition
temperature
704°C

5–15%
Gas in
air
mixture

Oxygen
(2 volumes of O_2
for complete
combustion)

Fuel gas
(methane)
CH_4

Figure 2.2   Essentials for combustion

Flame Speed

2.15   As fuel burns, the flame travels through the gas/air mixture at a set speed. This can be illustrated by watching a piece of paper burn and observing the speed at which the flame travels. In gases, this speed is not so easily observed, but nevertheless, the same phenomena occur. With gases such as hydrogen, the flame speed is so great it breaks the sound barrier with a loud bang. Burner pressures should be carefully adjusted to ensure a flame speed that keeps the flame at the head of the burner. If the pressure was too great, it would eject the gas faster than it could be consumed, resulting in a flame lifting off. On the other hand, insufficient pressure would result in the flame burning back inside the mixing tube to the injector.

Ignition Temperature

2.16   Fuel will not burn if the ignition temperature is not reached. For example, paper needs to be heated to around 80–100 °C before it can be consumed by fire. To quench a fire, the fire brigade pour water on it. For this, water at around 6 °C is used to cool the solid material. Gases also need to reach a sufficient temperature for combustion, and for natural gas, we can see from Table 2.2 that this temperature is around 704 °C. When the gas is ignited by a match or ignition spark, rapid heat transfer through the fuel soon enables it to reach the required ignition temperature.

System Operating Pressure

2.17   The system operating pressure is the pressure that the gas is regulated down to for all domestic dwellings and most commercial installations. The actual gas pressure of the regional grid network supplying the homes would be considerably higher (see Section 2.45).

Liquefied Petroleum Gas (LPG)

2.18   There are two main types of LPG: butane and propane. They are only used in their pure form for test purposes and are supplied as commercial butane or commercial propane, which are mixtures of the gas with other gases added to improve the working characteristics.

The key difference between LPG and natural gas is that when modest pressure is applied to the LPG it becomes a liquid. This makes it possible to store large quantities of fuel in specially constructed containers.

Gas Vaporisation

2.19   In order to convert a liquid to a gas, heat is required. You can see this when you boil a kettle of water to make a cup of tea. The liquid heats up and bubbles and steam rises. When the LPG cylinder or the pressurised LPG vessel valve is opened, the liquid released is subject to atmospheric pressure. It can be seen from Table 2.2 that butane, for example, boils at −2 °C. If the valve were opened in a very cold climate, nothing would happen. At a temperature of, say, 16–20 °C, on a typical UK day, the heat from the atmosphere would rapidly boil the liquid and the vapour would be drawn off. The greatest transference of heat is from the inside wetted metal surface that is in contact with the liquid gas, so air is free to circulate around the vessel or cylinder. The photographs (Figure 2.3a) and illustrations (Figure 2.3b) show liquid gas boiling.

Figure 2.3a   Inside view of vessel containing LPG

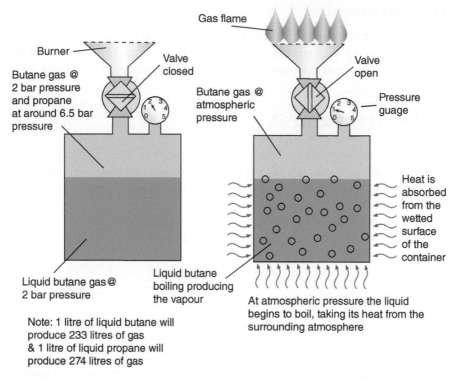

Figure 2.3b Illustration of LPG boiling at atmospheric pressure

When calculating the gas requirements for a particular supply, it is important to remember that it is the size of the vessel in which the LPG is contained and not the volume of liquid that determines the amount of gas that can be generated per hour. This is because larger vessels have a larger wetted surface area.

Container Pressure

2.20 The pressure within the vessel containing LPG varies, depending on the type of gas. Typical pressures for propane are 6–7 bar and for butane they are 1–2 bar. The containers are protected from overpressure by a safety pressure relief valve. Overpressure could occur due to excessive expansion caused by; for example, exposure to heat. The safety pressure relief valve forms part of the outlet valve on a cylinder where a cylinder is used.

The vessels in which the LPG is supplied are never filled to more than 80–87% and a void, called an 'ullage space', is left to allow for any expansion of the liquid caused by changes in atmospheric pressure or temperature. This space also allows for compressed vapour to form above the liquid level in readiness for use.

The Combustion Process

2.21 Combustion of fuel involves a chemical reaction; it produces heat as the fuel changes into a new compound, just as heat is produced within a compost heap. A compound is a substance with more than one kind of atom. For example, a hydrocarbon is a compound of carbon (C) and hydrogen (H). The proportion of carbon to hydrogen varies depending on the hydrocarbon: methane has 1 carbon atom to 4 hydrogen atoms (CH_4), whereas propane has 3 carbon atoms and 8 hydrogen atoms (C_3H_8).

Hydrocarbons react with oxygen in the presence of heat to undergo a chemical change, so producing a new by-product. When a piece of wood is burnt, we see that the wood, being a hydrocarbon, slowly disappears leaving nothing but a pile of ash. The process of combustion is misunderstood as being to destroy and consume. In reality, during the combustion process, the wood or gas, etc. is not consumed; it is simply converted to another form.

The Complete Combustion of Methane

2.22   Methane (CH_4) will burn in the presence of oxygen (O_2). Where the supply of oxygen is unlimited, methane uses 2 volumes of oxygen for every volume of fuel gas supplied. From Figure 2.4, you can see that during combustion, the molecular structure of the gases changes into a new form. They are still gases, and they still contain the same number and types of atoms. The difference is that these newly formed gases, in this case carbon dioxide (CO_2) and water vapour (H_2O), have new and different qualities; CO_2 and H_2O are both non-toxic gases. This is often expressed as:

$$CH_4 + 2\,O_2 = CO_2 + 2\,H_2O.$$

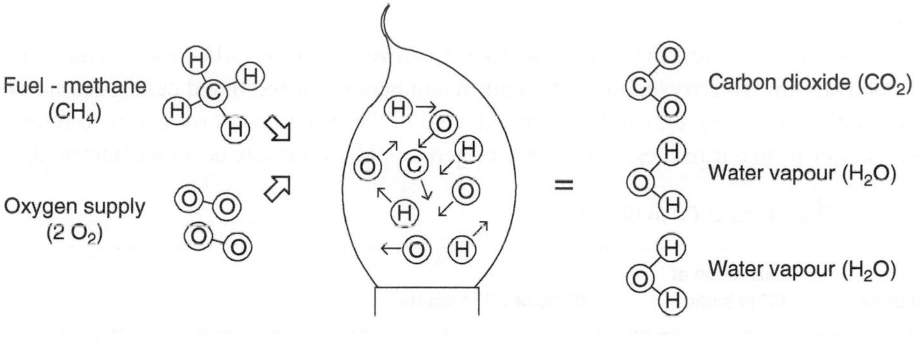

Figure 2.4   The combustion process

It should therefore be noted that for every 1 m^3 of natural gas consumed approximately 2 m^3 or 2000 litres of water vapour is produced. This equates to around 1.25 litres of condensate.

Incomplete Combustion

2.23   The gas flame is the visual chemical reaction that occurs when the fuel gas ignites with oxygen to form new compounds. As the oxygen and gas diffuse, the transition from the hydrocarbon to the resultant CO_2 and H_2O is not instantaneous. The heat generated by the flame causes the original constituents to break down, slowly forming different compounds, such as alcohols (CH_3OH), aldehydes (HCHO), free carbon (C) and carbon monoxide (CO).

In the presence of sufficient oxygen, these hydrocarbons will continue to burn until complete combustion has taken place. However, if the process is interrupted by a lack of oxygen, or the flame temperature falls below its ignition temperature, the unburnt gases will be discharged into the atmosphere. For example, in the combustion of methane, if the available oxygen is restricted by just 5% (i.e. 1.9 volumes O_2 instead of 2), we would begin to see levels of carbon monoxide (CO) being produced, and if the O_2 were further restricted, the level of CO would rise:

$$CH_4 + 1.9\,O_2 = 0.05\,CO + 0.9\,CO_2 + 9\,H_2O + 0.05\,CH_3OH.$$

Incomplete combustion can occur at an appliance when:

- The appliance is under or over-gassed.
- The air supply is inadequate.
- Flame impingement occurs on a surface cooler than the ignition temperature of the gas, resulting in flame chilling (see Section 2.35).
- Vitiation (see Part 3: Section 3.57).

Dangers of CO and the Effects of CO in the Air

2.24   CO in the environment can have a devastating effect on human health. For many of us, it would be the most dangerous gas we might ever experience. It is often referred to as the 'hidden gas' because it is very hard to detect. However, the effects of CO are quite dramatic, and in high concentrations, it will kill within a few minutes, as can be seen in Table 2.4.

When we breathe, our lungs absorb oxygen. However, CO is absorbed more readily into the blood stream than oxygen and when our haemoglobin, or red blood cells, become saturated with CO, no oxygen can be absorbed. One of the functions of oxygen is to remove waste matter from our tissues, so without oxygen, our blood rapidly becomes poisoned.

Table 2.4   The effects of CO in the air

CO in air	Saturation of CO in blood	Effects of CO in adults
0.01%	0–15%	Slight headache after 2–3 hours
0.02%	1–30%	Mild headache, feeling sick and dizziness after 2–3 hours
0.05%	30–50%	Strong headache, palpitations and nausea within 1–2 hours
0.15%	50–55%	Severe headache, nausea and dizziness within 30 minutes
0.3%	55–60%	Severe headache, nausea and dizziness within 10 minutes; increased breathing and convulsions, leading to collapse, possible death after 15 minutes
0.6%	70–75%	Severe symptoms within 1–2 minutes, death within 15 minutes
1.0%	85–90%	Immediate symptoms: death will occur within 1–3 minutes

Air Requirements for Combustion

2.25 The Table 2.5 gives an indication of the amount of oxygen required to consume completely 1 volume of a gas.

Table 2.5 Oxygen requirement to provide complete combustion of 1 volume of gas

Fuel		Volume of oxygen
Methane	CH_4	2.0
Ethane	C_2H_6	3.5
Propylene	C_3H_6	4.5
Propane	C_3H_8	5.0
Butane	C_4H_{10}	6.5

To calculate the oxygen requirements and, consequently, the air requirement for complete combustion of a specific fuel gas comprising a mixture of gases, such as natural gas, first list the constituents in order of percentage volume, taken from Table 2.1. Then multiply each percentage by the volume required for the individual fuel to obtain the oxygen required by each individual gas, from Table 2.5. Adding these gives the total amount of oxygen required.

Therefore, as can be seen in Table 2.6, 100 m³ of natural gas would require 206.15 m³ of O_2. This equates to 1 m³ of natural gas needing 2.062 m³ of O_2.

Table 2.6 O_2 requirement for natural gas

Constituent		%		Volume of O_2 required		
Methane	CH_4	90.0	×	2.0	=	180.0
Ethane	C_2H_6	5.3	×	3.5	=	18.55
Propane	C_3H_8	1.0	×	5.0	=	5.0
Butane	C_4H_{10}	0.4	×	6.5	=	2.6
Carbon dioxide	CO_2	0.6	×	0	=	0.0
Nitrogen	N_2	2.7	×	0	=	0.0
		100.0				206.15

Note: No consumption of the N_2 or CO_2 takes place as these are not fuel gases.

In round figures, 2 volumes of O_2 are needed to consume 1 volume of natural gas, see Figure 2.5.

In most cases, the oxygen supply is taken from the air surrounding the appliance. There is approximately 21% oxygen in the atmosphere; therefore, $2.06 \times (100 \div 21) = \underline{9.81 \text{ m}^3}$ of air would be needed for complete combustion of 1 m³ of natural gas.

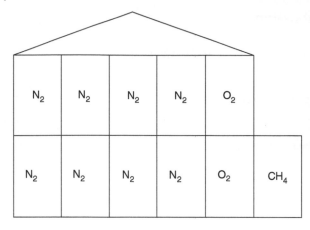

Imagine a house full of air. It would all be required to consume the volume of natural gas in the extension.
This same amount of air would consume less than half the gas if it was propane, hence the need for good ventilation.

Figure 2.5 Lots of air is required for combustion

In round figures, 10 volumes of air are needed to consume 1 volume of natural gas.

This equates to 1 m³ of propane needing 4.926 m³ of O_2 for complete combustion (Table 2.7).

Table 2.7 O_2 requirement for propane gas

Constituent		%				Volume of O_2 required
Ethane	C_2H_6	1.5	×	3.5	=	5.25
Propylene	C_3H_6	12.0	×	4.5	=	54.0
Propane	C_3H_8	85.9	×	5.0	=	429.5
Butane	C_4H_{10}	0.6	×	6.5	=	3.9
		100.0				492.65

In round figures, 5 volumes of O_2 are needed to consume 1 volume of propane gas.

This means that $4.93 \times (100 \div 21) = \underline{23.47 \text{ m}^3}$ of air are needed for combustion.

In round figures, 24 volumes of air are needed to consume 1 volume of propane gas.

Ventilation Requirements for Natural Gas and LPG Systems

2.26 It should be noted that LPG requires more air for combustion. Yet when calculating the size of the ventilation grille needed for a permanent dwelling, the same sized grille is used irrespective of the type of gas used. The size of the ventilation grille does not need to be larger for the LPG installation because, for the same heat input, less gas is consumed, i.e. the calorific value of natural gas is 38.5 MJ/m³, whereas that for propane is 95 MJ/m³.

To obtain 1 kW of heat from natural gas, the following amount of fuel is consumed:

$$1 \text{ kW} \times 3.6 \div CV = 1 \times 3.6 \div 38.5 = 0.094 \text{ m}^3 \text{ of gas.}$$

To obtain 1 kW of heat from propane gas, the following amount of fuel is consumed:

$$1 \text{ kW} \times 3.6 \div CV = 1 \times 3.6 \div 95 = 0.038 \text{ m}^3 \text{ of gas.}$$

So, although LPG requires more air, less gas is consumed!

Products of Combustion

2.27   The products of complete combustion, as previously stated, are carbon dioxide (CO_2) and water vapour (H_2O). The quantities are calculated in Table 2.8 for each of the fuel gases. Note: The percentage by volume for the constituents was given in the previous section.

Table 2.8   Products of combustion for natural gas with sufficient O_2 supplied

Constituent percentage by volume %			Ratio of CO_2 and H_2O produced by complete combustion of fuel		Total products of combustion	
			CO_2/vol.	H_2O/vol.	CO_2	H_2O
Methane	CH_4	90	1	2	90	180
Ethane	C_2H_6	5.3	2	3	10.6	15.9
Propane	C_3H_8	1	3	4	3	4
Butane	C_4H_{10}	0.4	4	5	1.6	2
CO_2		0.6			0.6	
N_2		2.7				
		100			105.8	201.9

For example: the total CO_2 for ethane is found thus 5.3 (%vol.) × 2 (CO_2/vol.) = 10.6; the total H_2O for ethane is found, thus 5.3 (%vol.) × 3 (H_2O/Vol.) = 15.9.

Therefore, if 1 m³ of natural gas was consumed, the total products of combustion, and ultimately the total volume of gases leaving the appliance, would be as follows:

Carbon dioxide	1.058
Water vapour	2.019
Nitrogen in gas	0.027
Nitrogen in air (9.81–2.06)*	7.750 * From Section 2.25
	10.854 or approx. 11 m³

'*' is indicating that (9.81 – 2.06) comes from Section 2.25.

With methane, shown in Figure 2.6, the volume of gases leaving the combustion chamber equals the volume of gases entering the combustion chamber. However, this is not always the case as can be seen when burning propane, where 23.5 air + 1 fuel enter yet 25.5 come out (Table 2.9).

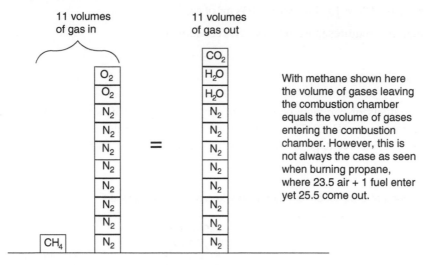

With methane shown here the volume of gases leaving the combustion chamber equals the volume of gases entering the combustion chamber. However, this is not always the case as seen when burning propane, where 23.5 air + 1 fuel enter yet 25.5 come out.

Figure 2.6 Gas volume in versus gas volume out

Table 2.9 Products of combustion for propane gas with sufficient O_2 supplied

Constituent percentage by volume %			Ratio of CO_2 and H_2O produced by complete combustion of fuel		Total products of combustion*	
			CO_2/vol.	H_2O/vol.	CO_2	H_2O
Propane	C_3H_8	85.9	3	4	257.7	343.6
Propylene	C_3H_6	12	3	3	36	36
Ethane	C_2H_6	1.5	2	3	3	4.5
Butane	C_4H_{10}	0.6	4	5	2.4	3
		100			299.1	387.1

For 1 m³ of propane gas consumed, the total products of combustion, and ultimately the total volume of gases leaving the appliance would be as follows:

Carbon dioxide	2.991
Water vapour	3.871
Nitrogen in air (23.47–4.926)*	18.544 * *From Section 2.25*
	25.405 or approximately 26 m³

'*' is indicating that (23.47 – 4.926) comes from Section 2.25.

Figure 2.7 illustrates how the number of atoms is equal before and after combustion has occurred. Note how the number of atoms on either side of the equal sign is the same.

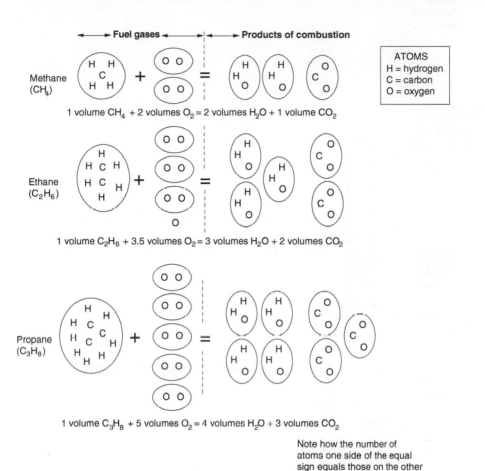

1 volume CH_4 + 2 volumes O_2 = 2 volumes H_2O + 1 volume CO_2

1 volume C_2H_6 + 3.5 volumes O_2 = 3 volumes H_2O + 2 volumes CO_2

1 volume C_3H_8 + 5 volumes O_2 = 4 volumes H_2O + 3 volumes CO_2

Note how the number of atoms one side of the equal sign equals those on the other

Figure 2.7 Typical combustion equations

The Gas Flame

2.28 The gas flame can have many different shapes and colours, each designed for a particular purpose. However, gas flames generally fall into one of two categories: post-aerated (Figure 2.8a) or pre-aerated (Figure 2.8b).

The Post-aerated Flame (Non-aerated Flame, Neat Flame or Luminous Flame)

2.29 A flame in which no air has been mixed with the gas prior to combustion. This type of flame is rarely used in modern gas supply systems because of its unstable characteristics.

Figure 2.8a Post-aerated flame

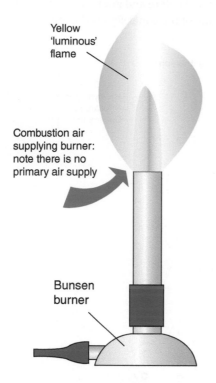

Yellow 'luminous' flame

Combustion air supplying burner: note there is no primary air supply

Bunsen burner

Figure 2.8b Pre-aerated flame

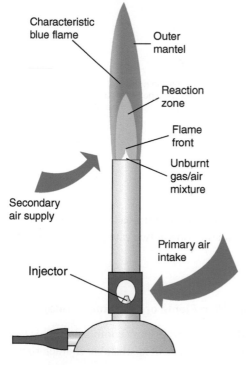

Characteristic blue flame

Outer mantel

Reaction zone

Flame front

Unburnt gas/air mixture

Secondary air supply

Primary air intake

Injector

The Pre-aerated Flame (Aerated Flame)

2.30   This is the type of flame and burner design used in most modern gas appliances. Air is drawn in by natural draught or by using a fan that mixes the air with gas prior to combustion. This pre-mixed air and gas is called primary air and it makes up some 40–50% of the air that is needed. The additional air that is needed to support complete combustion is obtained from the air that surrounds the flame. This air is known as secondary air. Forced draught premix burners draw all the air that they need prior to the combustion process, and no secondary air supply is required.

Burners that rely on natural draught draw the air into the mixing chamber. The flow of gas is restricted as it passes through an injector, forcing it to flow with increased velocity, and creating a negative pressure at the primary air port, which causes it to draw in the required air.

Post-aerated and pre-aerated flames produce the same amount of heat output per volume of gas consumed. However, the temperature of the pre-aerated flame is much higher due to the smaller intensity of the flame. When all the air is forced, as with the pre-mix burner, a smaller, intensely hot flame results.

The Flame Front

2.31   The gas/air mixture flows through the mixing chamber and is finally discharged at the burner ports where it is ignited. The point where the unburnt gas ends and the actual flame begins is referred to as the 'flame front'. At this point, the gas speed equals the flame speed, see Figure 2.9. The shape of the flame front varies slightly depending on the design of burner. It is often cone-shaped because the friction of the gas against the sides of the burner slows the gas, so that it travels faster at the centre, pushing the flame front away from the burner head. The primary air effects, the shape of the flame front, and a small amount of air produce a long flame front because the gas maintains a high speed at its centre. On the other hand, when large volumes of primary air are added, the flame front becomes flatter and noisier.

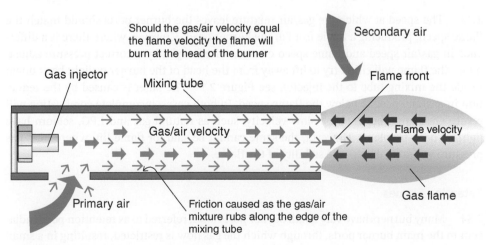

Figure 2.9   Gas/air speed = flame speed

Flame Pattern and Characteristics

2.32   A well-adjusted flame can be recognised by its colour: a bright blue-green inner cone and an outer flame that is slightly darker with a bluish-purple tinge. Such a flame is stable and there is a constant flow as the gas leaves the burner ports. Problems with a gas flame can often be identified by observing and understanding the flame pattern. Figure 2.10 illustrates the effect that a wrongly fitted or blocked injector will have on the flame pattern. Similarly, if the primary air port becomes blocked or damaged, it will also affect the flame characteristics. In both cases, the amount of primary air intake is affected, causing incomplete combustion of the fuel. A greater understanding of these simple observations illustrates the need for servicing and maintenance checks on appliances.

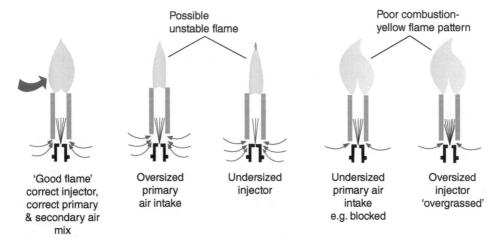

Figure 2.10   Flame characteristics caused by many variables

Flame Lift and Lighting Back

2.33   The speed at which the gas/air mixture leaves the burner ports should match the flame speed, producing a flame front at the burner head. However, where there is a difference in gas/air speed and flame speed caused, for example, by incorrect pressure adjustment, the flame will either try to lift away from the head of the burner or light back down inside the mixing tube to the injector, see Figure 2.11. The latter is caused by the reduction in gas/air speed to below the flame speed. In both cases, incomplete combustion will result. Gases that burn with a slow velocity, such as natural gas and LPG, seldom have lighting-back problems. However, the same cannot be said for flame lift, where high gas pressure will soon disrupt the stability of the flame.

Retention Flames

2.34   Many burners have very small burner port-holes, referred to as retention ports, adjacent to the main burner ports, through which the gas flow is restricted, resulting in a small stable flame, see Figures 2.12a and 2.12b. Should a main burner flame try to lift off from the burner head, these stable retention flames continually re-ignite the gas/air mixture, holding

Figure 2.11 Flame lift and lighting back

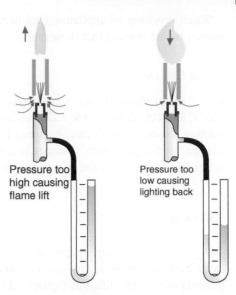

Pressure too high causing flame lift

Pressure too low causing lighting back

it securely in place. Cold air from the room feeds secondary air to the burner. This air is usually at about 21 °C, and it tends to cool the flame below the ignition temperature required (see Section 2.16). The retention flame also helps to increase the temperature of the gas/air mixture at this point, again assisting in keeping the flame stable.

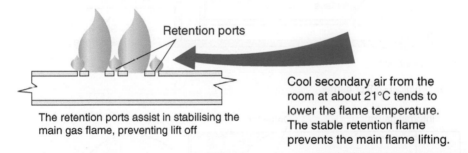

Retention ports

The retention ports assist in stabilising the main gas flame, preventing lift off

Cool secondary air from the room at about 21°C tends to lower the flame temperature. The stable retention flame prevents the main flame lifting.

Figure 2.12a Retention flames

Figure 2.12b Photograph showing main burner and retention ports

When servicing an appliance, care needs to be taken to maintain the condition of the burner head to ensure that these retention ports remain clear and undamaged.

Flame Chilling

2.35 Flame chilling occurs when the gas is cooled below its ignition temperature between 370 °C and 700 °C (see Table 2.2). This leads to the flame going out before complete combustion has occurred. It can be caused by flame impingement, where the flame touches the heat exchanger, which, although hot, is cooler than the ignition temperature of the gas. Where flame impingement occurs, extensive sooting or carbon build-up will result.

Atmospheric Burners

2.36 There are many designs of atmospheric burners, but they all operate with the same basic components: injector, primary air port, mixing tube and burner. These components can be identified in Figures 2.13a and 2.13b and are described in the following sections.

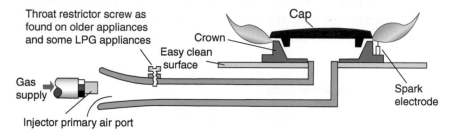

Figure 2.13a Typical cooker burner

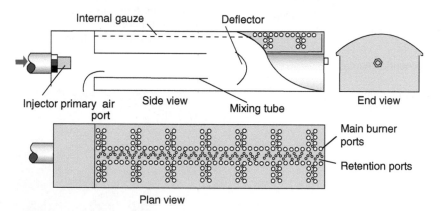

Figure 2.13b Typical bar burner

The Injector

2.37   The injector is usually made from brass with a central hole or series of holes through which the gas can pass. The pressure within the gas pipe is a few millibars above atmospheric pressure, so as the gas passes out through the relatively small hole, it moves at about 45 m/s, causing air trapped in its wake to be drawn in at the primary air port. The size and design of the injector control the amount of heat input to an appliance as indicated by the manufacturer; therefore, it is essential that the correct injector is selected for a specific burner as an incorrectly sized injector could have fatal consequences. During maintenance or when cleaning an injector, care needs to be taken to ensure that its hole does not become enlarged. For example, should an injector with a 5 mm diameter hole become enlarged to, say, 6 mm, it would supply over 40% more gas, which is nearly half as much again as intended.

Primary Air Port

2.38   The primary air port is located at the point where the injector leaves the pipeline, usually to the side or bottom. The size of this hole is generally fixed; however, sometimes some form of aeration control is incorporated to control the amount of primary air drawn into the burner by way of an air shutter or throat restrictor.

Mixing Tube

2.39   This is the chamber between the primary air port and the burner within which the gas and air can pre-mix prior to combustion. Some mixing tubes, particularly those where there may be resistance to flow, incorporate a venturi, which provides a more consistent pressure to the burner.

Burner

2.40   The burner provides the base on which the flames can burn with the required shape and heat input. The shape of the burner varies depending on the appliance in which it is installed, e.g. round for a cooker (see Figure 2.13a) and a long box or bar inside a boiler (see Figure 2.13b) or a tapered body (see Figure 2.14). The burner is designed to produce an even flame distribution and baffles are often put in to serve this purpose. Sometimes, a gauze is incorporated with bar burners to even out internal pressures. This also has the effect of preventing light back, as the flame will not travel back through the gauze. The burner ports are spaced to allow secondary air to reach all the flames yet must be close enough to allow cross-lighting from an igniter situated at one end. To prevent flame lift, retention ports are sometimes incorporated, as explained in Section 2.34. Some burners, referred to as ribbon burners, seen in Figure 2.15, use alternating strips of corrugated and flat metal.

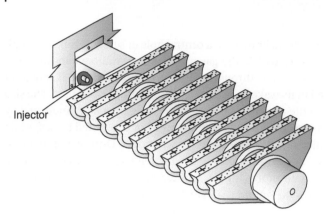

Figure 2.14 Tapered body burner

Figure 2.15 Ribbon burner

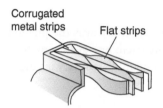

Corrugated metal strips   Flat strips

Pre-mix Burners

2.41   These burners, unlike atmospheric burners, do not have a primary and secondary air supply to the flame; all air is pre-mixed prior to combustion. As a result, the flame characteristics differ in that no outer mantle can be seen as all the gases are fully consumed by the air supplied through the point of intake. These flames tend to be more compact and, as a result, the temperature of the flame is higher. They can therefore have a smaller combustion chamber without risk of impingement and flame chilling (see Section 2.35). For a commercial application, it may be possible that all the air supply could be induced into the burner by raising the pressure of the gas supply. However, a fan that is adjusted to deliver a predetermined flow rate is usually used. The fan may be placed either upstream of the burner, in which case it is called a 'forced draught burner', or downstream, where it would be referred to as an 'induced draught burner'.

Forced Draught Burner

2.42   These burners are found typically on industrial appliances and in large domestic boilers (see Figures 2.16a and 2.16b). A centrifugal fan is used to blow air past the injectors, which are located centrally upstream of the burner ports. As the gas emerges from the injector it initially pre-mixes with the air before entering the burner ports. Here a flame pattern is created that interacts with the discharge from other burner ports, preventing flame lift. The position of the burner head in the housing can be changed to enable the length of the flame to be adjusted and to vary the rate of mixing to obtain the best possible combustion efficiency.

Figure 2.16a Appliance with forced draught burner.

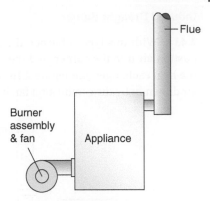

Figure 2.16b Typical forced draught burner.

Most modern systems package all the components in one unit called a package burner. However, the controls may be fitted onto the pipe that serves the burner, in which case it is called a gas train. This design of burner has an automatic control sequence that prevents unburnt gases from being pre-ignited within the combustion chamber and may follow a typical format as shown in Table 2.10.

Table 2.10 Typical ignition sequence for a forced draught burner

Stage	Ignition sequence
1	The air-proving device is checked, and all contacts confirmed to be in the correct position prior to any airflow
2	A pre-purge sequence begins and runs for 30 seconds to clear the combustion chamber of any remaining combustion products. Also, a check is made for the presence of a flame that, if detected, will cause the burner to go into lockout
3	A spark is produced at the burner head whereupon the control box confirms its presence for a minimum period of five seconds
4	Pilot gas is introduced, and the flame stability is checked for a further five seconds, whereupon the spark would go out
5	Main gas burner valve opens and full ignition takes place, the pilot gas terminates and flame monitoring continues throughout the run period

Note: At any time, if the correct sequence is not fully adhered to the burner will shut off the appliance. Some burners will conclude with a post-purge as the appliance shuts down.

Induced Draught Burner

2.43 With this type of burner, the fan serves two purposes: the first is to draw the combustion air into the burner, and the second is to remove the products of combustion from the heat exchanger (see Figure 2.16b). Gas enters the burner assembly and mixes with the air drawn in by the combustion fan before passing through the burner head (Figure 2.16c).

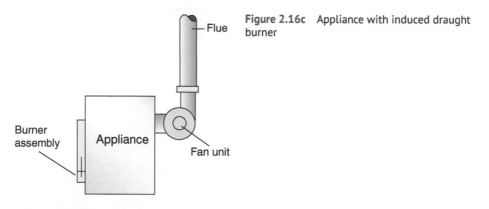

Figure 2.16c Appliance with induced draught burner

Tunnel Burners

2.44 These are specialist burners, associated with high-temperature work such as in a furnace or where rapid heating is required. The gas/air mixture burns within a tunnel of refractory material. The tunnel may form part of the burner design or may be part of the wall furnace itself. The flame is directed toward whatever is to be heated.

Gas Pressure and Flow

2.45 As previously stated, natural gas in the UK comes from underground gas fields beneath the North Sea. The gas pressure there is in the region of 100 bar. Large volumes of gas are distributed through a national gas transmission system and pass into the regional grid network at pressures, ranging from 7 bar down to 30 mbar. It arrives at customers' properties for use at pressures of 18.5 to 23 mbar and is finally used at pressures, ranging from 3 to 20 mbar at the domestic appliance. The pressures at the various stages of gas distribution are shown in Table 2.11.

Table 2.11 Pressures at the various stages of gas distribution.

	High pressure	Intermediate pressure	Medium pressure	Low pressure
Maximum	≤16 bar	≤7 bar	≤2 bar	≤75 mbar
Minimum	>7 bar	>2 bar	>75 mbar	30 mbar

Note: The low pressure may be as low as 19 mbar where the supply cannot be maintained. All pressure ≤75 mbar is regarded as low pressure.

The pressure of an LPG supply depends on the type of gas being used, i.e. propane or butane, and because it is supplied and stored on-site in liquid form, its pressure is subject to the ambient pressure and temperature conditions and is therefore variable. However, typical pressures within a propane vessel would be 6–7 bar, and for butane 1.5–2 bar. The gas supply within a building is reduced to a nominal pressure of 37 mbar for propane and 28 mbar for butane. There will be a variance in these pressures depending on the type of regulator that is installed (see Part 3: Table 3.2).

Gas Flow in Pipes

2.46   When the gas within a pipe is static, the pressure is the same throughout its entire length. However, as soon as the gas begins to flow, the pressure falls progressively from the entry point to the exit. This is called 'pressure absorption' or 'pressure loss' (pressure drop). This is a result of the friction that takes place as the gas rubs along the pipe wall. The rate at which the pressure drops is constant, provided that the pipe diameter and material are the same throughout the length. The illustration (Figure 2.17a) shows four pressure gauges fitted equidistantly along a 9 m pipe. When no gas is flowing, the pressure throughout the pipe is the 'standing pressure'. However, when the gas valve is opened, the gas pressure readings are seen to drop as the gas progresses through the pipe; this is the 'operating pressure', see Figure 2.17b. (See Section 2.56 for methods of taking standing and operating pressure readings.)

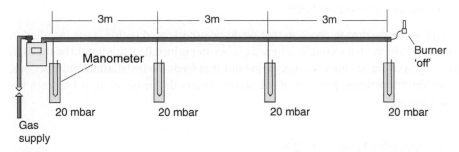

Figure 2.17a   Standing pressure

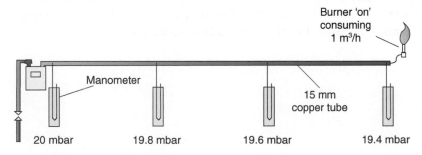

Figure 2.17b   Operating pressure

Factors Affecting Pressure Absorption

2.47   There are five factors that will affect 'pressure absorption' or 'pressure loss':

Length: With double the length of pipe, the pressure loss is doubled. Conversely, if its length is halved, the pressure loss would be halved.

Diameter: The pressure loss is inversely proportional to the fifth power of the pipe diameter (d^5). In simple terms, if you halve the diameter of a pipe, its pressure loss would increase 32 times.

Material: A smooth pipe wall will generate less friction than a coarse surface.

Quantity of gas: The amount of pressure loss is proportional to the square of the volume of gas flowing $(m^3)^2$. In simple terms, double the quantity and the pressure loss increases four-fold.

Specific gravity: Double the weight of the gas and the pressure loss will be doubled. Conversely, halve its weight and the pressure loss will be halved.

Gas Pressure Readings

2.48   The gas pressure in an installation varies and can be identified in several ways as shown in the following sections.

Maximum Incidental Pressure (MIP)

2.49   MIP is the maximum pressure that could be experienced within a gas installation under fault conditions. For example, where a gas booster is installed, this would be the MIP. Alternatively, it could be where a supply regulator that feeds the installation malfunctions. If the maximum operating pressure of the service enters the installation, it becomes the MIP of the supply.

Maximum Operating Pressure (MOP)

2.50   MOP is the maximum pressure experienced within a gas installation under normal operating conditions. This pressure is determined by the regulator that separates the installation from the service. For a typical building using natural gas, this pressure would not exceed 30 mbar, the pressure by which the regulator should lock up.

Operating Pressure

2.51 This is the usual pressure experienced within a gas installation under normal operating conditions. For a natural gas installation, at the point of entry to the building or at the meter, this would typically be 18.5 to 23 mbar. The pressure would progressively drop throughout the length of the system due to pressure absorption or pressure loss, described previously in Section 2.46.

Maximum Pressure Absorption Across the Installation (Pressure Loss/Drop)

2.52 To comply with the appropriate standards, the maximum pressure absorption in a gas installation, after the primary meter, must not exceed 1 mbar for natural gas installations or 2 mbar where LPG is to be used. Where these pressures are exceeded, the pipe would be considered undersized.

Pressure Absorption Across the Natural Gas Meter

2.53 The minimum pressure that the gas supplier must provide at the outlet of the emergency control valve, under peak flow conditions, is 19 mbar. The pressure absorption across the primary meter must not exceed 4 mbar at maximum rate of flow, which would provide a minimum of 15 mbar at the meter outlet. A further maximum of 1 mbar drop could occur across the pipework, leaving only 14 mbar at the appliance – a pressure at which manufacturing standards permit the appliance to operate safely. Ideally, we look for 18.5 to 23 mbar at the outlet of the meter and need to report any shortfall to the supplier. However, they may not be able to improve on the pressure of the gas supplied.

Appliance Burner Pressure

2.54 This is the pressure at which an appliance is designed to operate. This pressure may be the supply pressure, or it may be regulated down to a fixed or variable pressure, as deemed suitable by the manufacturer. Occasionally, appliances are range rated, which require the regulator to be adjusted up or down to suit the heating demand of the property.

Standing/Lock-up Pressure

2.55 The term 'standing pressure' relates to the pressure in a pipe when no gas is flowing. This pressure would not normally exceed the maximum operating pressure identified in Section 2.50. Lock-up pressure refers to the pressure at which a regulator will close to prevent over pressurisation of a system.

See Section 2.56 for a method of taking pressure readings.

Quick Reference Guide to Taking Pressures

Relevant Industry Documents
BS 6400 and IGEM/G/13

2.56 *Methods to Take and Record Pressures:*

Lock-up Pressure Test (see Figure 2.18a)

- Turn off the supply and connect a manometer to the system at a suitable test point. Ensure the gauge is calibrated to zero.
- 'Slowly' open the supply, allowing gas into the system.
- With no appliances operating the lock-up pressure can be read. Maximum lock-up pressure for natural gas is 30 mbar (for LPG see Part 3: Table 3.2).
- Close supply, remove the manometer and reseal the test point.
- Turn on the supply and test the joint with leak detection spray.

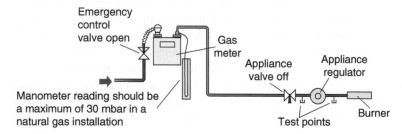

Figure 2.18a   Lock-up pressure (no gas flowing)

Natural Gas Meter Regulator Outlet Operating Pressure (see Figure 2.18b)

- Turn off the supply and connect a manometer to the test point at the meter. Ensure the gauge is calibrated to zero.
- 'Slowly' open the supply, allowing gas into the system.
- Operate all appliances to provide high operating load. This would typically be the highest output appliance running on maximum with all other appliances running at a 50% load. For example, a combination boiler would be put into high load by running hot taps at full flow, whilst simultaneously lighting 2 out of 4 burner rings on a hob.
- Once the appliances are running in high operating load, allow the pressure gauge to stabilise for one minute.
- The pressure should be between 18.5 to 23 mbar for natural gas (for LPG see Part 3: Table 3.2).
- Close supply, remove the manometer and reseal the test point.
- Turn on the supply and test the joint with leak detection spray.

Note that where natural gas is being supplied, if the pressure is not achieved you would need to contact the supplier for advice. You cannot make any adjustments without approval.

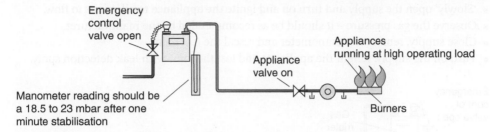

Emergency control valve open

Appliances running at high operating load

Appliance valve on

Manometer reading should be a 18.5 to 23 mbar after one minute stabilisation

Burners

Figure 2.18b   Meter regulator outlet operating pressure (gas flowing)

LPG Regulator Outlet Operating Pressure

- Turn off the supply and connect a manometer to the test point after the regulator.
- 'Slowly' open the supply, allowing gas into the system.
- Operate all appliances at full rate or ensure that there is a flow rate of at least 0.5 m³/h.
- The required pressure can be referenced from the Table 3.2 in Part 3.

Appliance Inlet Operating Pressure (see Figure 2.18c)

- Turn off the supply and connect a manometer to a suitable test point at the appliance, prior to any regulator. Ensure the gauge is calibrated to zero.
- 'Slowly' open the supply, allowing gas to flow to the appliance.
- Turn on and ignite the appliance and any other appliances on the system being tested. (For natural gas the pressure should be no more than 1 mbar below that read at the meter. For LPG the maximum drop should not exceed 2 mbar.)
- Close supply, remove the manometer and reseal the test point.
- Turn on the supply and test the joint with leak detection spray.

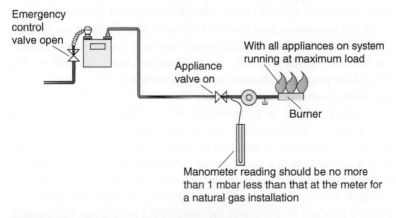

Emergency control valve open

Appliance valve on

With all appliances on system running at maximum load

Burner

Manometer reading should be no more than 1 mbar less than that at the meter for a natural gas installation

Figure 2.18c   Appliance inlet operating pressure

Appliance Burner Pressure (see Figure 2.18d)

- Turn off the supply and connect a manometer at the appliance, after any regulating device. Ensure the gauge is calibrated to zero.

- 'Slowly' open the supply and turn on and ignite the appliance to allow gas to flow.
- Observe the gas pressure – it should be as recommended by the manufacturer.
- Close supply, remove the manometer and reseal the test point.
- Turn on the supply, relight the appliance and test the joint with leak detection spray.

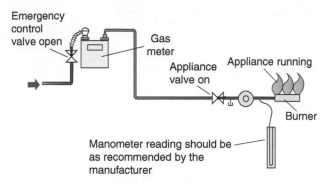

Figure 2.18d   Appliance burner pressure

Estimating a Suitable Pipe Size

Relevant Industry
Documents
BS 6891 and IGEM/UP/2

2.57   Basic pipe sizing is often undertaken using a computer package; however, you can make a simple calculation such as the one illustrated here to give you an indication as to a suitable size. When sizing pipes, the size chosen must take into account the maximum possible flow rate of gas for the appliances on the system. If there are potentially further appliances to be added at a later date, an allowance should be made for these also.

This method enables you to estimate pipe sizes and calculate whether the sizes you have estimated will achieve a pressure drop across the system of less than 1 mbar, when all appliances are running at full rate. Alternatively, you may be planning alterations to an existing system and calculating if the proposed extension to the pipework is viable.

The same method can be adapted to calculate using different pipe materials, such as copper or steel, as well as different gases, be it Natural Gas or LPG. You will need to use the appropriate tables to obtain the figures to carry out the calculation. For the system to be considered suitably sized, a natural gas installation pressure loss must not exceed 1 mbar, for systems using LPG a 2 mbar drop is acceptable, but a different table must be used.

The method described here needs some time and consideration but is quite simple to perform once the concept is understood. It consists of dividing the pipe installation into sections, each of which is to take a proportion of the maximum 1 mbar drop. For example, looking at the system shown in Figure 2.19; to supply gas from the meter at point A to the appliance at point G involves the gas passing through four sections, A–B, B–C, C–D and D–G. So, it follows, that when each of these sections' individual pressure losses are added together, the total must not exceed 1 mbar. An understanding of this principle is a key component of successful pipe sizing.

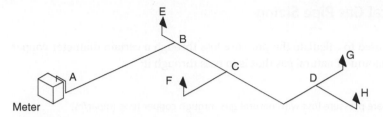

Figure 2.19 Internal installation pipework

2.58 Once the sections have been determined, there are four stages to calculating the pipe size for each section. Completed examples are shown from Section 2.61, Examples A and B, each stage being completed as follows:

Stage 1: Determine the total gas flow in m³/h, kW (*gross*) or kW (*net*) that could pass through the pipe section. This is found by totalling up the rating of all the appliances that are to be fed by the section.

Stage 2: Begin by estimating a suitable pipe size and then calculate the effective pipe length. This is the total measured length plus an allowance for fittings; this can be found in Table 2.12.

Stage 3: Determine the pressure loss per metre using Table 2.13 for copper pipe or Table 2.14 for steel pipe. Start with one of the columns on the left of the table, relevant to whether you are using m³/h, kW (*gross*) or kW (*net*).

Stage 4: Determine the total estimated pressure loss for the section. This is achieved simply by multiplying the adjusted length by the pressure loss per metre.

Table 2.12 Suggested pressure loss due to fittings

Nominal pipe size			Equivalent length (metres)				
			Bends		Fittings		
Steel		Copper	45°	90°	90° Elbow	Tee (flow entering the branch of the tee)	Tee (flow exiting the branch of the tee)
15 mm	R½	≤15 mm	0.15 m	0.20 m	0.40 m	0.75 m	1.20 m
20 mm	R¾	22 mm	0.20 m	0.30 m	0.60 m	1.20 m	1.80 m
25 mm	R1	28 mm	0.25 m	0.40 m	0.80 m	1.50 m	2.30 m
32 mm	R1¼	35 mm	0.30 m	0.50 m	1.00 m	2.00 m	3.00 m

Note 1: With regards to tees, these are to be calculated with the section of pipework that is connected to the branch of the tee. If a reducing tee is to be calculated, the size of the largest connection, not necessarily the branch size is to be used. For example, if you have a 22 mm x 1 5 mm x 15 mm tee, you will calculate using the 22 mm row and select the equivalent length based on whether the pipe is entering the branch or exiting the branch.

Note 2: For press end fittings, pliable corrugated stainless steel tube fittings, flexible connections, secondary meters, regulators etc. consult manufacturer's instructions.

Source: Data from BS 6891

Domestic Natural Gas Pipe Sizing

2.59 Table 2.13 is used to calculate the pressure loss through a certain diameter copper pipe related to the amount of natural gas that is to pass through it:

Table 2.13 Approximate pressure loss with natural gas through copper tube (mbar/m)

| Flow rate (m³/h) | Heat input kW (gross) | Heat Input kW (net) | Nominal pipe size (mm) with pressure loss per metre length | | | | | | |
			8	10	12	15	22	28	35
0.25	2.70	2.46	0.2675	0.0710	0.0255	0.0077	0.0014	0.0004	0.0001
0.50	5.40	4.91	0.8348	0.2188	0.0777	0.0231	0.0040	0.0011	0.0004
0.75	8.10	7.37	–	0.4285	0.1514	0.0447	0.0077	0.0022	0.0007
1.00	10.81	9.82	–	0.6940	0.2444	0.0719	0.0123	0.0035	0.0011
1.25	13.51	12.28	–	–	0.3553	0.1042	0.0178	0.0050	0.0016
1.50	16.21	14.73	–	–	0.4833	0.1414	0.0240	0.0067	0.0021
1.75	18.91	17.19	–	–	0.6276	0.1832	0.0311	0.0086	0.0027
2.00	21.61	19.65	–	–	0.7877	0.2296	0.0388	0.0108	0.0034
2.25	24.31	22.10	–	–	0.9630	0.2804	0.0473	0.0131	0.0042
2.50	27.01	24.56	–	–	–	0.3353	0.0565	0.0156	0.0049
2.75	29.72	27.01	–	–	–	0.3945	0.0663	0.0183	0.0058
3.00	32.42	29.47	–	–	–	0.4577	0.0769	0.0212	0.0067
3.25	35.12	31.93	–	–	–	0.5249	0.0880	0.0243	0.0077
3.50	37.82	34.38	–	–	–	0.5960	0.0998	0.0275	0.0087
3.75	40.52	36.84	–	–	–	0.6709	0.1123	0.0309	0.0097
4.00	43.22	39.29	–	–	–	0.7496	0.1253	0.0345	0.0108
4.25	45.92	41.75	–	–	–	0.8321	0.1390	0.0382	0.0120
4.50	48.63	44.20	–	–	–	0.9182	0.1533	0.0421	0.0132
4.75	51.33	46.66	–	–	–	–	0.1681	0.0462	0.0145
5.00	54.03	49.12	–	–	–	–	0.1836	0.0504	0.0158
5.25	56.73	51.57	–	–	–	–	0.1996	0.0548	0.0172
5.50	59.43	54.03	–	–	–	–	0.2162	0.0593	0.0186
5.75	62.13	56.48	–	–	–	–	0.2334	0.0640	0.0200
6.00	64.83	58.94	–	–	–	–	0.2511	0.0688	0.0215
6.25	67.53	61.40	–	–	–	–	0.2694	0.0738	0.0231
6.50	70.24	63.85	–	–	–	–	0.2882	0.0789	0.0247
6.75	72.94	66.31	–	–	–	–	0.3076	0.0842	0.0263
7.00	75.64	68.76	–	–	–	–	0.3275	0.0896	0.0280
7.25	78.34	71.22	–	–	–	–	0.3480	0.0952	0.0297

Source: Data from BS 6891

2.60 Table 2.14 is used to calculate the pressure loss through a certain diameter steel pipe related to the amount of natural gas that is to pass through it:

Table 2.14 Approximate pressure loss with natural gas through steel pipe (mbar/m)

Flow rate (m³/h)	Heat input kW (gross)	Heat input kW (net)	8 mm R 1/4	10 mm R 3/8	15 mm R 1/2	20 mm R 3/4	25 mm R 1	32 mm R 11/4
0.25	2.70	2.46	0.0621	0.0130	0.0039	0.0010	0.0004	0.0001
0.50	5.40	4.91	0.1908	0.0393	0.0115	0.0029	0.0010	0.0003
0.75	8.10	7.37	0.3731	0.0761	0.0221	0.0056	0.0019	0.0005
1.00	10.81	9.82	0.6037	0.1225	0.0354	0.0089	0.0030	0.0008
1.25	13.51	12.28	0.8792	0.1777	0.0512	0.0128	0.0043	0.0012
1.50	16.21	14.73	–	0.2414	0.0693	0.0172	0.0058	0.0016
1.75	18.91	17.19	–	0.3130	0.0897	0.0222	0.0075	0.0020
2.00	21.61	19.65	–	0.3924	0.1123	0.0278	0.0094	0.0025
2.25	24.31	22.10	–	0.4793	0.1370	0.0338	0.0114	0.0031
2.50	27.01	24.56	–	0.5735	0.1637	0.0404	0.0136	0.0037
2.75	29.72	27.01	–	0.6748	0.1924	0.0474	0.0159	0.0043
3.00	32.42	29.47	–	0.7831	0.2230	0.0549	0.0184	0.0050
3.25	35.12	31.93	–	0.8982	0.2556	0.0628	0.0210	0.0057
3.50	37.82	34.38	–	–	0.2901	0.0712	0.0238	0.0064
3.75	40.52	36.84	–	–	0.3264	0.0801	0.0268	0.0072
4.00	43.22	39.29	–	–	0.3645	0.0893	0.0269	0.0080
4.25	45.92	41.75	–	–	0.4044	0.0991	0.0331	0.0089
4.50	48.63	44.20	–	–	0.4461	0.1092	0.0364	0.0098
4.75	51.33	46.66	–	–	0.4896	0.1198	0.0400	0.0107
5.00	54.03	49.12	–	–	0.5347	0.1307	0.0436	0.0117
5.25	56.73	51.57	–	–	0.5816	0.1421	0.0474	0.0127
5.50	59.43	54.03	–	–	0.6302	0.1539	0.0513	0.0137
5.75	62.13	56.48	–	–	0.6804	0.1661	0.0553	0.0148
6.00	64.83	58.94	–	–	0.7323	0.1787	0.0595	0.0159
6.25	67.53	61.40	–	–	0.7859	0.1917	0.0638	0.0171
6.50	70.24	63.85	–	–	0.8410	0.2050	0.0682	0.0182
6.75	72.94	66.31	–	–	0.8978	0.2188	0.0727	0.0195
7.00	75.64	68.76	–	–	0.9562	0.2329	0.0774	0.0207
7.25	78.34	71.22	–	–	–	0.2474	0.0822	0.0220

Source: Data from BS 6891

Example A

2.61   Figure 2.20 shows an example of a typical natural gas installation; to illustrate, we will calculate the appropriate pipe sizes, using Tables 2.12 and 2.13 from the previous pages and input the data onto the following grid.

Section	Stage 1	Stage 2	Stage 3	Stage 4
	Find the total gas flow in m³/h, *Max.* kW (gross) or *Max.* kW (net)	Determine the adjusted pipe length from Table 2.12 *Actual length + allowance for fittings*	Determine the pressure loss per metre using Table 2.13	Determine total pressure loss for section
A–B Estimated 28 mm	All appliances are supplied, and information is in kW (gross) therefore:	Measured length = 5 m. 0.8 m is added for each 28 mm elbow of which	Heat input kW (gross) column look to find at least 53.8 row	Multiply adjusted pipe length by the pressure loss per metre $\therefore 8.2 \times 0.0504 = \mathbf{0.42}$
	$32 + 17.2 + 4.6 = 53.8$ kW (gross)	there are four $\therefore 4 \times 0.8 + 5m = $ **8.2 m**	Nearest is 54.03. Go along row to 28 mm column – 0.0504	$\therefore 8.2 \times 0.0504 = \mathbf{0.42}$ Rounded up to 2 decimal places
B–C Estimated 22 mm	The cooker and fire are supplied therefore: $17.2 + 4.6 = 21.8$ kW (gross)	Measured length = 2 m. There are no fittings $\therefore$ adjusted length is 2 m	Heat input kW (gross) column look to find at least 21.8 row Nearest is 24.31. Go along row to 22 mm column – 0.0473	Multiply adjusted pipe length by the pressure loss per metre $\therefore$ $2 \times 0.0473 = \mathbf{0.10}$ Rounded up to 2 decimal places
C–D Estimated 22 mm	Only the cooker, therefore, 17.2 kW (gross)	Measured length = 2 m. 0.6 m is added for each 22 mm elbow of which there are two $\therefore 2 \times 0.6 + 2 m = 3.2$ m	Heat input kW (gross) column, look to find at least 17.2 row Nearest is 18.91. Go along row to 22 mm column – 0.0311	Multiply adjusted pipe length by the pressure loss per metre $\therefore 3.2 \times 0.0311 = \mathbf{0.10}$ Rounded up to 2 decimal places
C–F Estimated 12 mm	Only the fire, therefore: 4.6 kW (gross)	Measured length = 2.5 m. 0.4 m is added for each 12 mm elbow of which there is one and for the tee (22 mm) which exits the branch $\therefore 0.4 + 1.8 + 2.5 m = 4.7$ m	Heat input kW (gross) column look to find at least 4.6 row Nearest is 5.4. Go along row to 12 mm column – 0.0777	Multiply adjusted pipe length by the pressure loss per metre $\therefore 4.7 \times 0.0777 = \mathbf{0.37}$ Rounded up to 2 decimal places

Section	Stage 1	Stage 2	Stage 3	Stage 4
B–E Estimated 22 mm	Only the boiler, therefore: 32 kW (gross)	Measured length = 1 m. 0.6 m is added for each 22 mm elbow of which there is one and for the tee (28 mm) which exits the branch ∴ 0.6 + 2.3 + 1 m = 3.9 m	Heat input kW (gross) column look to find at least 32 row Nearest is 32.42. Go along row to 22 mm column – 0.0769	Multiply adjusted pipe length by the pressure loss per metre ∴ 3.9 × 0.0769 = **0.30** Rounded up to 2 decimal places

As can be seen, the calculation was done in four stages the resultant figure at stage 4 is the pressure loss of the individual sections. The final part of the calculation is to confirm that the total pressure loss from the meter to the appliance inlets is no more than 1 mbar. So, one more sum is necessary for each pipe route, the sums are:

Meter to boiler: $(A - B)\,0.42 + (B - E)\,0.30 = \mathbf{0.72\ mbar}$

Meter to cooker: $(A - B)\,0.42 + (B - C)\,0.10 + (C - D)\,0.10 = \mathbf{0.62\ mbar}$

From meter to fire: $(A - B)\,0.42 + (B - C)\,0.10 + (C - F)\,0.37 = \mathbf{0.89\ mbar}$

Each pipe route is confirmed as having a pressure loss of less than **1 mbar**, therefore is an acceptable design.

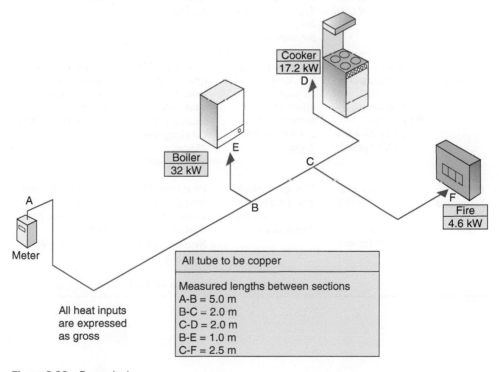

Figure 2.20 Example A

Domestic LPG Pipe Sizing

2.62   The same method of calculation that is used for natural gas can also be used for sizing LPG installations. Because of its higher specific gravity and its higher operating gas pressure, LPG requires smaller pipe diameters. Therefore, a natural gas system could easily be converted to LPG later without fear of any problems of pipe size. To prove this, we will now carry out a calculation with the same installation (Figure 2.20) that we have just calculated for natural gas but this time we will calculate for LPG instead.

2.63   The following Tables are for LPG, Table 2.15 is for copper and Table 2.16 is for steel. Table 2.12 is used for fittings as before.

Example B

2.64   To illustrate we will calculate the appropriate pipe sizes, using Tables 2.12 and 2.15 for the installation in Figure 2.20, that we have just calculated for natural gas but this time we will calculate for LPG instead,

Section	Stage 1	Stage 2	Stage 3	Stage 4
	Find the total gas flow in m³/h, *Max*. kW (gross) or *Max*. kW (net)	Determine the adjusted pipe length from Table 2.12 *Actual length + allowance for fittings*	Determine the pressure loss per metre using Table 2.15	Determine total pressure loss for section
A–B Estimated 22 mm	All appliances are supplied, and information is in kW (gross) therefore: $32 + 17.2 + 4.6 = 53.8$ kW (gross)	Measured length= 5 m. 0.6 m is added for each 22 mm elbow of which there are four ∴ $4 \times 0.6 + 5$ m = 7.4 m	Heat input kW (gross) column looks to find at least 53.8 row Nearest is 58.19. Go along row to 22 mm column – 0.0829	Multiply adjusted pipe length by the pressure loss per metre ∴ $7.4 \times 0.0829 = \mathbf{0.62}$
B–C Estimated 15 mm	The cooker and fire are supplied, therefore: $17.2 + 4.6 = 21.8$ kW (gross)	Measured length = 2 m. There are no fittings ∴ adjusted length is 2 m	Heat input kW (gross) column looks to find at least 21.8 row Nearest is 25.86. Go along row to 15 mm column – 0.1231	Multiply adjusted pipe length by the pressure loss per metre ∴ $2 \times 0.1231 = \mathbf{0.25}$

Section	Stage 1	Stage 2	Stage 3	Stage 4
C–D Estimated 12 mm	Only the cooker, therefore, 17.2 kW (gross)	Measured length = 2 m. 0.4 m is added for each 12 mm elbow of which there are two ∴ 2 × 0.4 + 2 m = 2.8 m	Heat input kW (gross) column looks to find at least 17.2 row Nearest is 19.4. Go along row to 12 mm column – 0.2590	Multiply adjusted pipe length by the pressure loss per metre ∴ 2.8 × 0.2590 = **0.73**
C–F Estimated 10 mm	Only the fire, therefore: 4.6 kW (gross)	Measured length = 2.5 m. 0.4 m is added for each 8 mm elbow of which there is one and for the tee (12 mm) which exits the branch ∴ 0.4 + 1.2 + 2.5 m = 4.1 m	Heat input kW (gross) column looks to find at least 4.6 row Nearest is 6.47. Go along row to 10 mm column – 0.1148	Multiply adjusted pipe length by the pressure loss per metre ∴ 4.1 × 0.1148 = **0.48**
B–E Estimated 15 mm	Only the boiler, therefore: 32 kW (gross)	Measured length = 1 m. From 0.4 m is added for each 15 mm elbow of which there is one and for the tee (22 mm) which exits the branch ∴ 0.4 + 1.8 + 1 m = 3.2 m	Heat input kW (gross) column look to find at least 32 row Nearest is 38.79. Go along row to 15 mm column – 0.2474	Multiply adjusted pipe length by the pressure loss per metre ∴ 3.2 × 0.2474 = **0.80**

The calculation was done in four stages and can be seen on the chart. The resultant figure at stage 4 is the pressure loss of the individual sections. As can be seen, the estimated pipework size is significantly smaller than the natural gas solution. The final part of the calculation is to confirm that the total pressure loss from the meter to the appliance inlets is no more than 2 mbar. So, one more sum is necessary for each pipe route, the sums are:

Meter to boiler: $(A - B)\,0.62 + (B - E)\,0.80 = $ **1.42 mbar**

Metre to cooker: $(A - B)\,0.62 + (B - C)\,0.25 + (C - D)\,0.73 = $ **1.60 mbar**

From meter to fire: $(A - B)\,0.62 + (B - C)\,0.25 + (C - F)\,0.48 = $ **1.35 mbar**

Each pipe route is confirmed as having a pressure loss of less than **2 mbar**, therefore, it is an acceptable design.

Table 2.15 Approximate pressure loss with LPG (propane) through copper tube (**mbar/m**)

Flow rate (m³/h)	Heat input kW (gross)	Heat input kW (net)	Nominal pipe size (mm) with pressure loss per metre length						
			8	10	12	15	22	28	35
0.25	6.47	5.88	0.4413	0.1148	0.0405	0.0119	0.0021	0.0006	0.0002
0.50	12.93	11.76	1.4367	0.3705	0.1298	0.0379	0.0064	0.0018	0.0006
0.75	19.40	17.63	–	0.7421	0.2590	0.0753	0.0127	0.0035	0.0011
1.00	25.86	23.51	–	1.2194	0.4246	0.1231	0.0206	0.0057	0.0018
1.25	32.33	29.39	–	1.7959	0.6243	0.1807	0.0302	0.0083	0.0026
1.50	38.79	35.27	–	–	0.8565	0.2474	0.0412	0.0113	0.0035
1.75	45.26	41.14	–	–	1.1200	0.3231	0.0537	0.0147	0.0046
2.00	51.72	47.02	–	–	1.4137	0.4074	0.0676	0.0185	0.0058
2.25	58.19	52.90	–	–	1.7369	0.5001	0.0829	0.0227	0.0071
2.50	64.65	58.78	–	–	–	0.6009	0.0995	0.0272	0.0085
2.75	71.12	64.65	–	–	–	0.7097	0.1174	0.0320	0.0100
3.00	77.58	70.53	–	–	–	0.8264	0.1365	0.0372	0.0116
3.25	84.05	76.41	–	–	–	0.9508	0.1569	0.0428	0.0133
3.50	90.51	82.29	–	–	–	1.0827	0.1786	0.0486	0.0151
3.75	96.98	88.16	–	–	–	1.2222	0.2014	0.0548	0.0170
4.00	103.44	94.04	–	–	–	1.3690	0.2254	0.0613	0.0191
4.25	109.91	99.92	–	–	–	1.5230	0.2506	0.0682	0.0212
4.50	116.38	105.80	–	–	–	1.6843	0.2770	0.0753	0.0234
4.75	122.84	111.67	–	–	–	1.8527	0.3045	0.0827	0.0257
5.00	129.31	117.55	–	–	–	–	0.3332	0.0905	0.0281
5.25	135.77	123.43	–	–	–	–	0.3630	0.0985	0.0305
5.50	142.24	129.31	–	–	–	–	0.3939	0.1069	0.0331
5.75	148.70	135.18	–	–	–	–	0.4259	0.1155	0.0358
6.00	155.17	141.06	–	–	–	–	0.4590	0.1245	0.0385
6.25	161.63	146.94	–	–	–	–	0.4932	0.1337	0.0414
6.50	168.10	152.82	–	–	–	–	0.5284	0.1432	0.0443
6.75	174.56	158.69	–	–	–	–	0.5648	0.1530	0.0473
7.00	181.03	164.57	–	–	–	–	0.6022	0.1631	0.0505
7.25	187.49	170.45	–	–	–	–	0.6406	0.1735	0.0537

Source: Data from BS 6891

Table 2.16 Approximate pressure loss with LPG (propane) through steel pipe (**mbar/m**)

Flow rate (m³/h)	Heat input kW (gross)	Heat input kW (net)	Nominal pipe size (mm/R) with pressure loss per metre length					
			8 mm R 1/4	10 mm R 3/8	15 mm R 1/2	20 mm R 3/4	25 mm R 1	32 mm R 11/4
0.25	6.47	5.88	0.0999	0.0203	0.0059	0.0015	0.0005	0.0001
0.50	12.93	11.76	0.3217	0.0647	0.0186	0.0046	0.0016	0.0004
0.75	19.40	17.63	0.6437	0.1288	0.0368	0.0091	0.0030	0.0008
1.00	25.86	23.51	1.0569	0.2107	0.600	0.0147	0.0049	0.0013
1.25	32.33	29.39	1.5558	0.3094	0.0878	0.0215	0.0072	0.0019
1.50	38.79	35.27	–	0.4239	0.1201	0.0294	0.0098	0.0026
1.75	45.26	41.14	–	0.5538	0.1567	0.0382	0.0127	0.0034
2.00	51.72	47.02	–	0.6985	0.1974	0.0481	0.0160	0.0043
2.25	58.19	52.90	–	0.8575	0.2421	0.0589	0.0196	0.0052
2.50	64.65	58.78	–	1.0307	0.2908	0.0707	0.0235	0.0063
2.75	71.12	64.65	–	1.2176	0.3432	0.0834	0.0277	0.0074
3.00	77.58	70.53	–	1.4180	0.3994	0.0970	0.0321	0.0086
3.25	84.05	76.41	–	1.6317	0.4594	0.1114	0.0369	0.0098
3.50	90.51	82.29	–	1.8584	0.5229	0.1267	0.0420	0.0112
3.75	96.98	88.16	–	–	0.5900	0.1429	0.0473	0.0126
4.00	103.44	94.04	–	–	0.6606	0.1599	0.0529	0.0140
4.25	109.91	99.92	–	–	0.7347	0.1778	0.0588	0.0156
4.50	116.38	105.80	–	–	0.8123	0.1965	0.0649	0.0172
4.75	122.84	111.67	–	–	0.8933	0.2159	0.0713	0.0189
5.00	129.31	117.55	–	–	0.9776	0.2362	0.0780	0.0207
5.25	135.77	123.43	–	–	1.0652	0.2573	0.0849	0.0225
5.50	142.24	129.31	–	–	1.1562	0.2791	0.0921	0.0244
5.75	148.70	135.18	–	–	1.2504	0.3018	0.0995	0.0263
6.00	155.17	141.06	–	–	1.3478	0.3252	0.1072	0.0284
6.25	161.63	146.94	–	–	1.4485	0.3494	0.1152	0.0305
6.50	168.10	152.82	–	–	1.5524	0.3743	0.1234	0.0326
6.75	174.56	158.69	–	–	1.6594	0.4000	0.1318	0.0348
7.00	181.03	164.57	–	–	1.7695	0.4264	0.1405	0.0371
7.25	187.49	170.45	–	–	1.8828	0.4536	0.1494	0.0395

Source: Data from BS 6891

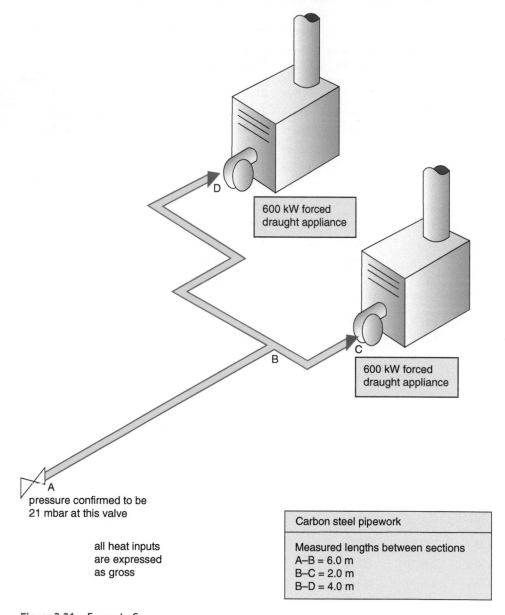

600 kW forced
draught appliance

600 kW forced
draught appliance

A
pressure confirmed to be
21 mbar at this valve

all heat inputs
are expressed
as gross

Carbon steel pipework
Measured lengths between sections A–B = 6.0 m B–C = 2.0 m B–D = 4.0 m

Figure 2.21 Example C

Commercial Natural Gas Pipe Sizing

2.65   This method is slightly different from the previous, as with many commercial installations, steel pipe has been selected and the appliances are much larger. The calculation for Example C is for the installation illustrated in Figure 2.21. You will notice that **kW** input is required to be converted to **m³/h**. This is done by the following sum:

$$kW \times 0.094 = m^3/h.$$

Reference is made to Tables 2.17 and 2.18.

Table 2.17   Suggested pressure loss due to fittings

Nominal size of pipe (mm)			Equivalent length (m)				
Carbon and stainless-steel pipe nominal bore	Copper nominal diameter	PE nominal diameter	45° bend, 90° long bend, bush & socket (one change of size)	90° bend, full bore valve, union, adapter, flange joint, through tee	90°elbow, bush & socket (more than one change of size)	Tee; flow entering the branch of the tee	Tee; flow exiting the branch of the tee
15	18	–	0.15	0.20	0.40	0.75	1.2
22	–	–	0.20	0.30	0.60	1.20	1.8
25	28	32	0.25	0.40	0.80	1.50	2.3
32	35	–	0.30	0.50	1.00	2.00	3.0
40	42	55	0.40	0.60	1.20	2.40	3.5
50	54	63	0.50	0.80	1.50	3.00	4.5
65	67	–	0.70	1.00	2.00	4.00	5.5
80	76	90	0.80	1.20	2.30	4.50	6.6
100	108	125	1.00	1.50	3.00	6.00	9.0
150	–	180	1.50	2.30	4.50	9.00	13.5
200	–	250	2.00	3.00	6.00	12.00	18.0
250	–	–	2.50	3.80	7.60	15.00	22.5

Source: Data from IGEM/UP/2

Table 2.18   Flow discharge through steel tube

Pipe size	Max. length: allowing 1 mbar pressure differential between each end									
	5	10	15	20	30	40	50	75	100	150
15	2.5	1.8	1.5	1.2						
20	6.5	4.3	3.4	2.9	2.3	2	1.7			
25	12	8	6	5.5	4.5	3.5	2			
32	19	15.5	13	11	8.5	7.5	6.5	5	4.5	
40	38	28	23	19	15	13	11	9	7	
50	75	52	42	35	28	24	22	18	15	12
65	140	95	80	65	55	45	40	32	27	22
80	240	170	135	115	90	85	75	60	50	40
100	500	370	250	210	170	150	130	110	90	70
150	1300	950	750	650	550	460	420	350	300	250

Discharge in m^3/h

Source: Data from IGEM/UP/2

Example C

Section	Stage 1	Stage 2	Stage 3	Stage 4
	Find the total gas flow in m^3 *Max.* kW $\times$ 0.094	Determine the effective pipe length using Table 2.17 *Actual length + Fittings*	Determine the pipe length for sizing purposes using Table 2.18	Go down column to required flow i.e. no lower than figure from stage 1. Correct pipe size will be in left-hand column
A–B Estimated size 80 mm	Both appliances are supplied, therefore: $600 + 600 =$ 1200 kW $\times$ 0.094 = 112.8 m^3/h	Measured length = 6 m + 4.5 m is added for the tee which the gas will enter $\therefore 6 + 4.5 = \underline{10.5\ m}$	Table shows no column for 10.5 m. You therefore use the figure from the 15 m column which is the next size up	Suggested pipe size of 80 mm confirmed suitable
B–C Estimated size 40 mm	Only one appliance supplied, therefore: 600 kW $\times$ 0.094 = 56.4 m^3/h	Measured length = 2 m + an amount is added for the one elbow (estimated size 40 mm) $\therefore 2 + 1.2 = \underline{3.2\ m}$	Table shows no column for 3.2 m. You therefore use the figure from the 5 m column which is the next size up	Suggested pipe size of 40 mm @ Stage 2 *not suitable* $\therefore$ ***Re-do calculations with larger size pipe!***

Section	Stage 1	Stage 2	Stage 3	Stage 4
B–C second Estimated size 50 mm	Only one appliance supplied, therefore: $600\,kW \times 0.094 =$ <u>$56.4\,m^3/h$</u>	Measured length = 2 m + an amount is added for the one elbow (estimated size 50 mm) $\therefore 2 + 1.5 =$ <u>$3.5\,m$</u>	Table shows no column for 3.5 m. You therefore use the figure from the 5 m column which is the next size up	Suggested pipe size of <u>50 mm</u> confirmed suitable
B–D Estimated size 65 mm	Only one appliance supplied, therefore: $600\,kW \times 0.094 =$ <u>$56.4\,m^3/h$</u>	Measured length = 4 m + an amount is added for the three elbows (estimated size 65 mm) $\therefore 4 + (3 \times 2) =$ <u>$10\,m$</u>	Table has column for 10 m	Suggested pipe size of <u>65 mm</u> confirmed suitable

Gas Rates and Heat Input

2.66   The gas rate refers to the amount of gas used in m^3, usually over one hour, whereas the heat input refers to the amount of heat generated in kW from burning a certain quantity of fuel.

To determine the gas consumption of an appliance, a gas meter is used to measure the amount of gas consumed in one hour in m^3. There are several types of gas meter, see Figure 2.22, providing readings in different units. Occasionally, one is found reading in

Figure 2.22   Typical gas meters

cubic metres per hour (m^3/h). However, for payment purposes, it only registers in either ft^3 or m^3 and not per hour.

It would be uneconomical to run an appliance for an hour just to determine its gas flow over this period; therefore, some kind of estimate is made, depending on the meter type. However, it is necessary to let the appliance run for about 10 minutes to allow it to warm up. This allows the injectors to heat up and expand and takes account of the effects of the physical laws relating to gas (see Sections 2.74–2.76).

Taking the Gas Rate with an Imperial Meter (Reading in ft^3)

2.67 Run the appliance and observe the test dial, see Figure 2.23, timing how long it takes, in seconds, to consume 1 ft^3 of gas. For the smaller U6 meter this would be one complete revolution of the test dial. For larger meters, 1 ft^3 may only be 1/10 of the test dial. For larger commercial appliances, which use large volumes of gas, a reading of, say, 10 ft^3 may be taken and the time taken for the gas to be consumed divided by 10, to give the time for 1 ft^3.

Figure 2.23 Test dial of an imperial meter

Knowing the time in seconds (t) to burn 1 ft^3, and with 3600 seconds in an hour, it is a simple process to calculate the gas rate per hour by understanding the following sum:

$$3600 \div t = ft^3/h$$

To convert ft^3/h into m^3/h divide by a conversion factor of 35.3.

Example: If we find that it takes 39 seconds to consume 1 ft^3, then consumption per hour is:

$$3600 \div 39 = 92.3 \, ft^3/h \div 35.3 = \underline{2.6 \, m^3/h}$$

Finding the Gas Rate from a Metric Meter (in m^3)

2.68 Record the reading from the meter reading index, see Figure 2.24, then take it again after two minutes. The amount of gas used in two minutes is the difference between the two readings (i.e. second reading minus the first). To find the gas rate over an hour in this example of a two-minute test you simply multiply by 30 (as $2 \times 30 = 60$ minutes).

Example Assume the index at the meter reads 07341.926 at the beginning of the two-minute test period and 07341.954 two minutes later. The consumption per hour is:

$$07341.954 - 07341.926 = 0.028 \, m^3 \text{ in 2 minutes} \times 30 = \underline{0.84 \, m^3/h}$$

Figure 2.24 A metric meter index

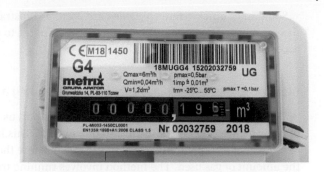

Finding the Gross Heat Input

Natural Gas

2.69 The gross heat input is expressed in kilowatt-hours and is found by multiplying the gas rate in m^3/h by the calorific value (CV) and then dividing by 3.6.

Example: To find the heat input where 0.84 m^3/h of gas is consumed. (Take the CV to be 38.5.)

$$\therefore 0.84 \times 38.5 \div 3.6 = 8.98 \text{ kWh}$$

Note: The 'h' is often omitted, giving 8.98 kW.

LPG

If we reconsidered the previous example with a CV of 95 (the CV for propane), the heat input would be:

$$0.84 \times 95 \div 3.6 = 22.16 \text{ kW}$$

This demonstrates the importance of using the correct calorific value for the fuel gas in use. Typical calorific values can be found by referring to Table 2.2.

Finding the Net Heat Input

2.70 The net heat input (see Section 2.10) is found by dividing the gross input by 1.1.

Example: Taking the gross input from the previous example as 22.16 kW, the net heat input will be:

$$22.16 \div 1.1 = 20.14 \text{ kW}.$$

Having found the heat input, one can check it against the manufacturer's data badge to confirm that the appliance is being supplied with the correct quantity of gas. When comparing your result against the manufacturer's figure from their technical data, there is an allowable tolerance of 5% above or 10% below. This is because your result may be based on a different CV from that used by the manufacturer.

Note: For older appliances, heat input may be indicated as gross, and for newer appliances, it should be recorded as a net.

Sometimes quick reference tables are used, as illustrated in Section 2.73, to determine the gas rate and heat input. However, they are limited to smaller-sized appliances and take account of only one particular CV.

Alternative Method for Gas Rate (Heat Input) Using a Metric Meter

2.71   An alternative method for gas rate (heat input) using a metric meter has been developed to reduce the time of the test to one minute. This has the advantage of reducing the likelihood of a larger appliance shutting down before the end of a test as well as reducing the amount of gas used. The method involves running the appliance at full rate, recording the reading from the meter index and then again after 60 seconds (The time taken is recorded as 60 seconds plus any additional seconds for the digits to change to a full unit). The amount of gas used in the time taken is the difference between the two readings (second reading minus the first). To find the heat input you apply the following formula:

$$(3600 \times m^3 \times 38.5 \div 3.6) \div (60 + \text{additional seconds for digits to change})$$

Worked Example: Assume the index at the meter reads 07341.943 at the beginning of the test period and 07341.954 at the end. The test took 61 seconds. The consumption per hour is:

$$(3600 \times 0.011 \times 38.5 \div 3.6)\ \div 61 = 6.94\,\text{kW gross}$$

The net heat input is found by dividing the gross input by 1.1.

Note: The relevant CV can be inserted for a different gas type.

Electronic and Smart Meters

2.72   Meters have evolved and there are several different manufacturers supplying various electronic smart versions, see Figure 2.25, all with differing methods to enable a gas rate to be performed. Therefore, the Gas Safe Register has published Technical Bulletin 112 which is a useful guide for an engineer when one is encountered.

Figure 2.25   Typical smart meter

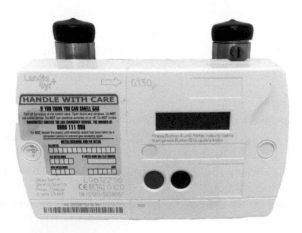

Quick Reference Tables: Gas Rate/Heat Input

2.73 The tables produced here are only suitable for domestic natural gas installations with a calorific value (CV) of 38.5. If the CV increases in value, the figures indicated would reduce. For a more accurate reading see Sections 2.66–2.70.

To determine the gas rate/heat input of an appliance, all other appliances in the system need to be turned off and the appliance in question run for about 10 minutes to enable it to warm up. The following procedure then needs to be adopted.

Meters Reading in m³, e.g. E6 and G4

Record the gas volume used over a two-minute period by recording a reading then take a second reading after two minutes, deducting the first figure from the second. Then refer to Table 2.19.

Meters Reading in ft³, e.g. U6

Observe the test dial on the meter and record the time in seconds that it takes to complete a full circle, i.e. 1 ft³. Then refer to Table 2.20.

Table 2.19 Gas rate and heat input for E6 and G4 type gas meters (average CV = 38.5)

2 min gas flow m³	Gas rate m³/h	Gross input (kW)	Net input (kW)	2 min gas flow m³	Gas rate m³/h	Gross input (kW)	Net input (kW)	2 min gas flow m³	Gas rate m³/h	Gross input (kW)	Net input (kW)
0.010	0.30	3.21	2.92	0.074	2.22	23.73	21.58	0.138	4.14	44.26	40.24
0.012	0.36	3.85	3.50	0.076	2.28	24.37	22.16	0.140	4.20	44.90	40.82
0.014	0.42	4.49	4.08	0.078	2.34	25.01	22.74	0.142	4.26	45.54	41.41
0.016	0.48	5.13	4.67	0.080	2.40	25.66	23.33	0.144	4.32	46.18	41.99
0.018	0.54	5.77	5.25	0.082	2.46	26.30	23.91	0.146	4.38	46.82	42.57
0.020	0.60	6.41	5.83	0.084	2.52	26.94	24.49	0.148	4.44	47.46	43.16
0.022	0.66	7.06	6.42	0.086	2.58	27.58	25.08	0.150	4.50	48.11	43.74
0.024	0.72	7.70	7.00	0.088	2.64	28.22	25.66	0.152	4.56	48.75	44.32
0.026	0.78	8.34	7.58	0.090	2.70	28.86	26.24	0.154	4.62	49.39	44.91
0.028	0.84	8.98	8.16	0.092	2.76	29.50	26.83	0.156	4.68	50.03	45.49
0.030	0.90	9.62	8.75	0.094	2.82	30.15	27.41	0.158	4.74	50.67	46.07
0.032	0.96	10.26	9.33	0.096	2.88	30.79	27.99	0.160	4.80	51.31	46.66
0.034	1.02	10.90	9.91	0.098	2.94	31.43	28.58	0.162	4.86	51.95	47.24
0.036	1.08	11.55	10.50	0.100	3.00	32.07	29.16	0.164	4.92	52.59	47.82
0.038	1.14	12.19	11.08	0.102	3.06	32.71	29.74	0.166	4.98	53.24	48.41
0.040	1.20	12.83	11.66	0.104	3.12	33.35	30.33	0.168	5.04	53.88	48.99
0.042	1.26	13.47	12.25	0.106	3.18	33.99	30.91	0.170	5.10	54.52	49.57

(Continued)

Table 2.19 (Continued)

2 min gas flow m³	Gas rate m³/h	Gross input (kW)	Net input (kW)	2 min gas flow m³	Gas rate m³/h	Gross input (kW)	Net input (kW)	2 min gas flow m³	Gas rate m³/h	Gross input (kW)	Net input (kW)
0.044	1.32	14.11	12.83	0.108	3.24	34.64	31.49	0.172	5.16	55.16	50.16
0.046	1.38	14.75	13.41	0.110	3.30	35.28	32.08	0.174	5.22	55.80	50.74
0.048	1.44	15.39	14.00	0.112	3.36	35.92	32.66	0.176	5.28	56.44	51.32
0.050	1.50	16.04	14.58	0.114	3.42	36.56	33.24	0.178	5.34	57.08	51.90
0.052	1.56	16.68	15.16	0.116	3.48	37.20	33.83	0.180	5.40	57.73	52.49
0.054	1.62	17.32	15.75	0.118	3.54	37.84	34.41	0.182	5.46	58.37	53.07
0.056	1.68	17.96	16.33	0.120	3.60	38.48	34.99	0.184	5.52	59,01	53.65
0.058	1.74	18.60	16.91	0.122	3.66	39.13	35.58	0.186	5.58	59.65	54.24
0.060	1.80	19.24	17.50	0.124	3.72	39.77	36.16	0.188	5.64	60.29	54.82
0.062	1.86	19.88	18.08	0.126	3.78	40.41	36.74	0.190	5.70	60.93	55.40
0.064	1.92	20.52	18.66	0.128	3.84	41.05	37.32	1.192	5.76	61.57	55.99
0.066	1.98	21.17	19.25	0.130	3.90	41.69	37.91	0.194	5.82	62.22	56.57
0.068	2.04	21.81	19.83	0.132	3.96	42.33	38.49	0.196	5.88	62.86	57.15
0.070	2.10	22.45	20.41	0.134	4.02	42.97	39.07	0.198	5.94	63.50	57.74
0.072	2.16	23.09	21.00	0.136	4.08	43.62	39.66	0.200	6.00	64.14	58.32

Table 2.20 Gas rate and heat input for U6 type gas meters (average CV = 38.5)

Seconds to burn 1 ft³	Gas rate m³/h	Gross input (kW)	Net input (kW)	Secs to burn 1 ft³	Gas rate m³/h	Gross input (kW)	Net input (kW)	Secs to burn 1 ft³	Gas rate m³/h	Gross input (kW)	Net input (kW)
400	0.255	2.73	2.48	207	0.493	5.27	4.79	77	1.324	14.16	12.87
396	0.258	2.75	2.50	204	0.500	5.34	4.86	76	1.342	14.35	13.04
392	0.260	2.78	2.53	201	0.507	5.42	4.93	75	1.360	14.54	13.22
388	0.263	2.81	2.55	198	0.515	5.51	5.01	74	1.378	14.73	13.40
384	0.266	2.84	2.58	195	0.523	5.59	5.08	73	1.397	14.93	13.58
380	0.268	2.87	2.61	192	0.531	5.68	5.16	72	1.416	15.14	13.77
376	0.271	2.90	2.64	189	0.540	5.77	5.24	71	1.436	15.36	13.96
372	0.274	2.93	2.66	186	0.548	5.86	5.33	70	1.457	15.57	14.16
368	0.277	2.96	2.69	183	0.557	5.96	5.42	69	1.478	15.80	14.37
364	0.280	3.00	2.72	180	0.567	6.06	5.51	68	1.500	16.03	14.58
360	0.283	3.03	2.75	177	0.576	6.16	5.60	67	1.522	16.27	14.80
356	0.286	3.06	2.78	174	0.586	6.27	5.70	66	1.545	16.52	15.02
352	0.290	3.10	2.82	171	0.596	6.38	5.80	65	1.569	16.77	15.25
348	0.293	3.13	2.85	168	0.607	6.49	5.90	64	1.593	17.03	15.49
344	0.296	3.17	2.88	165	0.618	6.61	6.01	63	1.619	17.31	15.74
340	0.300	3.21	2.92	162	0.630	6.73	6.12	62	1.645	17.58	15.99

Table 2.20 (Continued)

Seconds to burn 1 ft³	Gas rate m³/h	Gross input (kW)	Net input (kW)	Secs to burn 1 ft³	Gas rate m³/h	Gross input (kW)	Net input (kW)	Secs to burn 1 ft³	Gas rate m³/h	Gross input (kW)	Net input (kW)
336	0.304	3.24	2.95	159	0.641	6.86	6.23	61	1.672	17.87	16.25
332	0.307	3.28	2.99	156	0.654	6.99	6.35	60	1.700	18.17	16.52
328	0.311	3.32	3.02	153	0.667	7.13	6.48	59	1.729	18.48	16.80
324	0.315	3.36	3.06	150	0.680	7.27	6.61	58	1.758	18.80	17.09
320	0.319	3.41	3.10	147	0.694	7.42	6.74	57	1.789	19.13	17.39
316	0.323	3.45	3.14	144	0.708	7.57	6.88	56	1.821	19.47	17.70
312	0.327	3.49	3.18	141	0.723	7.73	7.03	55	1.854	19.82	18.02
308	0.331	3.54	3.22	138	0.739	7.90	7.18	54	1.889	20.19	18.36
304	0.335	3.59	3.26	135	0.755	8.08	7.34	53	1.924	20.57	18.70
300	0.340	3.63	3.30	132	0.773	8.26	7.51	52	1.961	20.97	19.06
296	0.345	3.68	3.35	129	0.791	8.45	7.68	51	2.000	21.38	19.44
292	0.349	3.73	3.39	126	0.809	8.65	7.87	50	2.040	21.80	19.83
288	0.354	3.79	3.44	123	0.829	8.86	8.06	49	2.081	22.25	20.23
284	0.359	3.84	3.49	120	0.850	9.08	8.26	48	2.125	22.71	20.65
280	0.364	3.89	3.54	117	0.872	9.32	8.47	47	2.170	23.20	21.09
276	0.370	3.95	3.59	114	0.895	9.56	8.70	46	2.217	23.70	21.55
272	0.375	4.01	3.64	111	0.919	9.82	8.93	45	2.266	24.23	22.03
268	0.381	4.07	3.70	108	0.944	10.09	9.18	44	2.318	24.78	22.53
264	0.386	4.13	3.75	105	0.971	10.38	9.44	43	2.372	25.35	23.05
261	0.391	4.18	3.80	102	1.000	10.69	9.72	42	2.428	25.96	23.60
258	0.395	4.23	3.84	99	1.030	11.01	10.01	41	2.487	26.59	24.18
255	0.400	4.28	3.89	96	1.062	11.36	10.33	40	2.550	27.26	24.78
252	0.405	4.33	3.93	93	1.097	11.72	10.66	39	2.615	27.95	25.42
249	0.410	4.38	3.98	91	1.121	11.98	10.89	38	2.684	28.69	26.09
246	0.415	4.43	4.03	90	1.133	12.11	11.01	37	2.756	29.47	26.80
243	0.420	4.49	4.08	89	1.146	12.25	11.14	36	2.833	30.28	27.54
240	0.425	4.54	4.13	88	1.159	12.39	11.26	35	2.914	31.15	28.32
237	0.430	4.60	4.18	87	1.172	12.53	11.39	34	3.000	32.07	29.16
234	0.436	4.66	4.24	86	1.186	12.68	11.53	33	3.090	33.04	30.04
231	0.441	4.72	4.29	85	1.200	12.83	11.66	32	3.187	34.07	30.98
228	0.447	4.78	4.35	84	1.214	12.98	11.80	31	3.290	35.17	31.98
225	0.453	4.85	4.41	83	1.229	13.14	11.94	30	3.399	36.34	33.04
222	0.459	4.91	4.47	82	1.244	13.30	12.09	28	3.642	38.94	35.40
219	0.466	4.98	4.53	81	1.259	13.46	12.24	26	3.922	41.93	38.13
216	0.472	5.05	4.59	80	1.275	13.63	12.40	24	4.249	45.43	41.30
213	0.479	5.12	4.65	79	1.291	13.80	12.55	22	4.636	49.55	45.06
210	0.486	5.19	4.72	78	1.307	13.98	12.71	20	5.099	54.51	49.56

The Physical Laws Relating to Gas

Graham's Law of Diffusion

2.74 This law is named after Thomas Graham (1805–1869), who discovered that gases would mix with one another quite readily due to the continuous movement of the molecules. However, the rate at which they mix depends on the specific gravity or density of the gases. Graham made a container, consisting of two separate compartments with a small hole in their dividing wall. He placed a different gas in each of the compartments, one having a higher specific gravity than the other. He found that faster, lighter molecules of the lighter gas passed through the hole rather than the slower, heavier molecules of the heavier gas. After many experiments, he discovered that the rates of diffusion, or mixing, varied inversely to the square root of the density of the gas. Thus:

$$\text{Diffusion rate} \propto 1 \div \sqrt{\text{Density}}$$

Which basically means that a light gas will diffuse twice as fast as a gas four times its density. (See Section 2.9 relating to specific gravity.) Graham's Law helps us understand the separation of gas with different densities (Figure 2.26).

Boyle's Law

2.75 This law is named after Robert Boyle (1627–1691), who discovered the relationship between volume and pressure of a gas (see Figures 2.27a and 2.27b). He found that the absolute pressure of a given mass of gas is inversely proportional to its volume provided that its temperature remains constant (Absolute pressure = Atmospheric pressure (1013 mbar + gauge pressure).

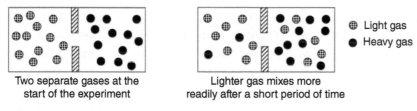

⊕ Light gas
● Heavy gas

Two separate gases at the start of the experiment

Lighter gas mixes more readily after a short period of time

Figure 2.26 Graham's Law of diffusion

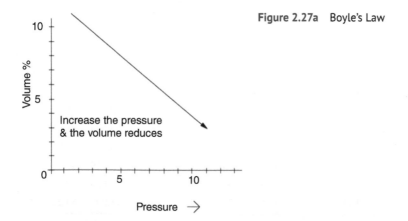

Figure 2.27a Boyle's Law

Volume %

Increase the pressure & the volume reduces

Pressure →

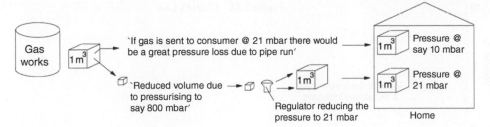

Figure 2.27b   Practical example of Boyle's Law

So:

$$\text{Pressure} \times \text{Volume} = \text{Constant}$$

To put it simply, if the absolute pressure (gauge pressure + atmospheric pressure) on a given quantity of gas decreases, then its volume will increase. So, if the absolute pressure is increased four-fold, the volume would be reduced to one-quarter. The formula can be redefined as:

$$P_1 V_1 = P_2 V_2$$

where P_1 = original pressure, V_1 = original volume, P_2 = final pressure and V_2 = final volume.

Practical example: Suppose that the supply pressure to a building is 80 mbar, and the total volume is 1 m^3, if the pressure was reduced to 20 mbar the new volume of the gas would be calculated as follows.

$$P_1 V_1 = P_2 V_2, \text{so } P_1 V_1 \div P_2 = V_2$$

$$\therefore (1013 + 80) \times 1 \div (1013 + 20) = \underline{1.06 \text{ m}^3} \text{ an increase of } \underline{6\%}.$$

Conversely, if the supply pressure is increased to 800 mbar when a new medium pressure regulator is fitted, supplying the same 1 m^3 volume of gas, and also reduced to 20 mbar, there is an interesting result:

$$P_1 V_1 \div P_2 = V_2$$

$$\therefore (1013 + 800) \times 1 \div (1013 + 20) = \underline{1.76 \text{ m}^3}, \text{ an increase of } \underline{76\%}$$

The effect of Boyle's Law is probably the reason why the gas supplier installed the regulator prior to the meter so that the gas is measured at the lower pressure.

Charles's Law

2.76   This law is named after Jacques Charles (1746–1823), a French physicist who discovered that all gases increase in volume by the same proportion if heated through the same temperature range, provided that the pressure remained constant (see Figure 2.28). This proportion is 1/273 of their volume at freezing point (0 °C or 273 K) for each 1 K rise above 273 K. Therefore, the temperature of a volume of gas would need to be increased from 0 °C to 273 °C to double its volume. *Note:* 1 K rise = 1 °C rise.

Where the pressure remains constant, Charles's law is expressed as:

$$\text{Volume} \div \text{Temperature} = \text{Constant}$$

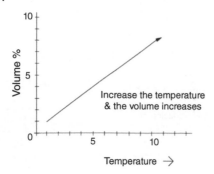

Figure 2.28   Charle's Law

In simple terms, if the temperature increases, so does the volume. As with Boyle's law, the formula can be redefined as:

$$V_1 \div T_1 = V_2 \div T_2$$

where V_1 = originai volume, T_1 = original temperature, V_2 = final volume and T_2 = final temperature.

Practical example: If 1 m^3 of gas enters a building from outside where the temperature is 2 °C and passes into a building where the temperature is 21 °C, the gas would increase in volume by:

$$V_1 \div T_1 = V_2 \div T_2, \text{so } (V_1 \times T_1) \div T_1 = V_2$$
$$\therefore 1 \times (21 + 273) \div (2 + 273) = \underline{1.07 \, \text{m}^3}, \text{an increase of } \underline{7\%}$$

This is why it is necessary to leave a period of time for pressure and temperature stabilisation during a tightness test.

Measurement of Gas

2.77   There are many different designs of gas metering devices, each using a different method to record the volume of gas flow. Some record the amount of gas used for billing purposes, whereas others give an indication as to the rate at which the gas is flowing, possibly for test purposes. Gas meters fall into two distinct classes: displacement and inferential.

Displacement Meters

2.78   These meters record a definite volume by the displacement of gas through one of the following methods:

- a bellows chamber or compartment;
- spaces between impellers or vanes.

Positive Displacement Meters (Diaphragm Meter)

2.79   This type of meter is a 'unit construction' (U-meter). It is a design that has been around since the 1800s. Since then, it has undergone many improvements, but it still operates along the same principles of the early design. The positive displacement meter uses diaphragms that are alternately inflated and deflated by the gas pressure flowing through. The movements of the diaphragms, via pivoting flag rods, opens and closes two valves located at the top of the meter. These valves control the passage of gas through the four measuring compartments. Figure 2.29 illustrates the cycle of opening and closing each compartment.

1. The gas flows into the inner chamber at the front of the meter, and in so doing allows gas to pass out into the system from the adjoining front chamber. As the diaphragm moves, it causes the flag rods to turn, which cause the valves at the top to move.
2. Owing to the movement of the flag rod, the inlet port to the front inner chamber eventually closes, as does the outlet from the outer front chamber, and gas can now only pass into the rear outer chamber, causing the gas to flow out from the inner rear chamber.
3. and 4. As the rear inner chamber empties, the movement of the flag rods continues and eventually the gas is allowed to flow into the front outer chamber and out from the front inner chamber filled during (1) above, and so the process continues.

As the flag rods pivot back and forth, opening and closing the valves, they cause a worm wheel to turn. This, via a gearbox, slowly turns the dial to register the quantity of gas used. The diaphragm used in the movement is made from a synthetic fibre called reinforced nitro-rubber. The typical U6 and U40 meters shown in Figure 2.30 are two of the many meters that use this design. The number following the U indicates the volume that the meter

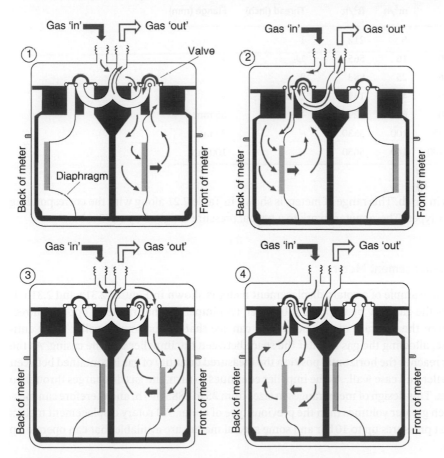

Figure 2.29 Operational cycle of a 'U' type positive displacement meter

Figure 2.30 Typical 'U' type diaphragm meters

Table 2.21 Displacement type meters

| | Badged rating | | Connection type | |
Type	m³/h	ft³/h	Thread (inch)	Flange (mm)
U6/G4	6	212	1	–
U16/G10	16	565	1¼	–
U25/G16	25	883	2	–
U40/G25	40	1412	2	–
U65/G40	65	2295	–	65 mm
U100/G65	100	3530	–	80 mm
U160/G100	160	5650	–	100 mm

can pass in m³/h. This range of meters is shown in Table 2.21 along with the corresponding 'G' meter types. These meters work to a typical pressure range of up to 75 mbar.

Rotary Displacement Meters

2.80　An example of a rotary displacement meter is shown in Figures 2.31a and 2.31b. It measures the gas flow by trapping it between two impellers that rotate in opposite directions. From the diagram of the meter, one can see that the bottom impeller turns anticlockwise, allowing the gas to enter the space between the impeller and the casing. As the impeller reaches the horizontal position the measured quantity of gas is contained between the impeller and case wall. As the impeller continues to turn, the gas discharges through to the outlet. This design of meter ranges in sizes from 25 to 2885 m³/h, and therefore can measure much greater volumes than the previous type of meter. All rotary displacement meters operate at pressures up to 10 bar and some special meters are available that can operate up to 38 bar.

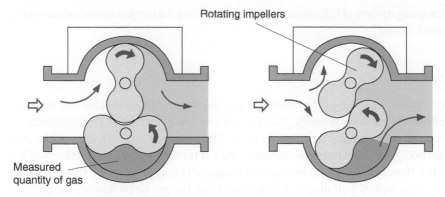

Rotating impellers

Measured
quantity of gas

Figure 2.31a Rotary displacement meter operation

Figure 2.31b Rotary displacement meter

With both the positive displacement and rotary displacement meter, the moving parts are activated by the gas flowing through the meter. Should the meter seize up, the gas supply will shut off. So, if the supply needs to be continuous, a bypass will need to be fitted or two meters would need to be fitted in parallel, to allow continuity of supply during shutdown, or for maintenance purposes.

Inferential Meters

2.81 Inferential meters record the gas flow by deducting one known quantity from another. These include:

- the rotation speed of a turbine;
- the measured differential across an orifice;
- the measured difference between static and kinetic pressures; and
- the measured temperature changes across a wire.

There are many designs of inferential gas meter, including the turbine meter, orifice meter and ultrasonic meter.

Turbine Meter

2.82   With this design, the gas impinges on specially shaped air-foil blades that rotate at high speeds as the gas flows through. The speed of rotation of the blades is proportional to the velocity and volume of gas passing over them. The meter records both the velocity and volume per hour. The drive from the turbine spindle is transferred by a magnetic coupling to record the flow onto an index mechanism counter. This type of meter should not be used to measure rapidly pulsating flows because when the gas stops flowing the turbine blades do not stop immediately, this can lead to significant errors in the measurements of gas flow. Pulsating flows can be caused by reciprocating compressors which, as well as affecting accuracy, can also damage the meter. Turbine meters are available in a full range of sizes from 65 to 25 000 m³/h with operating pressure up to 38 bar (Figures 2.32a and 2.32b).

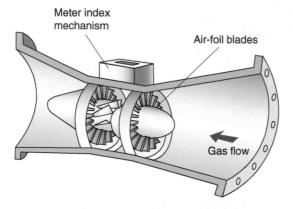

Meter index mechanism

Air-foil blades

Gas flow

Figure 2.32a   Workings of a turbine meter

Figure 2.32b   Turbine meter

Orifice Meter

2.83 This is a type of inferential meter and measures a pressure differential between points on each side of an orifice plate that has been positioned in the pipeline. This method of metering is only suitable where the flow is constant. Owing to its simple design and its ability to operate at very high pressures, it is often used for measuring the gas flow through transmission networks. The orifice plate itself is usually made of thin stainless steel or a material that would not be subject to corrosion. It is usually located between two flange joints allowing for regular maintenance to ensure that it does not become dirty, distorted or eroded. This access also allows the orifice plate to be changed for one with a different sized hole, so that different flow rates can be metered. Meter failure does not affect the gas flow as it is completely detached.

Figure 2.33 Orifice meter

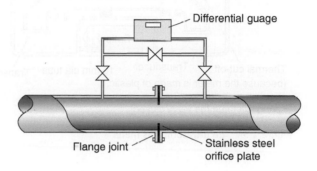

Ultrasonic Gas Meter

2.84 The first ultrasonic gas meters began to appear in about 1995. Initially, they were only used in the domestic market with meters operating up to 6 m^3/h, the meter was designated the E6. The meter consists of a tube with a transducer located at each end. A transducer is a device that converts a mechanical signal into an electrical impulse or vice versa; it can also act as a transmitter and receiver. A long-life battery, housed in the meter, provides power for the transducer to generate a sound wave. This is sent as an impulse through the moving gas flow, alternately, from each direction. The time of travel each way is measured and the difference between transmission and receiving in each direction is computed to provide an indication of the volume of gas consumed. The flow is displayed on a liquid crystal display (LCD) readout, located on the meter face. An example of an E6 ultrasonic gas meter is shown in Figure 2.25 and illustrated in Figure 2.34.

On the right-hand side of the meter is an optical communication port with a standard magnetic coupling from which information can be transmitted into or out of the meter, for purposes such as meter reading. There are several advantages to using this kind of meter including its compact size (it is smaller than the U series of meters); it is unaffected by air moisture and temperature, and it also includes a fraud detection system. Care needs to be taken when removing these meters as some, especially those that use a 'smartcard' credit system, have a tamperproof device fitted that will prevent gas flow until the problem has been investigated by the supplier.

More recently, non-domestic versions have evolved and have provided a more compact method of measuring gas on larger installations. The principle of operation is similar to the domestic version with the addition of multiple pairs of transducers to give further paths of measurement The units are called multipath and cartridge-type ultrasonic meters and are available in sizes up to 85 000 m^3/h and pressures up to 38 bar (Figures 2.33–2.36).

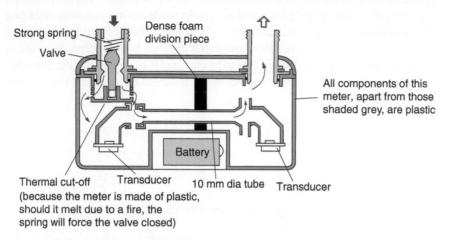

Figure 2.34   E6 ultrasonic gas meter

Figures 2.35 and 2.36 are examples of different capacity meters that are available.

Figure 2.35   G65 gas meter. Credit:Andy Lord

Figure 2.36 Size comparison of two gas meters both designed to measure $40\,\text{m}^3/\text{h}$. Credit:Andy Lord

3

Gas Controls

Quarter Turn Gas Control Valves

3.1 The following valves are examples of quarter-turn gas isolation valves. With this design of control, a change from the no-flow condition to full gas flow is achieved by a quarter turn (90°) of the operating handle. Often a groove is cut across the top of the spindle, which when lined up with the pipe, indicates the on position. Any lever fitted should be carefully positioned to align with this groove, so that it will indicate the on/off position. The lever, if fitted, should fall to off position as shown in Figure 3.1. This prevents it from opening accidentally should something fall onto the operating lever. There are three designs of quarter-turn valves which are shown in Figure 3.2. From left to right, the valves are ball, plug and butterfly. These three valves are described further in the following sections.

Figure 3.1 Correct fitting of a gas isolation valve in vertical pipework

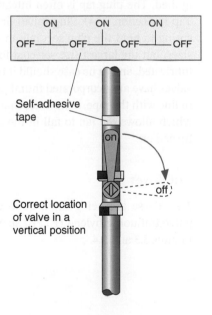

Gas Installation Technology, Third Edition. Andrew S. Burcham, Stephen J. Denney and Roy D. Treloar.
© 2024 John Wiley & Sons Ltd. Published 2024 by John Wiley & Sons Ltd.

Figure 3.2 Typical quarter turn valves – from left to right; ball valve, lubricated plug valve and butterfly valve

Plug Valve – Lubricated and Non-lubricated

3.2 This is the oldest and possibly the simplest form of gas control, see Figure 3.3. It consists of a tapered plug through which gas flows. There are two types, lubricated and non-lubricated. A lubricated plug valve can be lubricated whilst it is still in service, a non-lubricated valve needs to be dismantled to lubricate. The word 'valve' refers to a device fitted in the supply line. The word 'tap' refers to the terminal fitting, as will be found, for example, in a school science lab where it is opened to allow gas to flow through to be ignited. The plug tap is often integrated with the appliance (see also the Appliance Plug Tap in Section 3.14). Plug valves or taps when used with LPG incorporate an additional spring to hold the plug securely in the body of the valve. Plug valves are made in many sizes, but the largest size selected for gas pipework should not exceed 50 mm unless it is lubricated, and in no case should it be more than 100 mm in diameter. Many of the smaller valves have an incorporated thumbpiece or fan to indicate the on position. When the fan is in line with the pipe, it indicates that the supply is on. Sometimes a drop-fan design is used, which allows the fan to fall in the closed position, preventing it from being accidentally turned on.

Ball Valves

3.3 These are being fitted quite extensively these days; they incorporate a nylon or polytetrafluoroethylene (PTFE) seal to ensure adequate isolation of the gas supply, see Figures 3.3 and 3.4.

Figure 3.3   Plug valve shown in 'closed' position

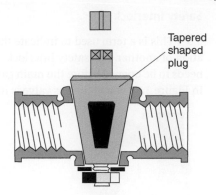

Tapered
shaped
plug

Figure 3.4   Ball valve shown in 'open' position

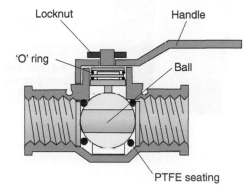

Locknut

Handle

'O' ring

Ball

PTFE seating

Butterfly Valves

3.4   These are available in many sizes; however, they are only found in commercial situations. These quarter-turn valves are different in that they allow for high volumes of flow because of the hinged disc, see Figure 3.5. To ensure complete isolation, a soft seating is incorporated which could be, for example, nitrile or neoprene.

Figure 3.5   Butterfly valve

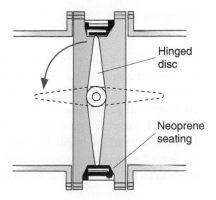

Hinged
disc

Neoprene
seating

Safety Interlock

3.5   This is a term used to indicate that one control cannot be operated without the operation of another. The safety interlock shown in Figure 3.6a, illustrates how a pilot gas line needs to be opened before the main gas line. Another design of safety interlock can be seen in Figure 3.6b where the gas valve is interlocked with the water supply pipe.

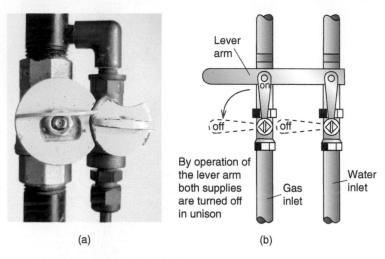

(a)   (b)

Figure 3.6   (a) Safety interlock linking main gas with pilot supply. (b) Safety interlock linking gas with water supply

Screw Down Gas Control Valves

The valves described in Sections 3.6–3.9 are examples of 'screw down' gas isolation valves. This design of valve makes use of the turning motion of a shaft, which slowly opens the gas supply by withdrawing the valve from its seating. This category of valves includes the following specific designs: disc, gate, diaphragm and needle valves.

Disc Valve

3.6   These valves are used for local gas isolation, see Figure 3.7. This valve design, using a disc that closes onto a seating, is more commonly found on automatic controls such as a solenoid or safety shut-off device. One design of disc valve used in the domestic sector is called the 'pillar-elbow restrictor', which is associated with the installation of gas fires (see Figures 3.8a and 3.8b). It consists of a brass- or chrome-plated body with a cover cap that needs to be removed in order to open or close the control.

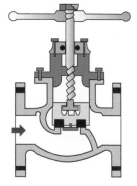

Figure 3.7   Disc valve

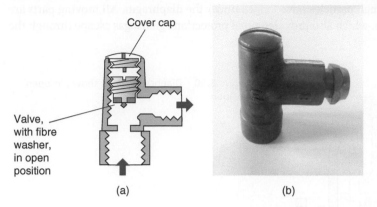

(a)   (b)

Figure 3.8 (a) Component parts of a pillar-elbow restrictor. (b) Pillar-elbow restrictor

Gate Valve

3.7 This design of valve is pictured in Figure 3.9a and illustrated in Figure 3.9b. In operation, the gate completely lifts out of the pipe run when fully open; therefore, it has minimal loss of flow and pressure.

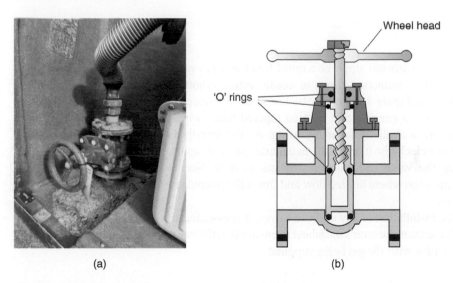

(a)   (b)

Figure 3.9 (a) Gate valve. (b) Gate valve shown in 'closed' position

Diaphragm Valve

3.8 This valve consists of a large flexible diaphragm that is raised off the valve seating by the operation of a wheel head, see Figure 3.10. As the wheel turns off the valve it compresses the diaphragm and forces the rubber firmly onto the seating, thus ensuring complete closure even where grit or small particles have lodged under the diaphragm. All moving parts are above the diaphragm, which also provides a real protection against gas escape through the valve head.

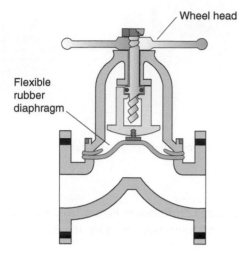

Wheel head

Flexible rubber diaphragm

Figure 3.10 Diaphragm valve shown in 'open' position

Needle Valve

3.9 The pressure test nipple on a multi-functional gas valve is one of the commonest forms of needle valve, although, as shown in Figure 3.11, it can be found as an independent control. It consists of a pointed tapered head, which locates into a similarly tapered seating. As the spindle is turned anticlockwise it draws the two mating surfaces apart, opening the valve and allowing the gas to flow. Needle valves are ideal where limited flow and fine adjustments are required.

Before installation of any of these valves, it is essential to consult the manufacturer's datasheet to ensure that the valve is compatible with the gas being supplied.

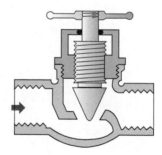

Figure 3.11 Needle valve

Valve Application

Relevant Industry Document
IGEM/UP/2

3.10   Table 3.1 gives an overview of the applications for some of the valves detailed in the previous sections with regard to acceptability and limitations in a commercial setting.

Table 3.1   Suitability of valves

Valve type	Valve application			
	Pipe section/plant isolation	Buried below ground	By-pass	AECV
Non-lubricated plug	Acceptable R – 1 and 8	Not acceptable	Acceptable R – 1 and 8	Acceptable R – 1 and 3
Lubricated plug	Acceptable R – 1, 8 and 9	Not acceptable	Acceptable R – 1, 2, 8 and 9	Acceptable R – 1, 2, 3 and 9
Ball	Acceptable R – 8 and 9	Acceptable R – 4 and 9	Acceptable R – 8 and 9	Acceptable R – 3 and 9
Wedge or parallel slide gate	Acceptable R – 6 and 8	Acceptable R – 4 and 6	Acceptable R – 6 and 8	Acceptable R – 3 and 6
Butterfly	Acceptable R – 5, 6, 7 and 8	Not acceptable	Acceptable R – 5, 6, 7 and 8	Not acceptable
Diaphragm	Not acceptable	Not acceptable	Not acceptable	Not acceptable
KEY TO CHART	Restrictions (R)			

1	Do not exceed 50 mm nominal bore
2	Except if upstream of safety shut-off valve
3	Fire resistance of valve to be checked (BS EN 1775 compliant) if required
4	Steel or iron only
5	Fully lugged
6	Dust/debris to be checked
7	Normally, MOP ≤ 100 mbar, above this pressure it may be appropriate to apply two valves in series with a valve test vent between them (double block and bleed) if gas-tight isolation is important for downstream operations
8	Can be sealed or locked in the closed position
9	Mechanical assistance (gearing) required above certain pressure and size

Source: Data from IGEM/UP/2

Spring-loaded Gas Control Valves

3.11   There are several designs of spring-loaded valves. Any valve that incorporates a spring to assist its operation is technically speaking 'spring loaded'. The valves identified here include the cooker safety cut-off valve, the plug-in quick-release valve, and the appliance plug tap, as found on most cookers and gas fires.

Cooker Safety Cut-off Valve

3.12   This type of spring-loaded valve is incorporated in devices such as a glass drop-down lid, as found on cookers. There are several designs: one, shown in Figure 3.12, uses a push rod to disengage a cut-off plunger. Another hotplate safety cut-off design for domestic gas cookers is illustrated in Figure 3.13a and pictured as 'open' in Figure 3.13b and 'closed' in Figure 3.13c.

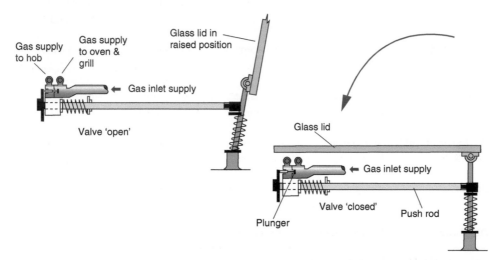

Figure 3.12   Cooker safety cut-off valve

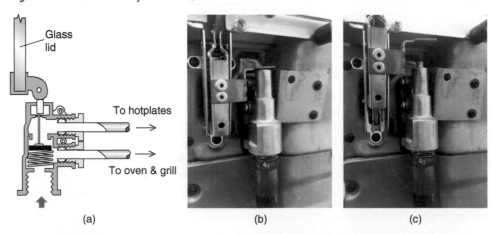

Figure 3.13   (a) Cooker safety cut-off device. (b) Cooker safety cut-off device lid open. (c) Cooker safety cut-off device lid closed

Plug-in Quick Release Valves

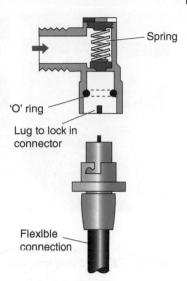

'O' ring

Lug to lock in connector

Spring

Flexible connection

Figure 3.14 Bayonet valve

3.13 These are used in conjunction with cooker and tumble dryer flexible gas connections, etc., and an assortment of leisure equipment such as barbecues and are designed to facilitate removal and maintenance of the appliance. They act as 'plug-in sockets', requiring a push and turning motion to make or break the connection, which is self-sealing. The two most common valves are the standard bayonet, illustrated in Figure 3.14 and the micropoint socket, used for leisure appliances, seen in Part 9: Section 9.135. The location of these fittings should be such that the length of flex is limited. Micropoint sockets can be obtained as independent fittings or can be supplied to be used in conjunction with a white plastic box. It should be noted that these valves can be regarded as suitable terminal fittings; no additional plugs are needed to seal off the pipe when removing the appliance and flex. However, good practice would include applying leak detection fluid (LDF) to ensure no gas is letting by. If an appliance is being disconnected permanently, it would be advisable to remove the bayonet and cap the pipe with an appropriate fitting.

Appliance Plug Tap

3.14 In order to turn on/off the supply of gas to an appliance, such as a gas fire or cooker, a tapered plug is used. Incorporated with the plug will be a spring, which is designed to prevent the control tap from accidentally being turned on. To operate the control, one must first push in the operating knob – this will then allow the tap to be turned. The appliances in which these taps are incorporated are often subject to high temperatures surrounding the valve and, as a consequence, they often dry out and become stiff to use. In this case, the valve will need to be re-greased. This is a simple operation involving removing the tapered plug from its housing, see Figure 3.15a–c. The grease used is usually a high-temperature grease such as 'Molykote'. When applying the lubricant, only the smallest smear is needed, and particular care should be taken on cooker taps to ensure that the very small simmer port is not blocked with excess grease. On re-assembly, after spraying with leak detection fluid, the tap should be checked for correct operation at all settings.

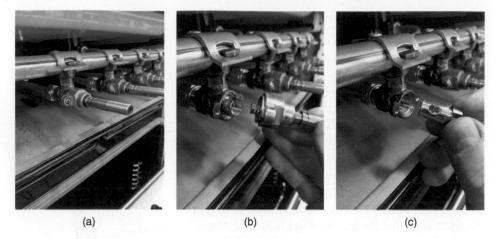

(a)   (b)   (c)

Figure 3.15 (a) Appliance plug tap in situ. (b) Appliance plug tap spindle removed. (c) Appliance plug tap – tapered plug removed

Electrically Operated Gas Control Valves

3.15 There are two electrically operated valves that are commonly encountered: the solenoid valve and the safety shut-off valve.

Solenoid Valve

3.16 This valve, also referred to as the magnetic valve, basically consists of a coil of wire through which electricity is passed, see Figure 3.16. This flow of electricity around the coil creates an electromagnetic field. This draws in an iron plunger that is free to slide inside a brass tube, referred to as the armature. The plunger is connected to a valve, which remains open while electricity is flowing, allowing gas to flow. When the electricity stops flowing

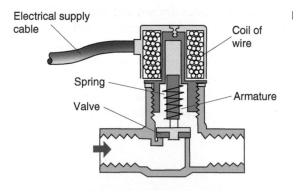

Figure 3.16 Solenoid valve

Electrical supply cable

Coil of wire

Spring

Valve

Armature

Valve shown in the 'closed' position
(i.e. no power supply)
when power is supplied to the coil the valve
lifts into the magnetic field created

through the coil, the valve closes due to the action of a spring, which pulls the plunger from the coil. Solenoids can operate on both alternating current (AC) and direct current (DC). With solenoids operating on AC, however, the valve sometimes hums because the armature vibrates within the coil as the current changes the magnetic flux back and forth within the coil. Solenoid valves are often incorporated within other more complicated controls, such as multi-functional gas valves and safety shut-off valves.

Safety Shut-off Valve

3.17 A safety shut-off valve is a gas valve that maintains a relatively slow opening opera-tion yet has a rapid closure. These valves are often installed on large automatic packaged burners or at the entrance of a large workshop or plant room. These valves are usually classified as 1, 2 or 3. The class number simply indicates the quality and the pressures to which the valve can safely operate, class 1 being the better valve. Two such designs are the Electro-hydraulic and the Electro-mechanical.

Electro-hydraulic

3.18 This device which can be seen in Figure 3.17, operates by using hydraulic power to force the valve open against a very strong spring. The oil pressure is obtained by the operation of a small electric motor. When power is supplied, a solenoid operates, closing off a small relief valve within the unit. At the same time, the motor operates pumping the oil from a reservoir through the non-return/check valve to force the diaphragm up, lifting the valve from its seating. Once fully open, a limit switch trips, breaking the circuit to the motor, and the valve is held open as the oil is trapped behind an energised solenoid valve. Switching off the power de-energises the relief solenoid, allowing the oil to flow back into the reservoir, assisted by the spring.

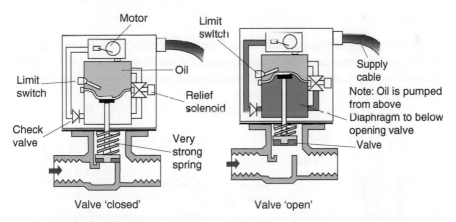

Figure 3.17   Electro-hydraulic safety shut-off valve

Electro-mechanical

3.19 With this device, illustrated in Figure 3.18, when a current is supplied it, energises a solenoid, pulling in a latch pedestal. This provides the fulcrum for a lever arm that forces the valve open as an electric motor turns a cam, forcing it down. As with the previous device, a limit switch stops the motor when the valve is fully open. Turning off the power de-energises the solenoid, which disengages the latch pedestal.

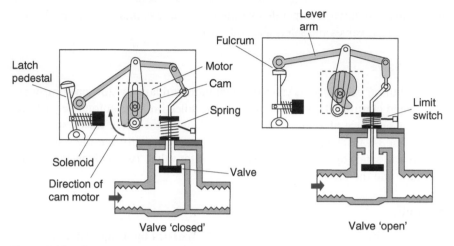

Figure 3.18 Electro-mechanical safety shut-off valve

Heat-operated Gas Control Valves

3.20 There are several types of heat-operated valves that you will encounter. *Note:* The use of drop-weight valves is no longer considered suitable.

Thermal Cut-off

3.21 This is a device designed to close off the gas supply to an installation in the event of a fire. Several designs have been developed over the years; the one shown in Figure 3.19a consists of a powerful spring that has been compressed to keep the valve open. A low-melting

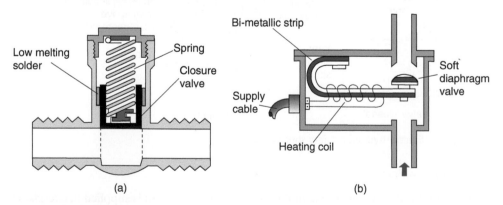

Figure 3.19 (a) Thermal cut-off. (b) Heat motor

solder has been used to secure it in this position. If a fire raises the ambient temperature to above 95°C, the solder melts, allowing the spring to force the valve closed. A thermal cut-off is incorporated in the design of an E6 gas meter (see Part 2: Figure 2.34 for an example).

Heat Motor

3.22   This gas valve is sometimes found incorporated in a gas cooker or multi-functional control. It consists of a small heater element, which is used to heat up a bi-metallic strip. One can see from the illustration in Figure 3.19b, as the element coil heats up, the strip bends because of the different expansion rates of the two metals, and slowly closes the supply. The pressure of the gas entering the valve also assists in fully closing the valve diaphragm. When no current is supplied to the element, the bi-metallic strip cools and re-straightens, causing the valve to snap back open. These controls can also be designed to have a normally closed position.

Pressure Stat

3.23   This device, pictured in Figure 3.20a, is designed to close down the flow of gas to appliances, such as pressure and steam boilers, when a predetermined pressure has been achieved. The illustration in Figure 3.20b shows a connection at the top of the control where steam enters. If the steam pressure generated is greater than the tension of the spring, the valve will close. Full closure may be obtained and sometimes a by-pass is incorporated to allow for a low fire rate. Other designs of pressure stat allow for pressure adjustment or the stat itself may be incorporated with a control such as a relay valve. Some pressure stats work by operating an electrical connection, which controls a safety shut-off valve.

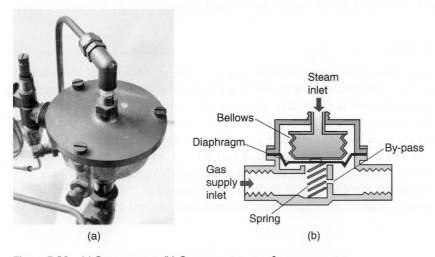

(a)   (b)

Figure 3.20   (a) Pressure stat. (b) Component parts of a pressure stat

Thermistor Thermostat

3.24   This is a form of temperature control that uses an electrical non-metallic thermal resistor. These thermistors are very sensitive to temperature change. As they heat up, the current flow through them varies. Thermistors are available in many shapes and are suitable for measurement control up to a useful temperature limit of about 300°C. Because of their size (they are no larger than a small bead), they are used to measure the temperature in the most inaccessible places (Figure 3.21).

(a)   (b)

Figure 3.21   (a) Flow temperature thermistor (red) and safety thermostat (black). (b) NTC thermistor

Pressure-operated Controls Valves

3.25   There are several types of controls that detect pressure.

Non-return Valve

3.26   These valves are also known as back-pressure valves or check valves. There are several designs. First, the most basic type, called a roll check valve, shown in Figure 3.22a, and is sometimes known as a flap valve. It consists of a hinged flap, which pivots at a fulcrum to close in a no-flow, or back-flow condition. A second design consists of a metal disc that is held open by slight pressure from a spring. As seen in Figure 3.22b, on top of the disc lie two leather or rubber washers that lift as gas flows through. When there is no flow or backflow the washers fall, sealing the outlet. Where excessive back pressure is encountered, the disc itself is forced down to overcome the spring pressure and close completely. Both of these non-return valves described need to be installed in the

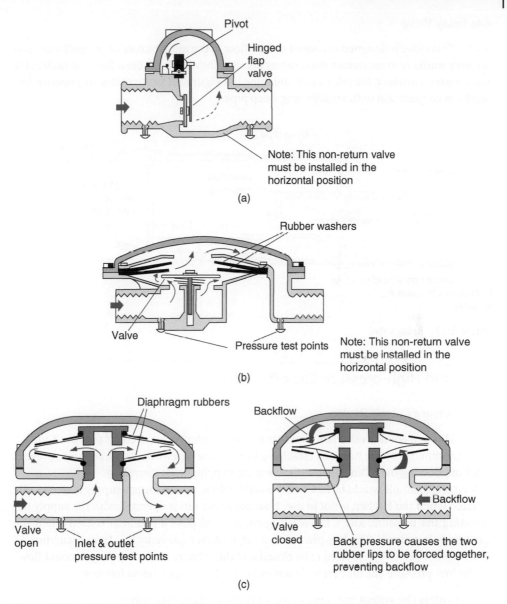

Figure 3.22 (a) Roll check valve. (b) Disc and washer valve. (c) Double diaphragm valve

horizontal position. A different design overcomes this problem and can be fitted in any position. Seen in Figure 3.22c, it consists of a pair of rubber diaphragms, housed inside a specially designed frame. Under normal gas flow, the diaphragms are pushed apart to allow the gas to flow freely, but under back-flow conditions the back pressure forces the two diaphragms together, stopping any flow.

Gas Relay Valve

3.27   This valve is designed to control the gas flow to the main burner of an appliance. You are very unlikely to encounter these valves because they have not been fitted on boilers for many years. However, the relay valve, illustrated in Figure 3.23, is operated by pressure and works in conjunction with an adjoining weep pipe.

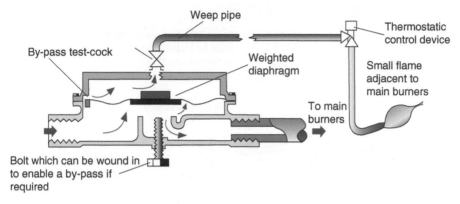

Figure 3.23   Relay valve

Low- and High-pressure Cut-off Devices

Low-pressure Cut-off Device

3.28   This is a special valve that prevents gas from entering a system of pipework until a predetermined pressure has been established within the pipework. Its purpose is to prevent gas from flowing out of any open ends when the supply is first turned on.

A typical example would be in a commercial kitchen, where the gas supply is electrically isolated when the kitchen is not in use. If an appliance tap is operated with the supply off, releasing the pressure, and a tap is left open, it could pose a potential hazard when the gas supply is restored. The low-pressure cut-off, however, prevents this from occurring: all downstream valves would need to be closed and the valve re-set before the gas could flow.

The low-pressure cut-off device shown in Figure 3.24a, operates as follows:

1. Gas enters the system but cannot progress beyond the closed valve.
2. The re-set plunger is held open. This allows gas to flow slowly, by-passing the valve into the installation pipework.
3. If all controls are closed downstream of this valve, the pressure will slowly build up below the main diaphragm to give a surface area pressure greater than that exerted by the spring and the valve will be lifted.
4. Should the supply be isolated upstream, the valve will remain open. If the pressure is lost, however, the valve will drop and need re-setting.

Nowadays, pressure switches are invariably used in conjunction with a safety shut-off valve to perform the same operation as shown in Figure 3.24b.

Figure 3.24 (a) Low pressure cut-off device. (b) Safety shut off with integrated proving system

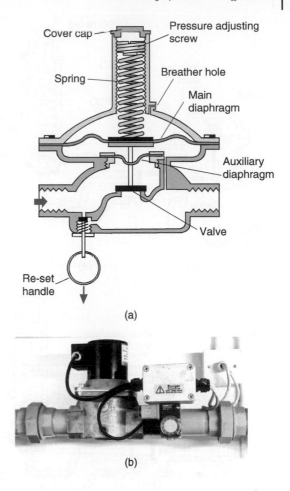

(a)

(b)

3.29 The following two controls are illustrated where they are combined into one unit. Figure 3.25a shows the valves under normal operation, whereas Figure 3.25b demonstrates fault conditions. The pressures at which these devices operate vary according to the type and design of regulator. For LPG, the pressures can be referenced by using Table 3.2.

Under Pressure Shut-off Valve (UPSO)

This control is not dissimilar to a low-pressure cut-off device and, in effect, performs the same task. Its mode of operation is slightly different, however, and it is invariably incorporated with a second-stage regulator, as found on an LPG installation. When the device senses an abnormally low outlet pressure between 25 and 32 mbar, it causes the diaphragm to drop to a particularly low position, resulting in the UPSO valve closing against its seating. Gas cannot therefore pass into the control valve until the re-set rod has been lifted to re-set the UPSO valve. This can only be achieved when all appliance controls are closed, and pressure can build up in the outlet pipework.

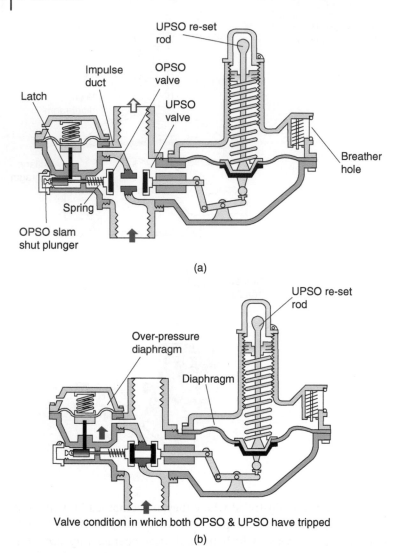

Figure 3.25 (a) Overpressure and underpressure shut-off device – in normal operation. (b) Overpressure and underpressure shut-off device – tripped

Overpressure Shut-off Valve (OPSO)

This control operates by detecting high pressure through an impulse duct, causing the over-pressure diaphragm to lift, which, in turn, lifts the latch holding the slam-shut plunger open. The slam shut is now free to move and a strong spring pushes the plunger, closing off the valve. Some OPSOs incorporate a sealed manual re-set. OPSO is often incorporated with a UPSO control. If the OPSO trips out for any reason, you should treat the situation as Immediately Dangerous and contact the supplier.

Both under and over-pressure shut-off valves are also incorporated within the medium pressure regulator now supplied on a natural gas installation, although the pressure tolerances would differ.

Pressure Regulators

3.30   There are constant pressure fluctuations within the gas supply pipe caused by various influences on the system, ranging from the operation of appliances to atmospheric conditions changing. The pressure in the distribution supply main also varies from district to district and, as more and more gas flows due to consumer use, so does the pressure. Therefore, it is essential to control the pressure entering a property and appliance to ensure safe and complete combustion of the fuel. This is achieved by the use of a regulator, or governor as it is also known, which automatically adjusts and opens or closes the gas line.

Constant Pressure Regulator

3.31   This regulator is simply constructed and may be found on the supply to an appliance. It works as follows:

1. With no gas pressure in the pipeline, as seen in Figure 3.26a, the spring pushes on the diaphragm, causing the valve to open.

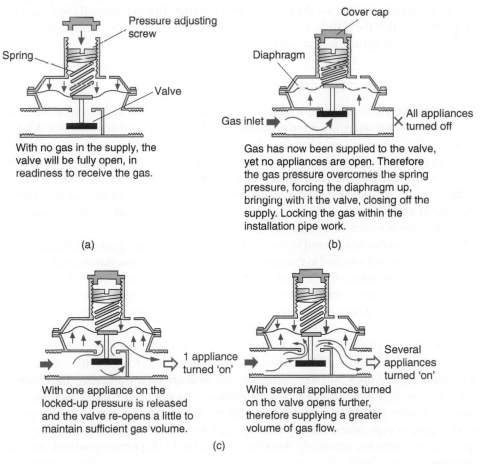

Figure 3.26   Operation of a constant pressure regulator. (a) with no gas flowing. (b) locked up. (c) with gas flowing

2. When connected to the supply, gas flows past the valve into the appliance or installation. If all the appliance controls are closed, the pressure in the outlet will build up until it equals the pressure of the spring. At this stage, the diaphragm will be in a state of equilibrium. Any further pressure rise will simply act on the diaphragm forcing it upwards and, as a result, the valve would equally be pulled up to close the supply, see Figure 3.26b. *Note:* When the valve is fully closed, it is referred to as 'locked-up' and the pressure within the system is referred to as the standing pressure.

3. Should an appliance be in operation, as illustrated in Figure 3.26c, the 'locked up' pressure would be released through the appliance injector, which would subsequently allow the spring to push on the diaphragm, so re-opening the valve. Thus, as the gas flows, the diaphragm monitors the pressure: continually rising and falling to meet the demand of the gas flow. The greater the volume of gas being consumed, i.e. the more burners in use, the more the valve will open.

3.32   Adjustment of the outlet pressure is made using a manometer when the gas is flowing; it is referred to as the operating pressure. With a manometer connected to the supply downstream of the regulator and with the cover cap removed, the spring pressure adjustor is turned clockwise to increase the pressure on the spring. This subsequently increases the gas pressure within the system, putting pressure on the diaphragm, and causing it to close. For pressure readings and adjustments, see Part 2: Section 2.56.

Compensated Constant Pressure Regulators

3.33   With the constant pressure regulator described above, the difference between the inlet and outlet pressures is not too great and therefore the upward and downward forces acting on the valve itself are virtually equal. However, as the difference in the pressures becomes greater, the force on the inlet side of the valve exceeds the force on the outlet side and, as a consequence, the valve tends to close prematurely, reducing the outlet pressure. A typical example would be where gas is taken from the district main with an inlet supply pressure of up to 75 mbar and then reduced to approximately 21 mbar. To overcome this problem, there is a need for compensation.

Compensation can be achieved by two basic methods. One is using a design that incorporates two valves, both attached to the same spindle. Figure 3.27 shows that in Example 1, the inlet pressure is trying to close one valve but is also trying to open the other, thus resulting in a zero-force influence. Smaller regulators tend not to favour this design because of the difficulty in adjusting the valves to close at exactly the same time. Instead, they use another method, which uses an auxiliary diaphragm instead of the second valve. In Figure 3.27, Example 2, an auxiliary diaphragm is used. It can be seen that the inlet pressure pushes on the valve and the underside of the auxiliary diaphragm. This pressure on the valve is trying to push it open but, simultaneously, the pressure on the auxiliary diaphragm is trying to close it; this results in a zero-force influence. As can be seen in the more detailed diagram, Figure 3.28, when gas passes through the valve seating, a small quantity passes up the impulse pipe, where it acts on the spring resistance, closing or reducing the valve opening, an operation previously described in the single diaphragm constant pressure regulator.

Example 1 shows how the gas pressure is trying to open one valve, yet is equally trying to close the other.

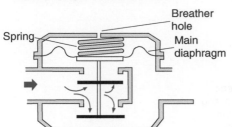

Example 2 shows how the gas pressure is trying to open the valve, yet is equally trying to close it by acting up onto the auxiliary diaphragm.

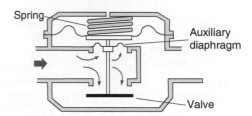

In both examples, it can be seen that zero force is being exerted on the valve allowing all influence for valve movement to be generated via the main diaphragm.

Figure 3.27   Principle of compensated pressure regulation

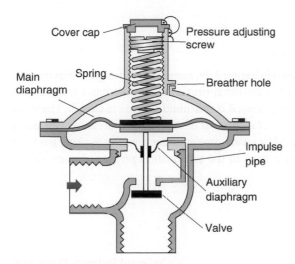

Figure 3.28   Compensating regulator

High-to-low Pressure Regulator

3.34   Nowadays, with more and more demand for gas for more houses, the pressure in many districts has been increased to above 75 mbar. Occasionally, a supply is taken from the high-pressure main prior to the locally regulated low-pressure supply district. In these cases, high-to-low-pressure regulators become necessary. *Note:* These are sometimes simply referred to as high-pressure service regulators. These regulators are designed to reduce the pressure from around 2 bar down to 21 mbar in one step. An overpressure shut-off device, as previously described, is incorporated with the regulator. A photo of this type of regulator can be seen in Part 5: Figure 5.10.

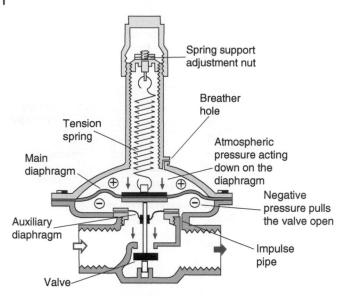

Figure 3.29 Zero governor

Zero Governor (Back-loaded Governor)

3.35 This design of regulator, illustrated in Figure 3.29, has a similar design to the compensating regulator, the difference being that a tension spring is used rather than a compression spring. Thus, instead of trying to open the valve, it is continually trying to close it. The valve will open only if the pressure downstream falls to that of atmospheric pressure, caused by suction. This negative pressure is caused by an air stream under pressure sucking gas from the pipeline and mixing it with air. It is used in the combustion process of many forced draught burners.

LPG Regulators

Relevant Industry Document
BS EN 16129 and BS 6891

Note: New regulators are made to BS EN 16129 and replace BS EN 12864, BS 3016, BS EN 13785 and BS EN 13786. Although regulators to these older standards may still be encountered, Liquid Gas UK advise that all new regulators should be marked BS EN 16129.

3.36 LPG regulators work on a similar principle to the natural gas regulators previously described, but because of the higher pressures generally involved, they are of a more robust design (Figure 3.30). They either use two regulators, one reducing the pressure to an intermediate pressure and the other to the final system working pressure; this system is referred to as first- and second-stage regulation and is illustrated in Figure 3.31a. Alternatively, a single-stage regulator is used; this reduces the pressure to that required in one step, see Figure 3.31b. Pressure variations and specifications for regulators are shown in Table 3.2.

(a)

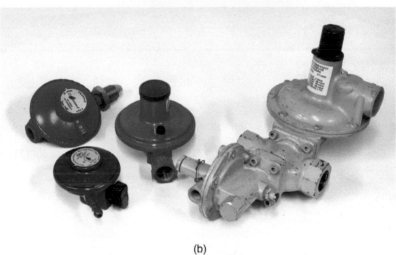

(b)

Figure 3.30 (a) LPG regulator to BS EN 16129. (b) Various older-style LPG regulators

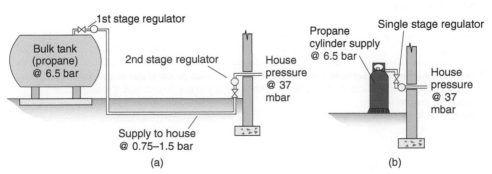

(a)

(b)

Figure 3.31 (a) First- and second-stage regulation. (b) Single-stage regulation

Table 3.2 LPG regulator set pressures (mbar) for the UK market

Pressures	Regulator set pressures (mbar)					
	Single-stage regulators		Final (2nd) stage regulator (low pressure)		Automatic changeover device (low pressure)	
	BS3016	BS EN 12864 or 16129	BS3016	BS EN 13785 or 16129	BS 3016	BS EN 13786 or 16129
Nominal outlet pressure						
Butane	28	29	–	–	–	29
Propane	37	37	37	37	37	37
Outlet pressure range						
Butane	23–33	22–35	–	–	–	22–35
Propane	32–42	32–45	32–42	32–45	32–42	32–45
Maximum lock-up pressure (above set pressure)						
Butane	+10	40	–	–	–	40
Propane	+15	50	+10	50	+15	50
Relief valve operating range						
Butane	–	48–150	–	–	–	48–150
Propane	–	60–150	50–62 Typical setting is 55 mbar	60–150	50–62 Typical setting is 55 mbar	60–150
UPSO operating range						
Propane	–	–	25–32	25–32	–	–
OPSO operating range						
Butane	–	48–150	–	–	–	48–150
Propane	–	60–150	70–80	60–150	70–80	60–150

Note: An on-site test of a relief valve, UPSO or OPSO is not normally practical due to the absence of a set procedure from the manufacturer. Therefore, it is only necessary to test these devices if manufacturer's instructions are provided.
Source: Data from BS 6891.

Gas pressures within the LPG sector depend on the gas being supplied: either butane or propane. The gas is supplied in liquid form and the gas pressure generated is dependent on the ambient atmospheric conditions and temperature. Typical supply pressures would be in the region of 1.5–2 bar for butane and 6–7 bar for propane. The distinguishing difference between the regulators used for these two gases is the gas connection to the high-pressure supply. With butane, the connection is either a plug-in bayonet-type connection, as found on a cabinet heater or a female gas connection, whereas propane connections are typically made via a left-hand male thread.

The regulator shown in Figure 3.32a and b, illustrates a typical single-stage LPG regulator and works as follows:

1. With the appliance taps closed, the pressure within the system pushes the diaphragm up, drawing the valve with it by means of a connecting lever, to the closed position, see Figure 3.32a.
2. Should an appliance valve be opened, the pressure on the underside of the diaphragm would drop, releasing the locked-up pressure and the valve would open to allow more gas to flow in through the valve, see Figure 3.32b. This operation continues until the pressure builds up, re-closing the valve.

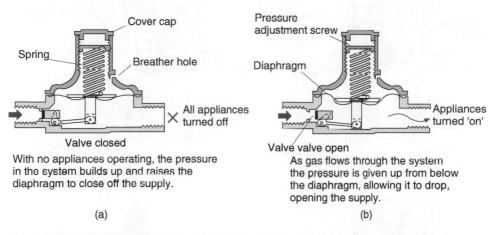

Figure 3.32 (a) Single-stage regulator locked up. (b) Single-stage regulator with gas flowing

Automatic Change-over Valves

3.37   This is a specially designed regulator that automatically closes the supply from one cylinder or bank of cylinders and opens another in order to ensure continuity of the gas supply, thus preventing any loss of supply when the cylinders have run out of gas. These can be used in conjunction with a second-stage regulator to give a high-pressure outlet supply or, as is more typically the case, they are used to reduce the pressure down to 37 mbar in a single stage. It will often be found that these controls have an overpressure shut-off (described in Section 3.29) incorporated with the valve.

The operation of the automatic change-over, illustrated in Figure 3.33a–c, is as follows:

1. When the manually operated lever control handle is turned to the on position, the angular base of the actuator applies pressure to the valve it faces. This overcomes the closure spring pressure and opens the valve. The high pressure then enters the valve and operates as previously described in the general operation of regulators.
2. When the supply pressure has finally all gone, the actuator spring pushes down further against the loss of pressure, and eventually, the valve from the second supply line is made to operate. At this time, a red indicator is displayed on the control handle informing the user that the cylinder supply has become exhausted. This indicator was previously white.
3. At this stage, the manually operated lever control handle can be turned to face the second gas supply. This causes the indicator to turn back to white but, more importantly, it allows the first exhausted supply valve to close. With the gas still flowing, the empty cylinders can now be exchanged for new full bottles.

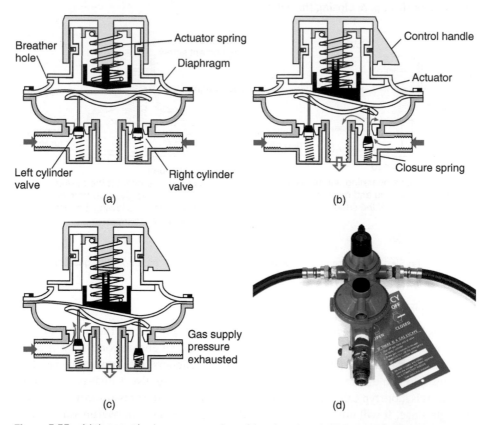

Figure 3.33 (a) Automatic change over valve with valve closed. (b) Automatic change over valve 'on position', taking supply from right cylinder. (c) Automatic change over valve right cylinder supply exhausted and now automatically taking gas from left cylinder. (d) Automatic changeover valve

Note: If the cylinder capacity was undersized, the change-over would occur prematurely with the result that the user would be unable to obtain the full quantity of gas from the cylinders. This is discussed in 'Vapour Off-take Capacity', Part 4: Section 4.65.

The location of an automatic change-over valve must be above the level of the cylinder outlet valves.

Single Stage Bayonet-type Low-pressure Regulator

3.38   This type of regulator is used where there is a minimal amount of gas flow required, such as to a small caravan or cabinet heater. The regulator, as seen in Figure 3.34, is simply offered to the compact valve located on the top of the gas cylinder. The locking device when turned a quarter turn clamps the regulator to the valve and with a second quarter turn allows gas to flow. It works on the same basic principle as the regulator described in Section 3.36. *Note:* The correct manufacturer's regulator must be selected as there are slight variances in valve connection sizes.

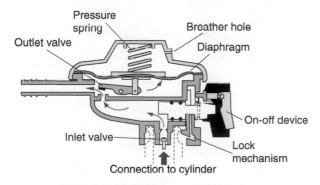

Figure 3.34   Bayonet-type butane regulator

Flame Supervision Devices

3.39   The term flame supervision device (FSD) refers to a gas control fitted into the pipeline. It is designed to turn off the main gas flow to an appliance in the event of flame failure. These devices are also sometimes referred to as flame failure devices (FFDs). The presence of the flame is detected by one of the following methods:

- using heat generated by the flame;
- allowing electricity to pass through a flame;
- detecting the light emitted from the flame.

Some of the devices identified below are no longer seen as common controls: future developments, aiming for improved efficiencies, will inevitably restrict their long-term use.

In compliance with the Gas Regulations, whenever you work on a gas appliance, you must test the operation of the safety control. This can be done by following the methods stated in the following sections or if available, specific instructions from the appliance manufacturer.

Bi-metallic Strip FSD

3.40 This device is no longer used as a modern gas control; it was found on older appliances, such as water heaters, cookers and warm air units. The device relies on the principle that a strip of two metals, with different expansion rates and bonded together, will bend when exposed to heat, see Figure 3.35a. This bending movement forces the gas valve to open, allowing the main gas to flow into the appliance, this is illustrated in Figure 3.35b. A pilot flame is used to supply the heat to bend the bi-metallic strip and ignite the main gas flow. Should the pilot flame be extinguished, the bi-metallic strip would cool, and the main gas valve will close. Problems with this design of FSD include the following:

- If the pilot light goes out, the gas freely discharges through the pilot tube. Thus, the pilot flame itself is unprotected.
- The continuously burning pilot flame is a waste of fuel.
- Distortion of the strip or valve spindle results in operational failure.

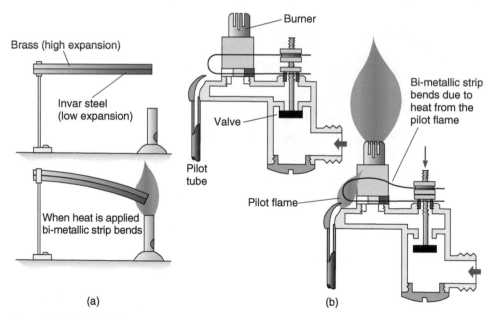

Figure 3.35 (a) Principle behind bi-metallic devices. (b) Bi-metallic strip FSD

Liquid Expansion FSD

3.41 As with the previous control, this device has no control over the pilot flame and, in the event of flame failure, gas continues to enter the appliance through an uncontrolled pilot or by-pass. Few appliances use this kind of FSD but possibly the best example would be that found in an older type of gas oven, as illustrated in Figure 3.36. A small phial containing a volatile fluid turns into a vapour when heated by a low flame on the burner. This vapour has a much greater volume than the liquid, therefore its expansion causes the bellows to increase in size, forcing the valve to open and thereby increasing the gas flow to produce a high flame at the burner. *Note:* The initial pilot flame that was used to generate the heat was supplied with gas through a by-pass hole. Should the supply to the unit be interrupted, the flame would go out. The vapour would then condense back into its liquid form and be forced back, by the spring-loaded valve, along the capillary tube to the phial.

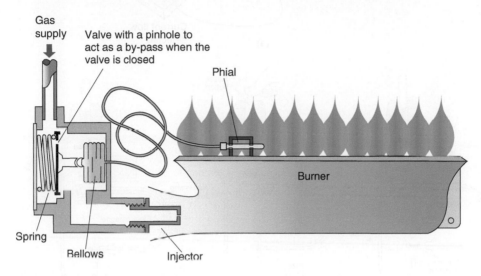

Figure 3.36 Liquid expansion FSD

Thermo-electric FSD

3.42 This device, certainly for the smaller domestic gas appliance, is possibly the most commonly encountered FSD. However, with greater improved efficiency now being sought from appliances, has now been replaced by units that rely on electronic ignition rather than a constant pilot. The biggest concern is that the pilot flame burns night and day, wasting fuel gas when the appliance is not in use. It is referred to as a thermo-electric device because it uses heat to generate electricity. The current produced is used to energise an electro-magnet, which holds the gas valve open. Thermo-electric valves come in all shapes and sizes and can be seen on boilers, cookers, water heaters, fires, etc. The device is often difficult to detect with the untrained eye and may form part of a larger multi-functional gas control. The giveaway is the location of a thermocouple connected between the FSD and pilot flame/burner.

The thermocouple is the device that generates the electrical current. In its simplest form, a thermocouple consists of two different metals connected together at one end, see Figure 3.37a. The other ends are connected together by a coil of wire. When heat is applied to the point where the two metals are joined, referred to as the hot junction, a small electromotive force (emf) of around 15–30 mV is generated. This causes a current

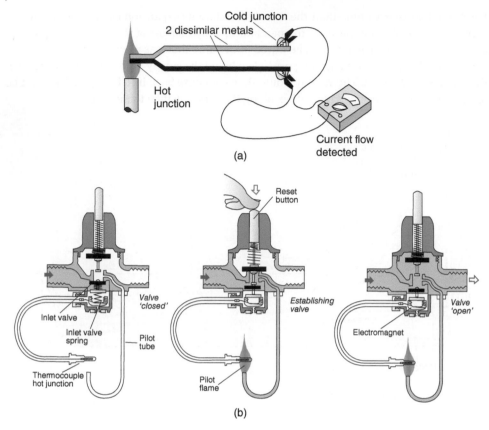

Figure 3.37 (a) Operating principle of a thermocouple. (b) Thermo-electric flame supervision device

to flow along one metal, around the coil, and back along the other metal to return to the hot junction, making a circuit. This continues as long as heat is supplied. The current flowing around the coil converts the electrical energy to magnetic energy and as previously stated, this magnetism is used to keep the gas valve open. Typical metals used for the thermocouple tip are copper and alloys such as constantan, chromel and alumel. The electrical conductor between the thermocouple tip and the electromagnet consists of a small copper tube with an insulated wire running through the centre.

The thermo-electric valve works as follows (see Figure 3.37b):

1. The re-set button is depressed. This closes off the outlet and then allows the inlet valve to open. Gas can now flow through and pass into the pilot tube.
2. Gas discharging from the pilot injector is ignited and the flame burns, playing onto the thermocouple hot junction.
3. After 10 seconds or so, a sufficient emf. has been generated to allow the current to flow around the electromagnet. The pressure can then be taken off the re-set button. The magnet continues to hold the inlet valve open, and the outlet valve returns to allow the gas to flow to the next stage in readiness to enter the appliance should heat be called for.

This main flow of gas would be ignited by the previously established pilot flame. Should the pilot flame go out, the thermocouple hot junction would cool, and the valve would fully close, assisted by the inlet valve spring.

Thermocouple Interrupter

3.43   A thermocouple interrupter is a method that allows a safety device to shut down the main gas supply to an appliance in the event of a safety-critical situation. An example of this could be the triggering of an atmosphere-sensing device, see Section 3.59. The interrupter forms part of the circuit of the thermocouple. When the safety device operates, it opens a switch and thus breaks the circuit of the current supplying the electromagnet in the thermo-electric valve. This causes the gas supply to shut down and will require to be manually reset. A very common fault with this device occurs when the connections are loose or dirty, this can cause an intermittent nuisance shut down. Furthermore, the device must never be left bridged out to get an appliance working, action such as this would be RIDDOR reportable because a safety device would have been over-ridden.

Flame Rectification FSD

3.44   As electricity can flow through a flame, its presence can be detected by positioning a probe in its path. This is located at a small distance off the burner head, as can be seen in Figure 3.38b. When a current is supplied, if a flame is present, it forms part of the electrical circuit. Should there be no flame, the circuit would be broken, and the appliance will fail safe.

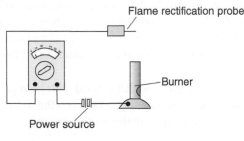

No flame therefore 'no' current detected on the meter

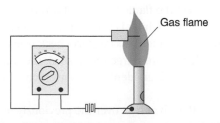

With a flame present current flow is detected at the meter

(a)

(b)

Figure 3.38   (a) Operating principle of flame rectification. (b) Flame rectification (right-hand probe) and spark ignition (left)

Early experiments using this concept used direct current (DC), which had a drawback in that, for example, should the probe be dislodged, or a conductive object fall on the probe, causing it to touch the burner head, it would complete the circuit, giving the false impression that a flame was present. Therefore, in order to avoid this, another method has been developed in which an alternating current (AC) is passed around the circuit (AC is the flow of infinitely small electrons first flowing in one direction and then, a split second later, in the other direction, continually back-and-forth throughout the circuit). Because the probe located in the flame is very small compared with the size of the total flame area, many of the electrons miss the probe, and the electron flow or current is not detected at the probe. Current is therefore only detected in one direction, i.e. when the electron flow is towards the burner head. This is referred to as a partially rectified AC current or, in simple terms, DC. Thus, AC is passed into the probe, but only DC is detected back at the control unit.

Possible faults

Should the probe become dislodged and touch the burner head, alternating current would be detected at the control unit, and it would then shut down the gas inlet valve. Excessive carbon build-up or a damaged lead or probe will cause the DC signal to break down and send the appliance into lockout. The control box could develop a fault, but it is recommended to examine the leads and probe first to establish that they are correct as they are more likely to be the source of a problem.

Ultraviolet (UV) Photo Cell FSD

3.45   When a flame burns, it gives off ultraviolet (UV) light. Therefore, by careful design, a UV detector would pick up its presence. Effectively, the UV detector is an electrical resistor; if the detector senses the presence of a flame, it allows electricity to flow in a circuit. Thus, as with the flame rectification FSD, if the circuit is maintained, it allows the gas valve to open. Where no flame is present, the valve will instantly close shutting off the flow of gas (Figures 3.39 and 3.40).

The detector consists of a UV transmitting glass, which is filled with inert gas. It has two electrodes positioned parallel to each other with a small gap between them. In the presence of UV light, electricity flows freely across the gap to maintain the circuit.

Possible Faults

A common fault is the failure of the cell to detect the flame because of a 'dirty or obscured cell'. Another fault is for the cell to detect a flame, which is not present. The name given to this fault is 'gone soft'. The control sequence includes a safe start check every time the burner starts and if the cell senses a flame during pre-purge (gone soft), the unit will go to lockout. UV cells should be exchanged yearly with an identical replacement of the same sensitivity.

Flame rectification and UV cell FSDs have the advantage over the previously used FSDs in that:

1. no permanent pilots are required;
2. complete gas isolation is maintained when the appliance is not in use;
3. they are very quick to respond to the presence of a flame.

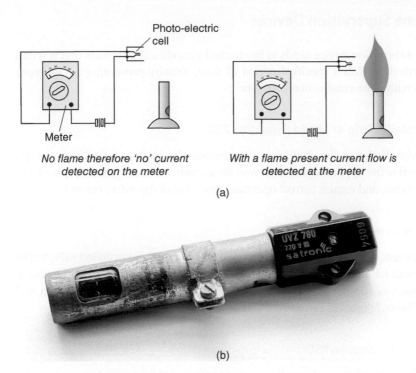

Figure 3.39   (a) Operating principle of ultraviolet (UV) flame detection. (b) UV photocell

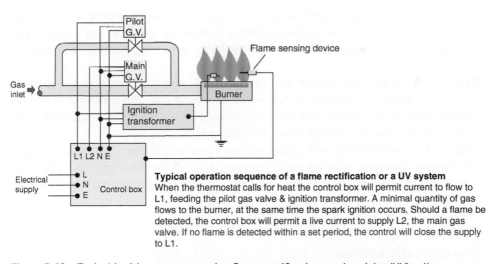

Typical operation sequence of a flame rectification or a UV system
When the thermostat calls for heat the control box will permit current to flow to L1, feeding the pilot gas valve & ignition transformer. A minimal quantity of gas flows to the burner, at the same time the spark ignition occurs. Should a flame be detected, the control box will permit a live current to supply L2, the main gas valve. If no flame is detected within a set period, the control will close the supply to L1.

Figure 3.40   Typical ignition sequence using flame rectification or ultraviolet (UV) cell

Testing Flame Supervision Devices

3.46 A flame supervision device needs to be checked periodically to ensure that it is still operating effectively within a specified period of time, thereby preventing a dangerous build-up of gas within the combustion chamber.

Testing a Bi-metallic Strip or Liquid Expansion FSD

To test this device, first, run the appliance with the main burner operating, turn the gas isolation valve off to the appliance and then wait 60 seconds. Turn the gas supply back on, relight the appliance, and ensure correct operation sequence of the safety control.

Testing a Thermo-electric FSD

Carry out a click test; with the main burner in operation, turn off the isolation valve to the appliance and wait for a period of up to 60 seconds or a time stated by the manufacturer. Within this time, you should hear a distinct click. Turn the isolating valve back on and ensure the appliance relights in the correct manner.

Testing a Flame Rectification FSD

The method of this test will vary depending on the appliance, if the manufacturer has provided a specific test, this should be carried out. However, there are three generic methods, any of which can be used, if suitable for the appliance. The three different methods are:

Method 1. With the main burner on, turn off the gas supply to the appliance and observe, that within 60 seconds, the appliance directly shuts down and locks out. Upon turning the gas supply back on, confirm that the appliance does not attempt to automatically relight.

Method 2. With the main burner on, turn off the gas supply and observe the appliance making a number of attempts to relight before locking out. This sequence of lock-out must last no more than two minutes. Upon turning the gas supply back on, confirm that the appliance does not attempt to automatically relight.

Method 3. There are some appliances that have flame rectification but are not designed to lockout. Therefore, with the main burner on, turn off the gas supply and observe the appliance continuing to spark while the gas is isolated. Re-establish the supply and confirm the normal ignition sequence.

Thermostatic Control

3.47 Thermostats are incorporated in many gas appliances to reduce the gas input, thereby decreasing the temperature. Some thermostats are designed to provide a gradual or modulating control over the flow of the gas, whereas others are designed to give an on/off control. For example, a gas oven would clearly need to have a variable control, whereas a typical conventional boiler may switch on and off as the thermostat calls for heat. Thermostats work on several principles including bi-metallic expansion, liquid expansion, and, in the more modern appliances electrical resistance, such as a thermistor in conjunction with an electronic control system.

Rod-type Bi-metallic Thermostat

3.48 These are only found on older gas equipment. The thermostat consists of two dissimilar metals, namely a brass tube, which has a high expansion rate and an Invar steel rod, which has a very low expansion rate. The Invar steel rod is secured at one end inside the brass tube. The brass tube is, in turn, secured to the body of the valve (see Figure 3.41). Thus, as the heat of the appliance warms the tube it expands and in turn pulls the Invar rod away from the valve seating allowing it to close, assisted by a spring, turning off the supply. To enable variable operational temperatures, a control knob can be adjusted to alter the distance the rod will need to travel in order for the valve to close fully.

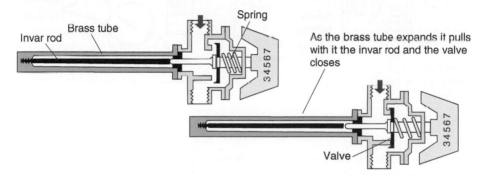

Figure 3.41 Rod-type thermostat

Note: Where this control is used in conjunction with a cooker, the valve body incorporates a by-pass hole to allow for the oven to drop to a simmer, and thus prevents the flame from going out.

Liquid Expansion Thermostat

3.49 These are far more common and are found on many gas appliances. There are two designs, both working on the same principle; however, they differ in that the expansion of the liquid or vapour could either open/close the gas line (typically found in cookers, see Figure 3.42) or could make/break a switch in an electrical circuit (typically found in boilers, see Figure 3.43). Electrical thermostats can also be designed to drop out completely, requiring manual intervention as in the case of high-limit thermostats.

The unit consists of a remote phial, which is housed high in the appliance, and a capillary tube that joins it to the bellows chamber. Within the phial is a volatile fluid that has a very rapid expansion rate. As the phial heats up the liquid expands, in some cases changing to a vapour, and passes through the small capillary tube to fill the bellows. The bellows in turn

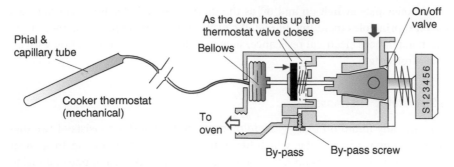

Figure 3.42 Liquid expansion cooker thermostat

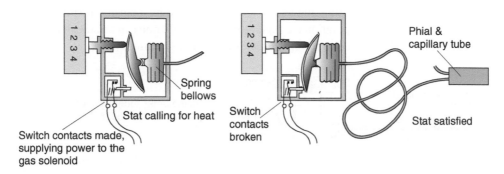

Figure 3.43 Liquid expansion boiler thermostat (electrical)

increase in size to take up this expansion and, in so doing, breaks the electrical circuit. In the case of a mechanical valve, it closes off the gas supply. On cooling, the fluid contracts and is forced back by the pressure of a spring acting on the bellows, through the capillary tube to the phial. As noted earlier with rod thermostats, when a mechanical thermostat is used in conjunction with an oven it requires a by-pass.

Multi-functional Gas Valve

3.50 As the name implies, this gas control has several functions. The valve was developed many years ago, incorporating several controls within one unit, thus saving space and allowing the appliance to be more compact and therefore smaller. The multifunctional control may include any or all of the following components:

- filter;
- safety shut-off valve (thermo-electric valve);
- regulator;
- solenoid valve (opens when appliance calls for heat);
- gas inlet pressure test point and burner pressure test point;
- pilot adjustment screw.

There are many different designs of multi-functional control, each offering a slight variation in operation. A typical version is shown in Figure 3.44a, which requires a permanent pilot flame. Figure 3.44b shows the valve at different stages of operation:

Stage 1: Shows the valve off.

Stage 2: Depressing the button opens the pilot valve, allowing gas to flow to the pilot burner, where it can be ignited.

Stage 3: After some 10–15 seconds, the thermo-electric valve will operate, allowing the latch lever to be drawn in, engaging the safety shut-off valve. Thus, when the pressure is removed from the button it rises, pulling with it the safety valve lever and opening the valve. *Note:* The pilot valve remains open as it is held open by the latch engaging the safety valve lever.

Stage 4: Should the thermostat be calling for heat, current is supplied to the solenoid. This opens the servo regulator valve and closes the weep valve outlet. Gas now flows to the underside of the servo regulator diaphragm and the working pressure diaphragm where it overcomes the valve spring pressure, forcing the main gas valve to open. Gas can now flow to the main burner and be ignited by the previously established pilot flame. Any adjustment to the outlet pressure adjustment screw alters the servo regulator position, which in turn adjusts the pressure to the working pressure diaphragm and so maintains a constant outlet pressure.

When the appliance temperature is sufficient, the solenoid will be de-energised, which will cause the servo regulator valve to close and allow the weep valve to re-open. The pressure within the regulating chamber will drop rapidly to zero, allowing the valve spring to close the main gas valve, as shown in Stage 3, Figure 3.44b.

Stage 5: To shut down, the control button is simply turned 15–20° and released. This allows the latch lever to disengage the safety valve lever and closes off the pilot valve. Note: It is impossible to re-establish the control until the magnetic valve is de-energised, re-lifting the latch lever in readiness to grab the safety valve trip.

Figure 3.44 (a) Multi-functional control valve

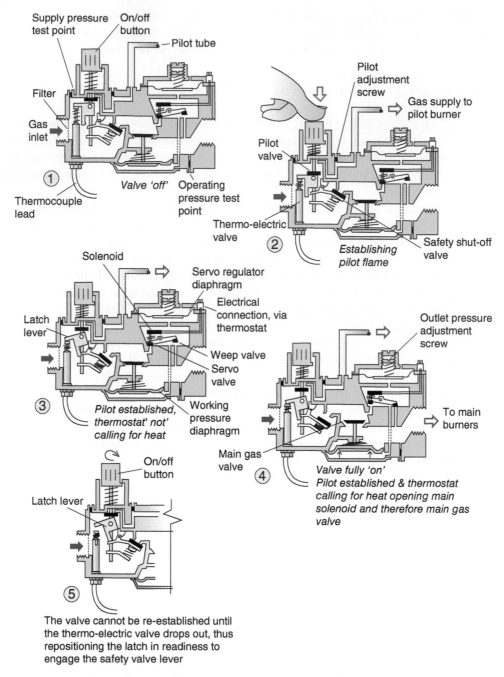

Supply pressure test point

On/off button

Pilot tube

Filter

Gas inlet

① Valve 'off'

Thermocouple lead

Operating pressure test point

Pilot adjustment screw

Gas supply to pilot burner

Pilot valve

Thermo-electric valve

② Establishing pilot flame

Safety shut-off valve

Solenoid

Servo regulator diaphragm

Electrical connection, via thermostat

Latch lever

Weep valve

Servo valve

③ Pilot established, thermostat' not' calling for heat

Working pressure diaphragm

Outlet pressure adjustment screw

To main burners

Main gas valve

④ Valve fully 'on'
Pilot established & thermostat calling for heat opening main solenoid and therefore main gas valve

On/off button

Latch lever

⑤

The valve cannot be re-established until the thermo-electric valve drops out, thus repositioning the latch in readiness to engage the safety valve lever

Figure 3.44 (b) Principle of operation of a multi-functional gas valve

Air/Gas Ratio Valve

3.51   The air/gas ratio valve shown in Figure 3.45c is typical of a design that is used in conjunction with an appliance that uses electronic ignition and a forced draught burner. The speed of the combustion fan is directly proportional to a variable modulating range of gas rates, thus the greater the fan speed the greater will be the flow of gas and consequently the heat input. As a result, there is a saving in the amount of gas consumed by the appliance because there is no permanent pilot flame burning 24 hours a day, every day.

The valve uses a servo pressure regulator that works in conjunction with the fan that supplies air to the appliance. This is shown in Figure 3.45a. The fan causes a negative pressure just in front of the gas injector, pulling the gas from the gas supply. This negative pressure is due to the action of the forced draught air supply whose velocity is increased as it passes through the restriction in the air supply line. It could be likened to the zero governor previously mentioned in Section 3.35. This gas valve eliminates the need for an air pressure switch as the gas valve cannot open without the fan running.

The valve shown in Figures 3.45a and b operates on the following principle:

1. With the thermostat calling for heat, both the safety shut-off and operator solenoids open. This allows gas to flow through the main safety shut-off valve, then into the servo inlet orifice, and eventually to the underside of the servo and working pressure diaphragms. This pressure alone would be insufficient to open the main gas valve to the burner.
2. If the fan is operational, the pressure differential on each side of the servo diaphragm causes the servo valve to move towards its seating. This causes the gas pressure to build up below the servo and in turn beneath the working pressure diaphragm. Eventually, it overcomes the main valve spring, allowing the valve to open and gas to flow to the burner. The greater the pressure on the servo, the more the main gas valve will open.
3. When the thermostat is satisfied, the two solenoids close and the fan ceases to run. With the pressure removed from the servo diaphragm, it lifts as the main valve spring pushes the gas back out from beneath the working pressure diaphragm. Any locked-up gas is discharged via the weep line into the appliance.
4. At (2) above, if the fan fails to operate during the initial operation of the appliance, the servo diaphragm will not be pulled down and therefore will not prevent the build-up of gas pressure below the working pressure diaphragm. Therefore, no gas will flow through to the burner injector. Any gas passing down through the weep line would be insufficient to establish a suitable flame and the appliance would go into lockout. This is because the flame is not detected by the flame rectification flame supervision device, and subsequently the solenoids would have to close the valve.

Figure 3.45a shows the valve in the open position, and Figure 3.45b shows the control fully closed. In addition to the components shown, there would be devices such as pressure test points and filters.

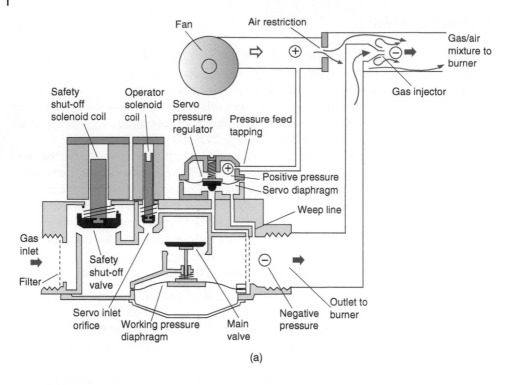

(a)

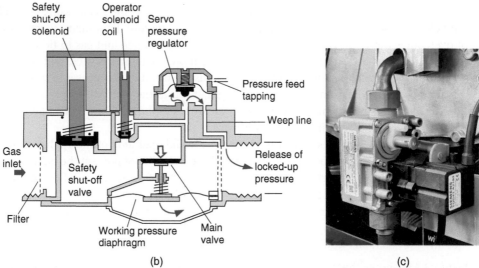

(b)   (c)

Figure 3.45 (a) Air/gas ratio valve in open position. (b) Air/gas ratio valve in closed position. (c) Air/gas ratio valve

Ignition Devices

3.52 There are several forms of ignition devices as described in the following sections.

Hot Surface Ignition

3.53 This design of ignition, seen in Figure 3.46, consists of a filament or coil of high-resistance metal, such as platinum, which glows red hot when a current of electricity is passed through it. As the gas flows past the glowing coil, it ignites. The power supply usually comes from a transformer. However, in some of the older filament igniters used on gas fires, a battery was often used. The glow coil igniter is not common but can still be found in some commercial warm air units.

Figure 3.46 Hot surface ignition

Piezo-electric Ignition

3.54 This type of igniter is commonly found on gas fires, water heaters and existing boilers. This spark ignition device, illustrated in Figure 3.47, consists of quartz, or similar crystals, which when exposed to pressure or stress produce a voltage of around 6000 V. The one shown uses a press button that, when depressed, allows the plunger head to strike the crystals, causing a spark to be generated. You can see from the diagram how the spark jumps to earth at the point where the electrode is positioned.

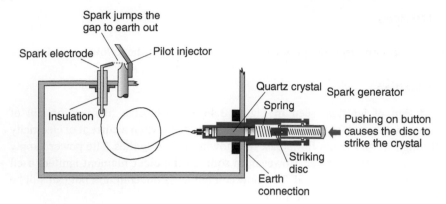

Figure 3.47 Piezo-electric ignition

Mains Ignition Transformer

3.55 This is simply a step-up transformer. It differs from the previous method of spark generation by supplying a continuous rapid spark that jumps or arcs between two electrodes or between one electrode and the earth. The voltage generated would be typically 5000–10 000 V. Higher voltages are obtained using two electrodes arcing together, see Figure 3.48.

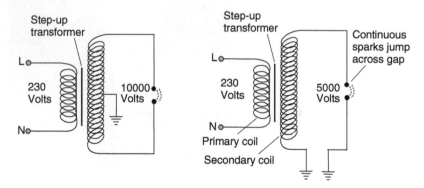

Figure 3.48 Mains ignition transformer

Electronic Pulse Ignition

3.56 With this method of spark generation, illustrated in Figure 3.49, instead of the continuous arcing across the electrodes as obtained by an ignition transformer, a controlled spark rate, typically of around 4–8 sparks per second is maintained. This is achieved by the use of electronics. The basic operating principle is as follows:

1. When the main switch is made, current flows through the rectifier and charges up the capacitor. *Note:* A rectifier acts as a non-return, allowing current to flow only in one direction. A capacitor acts like a storage unit, holding a charge or volume of electricity until it can be used or discharged.

2. After a predetermined period, and at set intervals, the timing switch closes, allowing the current to flow through the transformer. When this switch is closed the capacitor can discharge its contents and coupled with the normal electron flow, a high impulse flows through the step-up transformer causing a 15 000–20 000 V spark to jump across the electrode gap. The timing switch then breaks and so the capacitor re-charges in readiness for the next timed cycle a split second later.

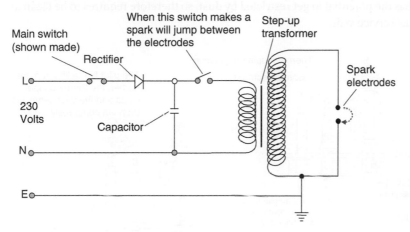

Figure 3.49 Electronic pulse ignition

Vitiation-sensing Devices

3.57 Where an open-flued or flueless appliance has been installed, it is possible that the oxygen within the room could become depleted as a result of inadequate ventilation or a poor flue arrangement. This lack of oxygen is called vitiation. Should vitiation occur, incomplete combustion of the fuel will result. A vitiation-sensing device shuts down the appliance when levels of CO_2 in the vicinity of the appliance reach between 1.5% and 2% (15 000–20 000 ppm), preventing its further use without manual intervention. There are two basic types of vitiation devices: those that monitor the oxygen supplying the pilot burner and those that sense heat, caused by spillage. These two devices are described in the following sections.

Oxygen Depletion System (ODS)

3.58 This method of vitiation detection, seen in Figure 3.50, works by sensing a depletion of the oxygen being supplied to the appliance for combustion of the pilot flame. During normal operation, primary air to supply the pilot flame is initially drawn in through an aeration port. This allows the flame to burn at the point of discharge in the typical stable fashion, characterised by a blue flame playing on the thermocouple. Should the environment lack sufficient oxygen, the amount of primary air would become depleted. This would have the effect of starving the flame of sufficient oxygen and the pilot flame would lift off the burner in search of oxygen and not play on the thermocouple. As a result, the thermocouple would cool, causing the thermo-electric valve to drop out. With appliances such as gas fires the aeration port is visible and during any service work it should be inspected to ensure that it is not blocked. A blockage would starve the flame, causing the pilot to reduce in size and burn yellow. Back boilers, on the other hand, have a sampler tube that passes from the aeration port up to a point above the heat exchanger. So where incomplete combustion occurs, due to spillage, vitiated air is drawn down through the tube and detected by the pilot jet. This tube also has the potential to get restricted by dust, so therefore requires to be cleaned during an annual service visit.

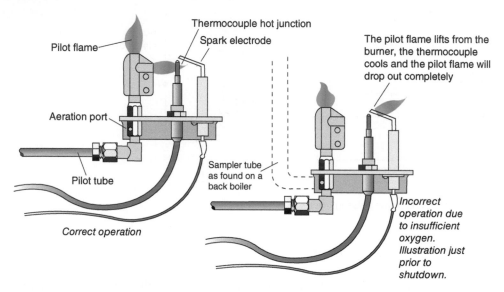

Figure 3.50 Oxygen depletion system

Atmospheric-sensing Device (ASD)

3.59 This device is usually fitted inside the down draught diverter and consists of a thermistor or similar heat-sensing device. Illustrated in Figure 3.51a, it has two wires that are connected and join the internal conductor wire of a thermocouple interrupter. Under normal operating conditions, dilution air is drawn over the heat sensor. However, during persistent and continued down draught and spillage the sensor heats up and breaks the circuit between the thermocouple and thermo-electric device. The valve then drops out and requires manual intervention to reset.

Many flueless appliances have this type of device fitted within the top third of the unit. It is designed to detect high temperatures that could otherwise damage the appliance, i.e. it senses the atmosphere, an example is shown in Figure 3.51b.

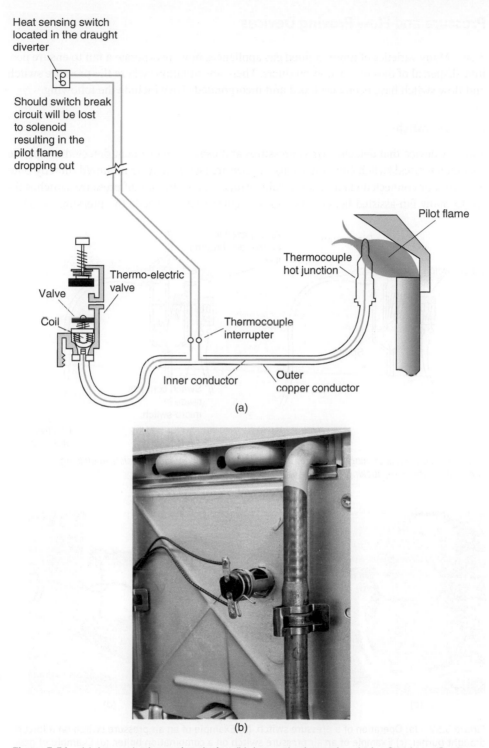

Heat sensing switch located in the draught diverter

Should switch break circuit will be lost to solenoid resulting in the pilot flame dropping out

Pilot flame

Thermocouple hot junction

Thermo-electric valve

Valve

Coil

Thermocouple interrupter

Inner conductor

Outer copper conductor

(a)

(b)

Figure 3.51 (a) Atmospheric sensing device. (b) Atmospheric sensing device fitted to a flueless water heater

Pressure and Flow Proving Devices

3.60 Many varieties of modern flued gas appliances now incorporate a fan to ensure positive dispersal of the combustion products. These are situations where the pressure switch and flow switch have been developed and incorporated. They include the following.

Pressure Switch

3.61 A device that detects air/gas pressures and uses the movement detected to operate a small electrical switch (micro-switch). A pressure switch may be mounted directly onto pipework or connected via a series of rubber tubes, as in the case of pressure switches fitted to many fan-assisted boilers. As seen in Figure 3.52a–d, inside the pressure switch is

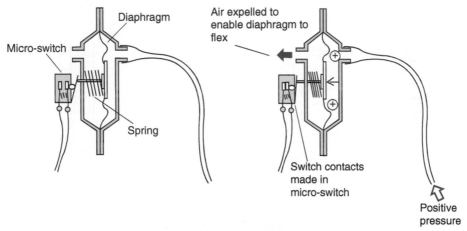

The switch shown is utilising a positive pressure; however, a variation consists in utilising a negative pressure, sucking the diaphragm towards the micro-switch.

(a)

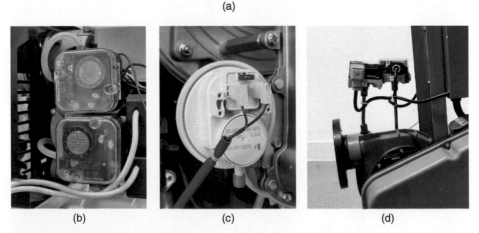

Figure 3.52 (a) Operation of a pressure switch. (b) Example of an air pressure switch on a forced draught burner. (c) Example of an air pressure switch on a combination boiler. (d) Example of gas pressure switches on a gas booster

a small diaphragm washer. On one side is the air/gas sensing tube and on the other the micro-switch. As the diaphragm moves in response to the pressure applied, it makes or breaks the electrical contacts.

Fan Flow Switch

3.62   This device, illustrated in Figure 3.53a and b, comprises a vein or paddle that moves in response to air movement. As the paddle moves it pivots at a fulcrum and, in so doing, pushes together the contacts of a micro-switch. As with the previous control switch, this electrical response can then be used to open a gas control valve.

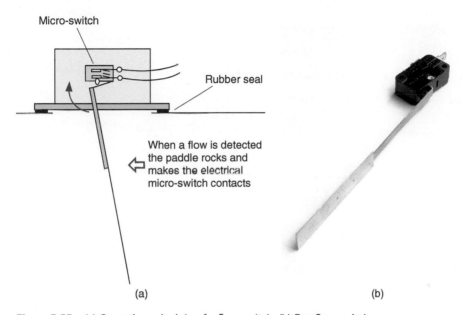

Micro-switch

Rubber seal

When a flow is detected the paddle rocks and makes the electrical micro-switch contacts

(a)

(b)

Figure 3.53   (a) Operating principle of a flow switch. (b) Fan flow switch

4

Installation Practices

Polyethylene Pipe Jointing

Relevant Industry Documents
BS 6891 and IGEM/UP/2

4.1 Polyethylene (PE) pipe can be used for both natural gas and LPG. It is available as either medium or high density. The maximum operating pressures need to be confirmed with the supplier. PE pipe must be fully protected, in most cases by burying below ground. It may rise above ground level, but it must be completely protected from damage, including that from ultraviolet light from the sun. It is always preferable to use metallic pipe entries into a building. However, if PE is used then the pipe entering the building must be placed inside a metal fireproof sheath to ensure that no gas escape from the plastic pipe could enter the building in the event of a fire. There are three methods of jointing to PE pipe: fusion welded joints, electro-fusion welded joints and compression joints. *Note*: Fusion welded joints may only be undertaken by companies or specialists who are competent and assessed for these jointing methods.

Fusion Welded Joint

4.2 This is a specialist joint that is undertaken using a fine stream of extremely hot air and a filler rod. With the heat directed at the pipe ends and the filler material applied, fusion occurs as the plastics melt together.

Electro-fusion Welded Joint

4.3 An electrical coil is incorporated into each joint (see Figure 4.1). This connects to the outside via two terminals. To make the joint, a special transformer that supplies a small voltage of 39.5 V is used for a set time of around 24–90 seconds, depending on the fitting size and manufacturer (see Figure 4.2). A label attached to the fitting will give precise data. First, the pipe end should be scraped clean using a special scoring tool (see Figure 4.3). Then the pipe is pushed fully into the fitting. At this point, a pencil mark to indicate the fully 'in'

Gas Installation Technology, Third Edition. Andrew S. Burcham, Stephen J. Denney and Roy D. Treloar.
© 2024 John Wiley & Sons Ltd. Published 2024 by John Wiley & Sons Ltd.

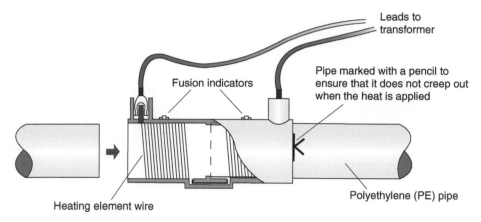

Leads to transformer

Fusion indicators

Pipe marked with a pencil to ensure that it does not creep out when the heat is applied

Heating element wire

Polyethylene (PE) pipe

Figure 4.1 Electro-fusion welded joint

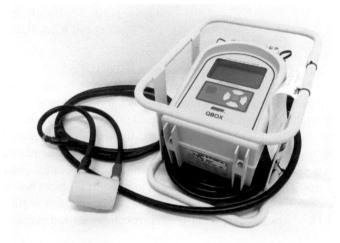

Figure 4.2 Electro-fusion welding transformer

position is made on the pipe. Failure to do this may lead to a problem, as the pipe tends to creep out from the fitting as it is heated. The mark will indicate any lateral movement so the pipe can be held in place. If the pipe did creep from the fitting, molten plastic would ooze inside and restrict the pipe. With the fitting fully assembled and the connections made, the supply is switched on for the set period. As the fitting melts, markers pop up from the fitting, indicating that sufficient weld temperature has been achieved. *Note*: There are two electrical connections to each fitting, thus with branches for example, the middle connection is a spigot on which another coupling would be required to complete the connection.

Compression Joint

4.4 These fittings are usually used where the PE is to join another material, such as copper or steel, and as such they are often referred to as transitional fittings (see Figure 4.4a and

Figure 4.3 Special scoring tool

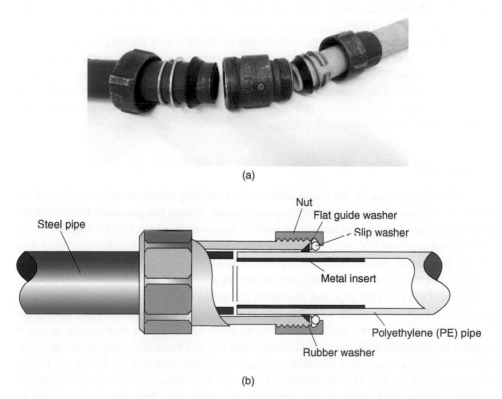

(a)

Steel pipe

Nut

Flat guide washer

Slip washer

Metal insert

Polyethylene (PE) pipe

Rubber washer

(b)

Figure 4.4 (a) Construction of a transitional compression joint. (b) Transitional compression joint

b). The fitting consists of a body into which the pipe spigot can enter and be clamped via a rubber compression ring. To prevent the rubber ring from twisting out from the fitting, as the lock nut is turned, a slip and guide washer is used. This allows the pressure to be spread evenly onto the rubber, forcing it squarely into the fitting. An insert needs to be placed inside the plastic pipe prior to connection; this maintains a solid true bore within the pipe, preventing leakage. The fittings used for an LPG installation may need confirmation with the manufacturer to ensure that the rubber ring is suitable for use.

Copper Pipe Jointing

Relevant Industry Documents
BS 6891 and IGEM/UP/2

4.5   Copper may be used for both natural gas and LPG, the pipe should be no larger than 108 mm in diameter. Where applicable, copper pipe needs to be suitably protected from mechanical damage, and in the situation where it is to be buried, it must be factory-sheathed. Buried copper pipe must not be connected to steel pipe and fittings as the surrounding groundwater will act as an electrolyte and cause electrolytic corrosion, eventually resulting in a leak. There are several jointing methods as described in the following sections.

Note: A manufacturer may produce alternative methods of pipe jointing that are not covered in the relevant standards, albeit they are stated as being of an equivalent level of safety, and an appropriate technical specification is provided for its use. The Gas Regulations do allow the use of these methods, provided that the procedures laid down by the manufacturer are followed and a suitable joint is made. Ultimately, the decision to use this alternative method would be for the engineer to make.

Capillary Joints

4.6   These joints require the use of solder with a melting temperature below that of the copper to be drawn into the fitting by capillary action. Where the pressures within the gas pipe will exceed 75 mbar, solders with a melting temperature above 600°C should be used. The type of capillary joint may be either the end feed or solder ring type (see Figure 4.5a and b). Whichever method is used, it is essential to examine the completed joint visually to confirm that the solder has run fully around the joint.

Method: First clean the pipe end and inside the fitting with wire wool and apply an inactive flux. It is possible to use an active (self-cleansing) flux that is active only during the heating process. However, caution needs to be observed to ensure that not too much is applied as the solder will rapidly flow everywhere the flux runs, including inside the pipe. It is therefore recommended that it be applied only to the pipe and not inside the fitting. The purpose of the flux is to maintain a clean oxide-free surface on which the solder can readily flow and stick. Heat is now applied to the joint until the solder melts; it will be seen to appear at the mouth of the solder ring fittings or as solder is applied with end feed

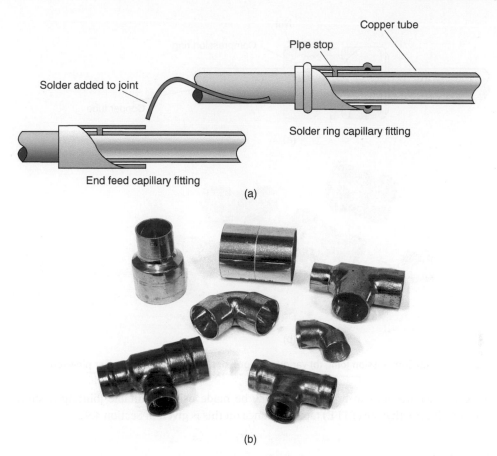

Figure 4.5 (a) Types of soldered joint used for copper pipework. (b) Various capillary fittings used for copper pipework

fittings. Where end feed fittings are used, the amount of solder required is no more than the diameter of the tube. So, for a 15 mm pipe, 15 mm of solder is required for each socket. Finally, the joint should be allowed to cool without movement, whereupon any remaining flux residue should be removed with a damp cloth.

Compression Joints

4.7 Compression joints cannot exceed 54 mm in diameter and where used, must be accessible; below floors or within voids without removable covers would not be deemed suitable locations. To make a sound joint, first, the pipe end needs to be cut square and de-burred, then the nut and compression ring (olive) are slid onto the pipe. The tube is now inserted fully into the fitting and the nut hand-tightened onto the thread, with an additional one-and-a-half turns made with a spanner to form a seal (see Figure 4.6a and b). However, when used within an LPG installation, soft copper compression rings instead of those made of brass must be used, due to the searching nature of the gas.

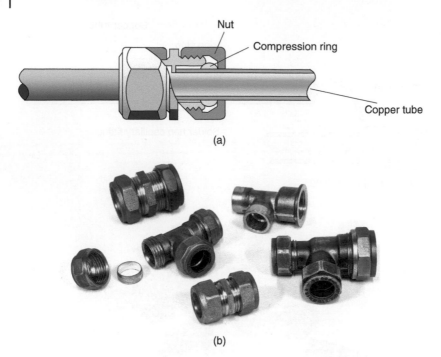

(a)

(b)

Figure 4.6 (a) Compression joint. (b) Various compression fittings used for copper pipework

Threaded connections to brass fittings may be made using a suitable jointing paste or polytetrafluoroethylene (PTFE) tape (guidance on this is given in Section 4.9).

Press Fitting Joints

4.8 This is a modern method of jointing that can be used where hot works are not permitted. It is essential that the correct design of press fitting is selected, as the seal used to make the joint is different from that used in many other fittings. A fitting suitable for gas is identified by a yellow/tan 'O' ring and a yellow product marking, whereas a fitting for water has a black 'O' ring, both types are pictured in Figure 4.7a. The joint is made by

(a)   (b)

Figure 4.7 (a) Press fittings – yellow 'O' ring for gas, black for water. (b) Making a press fitting joint

simply inserting the pipe into the fitting and using a special press fitting tool to compress the fitting tightly onto the pipe (see Figure 4.7b). The fittings are available in a range of sizes from 15 to 108 mm, with a maximum operating pressure of 5 bar.

Mild Steel Pipe Jointing

Relevant Industry Documents
BS 6891 and IGEM/UP/2

4.9   Medium- and heavy-grade mild steel are very versatile materials that can be used in the installation of gas pipework above ground without fear of damage. In fact, it is the only suitable material where large internal pipe diameters are sought. Pipes greater than 50 mm in diameter always need to be welded. However, depending on the maximum operating pressure (MOP), welding may be the required jointing method due to pressure; this can be seen in Table 4.1.

Table 4.1   Jointing of carbon and stainless-steel pipework

Pipe size in mm	Joint type					
	MOP ≤ 500 mbar		MOP > 500 mbar ≤5 bar		MOP > 5 bar	
	Screw	Weld	Screw	Weld	Screw	Weld
<25	✓	✓	✓	✓	x	✓
26–50	✓	✓	x	✓	x	✓
>50	x	✓	x	✓	x	✓

x not acceptable; ✓ acceptable.
Source: Data from IGEM/UP/2

Steel tube is suitable for LPG installations; however, galvanised pipe is often to be recommended. When galvanised pipe is used, galvanised fittings should also be used.

Jointing to threaded pipework is made using a suitable jointing paste, as indicated on the side of the tin/tube of compound used. Alternatively, PTFE tape or string/cord (see Figure 4.8a and b) may be used. Note: PTFE is not to be used with any other jointing medium or compound.

Cutting oil used to make the threads needs to be fully wiped from the pipe prior to use. The PTFE tape must be of the thicker design to BS EN 751 pt. 3 and applied with a 50% overlap as shown in Figure 4.8c.

Union connection joints, as seen in Figure 4.9a, must be in readily accessible positions. *Note*: Under floors and within ducts without removable covers are not considered to be accessible areas.

Where welded pipework is used, flange joints may be required to assist in the replacement of components (see Figure 4.9b). However, the number of flanged joints should be kept to a minimum, with the flanges welded to the pipes. All welding needs to be completed by a trained operative to ensure that the work meets the required standard.

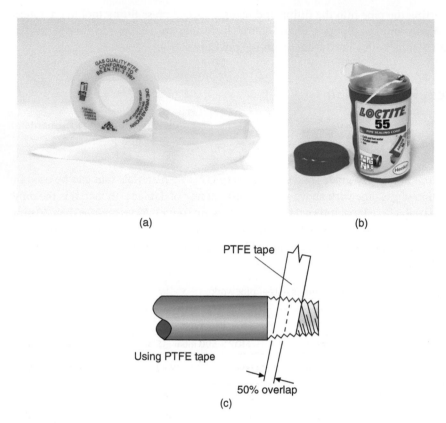

Figure 4.8 (a) Thicker one wrap PTFE tape for gas. (b) Pipe sealing cord. (c) Application of 'one wrap' PTFE

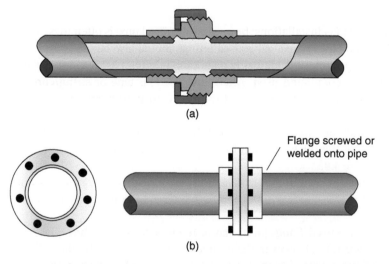

Figure 4.9 (a) Union connector – fitting used to join two fixed pipes. (b) Flange joint

Figure 4.10 Types of joint used for steel pipework

You can use compression joints when connecting to the smaller sizes, although these are generally restricted to making transitional connections to PE pipe (as seen in Figure 4.4b). Examples of threaded joints used for steel pipework can be seen in Figure 4.10.

Semi-rigid and Flexible Pipe Connections

Relevant Industry Documents
BS 6891 and IGEM/UP/2

Flexible Stainless Steel

4.10   The flexible stainless-steel connection, known as the anaconda, has been used at the gas meter for many years. However, a modern innovation is pliable corrugated stainless-steel tubing (PCSST) in sizes from 12 to 50 mm diameter, supplied in coils of up to 90 m in length. It is manufactured to the requirements of BS EN 15266 and limited to 500 mbar (0.5 bar) MOP. Previously, it was manufactured to BS 7838 and referred to as corrugated stainless-steel tube (CSST) and limited to 75 mbar MOP. The pipe can be easily manoeuvred to suit the design layout of the building, enabling it to overcome obstacles. The pipe has a factory-applied membrane coating on the outside to protect against corrosion and some manufacturers have included a second membrane designed for use in unventilated voids (see Figure 4.11a). However, you must consult the manufacturer of the selected product to confirm its suitability for use in a duct, below ground and in any other locations where protection is normally required. No special tools are needed as it is simple to cut using conventional tube cutters with a steel cutting wheel. The joints are made using two wrenches, clamping a threaded terminal connection to the pipe end, and flaring the pipe into the fitting, forming a metal-to-metal seal (see Figure 4.11b). The material may be used for both domestic and commercial installations. It is important, however, to ensure

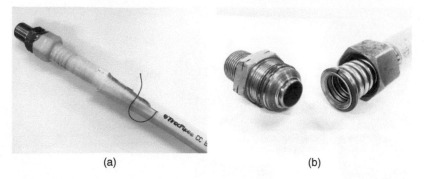

(a)   (b)

Figure 4.11 (a) CSST with second membrane and added protection tape. (b) CSST joint

that no flux is allowed to come into contact with this material, as it rapidly tends to corrode the metal, leading to pitting. The manufacturer provides tape to be wound around the exposed pipe and joint to further protect from corrosion damage. Flexible stainless steel can be used to accommodate structural movement when passing through a timber-framed wall. Some manufacturers of this product produce a specific installation programme and run courses to train and assess operatives to demonstrate competency in installation.

Flexible Hoses

4.11 These are used in the installation of appliances that need to be moved easily, for example, cookers and tumble dryers, or possibly for the high-pressure connections to an LPG cylinder (see Figure 4.12 for examples of hoses). The hose selected for each application needs to comply with the appropriate current standards, for example, the hose must be suitable for the gas being used such as LPG, and also for the pressure.

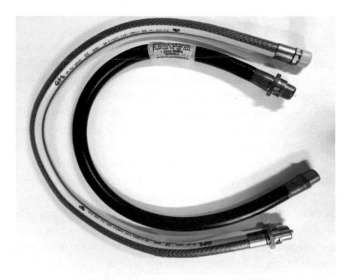

Figure 4.12 Examples of flexible hoses

Appliance Hoses

4.12   These are usually bought as a complete component, with all necessary brass threads and bayonet connections ready for use. LPG appliance hoses used to be characterised by a red stripe or band running along the tube, which itself was black. More recently, single hoses are being manufactured that are suitable for gas families 1, 2 and 3. It is important to ensure that any hose used, complies with the relevant standard. Furthermore, it is fitted as per manufacturer's specification with respect to location, especially the potential temperature of the area in which it will be installed. Appliances used in commercial situations invariably have a steel outer casing and, where used for catering purposes, are covered with an additional yellow plastic sheath.

LPG Supply Hoses

4.13   Where these are used to make the connection from the supply cylinders to the regulator, they are often referred to as pigtails. They are generally coloured black. All LPG hoses should be clearly marked with the manufacturer's name, BS specification, high or low pressure and date of manufacture or expiry. The service life for a hose would be no longer than 10 years and replacement at this age is recommended. Flexible hose ≥ 8 mm diameter and ≤ 50 mbar pressure can be secured with worm drive clips, swaged fittings, or crimp clips, whereas a hose with an internal diameter of <8 mm and pressure >50 mbar must be secured with swaged fittings or crimp clips; worm drive cannot be used. The hose should be no longer than is required and be in one continuous length. Maximum lengths vary depending on the application it is being used for and can be confirmed by referencing BS 6891. For example, in residential park homes; for the flexible pipe connecting from the standpipe to the home, a maximum length of 2 m is stated, this length is considered suitable to accommodate movement of the home. However, an exception is made if the home has a floatation device where a longer hose is permitted. This device allows the anchored home to rise if there is a flood and the hose should be long enough to allow for the maximum extension of the device. Also, due to the risk of rodent attack, hoses shall be armoured to provide additional protection.

Lead Pipes

4.14   Lead piping can no longer be installed; however, it may be found on some existing gas installations. Where found it can be left, providing it is gas tight and there is no sign of damage.

Gas Service and Installation Pipework

Relevant Industry Documents
BS 6891, IGEM/UP/2, IGEM/TD/3 and 4

4.15   The service connection from the main in the road to bring gas into the premises is undertaken by the supplier or one of their contractors and not by the gas installer/engineer.

In general, all new services today will be PE pipe and connections are made with a saddle tapping tee as shown in Figure 4.13. First, the tee is fused on to the main (see Section 4.3 for electro-fusion welded joint), then the service pipe is run to the desired location, usually just outside the building. Upon completion, it can be tested for tightness. Should the tightness test prove successful, a cutter in the top of the fitting is wound down to cut into the pipe. Owing to its design, the plug of PE is retained within the cutter and, when it is screwed up again, gas flows through into the service pipe. Where a steel main has been connected to a PE service (see Figure 4.14), a compression joint is made into a service tee, which has been cut into the top of the main. To provide additional support to the PE, an anti-shear sleeve is incorporated within the first 460 mm. The PE termination at the entry to the building is usually made using a transitional tee or a meter box adaptor.

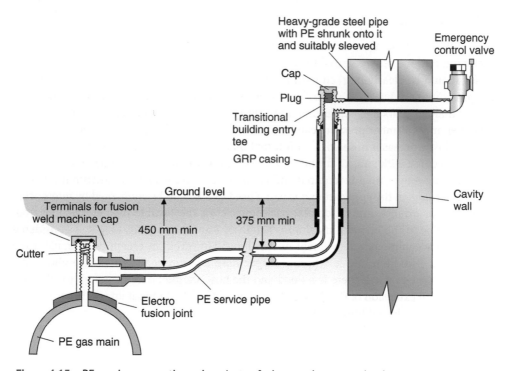

Figure 4.13 PE service connection using electro-fusion tapping tee and reducer

The service pipe should not pass through the wall below ground level; ideally, it should rise externally to enter the building above the floor level. Where the pipe is to pass through a wall, it must be suitably sleeved, sealing the internal surface of the sleeve with a flexible fire-resistant compound, so that any escaping gas can pass only to the external environment.

The route that a pipe is to take into a building should avoid unstable structures and ground that is liable to movement and subsidence. Where possible, the pipe should travel in a straight line at right angles to the building. If it is necessary to run it alongside a building, it should be kept a minimum distance from the wall. This distance can vary between 0.25 and 1 m for pipework up to 75 mbar, depending on whether the pipe is steel or PE. Where high pressure pipework is encountered, this distance needs to be increased.

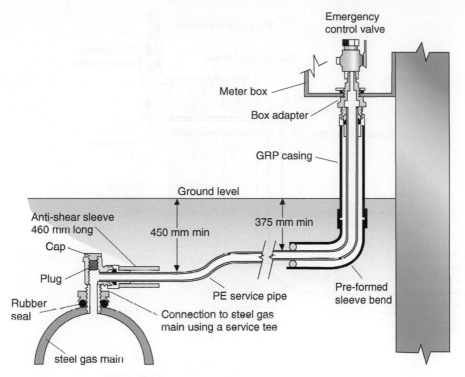

Figure 4.14 PE service connection into a meter box

Protection of Service or Installation Pipework Below Ground

4.16 The minimum depth of cover for new gas pipework depends on the location, nominal bore and pressure. For private gardens and pathways, a minimum depth of 375 mm is required. Where there is light vehicular traffic, this is increased to 450 mm. Where this cannot be achieved, additional protection must be provided such as concrete slabs laid 100 mm above the pipework. Should the pipe run under public roadway etc., then this depth would need to be increased.

Where a possible opening may occur in the protective coating surrounding a steel pipe below ground, the metal would be exposed to direct contact with the soil. Any stray electrical current passing through to earth would undoubtedly flow to earth at this point and cause electrolytic corrosion. To prevent this, a sacrificial anode is sometimes connected to the service pipe; this is destroyed in preference to the service pipe. The anode is simply screwed into a blind tee as shown in Figure 4.15a. The blind tee is a tee fitting with a branch connection that is blanked off. The tee is fully wrapped with a petroleum-impregnated woven bandage, with a minimum overlap of 50%, leaving the anode exposed. The protective anode is usually positioned just prior to the steel service turning upwards to enter the property. PE pipes should not be laid in ground that may be chemically reactive to the plastic material being laid; this includes ground that contains tar, oils, dry cleaning fluids, etc.

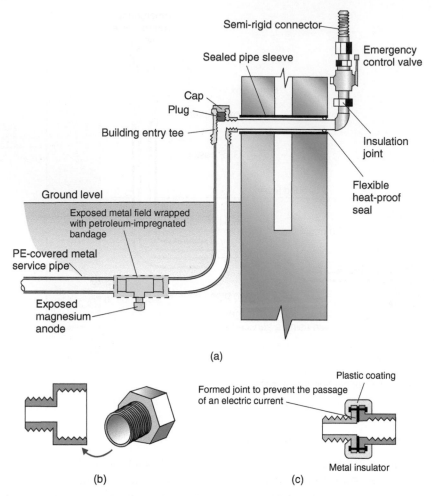

Figure 4.15 (a) Metal service with insulator and additional corrosion protection. (b) Fire retardant thermoplastic insulator. (c) Mica gasket, plastic collar and 'O' ring insulator

Protection to Service or Installation Pipework Above Ground

4.17 PE pipework brought to rise above the surface would be subject to ultraviolet light damage from the sun. It can be fully protected by surrounding it with a glass-reinforced plastic (GRP) sleeve.

Metal pipework needs to be securely fixed and, in general, a suitable coating of corrosion paint suffices to protect steel pipework. However, wrapping may be carried out in corrosive situations. All pipework should be kept at least 25 mm away from any electrical cables and 150 mm away from switchgear and consumer units.

Service Pipework Insulation

4.18 Electrical current flowing through steel pipework can cause corrosion, so to prevent stray electrical currents passing down metal service pipework, the service must be insulated from the internal installation. This is achieved by fitting an insulation joint just inside the building, as shown in Figure 4.15a. The insulation joint may be either of a fire-retardant thermoplastic (see Figure 4.15b) or of a metal design in which the two mating surfaces are prevented from directly touching each other by the use of a mica gasket, appropriate washers and 'O' ring, thus forming a solid sound fitting (see Figure 4.15c). The whole fitting is given a final protective coating to prevent electrical tracking due to moisture across its surface. Where a plastic insulator is used, it needs to be positioned directly on the semi-rigid connector at the meter.

Equipotential Bonding at the Meter

4.19 Within a distance of 600 mm from the primary gas meter (before the first branch) or at entry to the building, the installation pipe must have a suitable protective equipotential bonding conductor. It must be no less than 10 mm^2 and connected to the main earth terminal, which is located at the electrical consumer unit. A meter box must not be drilled or pierced for the earth bond to exit, the wire must be routed via the purpose provided rear entry spigot or by the bottom exit. The installation of this bonding is completed by a competent person, such as an electrician. The gas installer who initially connects the installation pipework to the gas meter has a duty to inform the responsible person for the property where this is required.

Emergency Control Valve

4.20 The person who supplies gas for use in a premises for the first time must install an emergency control valve at the point of entry to the building, in an accessible location (see Figure 4.16a). The valve should be fitted so that a union connector can be disconnected after the control valve. The valve must have a handle securely fitted along with a suitable indicator label showing the on/off position and instructions for the action to take in the event of an emergency. *Note*: the handle must be positioned so that when it is moved down to its lowest possible position the gas supply is off.

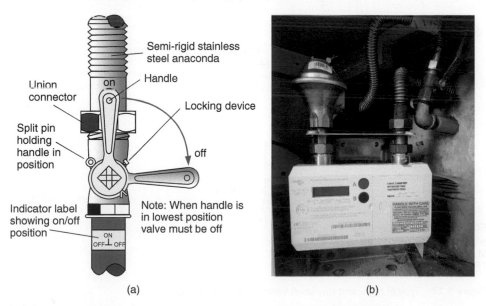

(a)   (b)

Figure 4.16   (a) Emergency control valve. (b) Typical internal meter installation

Connections to High-rise Buildings

Relevant Industry Document
IGEM/UP/2, IGEM/G/5 and BS 6891

4.21 For a multi-storey building, where the supply is to enter individual apartments, an additional emergency control valve (AECV) needs to be fitted inside each flat or in a suitable communal location, with a test point within 300 mm of the outlet of the valve. A notice must also be displayed giving instructions to the user on the action needed in the event of a gas escape, whilst also providing the telephone number of the emergency services.

It would be necessary to run the service up to the various levels or apartments. To do this, an internal fire-resistant shaft could be used (Figure 4.18). This shaft would, if possible, be located adjacent to an external wall to enable ventilation direct to the outside at its highest and lowest position. Where the gas supply pipe in a multi-occupancy building passes vertically through one dwelling in order to gain access to another (as shown in Figure 4.17),

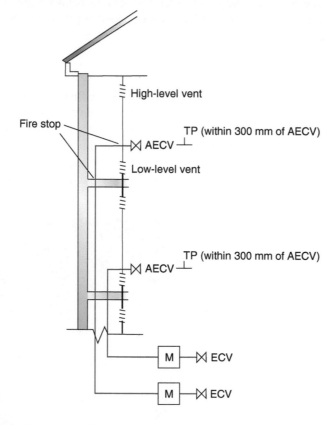

ECV – emergency control valve
AECV – additional emergency control valve
M – meter
TP – test point

Figure 4.17 Typical high- and low-level vents within enclosures

it must be installed within a purpose provided duct to prevent damage and meet current Building Regulations/Standards. These ducts and enclosures must be ventilated at high and low level at each floor where appropriate. The size of the vents is based upon the cross-sectional area of the duct as shown in Table 4.2.

Table 4.2   The required size of the ventilator openings for gas pipework in duct

Cross-sectional area of shaft (m^2)	Minimum "free air" ventilation size for each opening (m^2)
<0.01	0
0.01–0.05	Same size as cross-sectional area
0.05–7.5	0.05
>7.5	150th of the cross-sectional area

Example: The service riser shown in Figure 4.18 has a base measuring 0.450 m × 0.3 m, which gives a cross-sectional area of: 0.45 × 0.3 = 0.135 m^2. This clearly falls into the third row in the Table 4.2; therefore, the ventilation grille fitted at high and low level would each need to be 0.05 m^2. To convert m^2 to cm^2 multiply by 10 000.

$$\therefore 0.05 \times 10\,000 = 500\,cm^2$$

Where the riser passes through any wall or floor structures, it would need to be suitably fire-stopped to prevent the spread of fire. To provide an opening for maintenance purposes, a sealed half-hour fire resistant panel is required at each floor level.

Where the material to be used for a vertical gas riser is mild steel and, if >50 mm diameter, joints must be welded. If ≤50 mm diameter and ≤20 m in height, the joints can be screwed, or end load resistant fittings. If between 20 and 40 m in height, end load resistant fittings or welded, will be acceptable. If >40 m, in height, the riser must be welded. It is essential that the riser is adequately supported at its base; this is usually achieved by means of a concrete plinth with a duck foot bend. The riser should maintain a vertical rise throughout its length, without any change in direction and it should be adequately supported at various intervals to allow for thermal movement.

On lateral pipework connected to the riser at each floor level, a network lateral isolation valve can be fitted close to the riser, to allow for maintenance of the lateral. An expansion joint can be installed after this valve to allow for the differential thermal expansion between the riser and lateral service.

Protected Area

The term 'protected area' relates to an adequately ventilated, fire protected and compartmentalised space within a building such as a lobby or corridor, which would form a fire escape route. Consequently, there are restrictions to the installation of gas pipes within these areas and materials shall be one of the following:

From Approved Document B of the Building Regulation and BS 6891:

1. Steel pipe with screwed joints (≤50 mm).
2. Steel pipe with welded joints.

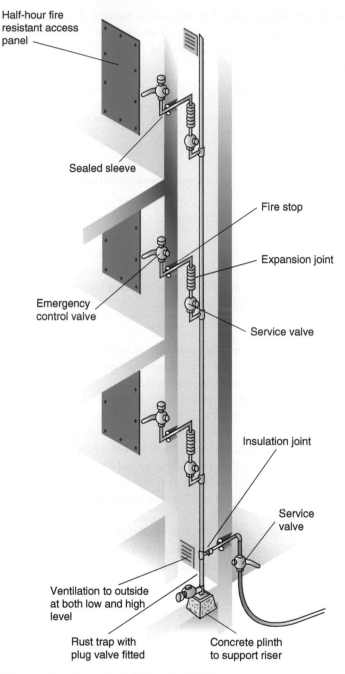

Half-hour fire resistant access panel

Sealed sleeve

Fire stop

Expansion joint

Emergency control valve

Service valve

Insulation joint

Service valve

Ventilation to outside at both low and high level

Rust trap with plug valve fitted

Concrete plinth to support riser

Figure 4.18 Gas service to high-rise building

In additions to 1 and 2 above, in certain situations, BS 6891 includes:

3. Copper pipe of a continuous length with no joints.
4. Pliable corrugated stainless-steel tube (PCSST) in a continuous length manufactured to satisfy Fire Test A from BS EN 1775, Annex A.

Note: PCSST manufactured to BS 7838 satisfies Fire Test A but PCSST manufactured to BS EN 15266 requires consultation with the manufacturer to confirm compliance.

A protected area containing gas pipework will require adequate ventilation to be provided at both high and low level direct to outside air, this can be calculated using Table 4.2. This must be natural ventilation, not mechanical. *Note*: Pipework is not considered to be within the protected area if it is contained within its own fire protected duct ventilated to outside air.

Gas Meter Installations

Relevant ACS Qualifications
MET1 – MET4

Relevant Industry Documents
BS 6400 and IGEM/GM/6

Meter Location

4.22   The gas meter should be located at a point as close as practical to the point of entry to the building and where the service pipe terminates, for example:

- in a purpose-made meter housing, to include a meter box, located outside the building either adjacent to the property or at a boundary enclosure;
- in a garage or outbuilding;
- inside the building.

The location must be easy to access for servicing, exchange and reading purposes or, in the case of electronic prepayment meters, be positioned to enable easy insertion of the payment card. The site must be well ventilated, and the gas meter must be mounted so that it is not in contact with the surrounding wall surfaces.

Fitted prior to the gas meter is the emergency control valve (ECV), with a label indicating the 'on' and 'off' positions; this is followed by the installation regulator. The regulator must not be situated exposed to the elements, water inside the valve can corrode the mechanism and freeze during winter. If the breather hole blocks, it will cause pressure regulation to become erratic.

Affixed to the meter, or adjacent to it, must be a completed and dated emergency control notice that tells the occupier what to do in the event of a gas escape, including turning off the supply. It should also identify the location of the ECV if it is greater than 2 m from the meter and give the telephone number of the emergency service contact. When gas is first supplied, where there is an existing label, this must be updated. Note: Where no meter bracket has been incorporated, ensure the meter is supported by a rigid steel pipe for at least a distance of 600 mm from the meter.

Meter Boxes

4.23 There are various types of meter box including built-in (Figure 4.19), surface mounted (Figure 4.20), and semi-concealed designs (Figure 4.21a–c). These boxes all have the following basic characteristics:

- They provide adequate ventilation to the external environment.
- They are suitably sized to enable installation, exchange and servicing of all components, e.g. meter, regulator, thermal cut-off, as appropriate.
- They are of non-combustible construction and have adequate fire resistance.
- They are constructed so that access can only be gained by the use of a special key, with which the consumer has been provided.

When installing the meter into these boxes, no drill holes should be made except those intended by the manufacturer and all exits should be suitably sealed so that any escape of gas can pass only to the external atmosphere and not into the building. The gas service

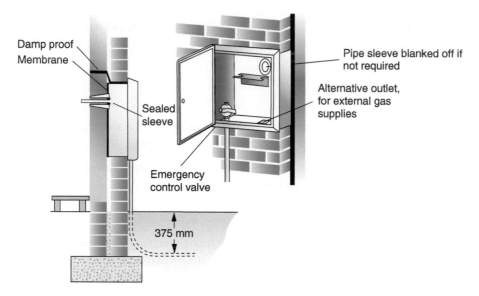

Damp proof Membrane

Sealed sleeve

Emergency control valve

375 mm

Pipe sleeve blanked off if not required

Alternative outlet, for external gas supplies

Figure 4.19 Built-in meter box

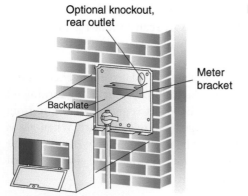

Optional knockout, rear outlet

Backplate

Meter bracket

Figure 4.20 Surface-mounted meter box

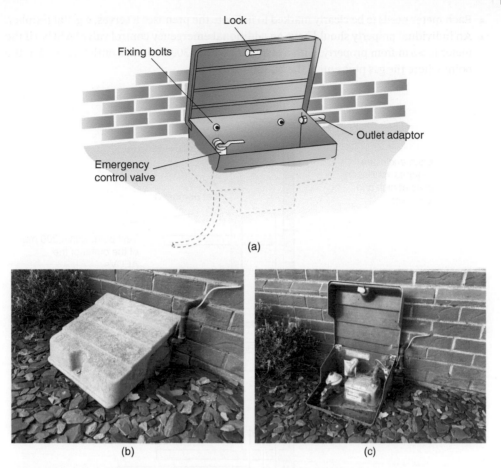

Figure 4.21   (a) Installation of a semi-concealed meter box. (b) Semi-concealed meter box.
(c) Semi-concealed meter box (open), including the usual spider webs

pipe is run to a point adjacent to the meter bracket, from which the gas meter is suspended. A semi-rigid stainless-steel connector (anaconda) connects to the emergency control valve and terminates at the regulator, which is, in turn, connected to the inlet of the meter. The outlet from the meter terminates with an elbow facing into the installation serving the building.

Note: If the meter installation is medium pressure, the outlet pipework must exit the box before entering the property, the rear sleeve must not be used.

Multi-occupancy Buildings

4.24   Should a number of primary meters, grouped together, be required to serve different parts of a building (see Figure 4.22), the following points should be observed:

- The meters need to be sited where access is available at all times.
- The meters should be enclosed in a lockable housing; this may be either as a large single enclosure or as individual meter boxes.

- Each meter needs to be clearly marked to indicate the premises it serves, e.g. flat number.
- An individual property should have an additional emergency control valve (AECV) if the meter is >6 m from property, with a test point within 300 mm of its outlet, located at the point where the gas pipe enters the building.

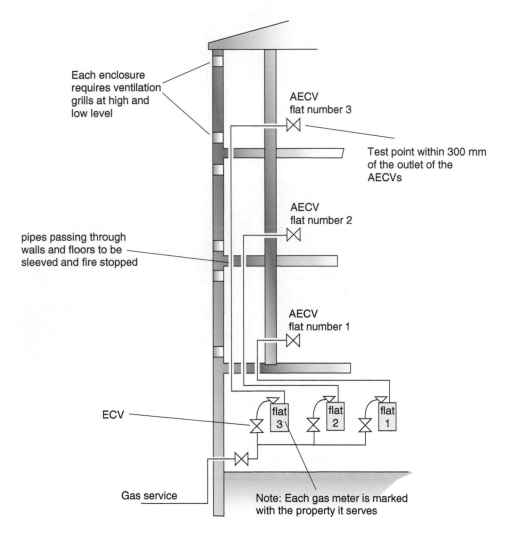

Figure 4.22 Pipework to multi-occupancy buildings

Locations Where the Meter Must Not be Sited

4.25 These include locations:

- subject to high temperatures and close to heat sources;
- where the meter could cause an obstruction or be liable to damage;
- that are damp or corrosive;

- in close proximity to electrical equipment;
- where food is stored;
- under a stairway or in a passageway of a building with two or more floors above ground floor (i.e. three storeys).

Although not preferable, if it is unavoidable, for buildings with only one floor above ground floor, a meter may be located along an escape route. However, the meter must be fire-resistant or be housed within a fire-resistant compartment with an automatic self-closing door or include a thermal cut-off valve, upstream, designed to cut off the gas flow should the temperature exceed 95 °C (see Part 3: Section 3.21).

Removal of the Gas Meter

4.26   When working on an installation that contains fuel gas, you need to be aware of the risks involved. For example, prior to performing any hot works on the installation pipework, the meter will need to be removed. Before a meter is removed, a temporary bond must be attached so that any stray electrical currents do not cause a spark to jump across to earth and to protect the operative from potential electrocution (see Part 15: Section 15.9). Upon removal, all open ends will need to be sealed with an appropriate fitting. This will include pipework and meter connections. Where an electronic pre-payment meter has been installed, care needs to be taken to prevent the tamper device becoming activated and closing off the supply. The meter must not be permanently removed without the authority of the meter owner. Where the gas pipe is to be left, plugged off, with a live supply remaining, it must be clearly marked to identify that it contains gas. Where permanent removal of the meter is to take place, the earth bond may also need to be made permanent.

Meter By-pass

4.27   Domestic meters are rarely by-passed. Where continuity of supply is needed, such as in factories or hospitals, a by-pass may be required, see Figure 4.23a and b. There are two

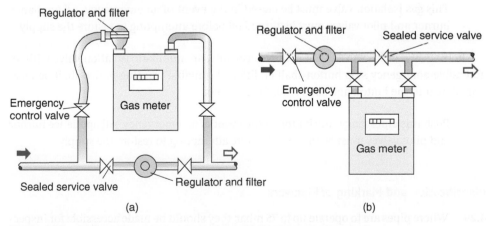

Figure 4.23   (a) Bypass to meter and regulator. (b) Bypass to meter only

methods; however, in both methods, it is essential that the supply is regulated. The valve fitted to the by-pass must be sealed by the supplier and only be opened in an emergency or during an exchange.

Commercial Gas Installations

Relevant Industry Documents
BS 6400 and IGEM/UP/2

(Specific points in addition to domestic gas installations)

Additional Isolation Valves and Line Diagrams

4.28   In compliance with Regulation 24 of Gas Regulations, where the service pipe is ≥50 mm for natural gas or ≥30 mm for LPG, additional isolation valves are to be installed in readily accessible locations where the pipework installation splits to different areas. Furthermore, a line diagram in permanent form must be attached to the building in a conspicuous and accessible position, such as at the primary gas meter or emergency control valve (ECV). In addition, other places could be included such as the security gatehouse or site engineer's office. The diagram is to show the location of all pipes of diameter greater than 25 mm, meters, ECVs, pressure test points and electrical bonding points. At operating pressures over 100 mbar, the line diagram should include pipework 15 mm and greater. Its purpose is to inform anyone, especially the emergency services, of the route of the gas supply, thus enabling them to identify and isolate it if necessary. A typical diagram is shown in Figure 4.24. An additional isolation valve shall also be installed in the pipework to any self-contained area such as a boiler room or a commercial kitchen, either outside the area or close to the exit. A suitably worded warning notice will be displayed to indicate the procedure in the event of an emergency. The notice shall read, for example:

> This gas isolation valve must be closed in the event of an emergency. All appliance burner and pilot valves must be turned off before attempting to restore the supply

Alternatively, if this valve is not readily accessible, an automatic isolation valve with an accessible emergency stop button shall be fitted. A suitable warning notice shall be positioned near to the button and shall read, for example:

> Push this emergency stop button in the event of an emergency. All appliance burner and pilot valves must be turned off before attempting to restore the supply

Identification and Marking of Pipework

4.29   Where pipes are to operate up to 75 mbar, they should be made accessible for inspection. Where pipework is accessible, it should be permanently marked in such a way that it

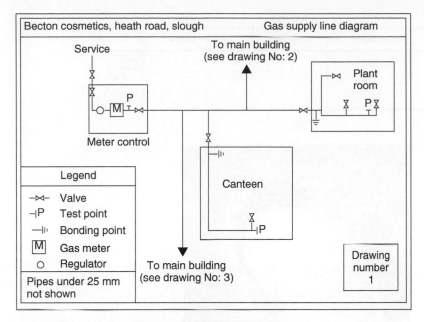

Figure 4.24 Typical gas supply line diagram

is easily recognisable as a gas pipe (see Figure 4.25a). This does not apply to pipework that is to be used in living accommodation. The method of identifying the gas pipe is by painting the pipe yellow ochre, in accordance with BS 1710. It is possible to simply apply banding at strategic points along the pipe, e.g. each side of valves, entry or exit through walls, or at branches. Where other gases, such as LPG, are used on the same site, an additional band of primrose yellow indicates that it is natural gas. Alternative additional labelling, e.g. 'Natural Gas' could be applied. In addition to marking, sometimes, particularly in confined spaces, a label identifying the direction of gas flow proves most useful. It is the installer's responsibility to ensure that the pipe is initially marked; however, the person responsible for the premises must ensure that the pipe remains recognisable as long as it is used to convey gas.

Purge Point

4.30 To facilitate purging the pipework, a purge point with a plugged-off valve should be located at each of the following positions:

- at every point where a section isolation valve is fitted;
- at the end of a pipe run;
- at the outlet on the primary gas meter;
- at each side of a secondary gas meter.

Secondary and Check Meters

4.31 The location of any secondary gas meters or check meters should be clearly identified on the line diagram. The location of these meters needs to be identified for several reasons:

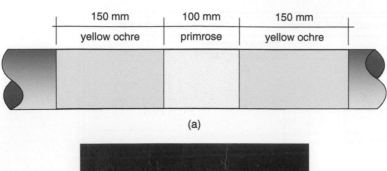

150 mm	100 mm	150 mm
yellow ochre	primrose	yellow ochre

(a)

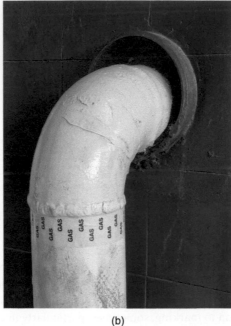

(b)

Figure 4.25 (a) One method of banding natural gas pipework in commercial premises. (b) Gas labelling

the emergency services need the information to isolate the supply and it also needs to be known for purging purposes after disconnection or reconnection. The secondary meter may only be fitted where a credit primary meter has been installed upstream and is used for private billing purposes or installed simply to monitor the gas consumption to an appliance. The responsibility for the labelling on the secondary meter rests with the gas installer.

Internal Installation Pipework

Relevant Industry Document
BS 6891

4.32 The common terminology used to identify the gas pipework to and within a building may vary across the country. However, Figure 4.26 identifies the British Standard terminology and symbols that should be used for universal recognition.

The materials that can be used for the pipework were identified earlier in the chapter, and include PE, copper and steel. As the pipework inside the building does not have the

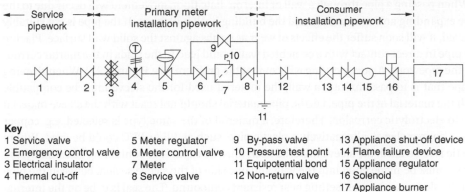

Key

1 Service valve
2 Emergency control valve
3 Electrical insulator
4 Thermal cut-off

5 Meter regulator
6 Meter control valve
7 Meter
8 Service valve

9 By-pass valve
10 Pressure test point
11 Equipotential bond
12 Non-return valve

13 Appliance shut-off device
14 Flame failure device
15 Appliance regulator
16 Solenoid
17 Appliance burner

(Note: Not all controls will be fitted to most premises)

Figure 4.26 British standard terminology and symbols for universal recognition

protection of being buried in the ground, it is continually subjected to the threat of damage from knocks, alterations or by further unexpected actions and building movement. Therefore, the gas installer must continually look ahead for possible problems, for example, you would use a steel pipe in a protected shaft containing a lift or stairwell used as a fire escape. Installing copper pipe runs in a school or college may prove useless if the pipes are soon damaged by students who continually knock against them and stand on them. It is important to ensure that the pipework is adequately supported, Table 4.3 should provide a suitable guide to the maximum interval between pipe supports.

Table 4.3 Identifying maximum interval between pipe support

Material and nominal bore			Maximum interval between pipe supports (m) H (Horizontal) V (Vertical)							
			Screwed steel		Welded steel		Copper		CSST	
Mild steel (mm)	CSST (mm)	Copper (mm)	H	V	H	V	H	V	H	V
15	≤15	15	2.0	2.5	2.5	3.1	1.2	1.8	1.5	2.0
20	22	22	2.5	3.1	2.5	3.1	1.8	2.4	2.0	2.5
25	28	28	2.5	3.1	3.0	3.7	1.8	2.4	2.0	2.5
32	32	35	2.7	3.3	3.0	3.7	2.4	3.0	2.0	2.5
40	40	42	3.0	3.7	3.5	4.3	2.4	3.0	2.0	2.5
50	50	54	3.0	3.7	4.0	5.0	2.7	3.0	2.0	2.5
65	–	67	–	–	4.5	5.6	3.0	3.6	–	–
80	–	76	–	–	5.5	6.8	3.0	3.6	–	–
100	–	108	–	–	6.0	7.5	3.0	3.6	–	–
150	–	–	–	–	7.0	8.7	–	–	–	–
200	–	–	–	–	8.5	10.6	–	–	–	–
250	–	–	–	–	9.0	11.2	--	--	–	–
300	–	–	–	–	10.0	12.0	–	–	–	–

Note: Ensure pipe clip and pipe are compatible to avoid corrosion.
Source: Data from IGEM/UP/2

When passing a pipe through a wall or intermediate floor, movement will occur due to the pipe expanding and contracting, and the building itself will move. If the pipe is not suitably sleeved, it will soon suffer the effects of wear as it rubs against the solid wall surface. Placing the pipe in direct contact with a cemented wall would lead to the acids in the mortar corroding the pipework. Again, sleeving will combat the problem (see Figure 4.27). When sleeving a pipe that is to pass through a wall, the material used for the sleeve must be compatible with the material of the pipe, i.e. the pipe material should not react with the sleeve material due to electrolytic corrosion. Therefore, a material of the same type is selected, e.g. copper pipe – copper sleeve, alternatively a plastic pipe, such as PE or PVC could be used. Where plastic is to be used, the sleeve material itself must be capable of containing gas. Finally, the sleeve must be made good into the wall and one side of the sleeve sealed, between the pipe and the sleeve, with a non-setting heat resistant compound. The seal is to be on the internal pipework. *Note*: No joints or parts of a joint should be within the sleeve section.

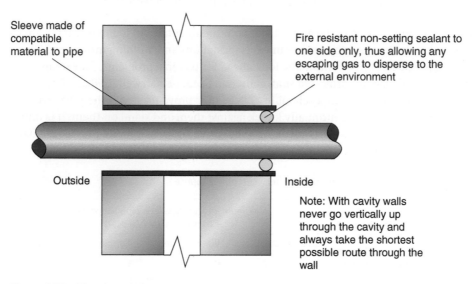

Figure 4.27   Sleeving of pipes passing through walls

Pipework Laid in Floors

Relevant Industry Document
BS 6891

Pipes Laid in or Under Wooden Joisted Floors

4.33   Pipes may be run suspended below timber joists or run parallel between joists. The location of the supports needs to be in accordance with Table 4.3. Should it become necessary to run the pipe perpendicular to the joist and there is no provision to secure it below the timber it may be possible to drill a hole or notch the timber as necessary. The maximum pipe diameter that could possibly pass through a joist would depend on the depth of the timber and whether a hole was drilled, or notch cut. This size is found by making the following calculation:

Maximum notch size = 0.125 × depth of joist

Maximum drilled hole size = 0.25 × depth of joist

Note: Joists ≤100 mm must not be cut or drilled and the maximum depth of joist to be considered is 250 mm; therefore, a 300 mm joist is only regarded as being 250 mm.

So, for example, where the joist is 200 mm deep, the largest pipe that could pass would be 25 mm where a notch is made or 50 mm where a hole has been drilled.

The location of the hole or notch also needs to be considered; it must only be positioned within the shaded areas as shown in Figure 4.28. Where more than one notch or hole is required, the two must be at least 100 mm apart horizontally. When refitting the floorboards, care needs to be taken to prevent damage. It is also a good policy to mark the floor, where possible, warning of the gas pipe below.

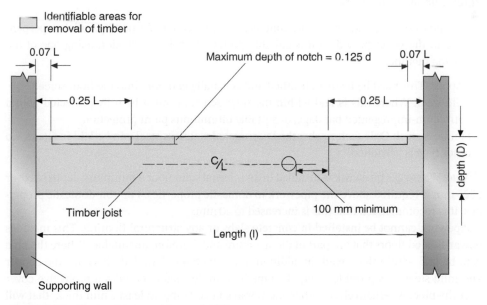

Figure 4.28 Timber joist notching and drilling

Modern construction methods have evolved, and the use of engineered joists is now common. These consist of timber flanges with either a fibreboard or a metal web construction between (see Figure 4.29a and b). It is most important that the timber flanges are never notched, and for metal web joists, the pipework is supported in such a way that it is not in contact with the metal webbing.

        (a)   (b)

Figure 4.29   (a) Examples of engineered joists – wooden. (b) Examples of engineered joists – Metal. Credit: mbolina\Adobestock

Pipes Laid in Solid Floors

4.34   Pipes may be laid within the concrete or floor screed of a floor, parallel or perpendicular to the walls, providing it is suitably protected against corrosion using one of the following methods:

- Copper: This must be factory sheathed and is usually laid only into the floor screed.
- Mild steel: Where this is laid within the floor screed, it must be fully protected with a petroleum-impregnated bandage, or suitable bituminous paint protection.
- Stainless steel: Only factory sheathed corrugated stainless steel can be laid in floors; no other type is permitted.

Note: Where gas pipework is installed in an internal solid floor a minimum depth of cover of 25 mm is required above the pipework in domestic properties. For non-domestic properties, it is recommended that this is increased to 40 mm.

Pipework cannot be installed in concrete slabs or any structural flooring. This includes power-floated floors that are part of the structure and therefore unsuitable. Where the pipe is to be laid within the screed, in addition to the method identified above for the copper and mild steel, it is possible to lay the pipe in a pre-formed duct, with a protective cover over the pipe. Alternatively, another additional soft covering, at least 5 mm thick, that will allow some movement may be used. See Figure 4.30a–d for examples of pipework installed in solid concrete floors.

In addition to the points identified, no compression joints or press end connections should be used below floor level and any joints used should be kept to a minimum. All joints are to be fully protected with a petroleum-impregnated bandage or similar after the completion of a successful tightness test. Where a pipe is simply to pass vertically through a floor, it should be suitably sleeved as previously described in Section 4.32.

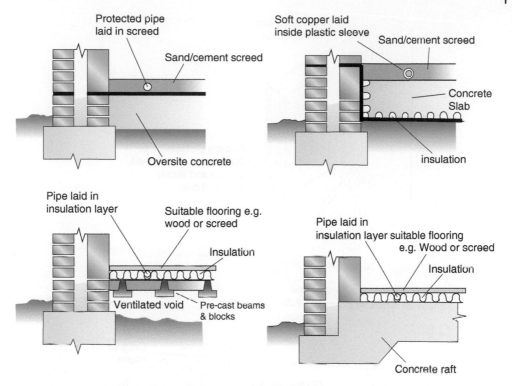

Figure 4.30 (a) Pipe laid in screed above oversite concrete. (b) Pipe laid in screed above a concrete slab. (c) Pre-cast beam and block floor. (d) Raft floor construction

Pipework in Walls

4.35 Whenever possible any pipework placed within the wall structure should run vertically, avoiding horizontal runs and be suitably protected against corrosion. The pipe ideally should be positioned within a duct; however, it is possible to chase the wall into which the pipe is placed. With the pipe secured in position, a tightness test needs to be completed before applying the final covering material. Joints within this section should be kept to a minimum and in no circumstances are compression joints or press end connections to be used. Where a pipe is to pass through a wall, it will need to be adequately sleeved as previously shown in Figure 4.27. The maximum permitted depth for a chase cut into the brick or block wall should not exceed:

Wall thickness ÷ 6 (*for horizontal pipes*);

Wall thickness ÷ 3 (*for vertical pipes*).

So, for example, for a 100 mm block wall, the maximum depth for a horizontal chase is 16.5 mm; this depth is increased to 33 mm for a vertical chase (see Figure 4.31a).

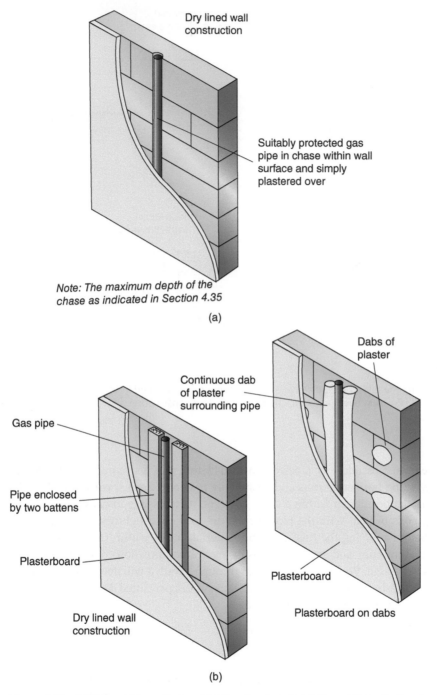

Dry lined wall construction

Suitably protected gas pipe in chase within wall surface and simply plastered over

Note: The maximum depth of the chase as indicated in Section 4.35

(a)

Dabs of plaster

Continuous dab of plaster surrounding pipe

Gas pipe

Pipe enclosed by two battens

Plasterboard

Dry lined wall construction

Plasterboard

Plasterboard on dabs

(b)

Figure 4.31 (a) Pipe within a plastered brick or block wall. (b) Pipes within dry lined walls and plasterboard on dabs.

Figure 4.31 (c) Pipe within timber stud wall

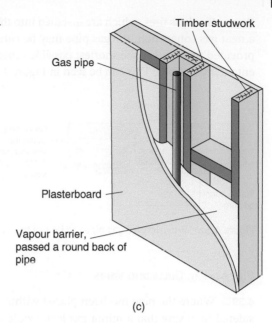

Timber studwork

Gas pipe

Plasterboard

Vapour barrier, passed a round back of pipe

(c)

Dry Lined Solid Walls

4.36 Where dry linings are to be used, one can install the gas pipe behind them. However, the pipe must be totally encased to prevent the void filling up with gas in the event of a gas escape. Two methods could be used here:

1. Place two battens, one each side of the gas pipe, along the entire length of the pipe run.
2. 'Dot and dab,' providing a continuous plaster or adhesive dab along either side of its length (see Figure 4.31b).

Timber Walls

4.37 Where gas pipework is to be contained within a timber stud wall, it should be adequately supported and run within a purpose-designed channel, protected from mechanical damage as appropriate (see Figure 4.31c). The number of joints should be kept to a minimum and, as previously shown, no compression joints are to be used in inaccessible places. The pipework should be installed more than 50 mm from the front face of the plasterboard.

Note: For both dry lined and timber walls, if the pipe is within 50 mm of the front face of the plasterboard, protection from penetration is required with a 1 mm thick steel plate covering the pipe.

Cavity Walls

4.38 It is not permissible to run a pipe down through a wall cavity and, when passing through a cavity wall, the pipe should take the shortest possible route, and will need to be sleeved. The Gas Safety (Installation and Use) Regulations have made provision for 'living flame effect' gas fires, which are installed into the cavity of a building structure. To provide a neat gas connection, the gas pipe may be run into the cavity through a suitable sleeve, provided that it takes the shortest possible route and is sealed at the point where the pipe enters the flue box. This can be seen in Figure 4.32.

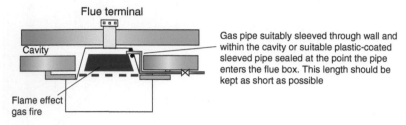

Figure 4.32 Provision for gas pipes to be run to a living flame effect, fan-assisted, gas fire

Pipework in Ducts and Voids

4.39   Where the pipe has been placed within a duct or void, ventilation should be considered to ensure that a minor gas leak would not create a dangerous situation. This has previously been discussed in Section 4.21.

Pipework Support and Allowance for Movement

Relevant Industry Documents
IGEM/UP/2

4.40   There are many designs of pipe support, some hold the pipe firmly in place and others allow for movement due to expansion and contraction or from external forces that move a structure or building. The clip must also prevent the pipes touching any surface that might cause corrosion. Pipe support spacing distances have previously been identified in Table 4.3.

Figure 4.33 shows a few of the many available designs of pipe support, some of which are fabricated on site to suit the needs of a particular situation. A well-designed system requires a mix of pipe supports and fittings that allow for movement yet maintain a firm sound system.

Expansion Due to Temperature Change

4.41   As a pipe heats up and cools down due to the changes in temperature of the environment, it expands and contracts. The amount of expansion or contraction is easily worked out using data from Table 4.4 with the following calculation:

Pipe length × Temperature change × Coefficient of linear expansion

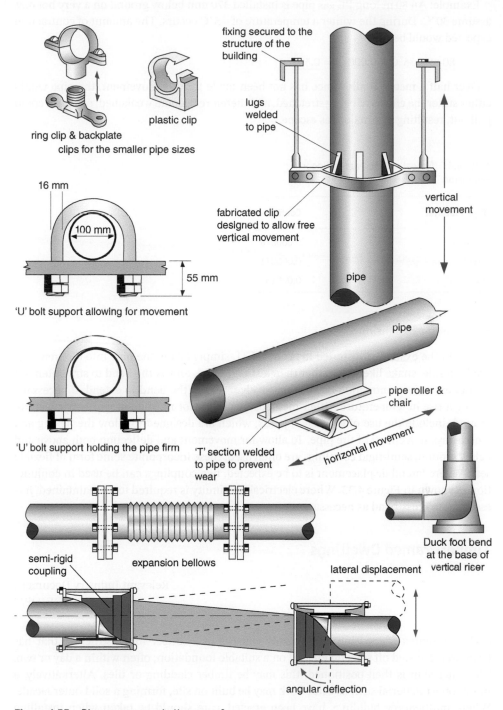

fixing secured to the structure of the building

lugs welded to pipe

vertical movement

pipe

fabricated clip designed to allow free vertical movement

plastic clip

ring clip & backplate
clips for the smaller pipe sizes

16 mm

100 mm

55 mm

'U' bolt support allowing for movement

pipe

pipe roller & chair

horizontal movement

'U' bolt support holding the pipe firm

'T' section welded to pipe to prevent wear

semi-rigid coupling

expansion bellows

lateral displacement

angular deflection

Duck foot bend at the base of vertical riser

Figure 4.33 Pipe support and allowance for movement

Example: An 80 m long PE gas pipe is installed 370 mm below ground on a very hot day, assume 30°C. During the winter a temperature of −5°C occurs. The amount of contraction expected would be:

$$80 \text{ m} \times 35°\text{C} \times 0.00018 = \underline{0.504 \text{ m}}$$

Over half a metre! If allowance has not been made for this movement, the pipe would either suffer the effects of being stretched, and therefore become weakened, or a joint could pull out, resulting in a major gas escape.

Table 4.4   Typical coefficients of linear expansion

Pipe Material	Coefficient
Copper	0.000016
Steel	0.000011
PE	0.00018

Provision for Expansion

4.42   In the example illustrated in Section 4.41, simply laying the pipe into the trench in a side to side, snake like way, rather than a straight line, allows the bend to straighten out should excessive contraction occur. Conversely it allows the bend to extend, if necessary, during a rise in temperature. Other methods that allow for movement, particularly above ground, include the use of expansion joints, which are designed to allow the flexing and lengthways movement of the pipe. To allow for movement and deflection both above and below ground, semi-rigid couplings are often used; these incorporate some form of flexible seal. Where lateral displacement is to be expected, two couplings can be used in conjunction as shown in Figure 4.33. Where electrical continuity is required to be maintained, it is essential to cross-bond as necessary.

Timber-framed Dwellings

Relevant Industry Document
IGE/UP/7

4.43   A timber-framed building usually consists of a timber inner structure that has been prefabricated off site and erected on a suitable foundation, often within a day or two. The outer skin is then positioned; this may be timber cladding or tiles. Alternatively, a more robust material such as brickwork may be built on site, forming a solid outer facade. Where multi-storey buildings have been erected, care should be taken when installing pipework through the external brickwork and timber frame at the highest level. This is because there is a greater variation of movement between the layers of the structure here compared with that experienced at a lower level. As stated in Section 4.10, flexible stainless steel can be used to accommodate structural movement when passing through a timber-framed wall.

Penetrating the Outer Wall

4.44   A vapour barrier can be seen immediately behind the plasterboard in the diagrams in Figure 4.34. This is to prevent any internal moist air passing through the wall where it may cool to the dew point and condense. This would lead to rotting in the timber structure. It is therefore imperative that this barrier is not unduly damaged or removed and care needs to be taken when penetrating it to pass pipes or a flue through the structure. In addition, moisture could pass from the outer skin, travelling across the cavity by clinging to horizontal pipe runs, including flues. As a consequence, the design of flue should include a moisture drip collar or some form of damp-proof membrane. Gas pipes should ideally pass through at a low level, below the level of the timber studwork.

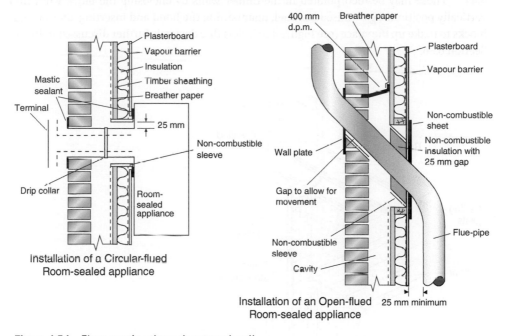

Installation of a Circular-flued
Room-sealed appliance

Installation of an Open-flued
Room-sealed appliance

Figure 4.34   Flues passing through external wall

Appliance and Pipework Installation

4.45   Plasterboard is technically combustible due to its paper surface; however, in practice most appliances can be installed on to or adjacent to combustible walls. However, the manufacturer's instructions must be consulted to see if any special precautions for fire-proofing are required. For new work, the builder should cater for the appliance fixing, incorporating a purpose-designed frame, timber stud or noggin. The plasterboard itself will have insufficient strength to support any significant load. Pipework contained within the walls should be run as shown in Figure 4.31c.

Flue Systems

Vertical Flue Pipes

4.46 Routing a vertical flue pipe through a timber-framed building poses the same problems as for a conventional dwelling; use sleeving of a non-combustible material and maintain a 25 mm air gap as shown in Part 6: Figure 6.27. Should the flue penetrate an external wall, the air gap should be packed with a non-combustible thermal insulation material and again care should be observed to ensure that moisture cannot get into the inner timber skin.

Pre-cast Concrete Flue Block Systems

4.47 These may be incorporated in the timber walls by enclosing the block work in a vertically positioned galvanised channel, maintaining the bond and inserting alternate cut blocks to make up the space (see Figure 4.35). This flue design is further discussed in Part 6: Section 6.36.

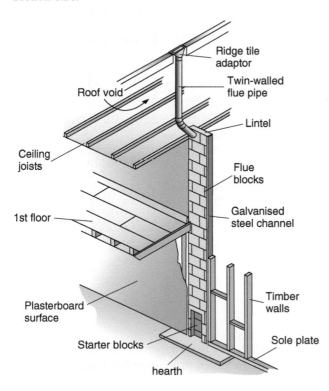

Figure 4.35 Incorporated pre-cast flue block system

False Chimneys

4.48 It is possible to include a false chimney breast as shown in Figure 4.36, enclosing the flue pipe and flue box; this is further discussed in Part 9: Section 9.22.

Figure 4.36 False chimney breast

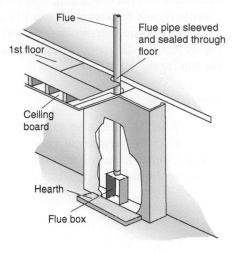

LPG Supply

Relevant Industry Documents
Liquid Gas UK Code of Practice 1, 22 and 24

4.49 LPG is supplied in liquid form and converted to its gaseous state as previously described in Section 2.19. It is stored either in cylinders, which need to be replaced as and when necessary, or in a large bulk tank on site that is refilled by a delivery tanker as necessary. Usually, with a bulk storage vessel, the LPG supplier is responsible for its installation, and periodic inspection and maintenance.

The same general principles apply to the storage of LPG, whether cylinders or a bulk tank are used. For protection against excessive pressure build-up within the storage vessel, a pressure relief valve is incorporated. For the bulk tank, this is an additional control that needs to be supplied, whereas it forms part of the outlet valve on the smaller cylinders.

LPG is a searching gas and will find the smallest of gas leaks; therefore, copper compression rings should be used in preference to brass as it is softer and compresses to form a tighter seal. Where steel pipe is to be used, this should be medium gauge galvanised pipe to BS 10255 and BS 10217 with malleable galvanised iron fittings to BS 143, 1256 or 10242.

Pressure Reduction

4.50 In order to reduce the pressure of the high-pressure gas (in the UK this would be typically in the region of 2–7 bar in the vessel), a regulator or series of regulators needs to be fitted. This may consist of a single regulator that reduces the pressure in a one-stage operation, or it may involve first reducing the pressure to an intermediate pressure of 0.75–1.5 bar.

A second-stage regulator would then reduce it further to a nominal operating pressure of 37 mbar (for propane) to be used within the building. This design of two-stage regulation is used particularly where the stored supply of LPG is some distance from the building, as it overcomes the problems arising from too great a pressure drop.

For bulk tank supplies of LPG, or where four or more cylinders are used for the supply, an under-pressure/overpressure (UPSO/OPSO) safety device needs to be included in the supply pipeline. This control was described in Part 3: Section 3.29, as was the operation of the regulators.

Blocks of Flats

4.51   Where LPG is used to supply a service riser on a multi-storey building (see Figure 4.37), the pressure supplied to the riser by the second-stage regulator must not exceed 75 mbar. In such a situation, it will be necessary to install a third-stage regulator within the dwelling to reduce the pressure down to 37 mbar.

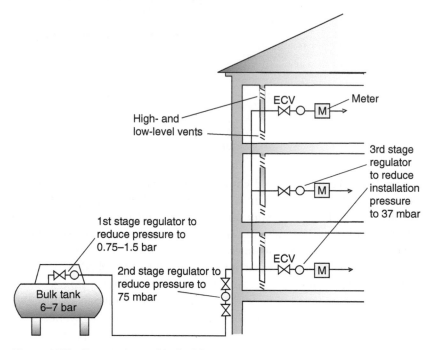

Figure 4.37   Gas supply to a block of flats

LPG Cylinder Installation

Relevant Industry Documents
Liquid Gas UK Code of Practice 22 and 24

4.52   Where cylinders of propane are to be used, in general, an automatic change-over valve would be provided, thereby ensuring that the supply is uninterrupted during use.

The automatic change-over valve (described in Part 3: Section 3.37) is a control valve that automatically switches the supply to a new cylinder when the gas runs out. The change-over valve also includes a regulator that reduces the gas pressure to that needed for use in the building. So that a cylinder can be replaced, the supply from the cylinder to the regulator is via a hose, or series of hoses. These hoses are referred to as the pigtails. The hoses are subject to the same high pressure as that found in the cylinder itself and, as a result, need close inspection during maintenance, etc. to ensure that they are not suffering stress from age or damage. Although BS 6891 recommends hoses are exchanged every 10 years, ultraviolet light from the sun slowly tends to perish these hoses. Therefore, certain manufacturers of changeover valves recommend that the pigtails are exchanged every five years.

Location of Cylinders

4.53 Propane cylinders must be located in an accessible position outside in the open air. They should be placed on a firm level surface and installed in the upright position, with the valve at the top so that only vapour may be drawn from the vessel when in use. The cylinders should ideally be located against a wall in a stable position where they are unlikely to fall or be knocked over and maintain minimum distances (see Figure 4.38). Where necessary, additional protection can be provided to secure the cylinders in place. At no time must any cylinder be used or stored in a cellar, basement or other low-lying area where escaping gas may accumulate. Butane cylinders incorporated with the appliance may be kept inside.

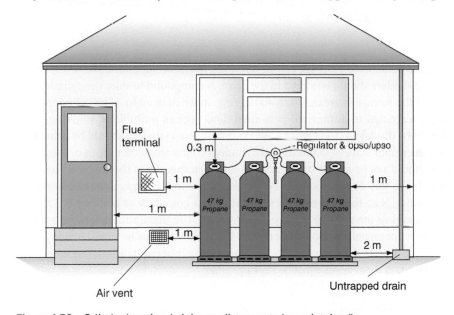

Figure 4.38 Cylinder location (minimum distances to be maintained)

Cylinders must not be positioned:

- at a distance of less than 1 m horizontally, or 0.3 m vertically, from the valve to any opening into the building, sources of ignition, heat sources or unprotected electrical equipment;

- closer than 2 m horizontally from an untrapped drain, unsealed gully or opening into a cellar unless a barrier wall, 250 mm in height is provided;
- within 3 m of any corrosive or toxic substances unless a fire barrier is provided;
- where they obstruct the access to and from the premises.

The associated equipment, e.g. regulators, should be located as close as practical to the cylinders, with the hoses kept as short as possible, allowing just enough room to provide the flexibility needed to exchange the cylinders. For exposed locations, some form of protection may be provided, such as a hood for weather protection. Where necessary, provision may need to be made to protect the cylinders from interference or damage by a third party but generally, they need to be as accessible as possible. Where a cylinder housing or compartment is used, the cylinder valves must be accessible and easily operated.

When a cylinder needs exchanging or disconnection, the cylinder valve and any other valves as applicable must be closed before undoing the supply pipe. This includes situations where the cylinders are empty. All sources of ignition must be extinguished.

Storage and Transportation of LPG Cylinders

Relevant Industry Documents
Liquid Gas UK Code of Practice 7 and 27

Storage of LPG

4.54   Specific guidance on storage can be obtained from the suppliers. However, in general, all LPG cylinders should be stored in the open. A compound to store the cylinders will be required if the storage is greater than 400 kg (e.g. more than eight 47 kg cylinders) with access being restricted to authorised personnel, and notices prohibiting smoking/naked flames should be displayed (see Figure 4.39a). All refillable cylinders should be considered as full whatever their state of contents and cylinders containing other gases or hazardous substances should not be stored in the vicinity. Small storage areas should be no less than 2 m from a drain; this distance should be increased to 3 m for larger storage compounds (see Figure 4.39b. Note that the minimum distance to a boundary or building is 1 m. This distance increases as the volume of gas exceeds 400 kg. The maximum height for stacking cylinders when not on pallets should not exceed 2.5 m. Butane may be stored inside providing it is no more than 15 kg; however, this must not be in a basement or near any emergency exits. The cylinders themselves must be kept upright.

Care with Empty Cylinders

4.55   When only a small amount or no more gas can be drawn from an empty cylinder, one could mistakenly think that it could be disconnected and left with the valve in the open position. However, the vessel will contain a small quantity of LPG, be it only vapour. During storage, the empty cylinders must be treated with great respect and the control valve kept securely closed with the plastic bung reinserted into the cylinder. Failure to do this could

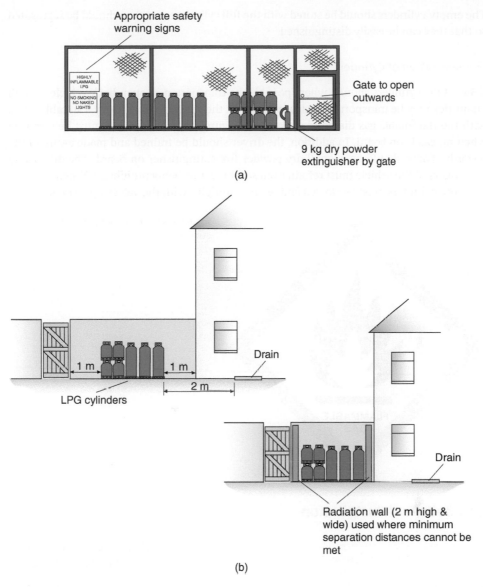

Figure 4.39 (a) Storage of LPG greater than 400 kg within a purpose built compound. (b) Open air storage for LPG supplies less than 400 kg

result in the volume of gas contracting during the evening when the temperature drops and, as a result, air would be drawn through the valve into the cylinder. There would then be a mixture of gas and air in the cylinder, which may be within the flammability limits for the gas. Should a heat source come within close proximity of the vessel, an explosion may result. While the cylinder was filled only with gas any heat applied would cause the pressure relief valve to open, expelling the gas. This could ignite as it issues from the release valve. The cylinder itself would not have expanded as there was no combustion within the vessel.

The empty cylinders should be stored with the full cylinders, but they should be segregated so that they can be easily distinguished.

Transportation of Cylinders

4.56 LPG cylinders should, where possible, be transported in open-back vehicles. Small quantities may be transported in closed vans, but the vehicle should carry a suitable label, with the flammable gas diamond sticker (see Figure 4.40). The sticker must be removed when no gas is on board. In addition, the driver should be trained and made aware of the hazards. There should be a small dry powder fire extinguisher on board. The driver and passengers of the vehicle must refrain from smoking, this being prohibited. When carrying LPG, the cylinders must be stowed and secured upright, with the valves uppermost.

Figure 4.40 Flammable gas diamond sticker

LPG Bulk Tank Installation

Relevant Industry Document
Liquid Gas UK CoP1

4.57 Bulk storage of LPG is usually undertaken by the supplier of the gas; it is a specialist operation undertaken by specially trained operatives. The gas stored in a bulk tank (an example of which can be seen in Figure 4.41) is propane and the size of the vessel selected will determine the total volume of gas that can be drawn from the tank in a given time. The colour of a bulk tank is usually white. This colour is selected to minimise the amount of heat absorbed on a hot day. This absorption would have the effect of expanding the gas and, where suitable expansion has not been allowed for, the pressure relief valve would open. Tanks painted darker colours would need to be much larger in size to hold the same quantity of fuel, thus allowing for a larger ullage space – the void above the level of the liquid LPG.

Figure 4.41 Typical bulk storage tank

Storage Vessel Siting

4.58   In general, the storage tank is sited above ground. It is possible to locate the tank below ground with access for maintenance and refilling as shown in Figure 4.42a. However the vapour take-off would be reduced due to the limited heat that can be taken from below ground level. Where the tank is to be sited above ground, it would need to be positioned on a concrete base. The location of a tank needs to be agreed with the supplier and user, thereby

Figure 4.42   (a) Below ground bulk storage tank access panels

(a)

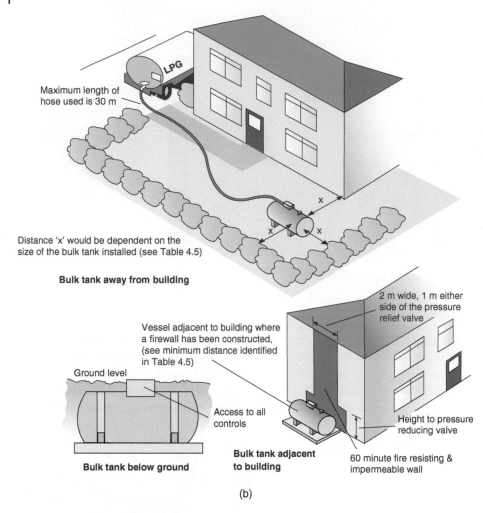

Maximum length of
hose used is 30 m

Distance 'x' would be dependent on the
size of the bulk tank installed (see Table 4.5)

Bulk tank away from building

Vessel adjacent to building where
a firewall has been constructed,
(see minimum distance identified
in Table 4.5)

Ground level

Access to all
controls

2 m wide, 1 m either
side of the pressure
relief valve

Height to pressure
reducing valve

Bulk tank below ground

**Bulk tank adjacent
to building**

60 minute fire resisting &
impermeable wall

(b)

Figure 4.42 (b) Storage vessel siting

ensuring that it is unobtrusive and easily filled from the road. There are minimal separation distances that need to be observed between the buildings and property boundaries, as indicated in the Table 4.5 and Figure 4.42b. The tank should not be located underneath any part of a building or overhanging structure. Overhanging tree branches and overhead power cables in close proximity also need to be avoided.

Firewall Protection

4.59 Where a firewall is constructed, it should be as tall as the vessel itself. It may form part of the boundary wall. For tanks ≤2500 litres, the building may form the firewall provided that it is impermeable to the liquid and is of 60 minute fire resistant construction for residential properties and 30 minutes elsewhere.

Warning Notices

4.60   The markings on a bulk tank must identify the contents and give a clear indication as to their highly flammable nature. There should be a durable, clearly visible and suitable sign affixed to an adjoining wall, fence or on to the vessel itself, prohibiting smoking or naked flames and, where the tank is segregated, a warning against unauthorised entry.

Storage Vessel Fittings

4.61   The gas controls fitted to a bulk storage vessel to enable safe filling (see Figure 4.45) are situated on top of the tank, normally under a hood or cover, which should be locked to prevent unauthorised entry.

Table 4.5   Minimum separation distances to be observed for bulk tank installations

Capacity litres	From buildings, boundaries and sources of ignition (m)	Distance if firewall is included (m)	Distance between vessels
≤500	2.5	0.3	1 m
>500–2500	3.0	1.5	1 m
>2500–9000	7.5	4	1 m
>9000–135 000	15	7.5	1.5 m
>135 000–337 500	22.5	11	$\frac{1}{4}$ sum of ϕ of tanks
>337 500	30	15	$\frac{1}{4}$ sum of ϕ of tanks

ϕ = diameter.

Gas Supplies from a Bulk Tank Installation

4.62   A bulk tank may serve an individual property, or it may serve several buildings. The service to a single building may be either low pressure or, where the supply is some distance from the building, a medium pressure installation may be found. Many of the controls listed were discussed in Part 3.

Connected to the outlet from a bulk tank will be a vapour take-off point and service control valve (as seen in Figure 4.43), and it is from here that the pipe is run to the building. This pipe is referred to as the service. Along its route will be the regulators and safety controls designed to allow the gas to flow safely and at the correct pressure. The first valve encountered is the first-stage regulator. This is a control regulator designed to reduce the gas pressure to an intermediate medium pressure of 0.75–1.5 bar. Following this valve may be fitted the second stage regulator, which reduces the gas pressure further to that required inside the building, usually 37 mbar. If the tank is sited some distance away from the building, this second-stage regulator may not be fitted close by the tank, but just prior to its entry

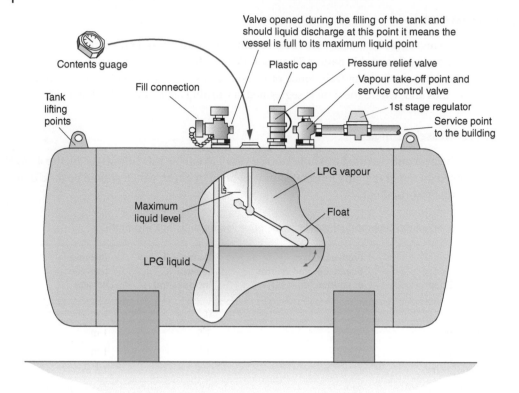

Figure 4.43 Typical bulk tank installation

to the building as shown in Figure 4.44. In this way, pressure loss within the pipework is reduced to a minimum. Also, along the gas service route to the building will be found the overpressure shut-off (OPSO) device and the under-pressure shut-off (UPSO) device. These two controls were described in Part 3: Section 3.29. They may be installed as separate controls or may be incorporated as a combined unit and include the second-stage regulator. *Note*: Where PE is used for the gas supply to the building, the OPSO will need to be installed close to the tank to ensure that any excessive pressure within the gas line does not cause a problem. Finally, before the entry into the building, an emergency control valve needs to be fitted.

Additional Storage Vessel Fittings

4.63 A pressure relief valve, designed to operate if the storage pressure becomes excessive, is fitted to the bulk tank. The pressure relief will discharge the vapour contents into the open air; therefore, consideration needs to be given as to a safe method of discharging the gas, which will otherwise flow to lower ground levels and accumulate. A liquid filling connection and its shut-off valve, along with a contents gauge, are also found.

Figure 4.44 LPG bulk supply entry to building

Figure 4.45 LPG bulk storage tanks being refilled

Sometimes, a combination valve is used; this comprises the filling connection and its shut-off control, the liquid-level indicator, and the vapour take-off with its appropriate service valve.

Multiple Building Installations

4.64 This is where the service is run from the bulk tank, through a distribution network to several different locations. Generally, the distribution service would be run at a pressure of 75 mbar with an OPSO/UPSO, second-stage regulator and emergency control fitted at the entry to the premises. A meter can be installed at this point to monitor the volume of gas used.

LPG Cylinder Sizing

Vapour Off-take Capacity

4.65   The storage vessels used must be suitably large to boil off the gas fast enough to provide the maximum hourly gas rate needed, with everything full on. This rate of gas production is known as the vapour off-take capacity. The off-take capacity is given in kilograms per hour (kg/h). To calculate the required capacity needed to provide a suitable supply, simply add all the heat input requirements for all the appliances connected to the system and refer to the Tables 4.6 and 4.7.

Table 4.6   Conversion of heat input **kW** (gross) to gravimetric calorific value **kg/h**

kW	kg/h	kW	kg/h	kW	kg/h	kW	kg/h	kW	kg/h
1	0.07	21	1.53	41	2.99	61	4.45	81	5.91
2	0.15	22	1.61	42	3.07	62	4.53	82	5.99
3	0.22	23	1.68	43	3.14	63	4.60	83	6.06
4	0.29	24	1.75	44	3.21	64	4.67	84	6.13
5	0.37	25	1.83	45	3.29	65	4.75	85	6.21
6	0.44	26	1.90	46	3.36	66	4.82	86	6.28
7	0.51	27	1.97	47	3.43	67	4.89	87	6.35
8	0.58	28	2.04	48	3.50	68	4.96	88	6.42
9	0.66	29	2.12	49	3.58	69	5.04	89	6.50
10	0.73	30	2.19	50	3.65	70	5.11	90	6.57
11	0.80	31	2.26	51	3.72	71	5.18	91	6.64
12	0.88	32	2.34	52	3.80	72	5.26	92	6.72
13	0.95	33	2.41	53	3.87	73	5.33	93	6.79
14	1.02	34	2.48	54	3.94	74	5.40	94	6.86
15	1.10	35	2.56	55	4.02	75	5.48	95	6.94
16	1.17	36	2.63	56	4.09	76	5.55	96	7.01
17	1.24	37	2.70	57	4.16	77	5.62	97	7.08
18	1.31	38	2.77	58	4.23	78	5.69	98	7.15
19	1.39	39	2.85	59	4.31	79	5.77	99	7.23
20	1.46	40	2.92	60	4.38	80	5.84	100	7.30

Note: Due to the similarity in gas rates of Propane and Butane this table covers both gases.
Source: Data from BS 5481-1

Table 4.7   Recommended off-take rates (cylinders)

Butane		Propane	
Cylinder size kg	Off-take kg/h	Cylinder size kg	Off-take kg/h
7	0.487	6	0.777
15	0.696	13	1.054
–	–	19	1.319
–	–	47	2.373

Source: Data from BS 5481-1

Example: A dwelling has a 28 kW combination boiler, 4.6 kW gas fire and 17 kW cooker (for cookers, we allow for 70% of its rating, due to it being unlikely that all the burners would be used at once). Therefore, the sum would be: $28 + 4.6 + (17 \times 0.7) = \underline{44.5\ kW}$.

We round this figure up to **45 kW** and refer to the Table 4.6 to find the appropriate gas rate in (**kg/h**). We can see this is **3.29 kg/h**.

We now want to find the most appropriate size cylinders for our installation. We do this by referring to Table 4.7. We divide our gas rate in **kg/h** by the off-take rate for our selected cylinder from Table 4.7. Therefore, for a **19 kg** cylinder, the sum would be: **3.29 ÷ 1.319 = 2.49**. We round this number up to the next whole number to get the number of cylinders required; in this case, it would be three. Let us try again with a **47 kg** cylinder: **3.29 ÷ 2.373 = 1.38**. This sum indicates that we would require two **47 kg** cylinders. Either of these options would correct, but the most appropriate would depend on a number of factors, including the location and amount of space available.

Typical propane and butane cylinders can be seen in Figure 4.46.

Figure 4.46   Typical propane and butane cylinders

Bulk Tank Specification

4.66   When specifying a bulk storage tank, it will be necessary to approach an LPG supplier to arrange its installation. As a guide, Table 4.8 gives an approximate vessel size in relation to heat input.

Table 4.8   Bulk tank size guide

Vessel capacity (l)	Equivalent heat input (kW)
1200	150
2000	187
3400	269

How Long Will a Cylinder Last?

4.67   Estimating the period of time that an LPG cylinder will last before it is required to be changed would clearly be dependent on how long it is used each day. However, it is possible to give an indication as to the continuous running time of an appliance. This is simply found by undertaking the following three steps:

Step 1. Calculate the total potential heat input from the available storage cylinders (*Note*: 1 kg = 14 kW).
Step 2. Calculate the total actual heat input of the appliance/s.
Step 3. Divide the potential input by the actual input (i.e. divide step 1 by step 2) to give the total hours' usage.

Example: Assuming there are no permanent pilot flames, how long will a 47 kg cylinder last when it supplies a 11 kW water heater and 4.7 kW fire.
Answer:

Step 1. $47 \times 14 = 658$ kW potential heat input;
Step 2. $11 + 4.7 = 15.7$ actual heat input; and
Step 3. $658 \div 15.7 = \underline{41 \text{ hours' usage.}}$

5

Testing and Purging

Testing and Purging

Relevant Industry Documents
IGE/UP/1, IGE/UP/1A and IGEM/UP/1B

Domestic or Commercial

5.1 The test procedures performed for domestic gas installations, be they for natural gas or LPG, are based on a general concept that the system pipework will not exceed 35 mm in diameter with a maximum volume of 0.035 m^3. This has allowed test times and purge volumes to be standardised for the gas used. On the other hand, with a commercial installation the size is in effect unlimited, and, as a result, more stringent test times and procedures need to be in place to allow for accuracy and safety in all aspects of the work. This includes testing, commissioning, and the de-commissioning of a supply. The flowchart shown in Figure 5.1 illustrates the relevant industry documents that should be complied with in order to ensure you are following the correct procedure. The following pages cover the use of natural gas and LPG at a range of pressures and pipe sizes. However, there are many variables to take into account and the relevant industry documents would need to be consulted for more in-depth information and advice.

Note: Where a large domestic property has pipework >35 mm in diameter or where the installation volume is > 0.035 m^3, the installation shall be tested to the procedures found in IGE/UP/1 or IGE/UP/1A by a gas engineer qualified in either TPCP1 or TPCP1A as appropriate.

In addition to tightness testing, all large gas installations require strength testing. Strength testing involves subjecting the pipework to the worst possible case scenario in terms of the pressures that the installation may experience. Strength testing is explained in Section 5.23 for IGE/UP/1A and Section 5.30 for IGE/UP/1.

When to Tightness Test

5.2 The Gas Regulations state that where you perform work on a gas installation that might affect gas tightness, you should test the pipework at least as far upstream and

Gas Installation Technology, Third Edition. Andrew S. Burcham, Stephen J. Denney and Roy D. Treloar.
© 2024 John Wiley & Sons Ltd. Published 2024 by John Wiley & Sons Ltd.

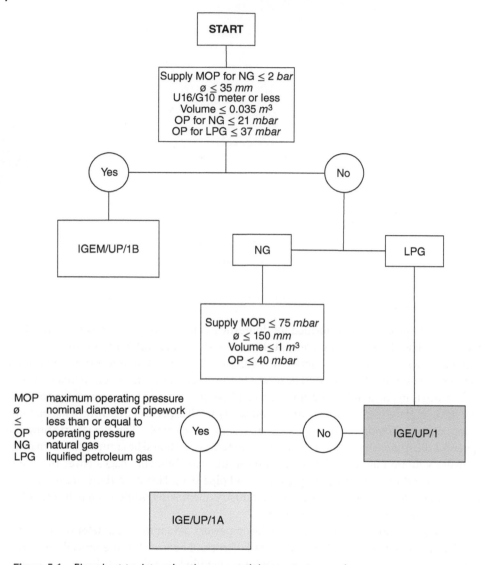

Figure 5.1 Flowchart to determine the correct tightness test procedure

downstream as the nearest valve. With this in mind, it is necessary to understand that you are not required to test the whole system, only that which you are working on. One can see from Figure 5.2 that where an extension has been run from a previously blanked end, the pipework needs only to be tested as far back as the valve marked 'X' as long as there is a suitable test point.

Tightness testing should also be carried out in the following circumstances:

- whenever installing new installations, and when altering, replacing or extending existing installations;
- before any work is performed on an existing installation;

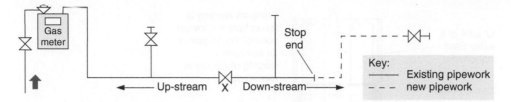

Figure 5.2 Responsibility for tightness testing

- where a gas leak has been reported, suspected or there is a smell of gas;
- where there has been a complete loss of system pressure;
- immediately before purging an installation;
- when performing routine inspection and testing of existing installations.

Testing Equipment

Pressure Testing Apparatus

5.3 *Pneumatic testing:* Testing with air can be achieved by using a simple hand pump where low pressures are required, as shown in Figure 5.8, or it may be necessary to use a compressed air supply for strength testing at high pressures (see Figure 5.23b).

Hydrostatic testing: Under certain circumstances when testing commercial pipework, it may be necessary to test with water. This may be undertaken by connection to the water supply main, with water authority approval and protection against backflow. Alternatively, it can be achieved using a purpose-designed hydraulic test pump.

Pressure Detecting Devices

5.4 *Manometer:* This instrument is used to determine the gas pressure within low-pressure pipework. Several designs will be encountered, the most common being the traditional water-filled 'U' gauge, which is available in various sizes. Other manometers include the 'J' gauge, which has the advantage of containing the fluid within an enclosed tube, so if the gauge falls over, the liquid will not run out. The J gauge also has the convenience of having only one leg of liquid from which to read the pressure. Battery-operated electronic gauges which give a digital readout are now commonplace; however, it is essential that where these are used, they are regularly checked for calibration.

Prior to using a manometer, it is essential that the gauge is zeroed to give an accurate reading. The water-filled gauge is read at the lowest point of the meniscus (see Figure 5.3). When reading a U gauge, both columns should always be viewed to ensure that they give the same reading. If they differ, re-set the gauge by adjusting the zero adjustor or average the readings by adding them together and dividing the total by 2.

5.5 *High specific gravity gauge:* This type of gauge is almost the same as the U gauge just described but the liquid used has twice the specific gravity of water, i.e. it is twice as heavy. Therefore, in effect, the high specific gravity gauge shows half the height of that of a water-filled manometer.

The sensitivity of a gauge will determine the gauge readable movement (GRM), see Table 5.17. Various types of manometers are shown in Figures 5.4a and b.

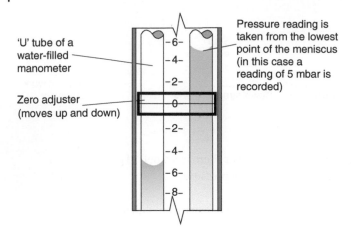

'U' tube of a water-filled manometer

Pressure reading is taken from the lowest point of the meniscus (in this case a reading of 5 mbar is recorded)

Zero adjuster (moves up and down)

Figure 5.3 Section from a water-filled manometer

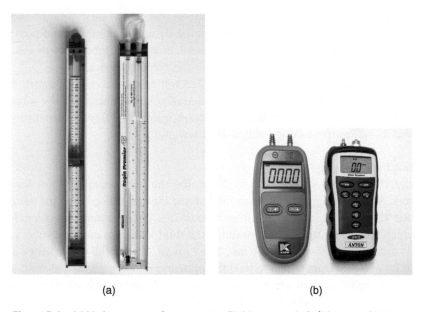

(a)   (b)

Figure 5.4 (a) Various types of manometer. Fluid gauges. Left: 'U' gauge. Right: specific gravity gauge. (b) Various types of manometer. Electronic gauges reading to different decimal points

Metering Devices

5.6   *Flow meters:* A flow meter measures the amount of liquid or gas that flows through it. Volumetric flow meters are often used in the gas industry and measure the volume of gas passing through the meter per unit of time, volume/time. The common unit of measurement in the gas industry is cubic meters per hour, which is shown as m^3/h. Some older versions of flow meters measure in cubic feet per hour (ft^3/h). An example of a volumetric flow meter is shown in Figure 5.5. The flow meter illustrated records the cumulative volume of gas passed in m^3 and the actual flow rate of the gas passing through the meter at any given time in m^3/h.

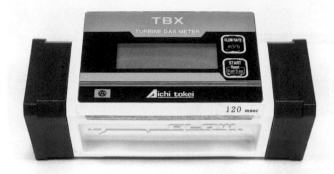

Figure 5.5   Volumetric flow meter

Gas Sampling Equipment and Detectors

5.7   There are many types and designs of gas detectors, including some that sample for flammable gas and give a warning bleep when gas is detected in the atmosphere, such as the gas 'sniffer' shown in Figure 5.6, while others detect only specific fuels. One such detector is commonly known as the 'Gascoseeker' and is shown in Figure 5.7. This particular device is designed to measure lower flammable limit (LFL) and volume flammable gas within the following ranges:

- LFL: 0–100%. Displays the flammability of the gas in the sample. For example, when a Gascoseeker is calibrated for methane, 100 % LFL equals 5 % gas in air.
- Volume gas: 0–100%. Displays the total volume of flammable gas with respect to air.

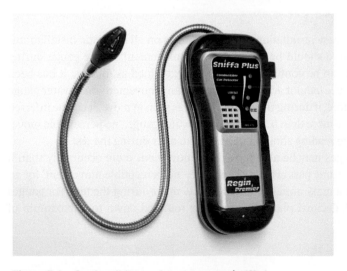

Figure 5.6   Combustible gas detector or gas 'sniffer'

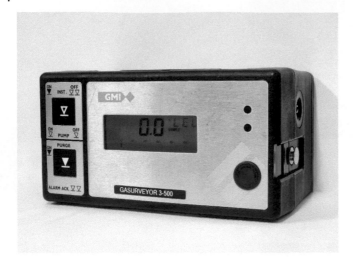

Figure 5.7   Gas sampling detector or 'Gascoseeker'

It should be noted that the Gascoseeker is designed to detect one specific fuel, e.g. methane, propane, etc., and therefore, it is important to ensure that the unit in use is correctly calibrated for the gas being sampled.

Electronic Equipment and Calibration Certificates

5.8   All test equipment used in conjunction with strength testing, tightness testing or purging operations must be intrinsically safe; therefore, it must not cause a spark to be generated where fuel gas is concerned. In addition, it must be maintained and tested prior to use and, where applicable, annually certified as correctly calibrated.

Perceptible Movement

5.9   For tightness tests of new installations, and let-by tests on all domestic installations, the pass criteria is that there should be 'no perceptible movement' of the gauge during the test. This terminology can be confusing but may be explained as follows: it has been suggested that the human eye cannot accurately observe any movement on a water gauge of 0.25 mbar or less. Therefore, if during the test the gauge is seen to move, it can be inferred that the gauge has moved by more than 0.25 mbar. For a water gauge, 'no perceptible movement' means that the gauge reading should not be seen to alter during the test.

Although electronic gauges may be able to register movement more accurately than a water gauge, they have the same pass criteria. Therefore, 'no perceptible movement' for an electronic gauge would be a maximum rise or fall of 0.25 mbar during the test. For gauges that can only register to 1 decimal place, this figure is rounded down to a maximum of 0.2 mbar.

Tightness Testing With Air (Domestic)

Installation Volume ≤0.035 m³

Relevant Industry Documents
IGEM/UP/1B

Natural Gas and LPG Pipework

5.10   When a new gas carcass or installation has been completed and before gas is distributed throughout the system, it is always a good practice to test for leaks with air, thereby avoiding problems with uncontrolled discharging gas entering the property. This procedure must be undertaken prior to protecting the pipework (e.g. wrapping, painting) and purging. The test is undertaken using either a water-filled manometer or an electronic gauge.

5.11   The methods used for tightness testing with air for natural gas and LPG systems are virtually the same, with the exception that different test pressures are used. During the test, a period of time is allowed for pressure and temperature stabilisation. This is the time allowed for the gas to expand or contract due to the temperature of the environment in which the gas is contained. For example, the air supply may be coming from a cold outside location, but the installation pipe itself is in a warm room. Following a successful air test, the system needs to be tested with the fuel gas that is to be used. The test described below is for installation pipework operating at the normal operating pressures that are encountered within domestic premises: 21 mbar for natural gas, 28 mbar for butane and 37 mbar for propane.

Tightness Test Procedure (Air)

1. Survey the system to ensure it meets the required standards (≤35 mm pipe size and ≤0.035 m³ volume, etc.). Make sure that all open ends are securely capped off, except one, to which is secured a test tee with associated pump, valve and manometer, as shown in Figure 5.8. Where appliances are connected, open all appliance isolation valves, turn off burner control taps, and if a cooker with a drop-down lid is fitted, ensure the lid is lifted fully up to open the safety cut-off valve.
2. Adjust the manometer to zero.
3. Operate the pump to increase the pressure, turning the valve off at the following pressures:
   - 20–21 mbar for natural gas installations;
   - 45–46 mbar for LPG installations.
4. Allow the pressure and temperature to stabilise for a period of one minute. During stabilisation, the pressure reading may rise or drop slightly. If necessary, at the end of the stabilisation period, readjust to the tightness test pressure.

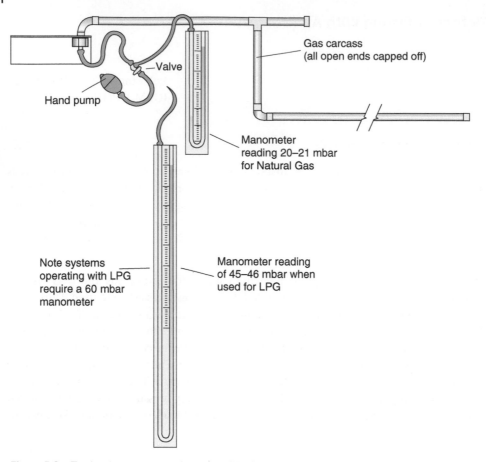

Figure 5.8 Testing low-pressure pipework with air

5. Then observe the gauge over the next two minutes. If there is no perceptible movement of the gauge, the system is deemed to be gas-tight.

   Note 1: For air testing LPG systems on boats, see Part 14: Section 14.23.

   Note 2: For strength and tightness testing of LPG service pipework, Liquid Gas UK Code of Practice 22 would need to be consulted.

Tightness Testing Natural Gas Installations (Domestic)

Installation Volume ≤0.035 m³

Relevant Industry Documents
IGEM/UP/1B

Domestic Natural Gas Tightness Testing

5.12 In order to undertake a tightness test on a large domestic or smaller non-domestic natural gas installation, firstly you must survey the system to ensure that the pipework is no larger than 35 mm in diameter and that the system volume is no greater than 0.035 m³. Larger systems would need to be tested in accordance with the commercial tightness testing procedures of IGE/UP/1 or 1A, described later in this chapter.

It would only be the largest of domestic houses that might exceed the 0.035 m³ volume. For example, where a typical U6/G4 gas meter has been installed, up to 45 m of 28 mm pipework could have been incorporated and you would still fall within the criteria of the 0.035 m³ maximum volume. However, where a U16/G10 meter is fitted, or any pipework greater than 28 mm diameter is installed, it will be necessary to estimate the installation volume to confirm it meets the domestic test criteria. This is a simple procedure with full details given in Section 5.18 and Example 5.1.

Tightness Test Procedure (Natural Gas)

5.13   The tightness testing procedure identified here is suitable for the majority of domestic gas installations; however, if a medium pressure regulator has been installed with no meter inlet valve (MIV), the procedure described in Section 5.14 should be adopted.

1. Survey the system to ensure that it meets the required standards (≤35 mm pipe size and ≤0.035 m³ volume, etc.). Make sure that all open ends are securely capped off with an appropriate fitting.
2. Ensure that any appliance isolation valves are in the open position, with cooker lids up and any appliance burner control tap or pilot flames turned off.
3. Turn off the gas supply using the appropriate supply control valve. Depending on the situation this could be an emergency control valve (ECV), additional emergency control valve (AECV) or meter inlet valve (MIV) (see Figure 5.9a–c).
   Note: An MIV is fitted to enable correct commissioning and testing of a medium-pressure meter installation and to assist in downstream tightness testing of the system. To prevent its operation by the consumer, it will not be fitted with a lever or key (see Figure 5.10).
4. Connect a manometer to the outlet of the gas meter, or other suitable test point, and zero the gauge.
5. Carryout a **let-by test** of the supply control valve by slowly opening the valve to increase the pressure to between **7 and 10** mbar, then close the valve. Observe the gauge reading over the next **one minute** to make sure the pressure reading does not indicate that the valve is letting-by. There should be **no perceptible movement** during this period. Should the pressure be seen to rise it could mean that the supply control valve is not fully closing off the supply and is letting-by. The valve would therefore need further inspection by disconnecting its outlet union and spraying leak detection fluid (LDF) into the valve to confirm let-by. If let-by is confirmed, the installation is made safe and the Gas Emergency Service Call Centre is immediately notified. The test is suspended until the issue is rectified.
6. Following a successful let-by test, slowly open the supply control valve and increase the pressure to between **20–21 mbar**, and close the valve. Now allow **one minute for pressure and temperature stabilisation**. If necessary, after one minute stabilisation readjust to between 20 and 21 mbar and close the valve.

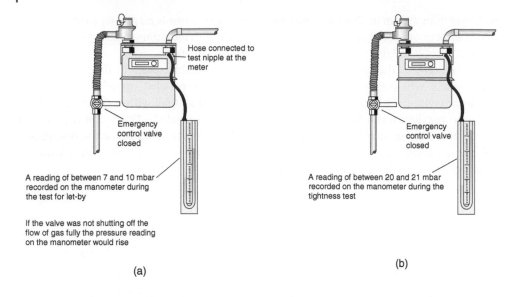

Hose connected to test nipple at the meter

Emergency control valve closed

A reading of between 7 and 10 mbar recorded on the manometer during the test for let-by

If the valve was not shutting off the flow of gas fully the pressure reading on the manometer would rise

(a)

Emergency control valve closed

A reading of between 20 and 21 mbar recorded on the manometer during the tightness test

(b)

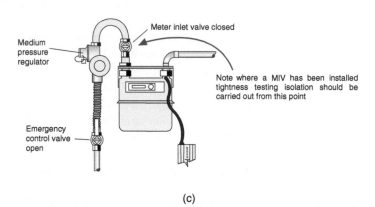

Medium pressure regulator

Meter inlet valve closed

Note where a MIV has been installed tightness testing isolation should be carried out from this point

Emergency control valve open

(c)

Figure 5.9 Tightness testing (a) Let-by test of the emergency control valve (ECV). (b) Using the emergency control valve (ECV) as supply control valve. (c) Using meter inlet valve (MIV) as supply control valve

7. Now observe the gauge over the next **two minutes**.
   - For **new installations** (with or without appliances connected) **and existing installations where no appliances are connected**, if there is **no perceptible movement** of the gauge reading **and no smell of gas**, proceed to step 8. If a drop is evident, or there is a smell of gas, the test has failed. If the test fails, the fault should be investigated and when rectified the test repeated.
   - For **existing installations with appliances connected**, if the pressure drop does not exceed those given in Table 5.1, **and there is no smell of gas**, proceed to step 8.

Figure 5.10 Medium-pressure regulator assembly showing (a) re-set lever, (b) meter inlet valve (MIV) and (c) vent pipe

If a drop exceeds those given in Table 5.1, or there is a smell of gas, the test has failed. If the test fails, the fault should be investigated and when rectified the test repeated.

8. Upon a satisfactory test, the manometer is removed and the test point re-sealed. The supply control valve is now fully re-opened and LDF applied to the test point. In addition, ensure that all connections between the ECV and the regulator are tested with LDF or a gas detector. This is important as once the ECV is turned on, the section of pipework between the ECV and the regulator will be at a higher pressure than the test pressure and a leak may only be evident at this higher pressure.

9. Finally, there must be no detectable smell of gas anywhere for the system to pass the tightness test.

Table 5.1 Maximum permissible pressure loss for existing installations with appliances connected and no smell of gas

Pipe diameter	Ultrasonic ≤6 m³ (E6 type meter)	Diaphragm ≤6 m³ (U6/G4 meter)	Diaphragm ≤16 m³ (U16/G10 meter)	No Meter AECV only (e.g. a flat)
≤28 mm	8 mbar	4 mbar	1 mbar	8 mbar
>28 mm ≤ 35 mm	4 mbar	2.5 mbar	1 mbar	4 mbar

Tightness Test Procedure For Testing Medium Pressure Systems with No MIV

5.14 Systems supplied with a medium-pressure regulator should be fitted with a meter inlet valve (MIV). Where this has not been installed, as may be found on some older systems, the responsible person for the premises should be informed and advised to contact their gas supplier. However, it is still possible to test the installation for tightness with the additional tests identified here to ensure the emergency control valve and medium pressure regulator are not letting-by. The procedure that follows is one way that the test can be conducted with further guidance found in IGEM/UP/1B Appendix 4.

1. Survey the system to ensure that it meets the required standards (≤ 35 mm pipe size and ≤ 0.035 m^3 volume, etc.). Make sure that all open ends are securely capped off with the appropriate fitting.
2. Ensure that any appliance isolation valves are in the open position, with cooker lids up and any appliance burner control tap or pilot flames turned off.
3. Turn off the supply at the emergency control valve (ECV). Remove the test nipple on the meter to release the gas pressure from the system. This will require you to hold open the re-set lever on the regulator (see Figure 5.10).
4. Connect a manometer to the outlet of the gas meter and zero the gauge.
5. Hold open the re-set lever and **slowly** open the ECV to increase the pressure to between 7 and 10 mbar, then close the ECV. Continue to hold open the re-set lever for one minute and observe the gauge (see Figure 5.11a). There should be no perceptible movement during this period. This confirms the ECV is not letting-by.
6. The re-set lever is allowed to return to its rest position and the ECV is opened and the gauge is observed for another one minute (see Figure 5.11b). Again there should be no perceptible movement. This confirms the regulator mechanism itself is not letting-by.

   If let-by is apparent, and confirmed, the installation should be made safe and the Gas Emergency Service Call Centre immediately notified. The test is suspended until the issue is rectified.
7. Following successful let-by tests, re-close the ECV and once again open the re-set lever and release any locked-up pressure via the test nipple, and then reconnect gauge.

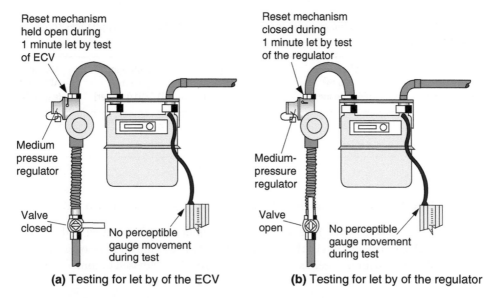

(a) Testing for let by of the ECV   **(b)** Testing for let by of the regulator

Figure 5.11   Testing for let-by on a medium pressure regulator where no MIV is fitted (a) testing for let-by of the ECV. (b) Testing for let-by of the regulator

8. With the re-set lever held open, **slowly** open the ECV and adjust the pressure to between 18 and 19 mbar, and close the ECV and the re-set lever. Wait for a period of one minute for pressure and temperature stabilisation. If necessary, after one minute of stabilisation readjust to the tightness test pressure and close the valve.
   Note: The system is only tested at 18–19 mbar thereby ensuring the regulator does not lock up.

9. Now observe the gauge over the next two minutes.
   - For **new installations** (with or without appliances connected) **and existing installations where no appliances are connected**, if there is **no perceptible movement** of the gauge reading **and no smell of gas**, proceed to step 10. If a drop is evident, or there is a smell of gas, the test has failed. If the test fails, the fault should be investigated and when rectified the test repeated.
   - **For existing installations with appliances connected**, if the pressure drop does not exceed those given in Table 5.1, **and there is no smell of gas**, proceed to step 10. If a drop exceeds those given in Table 5.1, or there is a smell of gas, the test has failed. If the test fails, the fault should be investigated and when rectified the test repeated.

10. Upon a satisfactory test, the manometer is removed and the test point re-sealed. The supply control valve is now fully re-opened and leak detection fluid (LDF) applied to the test point. In addition, ensure that all connections between the ECV and the regulator are tested with LDF or a gas detector. This is important as once the ECV is turned on, the section of pipework between the ECV and the regulator will be at a higher pressure than the test pressure and a leak my only be evident at this higher pressure.

11. Finally, there must be no detectable smell of gas anywhere for the system to pass the tightness test.

Tightness Testing LPG Installations (Domestic)

Installation Volume ≤0.035 m³

Relevant Industry Documents
IGEM/UP/1B

Domestic LPG Tightness Testing

5.15   Before tightness testing with gas, the system needs to have satisfactorily passed an initial tightness test with air as described in Section 5.11 and then be fully purged with the fuel gas. This is because LPG is a searching gas with a molecular structure that can find the smallest of holes within a pipeline, including those that air testing would not detect. The tightness test is then repeated with fuel gas as the test medium using the following procedure.

Tightness Test Procedure (LPG)

5.16   When tightness testing, first establish that the supply valve feeding the system is operating correctly. Failure to check this control valve may mean that gas can 'let-by' and pass into the system, thereby making up any losses during the test period. The valve is tested for let-by as follows:

1. Survey the system to ensure it meets the required standards ($\leq$35 mm pipe size and $\leq$0.035 m^3 volume, etc.). Make sure that all open ends are securely capped off with an appropriate fitting. Ensure any appliance isolation valves are in the open position with cooker lids up and that any appliance burner control taps or pilot flames are turned off.
2. Turn off the gas supply with the appropriate valve; this would typically be the emergency control valve or the cylinder valve as shown in Figure 5.12.
3. Connect a manometer to the outlet side of this valve and zero the gauge.
4. Slowly open the supply valve and allow the pressure to rise to between 7 and 10 mbar and close the valve. If an under pressure shut-off valve (UPSO) is installed on the outlet side of the supply control valve it will be necessary to operate the UPSO to allow gas to flow at these low pressures (see Figure 5.13).
   Important Note: If the pressure needs to be reduced at this or any other point during the test, gas cannot be simply vented into an enclosed space. It must either be safely vented to outside air or burnt off through a burner as described in Section 5.22.
5. Observe the gauge reading over the next one minute. If it was necessary to operate the UPSO at the beginning of the test, then the UPSO should be operated again at the end of the one minute before the final reading is taken. There should be no perceptible movement of the gauge reading during this period. Should the pressure be seen to rise it could mean that the supply control valve is not fully closing off and is letting-by. The valve would therefore need further inspection by disconnecting its outlet union and spraying leak detection fluid into the valve to confirm let-by. If let-by is confirmed, the test is suspended and made safe until the issue is rectified.

With no let-by, the test can proceed as follows:

6. Slowly re-open the supply control valve and raise the pressure to the correct tightness test pressure given in Table 5.2a or 5.2b depending on the gas in use. You may need to operate the UPSO again to let gas into the system.
   The tightness test pressure is determined by the operating pressure and whether or not the regulator is included in the test, see also Figure 5.14A,B.
7. Allow a period of one minute for pressure and temperature stabilisation. If necessary, after one minute stabilisation, readjust to the correct tightness test pressure and close the valve.
8. Now observe the pressure gauge over the next two minutes during which time there must be no perceptible movement of the gauge reading and no smell of gas.
   Note: It may be permitted to have a higher leakage rate but only for **existing installations with appliances connected**. For this to be acceptable, the installation volume will need to be calculated and compared to the maximum permissible pressure drops given in Table 5.3. As long as the pressure drop does not exceed the values given in Table 5.3 **and there is no smell of gas**, the installation may be deemed to have passed the test.
9. Upon completion of a satisfactory test, the manometer is removed and the test point re-sealed. The supply control valve is now slowly fully re-opened and leak detection fluid (LDF) applied to the test point. In addition, ensure that all connections between the ECV

and the regulator are tested with LDF or a gas detector. This is important as once the ECV is turned on, the section of pipework between the ECV and the regulator will be at a higher pressure than the test pressure and a leak may only be evident at this higher pressure.

Table 5.2a Tightness test pressures for installations supplied by propane

Type of installation	Operating pressure (mbar)	Tightness test pressure propane (mbar)
Regulator not included in the test	37 30 (see note)	36–37 29–30
Regulator included in the test	37 30 (see note)	30–31 28–29

Note: 30 mbar *operating pressure is used in leisure accommodation vehicles (LAV) such as caravans and motor caravans designed for use on the road in accordance with BS EN 1949.*
Source: Data from IGEM/UP/1B

Table 5.2b Tightness test pressures for installations supplied by butane

Type of installation	Operating pressure (mbar)	Tightness test pressure butane (mbar)
Regulator not included in the test	28 30 (see note)	27–28 29–30
Regulator included in the test	28 30 (see note)	20–21 28–29

Note: 30 mbar *operating pressure is used in leisure accommodation vehicles (LAV) such as caravans and motor caravans designed for use on the road in accordance with BS EN 1949.*
Source: Data from IGEM/UP/1B

Table 5.3 Maximum permitted pressure drop for existing LPG installations with appliances connected and no smell of gas

Installation volume (m³)	Maximum permitted pressure drop (mbar)
≤ 0.0025	2
> 0.0025 ≤ 0.005	1
> 0.005 ≤ 0.01	0.5
> 0.01 ≤ 0.035	No perceptible movement allowed
≤ *less than or equal to* > *greater than*	

Source: Data from IGEM/UP/1B

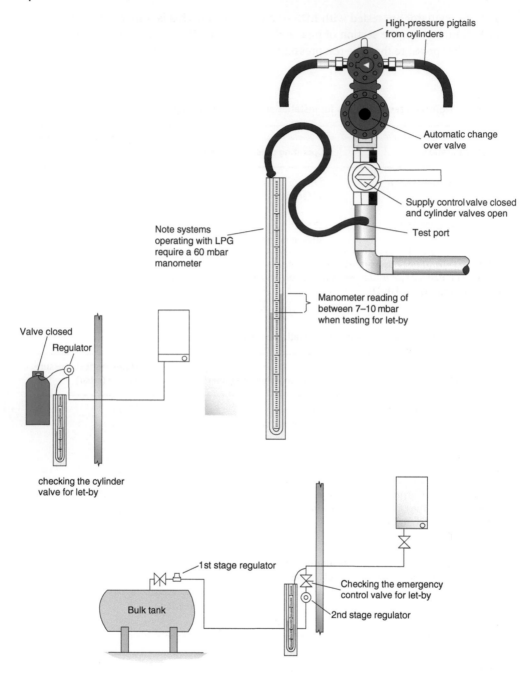

High-pressure pigtails from cylinders

Automatic change over valve

Supply control valve closed and cylinder valves open

Test port

Note systems operating with LPG require a 60 mbar manometer

Manometer reading of between 7–10 mbar when testing for let-by

Valve closed

Regulator

checking the cylinder valve for let-by

1st stage regulator

Bulk tank

Checking the emergency control valve for let-by

2nd stage regulator

Figure 5.12 Testing for let-by using various supply control valves depending on installation type

Figure 5.13 LPG regulator showing UPSO and OPSO resets

UPSO reset button

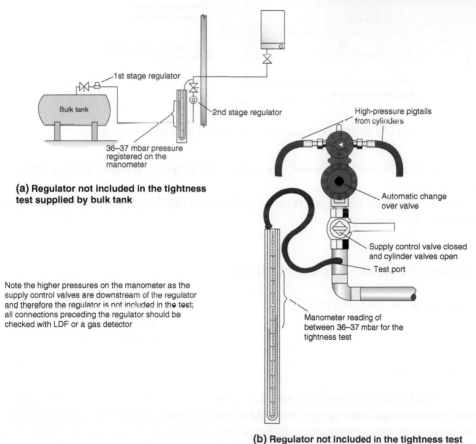

1st stage regulator

Bulk tank

2nd stage regulator

36–37 mbar pressure registered on the manometer

(a) Regulator not included in the tightness test supplied by bulk tank

High-pressure pigtails from cylinders

Automatic change over valve

Supply control valve closed and cylinder valves open

Test port

Manometer reading of between 36–37 mbar for the tightness test

Note the higher pressures on the manometer as the supply control valves are downstream of the regulator and therefore the regulator is not included in the test; all connections preceding the regulator should be checked with LDF or a gas detector

(b) Regulator not included in the tightness test supplied by cylinder changeover valve assembly

(A)

Figure 5.14 (A) LPG propane tightness test – regulator not included in the test

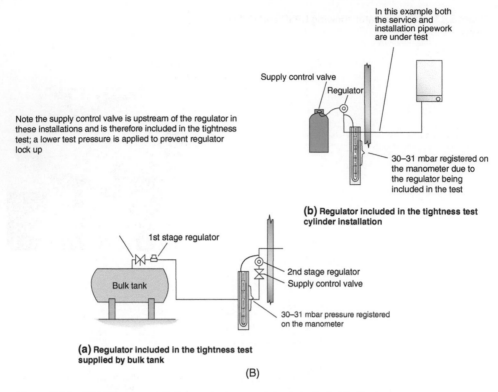

In this example both
the service and
installation pipework
are under test

Supply control valve

Regulator

Note the supply control valve is upstream of the regulator in
these installations and is therefore included in the tightness
test; a lower test pressure is applied to prevent regulator
lock up

30–31 mbar registered on
the manometer due to
the regulator being
included in the test

(b) Regulator included in the tightness test
cylinder installation

1st stage regulator

Bulk tank

2nd stage regulator
Supply control valve

30–31 mbar pressure registered
on the manometer

(a) Regulator included in the tightness test
supplied by bulk tank

(B)

Figure 5.14 (B) LPG propane tightness test – regulator is included in the test

If the installation fails the test, the leak should be traced, repaired and the tightness test repeated.

Note: For tightness testing of LPG service pipework, Liquid Gas UK Code of Practice 22 would need to be consulted.

Purging Domestic Gas Installations

Installation Volume ≤ 0.035 m³

Relevant Industry Documents
IGEM/UP/1B

Purge Volume

5.17 Purging relates to the discharge of air or gas from the pipework in order to commission or decommission the system. It should be noted that the tightness test always precedes the purging operation. To successfully purge a gas installation first you need to determine the purge volume. You will see from Table 5.4 that the purge volume is dependent on the meter type and the installation pipework diameter. For example, if a G4 or E6 meter is fitted along with pipework up to and including 28 mm in diameter, the purge volume is always 0.01 m³. Where an older U6 meter is installed, the equivalent volume is passed but is measured in cubic feet per hour (0.35*ft³*). However, if there is any pipework greater than 28 mm diameter on the installation, or a U16/G10 meter fitted, the installation volume must be estimated and this figure multiplied by 1.5 to calculate the purge volume. For guidance on how to estimate the installation volume, see Section 5.18 and Example 5.1. For meter installation volumes see Table 5.5.

Table 5.4 Domestic purge volumes

	Installation type	
Meter type	**Pipe diameter**	**Purge volume**
G4 or E6	≤28 mm	0.01 m³
U6		0.35 ft³
U6, G4, E6, U16 or G10	>28 mm ≤ 35 mm	Calculate the installation volume and multiply by 1.5

≤ less than or equal to > greater than

Table 5.5 Meter volumes

Meter volume (m³)	
U6/G4	0.008
E6	0.0024
U16/G4	0.025

Note: For the meter purge volume, multiply the meter volume by 1.5.

Calculation of the Installation Volume and Purge Volume

5.18   Where pipework on the installation has a diameter greater than 28 mm, or a U16/G10 meter is fitted, the installation volume must be estimated in order to determine the purge volume. This is calculated by adding together the volumes of the meter, pipework and fittings on the installation, and then multiplying this figure by 1.5 (see Figure 5.15). The final figure will give the total volume in cubic metres to be purged through the system. However, if purging through a U6 or U16 meter using the test dial, this figure will need to be converted to cubic feet. This is achieved by multiplying cubic metres by 35.31.

Material	Diameter (mm)	Volume of pipe per meter (taken from Table 5.8 column B)	Total length (m)	Internal volume of pipework (m³)
		×	=	+

Notes:	
To find **total pipework volume,** add together the internal volumes worked out for pipework in the columns above.	Total pipework volume =
	Meter Volume taken from Table 5.5 +
	Installation volume =
	Installation volume × 1.5 = Purge Volume
The **meter volume** is found by referring to Table 5.5	Where the installation volume is up to 0.035 m³, follow **IGEM/UP/1B**
For **installation volume,** add together the pipework volume and the meter volume.	Where the installation volume total is greater than 0.035 m³, follow **IGEM/UP/1 or 1A**
	Installation volume **up to 0.02 m³** - purge to a well-ventilated area
For **purge volume** multiply the installation volume by 1.5	Installation volume **greater than 0.02 m³** - purge to a well-ventilated area and apply ignition

Note: Column B of Table 5.8 includes an allowance of 10% for fittings with the pipe volume

Figure 5.15 Purge volume calculation template

Example 5.1 Assume a copper pipe installation installed with a G4 gas meter has 5 m of 35 mm pipe, 12 m of 28 mm pipe and 10 m of 15 mm pipe. The purge volume would be calculated as shown in Figure 5.16.

In this example, the installation volume exceeds 0.02 m³, therefore the purge should be completed by discharging the gas in a well-ventilated area through a burner, where there is a permanent source of ignition, see Section 5.20.

Purging Procedure for Natural Gas Systems

5.19 Small domestic-sized systems with an installation **volume of 0.02 m³ or less** may simply be **purged internally into a well-ventilated area**, opening any windows and/or doors as necessary. Other persons in the area should be notified of your intention to purge and that they may smell of gas. Ensure there is no smoking, no naked flames or sources of

Material	Diameter (mm)	Volume of pipe per meter (taken from Table 5.8 column B)		Total length (m)		Internal volume of pipework (m³)
	35	0.000924	×	5	=	0.00462
Copper	28	0.000594	×	12	=	0.007128
	15	0.000154	×	10	=	0.00154

Notes:	
To find **total pipework volume,** add together the internal volumes worked out for the 35 mm, 28 mm and 15 mm pipes in the columns above.	Total pipework volume = 0.013288
	Meter Volume taken from Table 5.5 + 0.008
	Installation volume = 0.021288
	Installation volume × 1.5 = Purge Volume 0.031932
The **meter volume** is found by referring to Table 5.5	Where the installation volume is up to 0.035 m³, follow **IGEM/UP/1B**
For **installation volume,** add together the pipework volume and the meter volume.	Where the installation volume total is greater than 0.035 m³, follow **IGEM/UP/1 or 1A**
	Installation volume **up to 0.02 m³** - purge to a well-ventilated area
For **purge volume** multiply the installation volume by 1.5	Installation volume **greater than 0.02 m³** - purge to a well-ventilated area and apply ignition

Note: Column B of Table 5.8 includes an allowance of 10% for fittings with the pipe volume

Figure 5.16   Purge volume calculation for Example 5.1

ignition present and do not operate electrical switches during the purge. Begin by slowly turning on the gas supply and make a note of the test dial position or index reading on the meter (see Figures 5.17 and 5.18). From a suitable purge point (appliance burner control tap or loosened union) allow gas to flow in a controlled way until gas is detected by smell or ignition of the burner. Turn off the control tap, or retighten the fitting, and return to the meter to see if the correct amount of gas has passed through the meter, see Table 5.4. If not, continue the process until the purge is complete ensuring every branch is purged and a stable flame is present at each burner. *Important Note: The area where gas is being released must not be left unattended during the purging procedure and any joint disconnected for this purpose needs to be sprayed with leak detection fluid to confirm it is gas-tight before being left unattended and again on completion of the purge.*

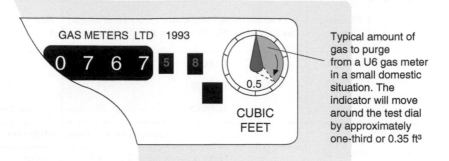

Figure 5.17 U6 gas meter test dial

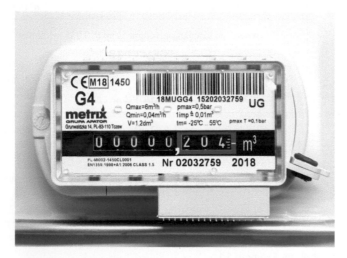

Figure 5.18 G4 index meter reading. On this meter to confirm a purge of 0.01 m³ has been completed, the meter index should move from 00.204 to 00.214

5.20 Where the **installation volume is greater than 0.02 m³**, the gas should not be permitted to discharge freely into the room. In this case, **the gas must be purged in a well-ventilated area through an appliance burner with a permanent source of ignition** to ignite the gas when it is present. Where no appliances are fitted, where appliances have flame supervision devices, or possibly there are no open burners, it may be necessary to connect a temporary burner to a suitable purge point to effectively complete the purge. Purging is only completed when all branch runs have been adequately purged and all appliances have been reinstated with a stable flame.

Any appliance that is purged with gas must be commissioned or otherwise disconnected from the supply.

5.21 Where a U6/G4 or E6 meter is to be permanently removed from an installation, such as in a meter exchange, the removed meter must be capped and sealed with an appropriate fitting. For domestic-sized meters with a badged capacity in excess of 6 m³/h (U16/G10) that are to be permanently removed, these meters must be purged of any fuel gas in accordance with IGE/UP/1 or 1A.

Purging Procedure LPG Systems

5.22   The process of purging LPG is very similar to that of natural gas, and you can follow the general principles of the procedure described in Sections 5.19 and 20 but with one very important exception. LPG must not be allowed to flow freely from an open end or unlit burner. LPG, being heavier than air, would fall to a low level and, where undetected, may accumulate and potentially lead to an explosion. Therefore, the gas must only be discharged to a well-ventilated area through an appliance burner, such as a cooker, whilst holding a permanent source of ignition close to the burner to ignite the gas immediately as it begins to issue from the burner head. It is essential that all branches are fully purged in this manner and that a stable flame is observed at each appliance burner upon completion. Where a meter is fitted, also ensure that the correct purge volume has passed as given in Table 5.4. Where no appliances are fitted, where appliances have flame supervision devices, or possibly there are no open burners, it may be necessary to connect a temporary burner to a suitable purge point to safely complete the purge.

Commercial Strength and Tightness Testing (TPCP1A)

Installation Volume ≤ 1 m³

Relevant ACS Qualifications	Relevant Industry Document
TCPC1A	IGE/UP/1A

Strength Testing for Commercial Pipework to IGE/UP/1A

5.23   In order to test in accordance with IGE/UP/1A, it is first necessary to use the flowchart shown in Figure 5.1 to check that the installation specifications fall within the criteria for the documented procedures. If the installation you are to test is outside of the parameters, reference must be made to IGE/UP/1, which is covered in Section 5.30.

5.24   Where a new system or extension has been installed, the pipe needs to be tested not only for tightness but also to ensure that it can withstand the stresses that may occur during its life without leakage. This is referred to as a strength test. Existing pipework should not be subjected to strength testing. An exception to this would be where new sections of pipework have been installed onto existing pipework and where testing them separately would be impractical. A detailed risk assessment would be necessary beforehand.

Strength Test Pressures, Test Criteria and Method

5.25   The actual pressure to be applied to the system under test depends on the maximum operating pressure of the system. Table 5.6 gives the test criteria to be observed. All components such as meters, governors or a pre-tested skid unit, that could be damaged, should be removed or disconnected, inserting bridging pieces, or the section should be spaded off. Do not rely on valves to isolate. These sub-assemblies need to be tested separately to an appropriate standard.

To carry out a strength test, firstly ensure that all valves are open within the section to be tested and, as far as possible, all pipework should be exposed. The pressure in the system is then 'slowly' raised to the required strength test pressure. Once the test pressure is reached, the required stabilisation period is observed. After this period, the test pressure is recorded. At the end of the strength test duration, the pressure is recorded again to identify any drop. Finally, a calculation is made and checked against Table 5.6 to ensure that any pressure loss does not exceed the maximum permitted drop. For a worked example, see Section 5.26.

Table 5.6   Test criteria for strength testing to IGEM/UP/1A

Strength test pressure (STP)	Stabilisation time	Strength test duration (STD)	Maximum permitted drop
82.5 mbar or 2.5 x MOP (whichever is the greater)	5 min	5 min	20% of the STP

MOP = maximum operating pressure; The MOP is the pressure at which the system is designed to operate.
Source: Data from IGEM/UP/1a

Worked Example of Strength Testing

5.26   A gas installation of $0.68 \, m^3$ has been installed. The maximum operating pressure (MOP) is 30 mbar. By referring to Table 5.6 the strength test pressure to be applied would be whichever is the greater of:

82.5 mbar or 2.5 x MOP

With a MOP of 30 mbar the calculation is 2.5 x 30 mbar = 75 mbar

∴ *Strength test pressure* = 82.5 mbar

The stabilisation time, taken from Table 5.6, is five minutes; similarly, the test duration is also five minutes. At the end of the five-minute strength test, it must be confirmed that the maximum permitted drop of 20% has not been exceeded:

∴ 82.5 x 0.2 = 16.5 mbar maximum permitted drop

Following a satisfactory test, the pressure can be reduced and a tightness test may be carried out immediately, subtracting the time used so far in strength testing from the tightness test stabilisation period.

Commercial Tightness Testing to IGE/UP/1A

5.27   The method described here may be used on commercial systems within the parameters of IGE/UP/lA as shown in Figure 5.1. In order to perform a tightness test on an installation, one needs access to several tables, which can be found in Section 5.28.

Tightness Test Procedures

1. A calculation is made to find the tightness test duration. This involves calculating the installation volume taking data from Table 5.7 for meters and Table 5.8 for pipework before referencing Table 5.9 for a new installation and extensions or Table 5.10 for an existing installation.

2. All components that may trap the gas pressure, e.g. non-return valves and regulators, must be temporarily by-passed. On a **new installation**, all valves on the system must be in the open position, thus ensuring all necessary pipework is subjected to the test. However, on an **existing installation**, it is recommended to turn off all appliance isolation valves.

3. Open ends need to be blanked off and all valves to and from the section under test spaded or capped.

4. A review of the ambient temperature and barometric conditions needs to be made to ensure that they will remain stable throughout the test.

5. Connect the test apparatus to the system. For existing installations, or where the gas supply is being used to pressurise the system, a test for 'let-by' will be required. This is achieved by slowly pressurising the system to 50% of its test pressure and then closing the supply valve. A period, obtained from Table 5.11 is waited, during which there must be no perceptible rise in pressure (perceptible movement is explained in Section 5.9), thus confirming that the valve is shutting off correctly. After completion of a satisfactory let-by test, the supply is raised to the test pressure (the system operating pressure). *Note:* a let-by test is not required when testing with air.

6. The pipework system is now allowed to stabilise for a minimum period of six minutes or the length of the tightness test duration, whichever is longer. At the end of the stabilisation period, the pressure must be adjusted to the tightness test pressure.

7. The tightness test duration is then waited.

8. For a **new installation**, no pressure drop shall be shown on the gauge that exceeds gauge readable movement (GRM) for the system to be deemed gas-tight. GRM, referenced from Table 5.12, is the lowest change in pressure that it is deemed possible to read on a gauge.

   For an **existing installation**, a small pressure drop is permitted, provided there is no smell of gas. This maximum permitted drop is referenced in Table 5.13. The data required to obtain the permitted drop is the installation volume of the section being tested and the smallest occupied space volume (the smallest occupied space that the gas pipework passes through). *Note:* If any of the pipework system passes through an inadequately ventilated area, it will be a good practice to test the whole installation as if it were new. Alternatively, you can isolate such a section and test it as new, or at the end of the test, access all joints in the section to test with LDF or a gas detector.

9. After proving tightness, the pipework, where applicable, should be purged and fuel gas introduced.

10. Finally, the tightness test should be documented on a certificate, which should be given to the person responsible for the property.

11. Following a successful tightness test, an appliance connection test, as detailed in Section 5.44, shall be carried out on the pipework after the appliance isolation valve. This is necessary before an appliance can be commissioned or re-commissioned.

Tightness Test Tables (IGE/UP/1A)

5.28   In order to perform a tightness test on an installation, one needs access to several tables from IGE/UP/1A. These have been reproduced but to our own design.

Table 5.7 Volume of gas meters

Meter type	Total volume (m^3)	Badge rating (m^3/h)	Badge rating (ft^3/h)	Capacity/revolution Cyclic volume (m^3)
G4/U6	0.008	6	212	0.002
G10/U16	0.025	16	565	0.006
G16/U25	0.037	25	883	0.01
G25/U40	0.067	40	1412	0.02
G40/U65	0.100	65	2295	0.025
G65/U100	0.182	100	3530	0.057
G100/U160	0.304	160	5650	0.071
E6	0.0024	N/A	N/A	N/A
Turbine	0.79$d2l$ or equivalent pipe length	N/A	N/A	N/A
Rotary	0.79d^2l or equivalent pipe length	N/A	N/A	N/A

d = diameter, l = length.

Table 5.8 Volume of 1 m of pipe

Material and nominal size		(A) Volume of 1 m pipe (pipework only) (m^3)	(B) Volume of 1 m pipe (including 10% for fittings) (m^3)
Steel			
(mm)	(in)		
15	$\frac{1}{2}$	0.00024	0.000264
20	$\frac{3}{4}$	0.00046	0.000506
25	1	0.00064	0.000704
32	$1\frac{1}{4}$	0.0011	0.00121
40	$1\frac{1}{2}$	0.0015	0.00165
50	2	0.0024	0.00264
65	$2\frac{1}{2}$	0.0038	0.00418
80	3	0.0054	0.00594
100	4	0.009	0.0099
125	5	0.014	0.0154
150	6	0.02	0.022
Copper:	**(mm)**		
	15	0.00014	0.000154
	22	0.00032	0.000352
	28	0.00054	0.000594
	35	0.00084	0.000924
	42	0.0012	0.00132

Table 5.9 Tightness test duration (TTD) for a new installation or extension

Installation volume (m³)	Tightness test duration (min) by gauge type using air		
	Water gauge or electronic gauge reading to one decimal place (Any movement greater than 0.5 mbar means the test has failed)	Electronic gauge reading to two decimal place (Any movement greater than 0.1 mbar means the test has failed)	Water gauge with no perceptible movement (Any movement of the gauge means the test has failed)
≤0.06	2	2	2
>0.06 ≤ 0.09	3	2	2
>0.09 ≤ 0.12	4	2	2
>0.12 ≤ 0.15	5	2	3
>0.15 ≤ 0.18	6	2	3
>0.18 ≤ 0.21	7	2	4
>0.21 ≤ 0.24	8	2	4
>0.24 ≤ 0.27	9	2	5
>0.27 ≤ 0.30	10	2	5
>0.30 ≤ 0.33	11	3	6
>0.33 ≤ 0.36	12	3	6
>0.36 ≤ 0.39	13	3	7
>0.39 ≤ 0.42	14	3	7
>0.42 ≤ 0.45	15	3	8
>0.45 ≤ 0.48	16	4	8
>0.48 ≤ 0.51	17	4	9
>0.51 ≤ 0.54	18	4	9
>0.54 ≤ 0.57	19	4	10
>0.57 ≤ 0.60	20	4	10
>0.60 ≤ 0.63	21	5	11
>0.63 ≤ 0.66	22	5	11
>0.66 ≤ 0.69	23	5	12
>0.69 ≤ 0.72	24	5	12
>0.72 ≤ 0.75	25	5	13
>0.75 ≤ 0.78	26	6	13
>0.78 ≤ 0.81	27	6	14
>0.81 ≤ 0.84	28	6	14
>0.84 ≤ 0.87	29	6	15
>0.87 ≤ 0.90	30	6	15
>0.90 ≤ 0.93		7	16
>0.93 ≤ 0.96		7	16
>0.96 ≤ 1		7	17

Source: Data from IGEM/UP/1a

Table 5.10 Tightness test duration for an existing installation

Installation volume (m³)	Tightness test duration with air (min)	Tightness test duration with gas (min)
≤0.15	2	2
>0.15 ≤ 0.30	3	2
>0.30 ≤ 0.45	5	3
>0.45 ≤ 0.60	6	4
>0.60 ≤ 0.75	8	5
>0.75 ≤ 0.90	9	6
>0.90 ≤ 1	10	6

Source: Data from IGEM/UP/1a

Table 5.11 Let-by test period

Installation volume (m³)	Let by test (min)
≤0.5	2
>0.5 ≤ 0.8	3
>0.8 ≤ 1.0	4

Source: Data from IGEM/UP/1a

Table 5.12 Readable movement of pressure gauges

Gauge type	Gauge readable movement (GRM) (The lowest change in pressure that it is deemed possible to read on a gauge)
Water	0.5 mbar
Electronic to 1 decimal place	0.5 mbar
Electronic to 2 decimal places	0.1 mbar

Source: Data from IGEM/UP/1a

Table 5.13 Maximum permitted drop on an existing installation with no smell of gas

Smallest Occupied Space Volume	Installation volume (m³)																	
	≤0.15	>0.15 ≤0.2	>0.2 ≤0.25	>0.25 ≤0.3	>0.3 ≤0.35	>0.35 ≤0.4	>0.4 ≤0.45	>0.45 ≤0.5	>0.5 ≤0.55	>0.55 ≤0.6	>0.6 ≤0.65	>0.65 ≤0.7	>0.7 ≤0.75	>0.75 ≤0.8	>0.8 ≤0.85	>0.85 ≤0.9	>0.9 ≤0.95	>0.95 ≤1
≤10	0.7	0.7	0.7	0.7	0.7	0.6	0.6	0.6	0.6	0.6	0.6	0.6	0.5	0.5	0.5	0.5	0.5	0.5
11	0.8	0.8	0.7	0.7	0.7	0.7	0.7	0.7	0.6	0.6	0.6	0.6	0.6	0.5	0.5	0.5	0.5	0.5
12	0.8	0.8	0.8	0.8	0.8	0.8	0.7	0.7	0.7	0.7	0.7	0.7	0.6	0.6	0.6	0.6	0.6	0.6
13	0.9	0.9	0.9	0.9	0.9	0.8	0.8	0.8	0.8	0.7	0.7	0.7	0.7	0.7	0.6	0.6	0.6	0.6
14	1	0.9	0.9	0.9	0.9	0.8	0.8	0.8	0.8	0.8	0.8	0.8	0.8	0.8	0.7	0.7	0.7	0.7
15	1	1	0.9	0.9	0.9	0.9	0.9	0.9	0.9	0.8	0.8	0.8	0.8	0.8	0.8	0.7	0.7	0.7
16	1.1	1.1	1	1	1	1	1	1	1	0.9	0.9	0.9	0.9	0.9	0.8	0.8	0.8	0.8
17	1.2	1.2	1.1	1.1	1.1	1	1	1	1	1	1	1	0.9	0.9	0.9	0.9	0.9	0.8
18	1.2	1.2	1.2	1.2	1.2	1.1	1.1	1.1	1.1	1.1	1.1	1	1	1	1	0.9	0.9	0.9
19	1.3	1.3	1.3	1.2	1.2	1.2	1.2	1.2	1.1	1.1	1.1	1.1	1.1	1.1	1	1	1	0.9
20	1.4	1.4	1.3	1.3	1.3	1.3	1.3	1.2	1.2	1.2	1.2	1.2	1.1	1.1	1.1	1	1	1
21	1.4	1.4	1.4	1.3	1.3	1.3	1.3	1.3	1.2	1.2	1.2	1.2	1.1	1.1	1.1	1.1	1	1
22	1.5	1.5	1.4	1.4	1.4	1.3	1.3	1.3	1.3	1.3	1.3	1.2	1.2	1.2	1.2	1.1	1.1	1.1
23	1.6	1.6	1.5	1.5	1.5	1.5	1.4	1.4	1.4	1.4	1.4	1.3	1.3	1.3	1.2	1.2	1.2	1.1
24	1.7	1.7	1.6	1.6	1.6	1.5	1.5	1.5	1.5	1.4	1.4	1.4	1.3	1.3	1.3	1.2	1.2	1.2
25	1.7	1.7	1.7	1.6	1.6	1.6	1.5	1.5	1.5	1.5	1.4	1.4	1.4	1.4	1.3	1.3	1.3	1.3
26	1.8	1.8	1.7	1.7	1.6	1.6	1.6	1.5	1.5	1.5	1.5	1.5	1.4	1.4	1.4	1.3	1.3	1.3
27	1.8	1.8	1.8	1.7	1.7	1.7	1.7	1.6	1.6	1.6	1.5	1.5	1.5	1.5	1.4	1.4	1.4	1.3
28	1.9	1.9	1.8	1.8	1.7	1.7	1.7	1.7	1.6	1.6	1.6	1.6	1.5	1.5	1.5	1.5	1.4	1.4
29	2	2	1.9	1.8	1.8	1.8	1.8	1.7	1.7	1.7	1.7	1.6	1.6	1.6	1.5	1.5	1.5	1.4
30	2.1	2.1	2	2	1.9	1.9	1.9	1.9	1.8	1.8	1.8	1.7	1.7	1.7	1.6	1.6	1.6	1.5
35	2.4	2.3	2.2	2.1	2.1	2	2	2	2	1.9	1.9	1.9	1.9	1.9	1.8	1.8	1.8	1.7
40	2.8	2.8	2.7	2.7	2.6	2.6	2.5	2.5	2.4	2.4	2.3	2.3	2.2	2.2	2.1	2.1	2	2
45	3.1	3	2.9	2.9	2.8	2.8	2.7	2.7	2.6	2.6	2.5	2.5	2.4	2.4	2.3	2.3	2.2	2.2
50	3.5	3.4	3.3	3.2	3.2	3.1	3.1	3	3	3	2.9	2.9	2.9	2.8	2.8	2.6	2.6	2.5
55	3.9	3.8	3.7	3.6	3.6	3.5	3.4	3.4	3.3	3.3	3.2	3.2	3.1	3.1	3	3	2.9	2.8
≤60	4.2	4.1	4.0	3.9	3.9	3.8	3.7	3.7	3.6	3.6	3.5	3.5	3.4	3.3	3.2	3.1	3.1	3.0
	Maximum permitted pressure drop (mbar)																	

Note: If the smallest occupied space volume is between two volumes, round down to the lower room volume to reference pressure drop.

Worked Example

5.29 Figure 5.19 shows a schematic pipe layout of a welded steel natural gas installation. Specific Details:

Section A – B = 50 mm dia and runs through an outside area
Section B – C = 25 mm dia and runs through an occupied room of 52 m^3
Section B – D = 50 mm dia and runs through a plantroom

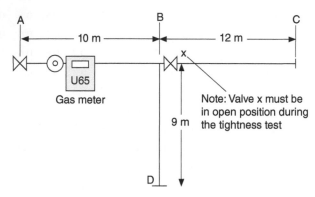

Figure 5.19 Worked example (TPCP1A)

Calculating the Installation Volume

Figure 5.20 provides a useful method to calculate installation volume using data from Table 5.7 for meter and Table 5.8 (column B) for pipework volumes.

Calculated Tightness Test Time:

For the installation shown in Figure 5.19, if it was new and to be tested using air, Table 5.9 would be referenced to determine the tightness test duration (TTD). For example, with the installation volume of 0.16 m^3 and using an electronic gauge to one decimal place, it can be seen that the TTD would be six minutes. Alternatively, if it were an existing installation Table 5.10 would be referenced to find a TTD. With gas as the medium the TTD would be two minutes, and with air, three minutes.

Allowable Pressure Drop

On a new installation, for the test to pass, there must not be a drop on the gauge that exceeds GRM as indicated in Table 5.12. Whereas, if our worked example was an existing installation Table 5.13 would be referenced using our previously calculated figures for installation volume (0.16 m^3) and the smallest occupied space volume (52 m^3, rounded

1	2	3	4	5
Section	Diameter	Length	Volume per metre	Section volume
	(mm)	(m)	(m³/m)	(m)
A-B	50	10	0.00264	0.0264
B-C	25	12	0.000704	0.008448
B-D	50	9	0.00264	0.02376
Meter volume: (U65)				0.100
Total installation volume:				0.1586 (**0.16 m^3**)

Figure 5.20   Installation volume calculation chart

down to 50 m³). Using these figures, Table 5.13 tells us that a maximum drop over the two-minute period would be **3.4** mbar.

Commercial Strength and Tightness Testing (TPCP1)

Installation Volume > 1 m³

Relevant ACS Qualifications   Relevant Industry Document
TCPC1   IGE/UP/1

Strength Testing for Commercial Pipework to IGE/UP/1

5.30   Testing to this standard is required if the installation is outside the scope of IGEM/ UP/1A and 1B. This can be confirmed by referencing the flowchart shown in Figure 5.1.

5.31   Where a new system or extension has been installed, the pipe needs to be tested not only for tightness but also to ensure that it can withstand the stresses that may occur during its life without leakage. This is referred to as a strength test. Existing pipework should not be subjected to strength testing. There are two types of strength tests: those that use air or nitrogen, called pneumatic tests, and those that use water, called hydrostatic tests. The method selected depends on location, system size and design.

Strength Test Pressures, Test Criteria and Method

5.32   The actual pressure to be applied to the system under test depends on the material used and the maximum operating pressure of the system. Table 5.14 gives the test criteria to

Table 5.14 Test criteria for strength testing

Maximum operating pressure	Diameter (mm)	Test Medium Pneumatic (P) Hydrostatic (H)	Strength test pressure Simply calculate values in the brackets and select the largest result as your test pressure	Stabilisation period (min)	Strength test duration (min)	Permitted pressure loss percentage Pneumatic (P) Hydrostatic (H)	
						P	H
Metallic pipes							
≤100 mbar	All	P or H	(MOP×2.5) or (MIP×1.1)	5	5	20	5
>100 mbar ≤1 bar	All	P or H	(MOP×2.0) or (MIP×1.1)	10	5	20	5
>1 bar ≤2 bar	All	P or H	(MOP×1.5) or (MIP×1.1)	10	5	20	5
>2 bar ≤16 bar	≤25	P or H	(MOP×1.5) or (MIP×1.1)	15	30	20	5
>2 bar ≤7 bar	>25 ≤150	P or H	(MOP×1.5) or (MIP×1.1)	30	30	20	5
>2 bar ≤7 bar	>150	H	(MOP×1.5) or (MIP×1.1)	30	30	NA	5
>7 bar ≤16 bar	>25	H	(MOP×1.5) or (MIP×1.1)	30	30	NA	5
PE pipes							
≤100 bar	All	P or H	(MOP×2.5) or (MIP×1.1)	5	5	20	5
>100 mbar ≤200 mbar	All	P or H	(MOP×1.75) or (MIP×1.1)	10	15	20	5
>200 mbar ≤1 bar	All	P or H	(MOP×1.5) or (MIP×1.1)	15	15	20	5
>1 bar ≤3 bar	All	P or H	(MOP×1.5) or (MIP×1.1) or (3 bar)	30	15	20	5

Note 1: If the calculated strength test pressure exceeds 10.5 bar for pipework diameter ≤150 mm or 3.5 bar for pipework diameter >150 mm, pneumatic testing is not allowed.

Note 2: After strength testing, PE pipe should be allowed to relax at atmospheric pressure for at least three hours before a tightness test is attempted. This is due to creep.

MOP = maximum operating pressure; MIP = maximum incidental pressure (e.g. supply pressure).

The MOP is the pressure at which the system is designed to operate. The MIP is the pressure under fault conditions, such as when a regulator fails or a gas booster is incorporated, etc. These terms are further defined in Part 2, Sections 2.49 and 2.50.

See IGE/UP/1 for pipes/pressures outside those listed above.

be observed. Strength testing must be considered at the design stage and it should be noted that applying high pressures to pipework needs considerable care. Therefore, it is essential that a suitable risk assessment is completed and that the pipework is suitably anchored to withstand the strength test pressure. All components that could be damaged should be removed or disconnected, inserting bridging pieces or the section should be spaded off. Do not rely on valves to isolate. These sub-assemblies need to be tested separately to an appropriate standard.

If the strength test is pneumatic and the pressure is calculated to be above 1 bar Table 5.15 must be referenced to determine the required exclusion zone that must be applied (see Section 5.34). To carry out a strength test, firstly ensure that all valves are open within the section to be tested and, as far as possible, all pipework and joints should be exposed. The pressure in the system is then 'slowly' raised to the required strength test pressure. If the strength test pressure has been calculated at above 2 bar, it should initially be raised to 350 mbar to check for general integrity, such as an open end. Following this, raise the pressure to 2 bar and then increase in 10 % stages to the actual test pressure. Once the test pressure is reached the required stabilisation period is observed. After this period, the test pressure is recorded and then recorded again at the end of the strength test duration to identify any drop. Finally, if necessary, a calculation is made to confirm that any drop in pressure is within permissible limits as stated in Table 5.14. For a worked example see Section 5.35.

Table 5.15 Exclusion zone for pneumatic tests exceeding 1 bar pressure

Strength Test pressure (bar)	Installation volume (m³)									
	1	2	3	4	5	6	7	8	9	10
>1≤2	1.0	1.3	1.6	1.8	2.0	2.2	2.4	2.7	2.7	2.9
>2≤3	1.5	1.9	2.4	2.7	3.0	3.3	3.6	3.8	4.1	4.3
>3≤4	1.75	2.4	3.0	3.5	3.9	4.2	4.6	4.9	5.2	5.5
>4≤5	2.0	2.9	3.7	4.1	4.6	5.0	5.4	5.8	6.2	6.5
>5≤6	2.4	3.3	4.1	4.7	5.2	5.7	6.2	6.6	7.0	7.4
>6≤7	2.6	3.7	4.5	5.2	5.8	6.4	6.9	7.4	7.8	8.3
	Distance to be maintained between operatives and centre-line of pipework									

The method for estimating the installation volume is described in 'Commercial Tightness Testing' in Section 5.36. For volumes in between, round up to the next highest figure.

Hydrostatic (Water) Testing

5.33 Hydrostatic testing would never be the first choice due to the obvious issues raised by filling a gas pipe with water to test it. Ultimately, you would have to get the water out and dry the pipework afterwards. However, if the size of the pipe related to the pressure falls within the parameters that require hydrostatic testing (as set out in Note 1, Table 5.14) the following points would need to be considered:

1. Ensure the pipework support can withstand the weight of the water in the system.

2. Acquire a hydraulic testing pump (see Figure 5.21).

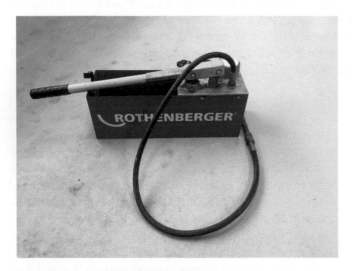

Figure 5.21 Hydraulic testing pump

3. Make sure there is provision to remove the water from the low points at the end of the test and to remove the air at the start (see Figure 5.22). Warm air can be circulated through the pipework to help dry it out.

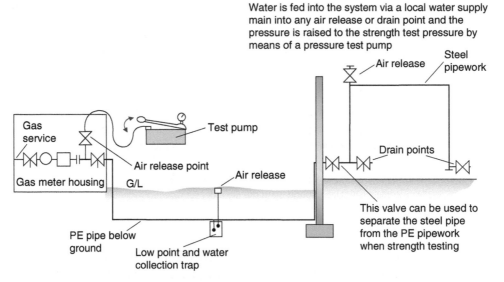

Figure 5.22 Hydrostatic strength testing

4. Take pressure readings at the highest point. Allowance needs to be made for the effects of the weight of water at the lowest points.

Pneumatic (Air) Testing

5.34   Specific notes: Where test pressures are to exceed 1 bar, an exclusion zone where no operative is permitted to enter must be maintained, as shown in Figure 5.23a and calculated using Table 5.15. This is to protect against fittings, etc. shooting off from the pipe with explosive force, causing injury. When hydrostatic testing, this exclusion zone is not required because, unlike air, water cannot be compressed.

If the pipework is overhead, this exclusion zone will extend to each side of a vertical line down to ground level. Leaking pipework is traced using leak detection fluid, with the pressure reduced to <1 bar. A pressure relief valve, suitably adjusted, to open just above the correct strength test pressure, is included in the test apparatus (as seen in Figure 5.23b).

Worked Example of Strength Testing

5.35   A gas installation of 4 m³ has been installed. The maximum operating pressure (MOP) is to be 30 mbar, and the system is to be supplied from a medium-pressure supply with a maximum incidental pressure (MIP) of 2 bar.

By referring to Table 5.14, with an MOP of 30 mbar, the strength test pressure to be applied would be whichever is the greater of:

MOP × 2.5 or MIP × 1.1

With a MOP of 30 mbar the calculation is 2.5 × 30 mbar = 75 mbar

With a MIP of 2 bar the calculation is 1.1 × 2 bar = 2.2 bar

∴ Strength test pressure = 2.2 bar

The stabilisation time, based on the MOP, would be five minutes and the test duration would also be five minutes.

When hydrostatic testing, the system should be slowly filled with water, observing the guidelines previously given for hydrostatic testing, and allowing a maximum pressure loss of 0.11 bar (5% of 2.2 bar) over the test period. However, where pneumatic testing is undertaken, when the air pressure exceeds 1 bar all operatives should be excluded from the work area to a distance of 2.7 m, as identified in Table 5.15. The maximum pressure loss during the test, with reference to Table 5.14 is 0.44 bar (20% of 2.2 bar).

Where a pneumatic test is satisfactory, the operating pressure can be reduced and a tightness test may be carried out immediately, subtracting the time used so far in strength testing from the tightness test stabilisation period.

Commercial Tightness Testing to IGE/UP/1

5.36   The method described here may be used on all commercial systems, using various gases up to 16 bar pressure. It is possible to use another test method as identified in IGE/UP/lA; however, this is restricted to 40 mbar natural gas supplies of ≤1 m³ volume and up to 150 mm in diameter. The procedure for IGE/UP/1A is covered in Section 5.27.

Commercial Tightness Test Procedures

5.37   In order to perform a tightness test on an installation, one needs access to several tables from IGE/UP/1. These have been reproduced in Section 5.38, but to our own design.

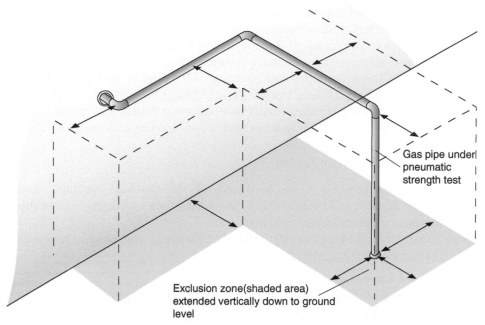

Should the pipe illustrated be pneumatically tested to 3 bar and assume the installation volume was 1.2 m^3, the exclusion zone would need to be 1.9 m

(a)

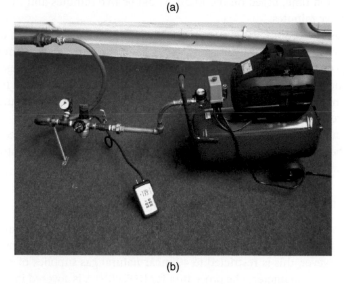

(b)

Figure 5.23 (a) Exclusion zones when pneumatic pressure testing. (b) Set up for pneumatic pressure testing

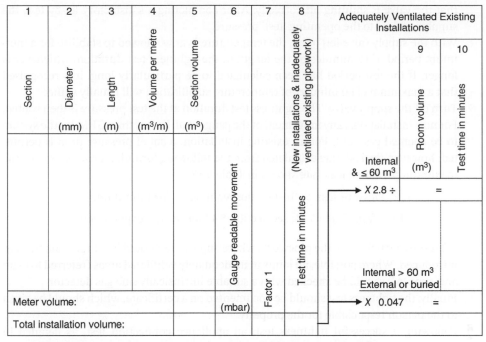

1	2	3	4	5	6	7	8	Adequately Ventilated Existing Installations		
Section	Diameter (mm)	Length (m)	Volume per metre (m³/m)	Section volume (m³)	Gauge readable movement (mbar)	Factor 1	(New installations & inadequately ventilated existing pipework) Test time in minutes	Internal & ≤ 60 m³ X 2.8 ÷	9 Room volume (m³) =	10 Test time in minutes
								Internal > 60 m³ External or buried X 0.047	=	
Meter volume:										
Total installation volume:										

Note: 1: Round up to the nearest minute & never test for less than two minutes
Note: 2: Ensure test time does not exceed test duration for gauge selected from Table 5.17
Note: 3: For test durations outside those listed in Table 5.17 and extended test periods, see IGE/UP/1

Figure 5.24   Tightness test table

Furthermore, Figure 5.24 provides a useful template on which to calculate the required test criteria.

The following ten points explain the procedure:

1. A calculation is made to find the tightness test duration on a template found in Figure 5.24, which uses the formulas obtained from Table 5.19 (see Section 5.41 for an example).
2. All components that may trap the gas pressure, e.g. non-return valves and regulators, must be temporarily by-passed, and all valves on the system must be in the open position, thus ensuring all necessary pipework is subjected to the test.
3. Open ends need to be blanked off and all valves to and from the section under test spaded or capped off. Appliance isolation valves should be closed.
4. A review of the ambient temperature and barometric conditions needs to be made to ensure that they will remain stable throughout the test.
5. With the test apparatus connected, slowly raise the pressure within the system to the required test pressure, indicated in Table 5.16, which is generally the system operating pressure. Table 5.17 can be used to select the most appropriate gauge. *Note:* For existing installations, or where the gas supply is being used to pressurise the system, a test for 'let-by' will be required. This is achieved by slowly pressurising the system to 50% of its test pressure and then closing the supply valve. A period is waited, equal to the length of time that will be used for the duration of the tightness test, during which there must be no rise in pressure in excess of GRM, thus confirming

that the valve is shutting off correctly. After completing a satisfactory let-by test, the supply is raised to the operating 'test' pressure.

6. With the supply valve left open, the temperature is now allowed to stabilise for a minimum period of 15 minutes or the length of the tightness test duration, whichever is longer. If the test period has been calculated to be particularly long, it is recognised that a maximum of 60 minutes of temperature stabilisation will usually suffice.

7. Turn off the supply valve. The tightness test duration at the test pressure is then waited. For new installations, any movement of the gauge shall be less than GRM for the system to be deemed gas-tight. For an existing installation, a small pressure drop is permitted; however, the leak rate calculated from the following formula, must not exceed the maximum permitted leak rate shown in Table 5.18:

F3 × Pressure loss × Installation volume ÷ Test duration.

(F3 is from Table 5.21; see Section 5.42 for further explanation)

8. After proving tightness, the pipework, where applicable, should be purged and fuel gas introduced. Where possible, all joints in inadequately ventilated areas (referred to as an area type 'A') should be checked with a suitable intrinsically safe gas detector.

9. Finally, the tightness test should be documented on a certificate, which should be given to the person responsible for the property.

10. Following a successful tightness test, an appliance connection test, as detailed in Section 5.44, shall be carried out on the pipework after the appliance isolation valve. This is necessary before an appliance can be commissioned or re-commissioned.

5.38   Tightness Test Tables (IGE/UP/1)

Table 5.16   Tightness test pressures

Pipe type	System pressure	Tightness test pressure
Metallic	All	@ Operating pressure
PE Pipe	≤1 bar	@ Operating pressure
	>1 bar	See IGE/UP/1

Table 5.17   Selection of pressure gauge

Gauge type	Range (mbar)	Gauge readable movement (GRM) (The lowest change in pressure that it is deemed possible to read on a gauge) (mbar)	Maximum test duration (min)
Water (SG = 1.0)	0–120	0.5	30
High SG (SG = 1.99)	0–200	1.0	45
Electronic to 1 decimal place	0–2000	0.5	30
Electronic to 2 decimal places	0–200	0.1	15
Electronic to 0 decimal places	0–20000	5.0	60

SG = specific gravity
For other gauges and longer test durations, see IGE/UP/1.

Table 5.18 Maximum permitted leak rate (MPLR)

Gas type	New installations where TTD is greater than maximum for gauge Existing installations 'inadequately vented' (Area type A)	Existing installations ≤60 m³, 'adequately vented' Rate per m³ of Smallest Space Volume (SSV) (Area type B)	Existing installations 'adequately vented' Volume >60 m³ or externally exposed or buried (Area type C and D)
		m³/h	
Natural	0.0014	0.0005 × SSV	0.03
Propane	0.00044	0.00016 × SSV	0.0098
Butane	0.00057	0.0002 × SSV	0.0123

Table 5.19 Formulas for calculation of the tightness test duration

	New installations and existing installations 'inadequately vented' (Area type A)	Existing installations ≤60 m³, 'adequately vented' (Area type B)	Existing installations 'adequately vented' Volume >60 m³ or externally exposed or buried (Area type C and D)
Tightness test duration in minutes	$GRM \times IV \times F1$	$IV \times GRM \times F1 \times 2.8 \div RV$	$IV \times GRM \times F1 \times 0.047$

IV = Installation Volume, GRM = Gauge Readable Movement, F1 = Factor from Table 5.20, RV = Room Volume
Source: Data from IGEM/UP/1

Table 5.20 Factor (F1) to apply when calculating tightness test

Gas type	F1 if using fuel gas to test	F1 if using air or nitrogen to test
Natural	42	67
Propane	102	221
Butane	128	305

Table 5.21 Factor (F3) to apply when calculating leak rate

Gas type	F3 if using fuel gas @ operating pressure	F3 if using air or nitrogen @ operating pressure
Natural	0.059	0.094
Propane	0.059	0.126
Butane	0.059	0.134

Tightness Test Duration and the Completion of the Calculation Chart (Figure 5.24)

5.39 A worked example can be seen in Figure 5.25 with a completed chart in Figure 5.26.

Column 1: Make an accurate survey of the system and divide it into sections, i.e. at branch connections and changes in diameter. Test different materials separately, e.g. steel and PE.

Column 2: Enter the pipe diameter for each section.

Column 3: Enter the actual length of the section.

Column 4: Enter the volume of 1 m of pipe taken from Table 5.8 (column B), depending on the pipe size. *Note:* For ease, column (B) of Table 5.8 in this book includes the 10% allowance for fittings, this column should be used when using the chart in Figure 5.24 to calculate. Column (A) of Table 5.8 is the volume without the allowance for fittings.

Column 5: Multiply column 3 by column 4 to give the total volume of the section. Add the meter volume, taken from Table 5.7, then total up all the sections to give a total installation volume and record it in the box at the base.

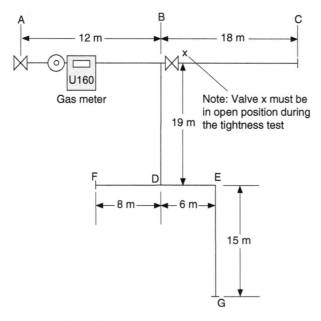

Figure 5.25 Worked example (TPCP1)

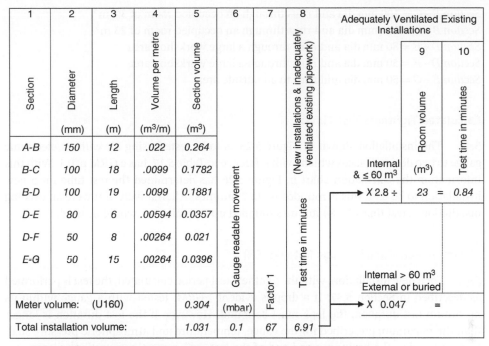

1	2	3	4	5	6	7	8	Adequately Ventilated Existing Installations		
									9	10
Section	Diameter	Length	Volume per metre	Section volume	Gauge readable movement	Factor 1	Test time in minutes (New installations & inadequately ventilated existing pipework)		Room volume	Test time in minutes
	(mm)	(m)	(m³/m)	(m³)					(m³)	
A-B	150	12	.022	0.264						
B-C	100	18	.0099	0.1782				Internal & ≤ 60 m³		
B-D	100	19	.0099	0.1881				→ X 2.8 ÷	23 =	0.84
D-E	80	6	.00594	0.0357						
D-F	50	8	.00264	0.021						
E-G	50	15	.00264	0.0396				Internal > 60 m³ External or buried		
Meter volume: (U160)				0.304	(mbar)			→ X 0.047 =		
Total installation volume:				1.031	0.1	67	6.91			

Note: 1: Round up to the nearest minute & never test for less than two minutes
Note: 2: Ensure test time does not exceed test duration for gauge selected from Table 5.17
Note: 3: For test durations outside those listed in Table 5.17 and extended test periods, see IGE/UP/1

Figure 5.26 Tightness test calculation chart

Column 6: Enter the gauge readable movement, taken from Table 5.17, which depends on the gauge used.

Column 7: From Table 5.20, enter a factor number based on the gas type to be used as the test medium.

Column 8: Complete the calculation along the bottom row by multiplying column 5 by column 6 by column 7 to give the test time in column 8. This test time is to be selected for all 'new installation work' and existing situations where the environment is inadequately ventilated, and it is not possible to test all joints with leak detection fluid.

For existing installations that are adequately ventilated, a further calculation needs to be made. This will depend on the degree of ventilation:

- for internal rooms less than 60 m³ in volume, multiply the test time above by 2.8 and divide by the room volume, columns 9–10;
- for external environments or internal rooms greater than 60 m³, the test time is multiplied by 0.047.

Worked Example:

5.40 Figure 5.25 shows a schematic pipe layout of a welded steel natural gas installation. Specific Details:

Section A – B = 150 mm dia and runs through an outside area

Section B – C = 100 mm dia and runs through an occupied room of 30 m³
Section B – D = 100 mm dia and runs through an occupied room of 23 m³
Section D – E = 80 mm dia and runs through a large workshop area
Section D – F = 50 mm dia and runs through a large workshop area
Section E – G = 50 mm dia and runs to an outside area

Calculated Tightness Test Time:

5.41   The installation shown in Figure 5.25, is to be tested using air with an electronic gauge to two decimal places which, with reference to Table 5.17, has a GRM of 0.1. With this information the calculation chart in Figure 5.26 has determined the test time for a new installation to be 6.91, this is rounded up to 7 minutes; Alternatively, if it were an existing installation, a test time of 0.84 minutes rounded up to 2 minutes, would be used.

Maximum Permitted Leakage Rate (MLPR)

5.42   For a new installation, with the tightness test period calculated, the test is performed as described in Section 5.37. If a drop is evident $\geq GRM$, tested with a gauge within its maximum test duration, the leak must be found. However, if the test duration is longer than the maximum prescribed for the gauge, temperature and atmospheric pressure shall be monitored at the beginning and end of the test, and corrections applied as necessary (see IGE/UP/1 Appendix 6 for details). Further calculations will be necessary to confirm whether the actual leak rate is less than the maximum permitted leak rate for a new installation.

The leak rate is determined from the following calculation:

$F3$ × Pressure loss × Installation volume ÷ Test duration

(F3 is a factor taken from Table 5.21).

The result of this calculation will be in m³/h and is compared with the figures in Table 5.18. If the calculated leak rate is greater than the maximum permitted leak rate, then the leak must be found.

From the example in Section 5.41, where an existing gas installation is tested with air, if a drop in pressure of 0.26 mbar is experienced during the test time, the above calculation would need to be made to determine whether the system could be left in service. Hence:

$$0.094 \times 0.26 \times 1.031 \div 2 = 0.0126 \text{ m}^3/\text{h}$$

This quantity (0.0126 m³/h) is now compared with the maximum permitted leak rate given in Table 5.18. In this example, the middle column of the table would need to be selected because our smallest space volume is <60 m³, which suggests that the leak should not exceed $0.0005 \times$ *Smallest space volume*. Our smallest space is 23 m³, therefore, $0.0005 \times 23 = 0.0115$ m³/h, which will be the maximum leak rate allowed. Our value of 0.0126 exceeds this figure and, as a consequence, the leak will need to be found.

Successful Installation Tightness Test and Completion

5.43 Immediately following the tightness test and when the pressure is removed from the system, any tools such as spades should be removed. Disturbed joints should be checked with leak detection fluid and the system should be purged to fuel gas. If this is not done immediately, a further tightness test will be required prior to purging. Once fuel gas is contained within the pipework, all accessible joints in inadequately ventilated areas will need to be checked with a suitable intrinsically safe gas detector set on the 0–10% LFL scale, which should detect no gas. Finally, the tightness test results should be recorded on a formal certificate, such as shown in Figure 5.27, and a copy given to the owner or person who is responsible for the property.

Figure 5.27 Sample strength testing, tightness testing and purging certificates

Appliance Connection Test

5.44 It would have been noted that the appliance connection was isolated during the tightness test. Therefore, after a successful test, this section of pipework now needs to be checked for tightness prior to commissioning or re-commissioning the appliance. To ensure that the whole section is tested, any appliance regulator must either be screwed down or bypassed to prevent lock-up, a non-return valve should also be bypassed. There must be no perceptible movement of the gauge over a two-minute test period. This test is limited to a maximum installation volume of 0.12 m³. For installation, volumes over 0.12 m³ IGE/UP/1 or 1A should be followed as appropriate (see Section 5.27 or 5.37). However, where the operating pressure, pipe size and volumes permit, this test may be achieved by undertaking a tightness test in accordance with IGE/UP/1B as previously described in Section 5.13.

Direct Purging Commercial Pipework

Relevant ACS Qualifications   Relevant Industry Documents
TCPC1 and TPCP1A   IGE/UP/1 and IGE/UP/1A

5.45 Direct purging refers to the exchanging of air in a pipeline for fuel gas or, conversely, removing the gas by replacing it with air. It differs from 'indirect purging' in which nitrogen (N_2) is introduced into the pipe between the full gas or air condition. Both direct and indirect purging processes have a certain degree of danger attached, for example, the danger of an explosion in the case of a direct purge and asphyxiation when nitrogen is used; therefore, many safety checks are needed before you start. A thorough risk assessment should also be documented.

General Considerations when Purging

1. The system to be purged should be surveyed and a tightness test carried out immediately prior to purging. Where a 'ring main' or looping system exists, the system needs to be divided to enable a complete purge.
2. The environmental impact of releasing fuel gas into the air should be considered and, where practical, the gas should be flared or burnt off. Gases such as LPG should be flared in any case, preventing their accumulation in low-lying areas.
3. An adequate provision of fire extinguishers of an appropriate type will need to be situated near the vent position, which is further explained in Sections 5.52 and 5.53.
4. The pressure within the pipe during the purge should not exceed the maximum operating pressure. Conversely, where the fuel gas is being drawn from a connecting supply, this too should be prevented from dropping to a pressure that could affect the operation of appliances installed upstream.
5. Purging of meters should only be undertaken with the agreement of the meter owner.
6. All pipework not forming part of the purge should be capped as appropriate, or if impractical the section should be locked off to prevent unauthorised use.
7. Sufficient gas operatives need to be available, each with their specific duty.

8. In all but the simplest purge operations, prepare a method statement and risk assessment.

9. In the event of not achieving the required purge rate (see Figure 5.28) the purge should be aborted and re-started if the cause can be determined. Alternatively, the system should be indirectly purged with nitrogen. Provision for it should therefore be included at the planning stage.

10. Purging needs to be done progressively and if there are branches these should be done in order of volume.

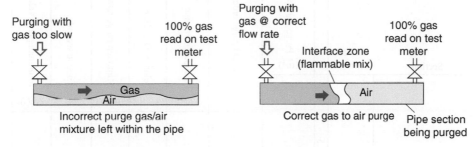

Figure 5.28   How the flow rate affects the success of the purge

Purging Internally

5.46   It is possible to purge small volumes of gas directly into a well-ventilated area, providing the following criteria are met:

- The operating pressure must not exceed 21 mbar.
- The volume of the ventilated internal space is not less than 30 m³.
- The installation volume of the pipework does not exceed 0.02 m³ and not include a diaphragm meter.
- There are no sources of ignition within a distance of 3 m.
- Gas concentrations in the room are monitored and where the level reaches 10% LFL, the purge should be stopped immediately.

Purge Velocity

5.47   The rate at which the gas or air is dispelled from the pipe determines whether the purge would be successful or not. If the velocity is insufficient, the two gases may stratify and mix together to form a combustible mixture within the pipe, see Figure 5.28. Two methods can be employed to verify that the purge velocity is maintained including:

1. using a suitable flow meter, as can be seen in Figure 5.5, which can directly read the volume of the gas/air flowing through the pipe (e.g. in m³/h), or

2. using a suitably sized 'volume' meter, either already fitted in the section to be purged, or incorporated for the test. The volume meter is used in conjunction with a timer so that the flow rate can be calculated.

In addition to maintaining the velocity, the discharging gas/air needs to be tested to confirm that the percentage of gas is greater than 90%, when inserting the fuel gas; or less than 1.8% gas when purging to air. *Note:* LPG is indirectly purged with N_2.

One method of finding the purge velocity is to complete a sheet as shown in Figure 5.29, carrying out each step as described below.

A			B	C	D	E	F	G	H
Section	Volume (m³)		Purge volume (m³)	Sequential purge order	Total purge volume (m³)	Column 'D' × 3600	Minimum purge rate (m³/h)	Time in seconds (s)	Minimum purge velocity (m/s)
		Multiply by 1.5							
Purge equipment									
Diaphragm meter capacity/rev	× 5 =								

Figure 5.29 Purge velocity and timed flow calculation sheet

Purge Velocity and the Completion of the Calculation Sheet

5.48 The calculation sheet is completed as follows:

(A worked example is shown in Section 5.49)

Column A Firstly, enter the installation volume for the various sections to be purged, taken from the tightness test table in Section 5.39, Figure 5.26 (column 5). Next, enter the meter volume for the ultrasonic, rotary or turbine meter from Table 5.7. Then, enter the volume of the purge equipment; this is calculated in the same way as the installation pipe was determined for tightness testing. To calculate the installation volume of the purge hose, you can use the comparable steel pipework size from Table 5.8.

Sub-section A Where a diaphragm meter is installed, enter the capacity/revolution (cyclic volume) taken from Table 5.7 into the appropriate box.

Column B Multiply the figures in column 'A' by 1.5 and any diaphragm meter by 5 to give the required purge volume for each part of the supply.

Column C Referring to the schematic drawing made in the survey, total up the purge volumes of each individual run and determine which has the highest total purge volume. This is usually the section to purge first. List the pipe sections in sequential order on the table.

Column D Enter the total purge volume for each of the sections shown at C. Do not forget that the purge equipment is added to each individual section.

Column E Multiply Column 'D' by 3600.

Column F Using Table 5.22, column 3, enter the minimum purge flow rate, based on the largest pipe within the section to be purged.

Column G Divide column E by column F to give the maximum purge time in seconds.

Column H Enter the minimum purge velocity, taken from Table 5.22, column 2.

Table 5.22 Minimum flow rates and associated purge stack dimensions

1	2	3	4	5	6	7
Nominal pipe diameter (mm)	Minimum purge velocity (m/s)	Minimum purge flow rate (m³/h)	Maximum purge length (m)	Purge point nominal bore (mm)	Purge hose and vent stack size (mm)	Flame arrestor nominal size (mm)
20	0.6	0.7	N/A	20	20	20
25	0.6	1.0	N/A	20	20	20
32	0.6	1.7	N/A	20	20	20
40	0.6	2.5	N/A	20	20	20
50	0.6	4.5	N/A	25	40	50
80	0.6	11	N/A	25	40	50
100	0.6	20	N/A	25	40	50
125	0.6	30	N/A	40	50	50
150	0.6	38	N/A	40	50	50
200	0.7	79	500	50	100	200
250	0.8	141	500	80	100	200
300	0.9	216	500	80	150	200
400	1.0	473	500	100	150	200

For larger pipe diameters see IGE/UP/1

With the table completed, sufficient information is now available for either of the purge methods, mentioned in Section 5.47, to be adopted.

Worked Example of Calculating the Purge Sequence and Velocity

5.49   We refer back to the worked example, as completed in Section 5.40 and assume that this system, as illustrated in Figure 5.30, has now to be purged. It will be seen that valve 'X'

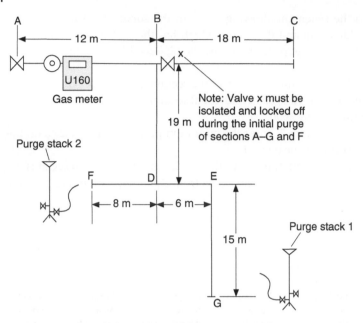

Figure 5.30 System from earlier tightness test worked example in Section 5.40

within the pipework could be isolated; this would therefore allow the initial purge operation to be completed from A through to points F or G. Either route could be selected for the initial purge; however, ideally, the route with the largest installation volume should be chosen. Totalling up each individual section it is found that the greater volume passes through to G. Thus the first purge would be from A to G.

When gas is detected at G, a purge from point F should begin immediately. Therefore, ideally, two purge stacks would be required. With most of the installation purged all that remains to be completed now is the section B–C, which was previously isolated. For this, a stack from the first purge could be re-used.

Installation Volume of the Purge Stack

5.50 The size/diameter of the purge stack and hose to be used is based on the largest diameter of pipework in the section being purged and is taken from Table 5.22. With the size selected, in this case, 50 mm, and a measurement of the total length of hose and stack (assume the hose to be 10 m in length), the installation volume can be determined by simply following the same format as that used to find the volume of the pipework. *Note:* It is acceptable to use the comparable steel pipework size from Table 5.8 for this purpose.

1	2 (size)	3 (length)	4 (volume)	5 (section volume)
Purge equipment	50 mm	10 m	0.00264 m^3	0.0264 m^3

If a meter is included with the purge equipment, it would need to be added to this volume.

Completion of Purge Velocity and Timed Flow Calculation Chart

5.51 The chart is completed by following the steps laid down in Section 5.48. The information is simply added to the chart, as seen in Figure 5.31, and where necessary, the calculations are carried out.

A		B	C	D	E	F	G	H
Section	Volume (m³)	Purge volume (m³)	Sequential purge order	Total purge volume	Column 'D' × 3600	Minimum purge rate	Time in seconds	Minimum purge velocity
				(m³)		(m³/h)	(s)	(m/s)
A-B	0.264	0.396						
B-C	0.1782	0.2673	A-G	1.19	4284	38	113	0.6
B-D	0.1881	0.2822						
D-E	0.0357	0.0536						
D-F	0.021	0.0315	D-F	0.072	259.2	4.5	58	0.6
E-G	0.0396	0.0594						
Purge equipment	0.0264	0.0396	B-C	0.31	1116	20	56	0.6
Diaphragm meter capacity/rev	0.071 ×5 =	0.355						

(Note in column B: *Multiply by 1.5*)

Figure 5.31 Completed purge velocity and timed flow calculation chart

To assist in understanding how the total purge (row D) was determined, the following separate calculations needed to be made, totalling up the volumes through which the gas flow would pass.

Purge volume for A–G:

Gas meter	0.355
A–B	0.396
B–D	0.2822 +
D–E	0.0536
E–G	0.0594
Purge equipment	0.0396
	1.1858

Purge volume for D–F:

D–F	0.0315 +
Purge equipment	0.0396
	0.0711

Purge volume for B–C:

B–C	0.2673 +
Purge equipment	0.0396
	0.3069

With the calculation chart completed, all risk assessments undertaken, and purge equipment erected, the purging operation can be undertaken. This consists of allowing the gas/air to flow through into the system and simultaneously starting the timer, measuring the flow of purge gas by the chosen method, either:

- using a suitably sized 'flow' meter that passes a minimum purge flow rate (in column F); or
- using a 'volume' meter, passing a total volume (in column D).

For the full purge procedure and method, see Section 5.53 and 5.54.

Purge Equipment and Vent Stack

5.52 Examples of purge equipment configuration can be seen in the illustration in Figures 5.32 and 5.33. For any purge to succeed, the purge points, hoses and vent stack need to be of the correct size as previously listed in Table 5.22, columns 5–7. Where the purge gas is to be supplied by cylinders, it is essential to confirm their contents as correct and that there are a sufficient number to complete the purge. Cylinder gas would need to be supplied through a high-capacity regulator and any valves used in the purging operation would need to be of the full-bore type. If these sizes cannot be achieved, it is possible to use multiple vent stacks simultaneously, providing each point is adequately supervised. The hose used to convey the gas to the vent stack needs to be suitable for the gas used and be secured firmly to the pipework and vent. In order to avoid sparking where an externally armoured hose is used, the hose needs to be suitably earthed. PE hoses should not be used as they may generate static electricity.

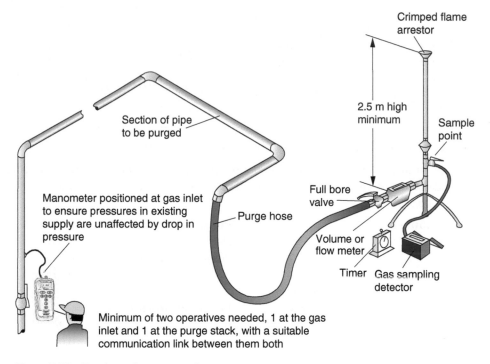

Figure 5.32 Purging using vent stack

The vent stack itself should include a flame arrestor; it may be in-line, where the gas is to be burnt off, or as a termination fitting where the gas is not flared. In addition, a sampling point is needed to check for the presence of gas. Where flaring or burning off the gas, as can be seen in Figure 5.33, the in-line arrestor should be fitted at least 2 m from the gas discharge point. A source of ignition is also required at the burner, and care needs to be taken with

Figure 5.33 Purging using flare stack

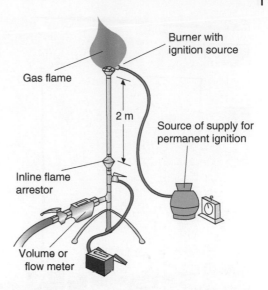

the issuing flame. The outlet should be in the open air and terminate at a distance of at least 2.5 m above ground. Where venting, the outlet needs to be located at least 5 m downwind of any sources of ignition and any drifting into buildings should be prevented. Natural gas should be flared off where practical, but LPG should always be flared.

Purging Procedure

5.53   Before beginning the purge, ensure that all the general considerations for purging have been met, including the completion of the appropriate risk assessment. Ensure safety barriers and signs are in place and that sufficient fire-fighting apparatus is available, see Figure 5.34. All radios used should be intrinsically safe and their operation tested. The communication procedure should be practised by all operatives involved in the purge operation to ensure that they know their specific duties.

5.54   Method:

1. A manometer should be located at the inlet to the system and an operative stationed at this point. Their job is to keep the supervisor informed should the pressure rise dramatically when supplying air, via an external source, when supplying fuel gas from an existing pipeline, or fall to a point that could affect upstream supply pressure. In addition, they should turn off the supply if instructed to do so by the supervisor.
2. At the purge point/vent stack, as seen in Figure 5.35, the valve should be opened to allow the gas to flow and at the same time, the timer should be started. At about halfway through the purge time, samples of gas/air should be taken to test the gas concentration using a gas detector looking for:
   - 90% fuel gas when purging to natural gas; or
   - 1.8% fuel gas when purging from natural gas to air or N_2.

Figure 5.34 Purge equipment and signage

Note: It is also possible to select the 100% LFL scale when purging to air looking for a reading of 40% LFL, which is equivalent to 1.8% gas. For LPG, see Section 5.55.

The purge should be aborted if the gas/air concentration is not achieved within the purge time. It may be re-attempted if the reason can be determined (e.g. blocked hose), alternatively, an indirect purge using nitrogen should be undertaken.

3. On completion of a satisfactory purge, the equipment should be removed and any disturbed joints checked with leak detection fluid. Inadequately, ventilated areas also need to have all accessible joints checked using a suitable gas detector. All connected appliances must be commissioned as necessary, or disconnected from the supply. The purge apparatus should also be purged with air before storage. Finally, an appropriate purging certificate, as shown in Figure 5.27, should be completed and given to the owner or person responsible for the property.

Figure 5.35 Gascoseeker and flow gauge at vent stack

Indirect Purging Using Nitrogen (N$_2$)

5.55   Indirect purging, as illustrated in Figure 5.36a, would only be carried out if:

- the risk assessment deemed a direct purge to be unsafe, such as when purging LPG;
- insufficient purge points are available;
- a direct purge had been unsuccessful and aborted.

Note: It is possible to re-attempt a direct purge if the cause for its abandonment can be determined, such as an error in the calculation or a restriction in the purge line. However, when such repeated purges are undertaken particular care needs to be observed due to the nature of a gas/air mixture being expelled from the pipe.

When purging to N_2 particular attention needs to be paid to avoid asphyxiation, especially in basements or confined places. It is also important to monitor the situation and pressure to ensure that no purge gas enters the distribution network supplying the installation. When purging with N_2 gas, the aim is to provide a complete displacement based on installation volume and sampling at the vent stack and, where possible, maintain the purge velocity

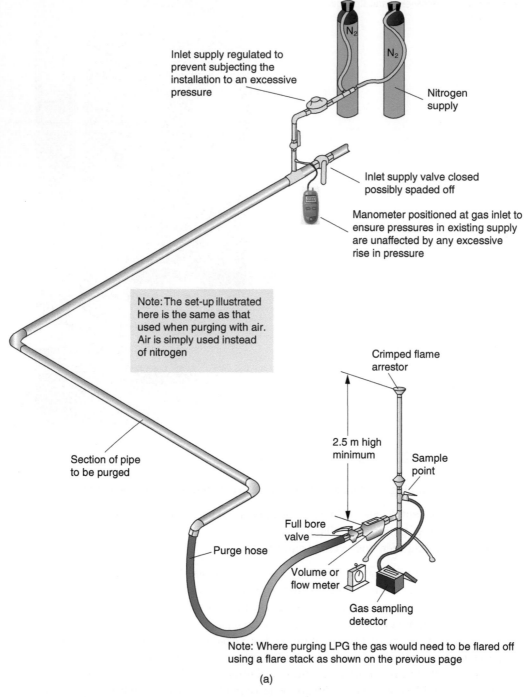

Inlet supply regulated to prevent subjecting the installation to an excessive pressure

Nitrogen supply

N₂

N₂

Inlet supply valve closed possibly spaded off

Manometer positioned at gas inlet to ensure pressures in existing supply are unaffected by any excessive rise in pressure

Note: The set-up illustrated here is the same as that used when purging with air. Air is simply used instead of nitrogen

Crimped flame arrestor

2.5 m high minimum

Sample point

Section of pipe to be purged

Full bore valve

Purge hose

Volume or flow meter

Gas sampling detector

Note: Where purging LPG the gas would need to be flared off using a flare stack as shown on the previous page

(a)

Figure 5.36   (a) Indirect purging using vent stack

(b)

Figure 5.36 (b) Example of a nitrogen cylinder

for direct purging. It is important to ensure that any dead legs are adequately purged. N_2 is an inert gas and will therefore not readily mix with the fuel gas; where any mixing does occur no combustible mixture will be formed. Thus with the installation volume worked out (as previously described in the section on direct purging and tightness testing), gas is released into the system to the volume calculated. Sample gas readings are taken at the various vent points to give the following safe purge quantities/gas concentrations:

Natural gas

- 7.5% gas (on the 100% gas scale) if purging to N_2; or
- 90% gas (on the 100% gas scale) if purging to fuel from N_2.

Propane gas

- 3.5% gas (on the 100% gas scale) if purging to N_2; or
- 90% gas (on the 100% gas scale) if purging to fuel from N_2.

Butane gas

- 3.0% gas (on the 100% gas scale) if purging to N_2; or
- 90% gas (on the 100% gas scale) if purging to fuel from N_2.

As a guide to the calculation of the total number of N_2 cylinders required for adequate inert purging, it should be noted that a standard 1.5 m high bottle contains approximately 6.5 m^3 of N_2. This could be discharged at a rate of approximately 60 m^3/h through a standard single-stage regulator. Where inert N_2 purging has been completed to allow for hot work to be undertaken, it is essential to take special care. Possibly consider localised monitoring for traces of fuel gas, as small pockets may remain as a result of the effects of stratification, particularly in dead legs.

6

Flues and Chimneys

Flue/Chimney Classification

Relevant Industry Documents
BS 5440, BS 6644, IGEM/UP/10 and BS EN 1443

6.1 BS 5440 now recognises the term flue to mean the passage through which the combustion products pass and the term chimney to be the structure or wall of the material enclosing the flue, to include masonry, metal or plastic materials.

6.2 Appliances generally fall into one of the following classifications: Type 'A': Flueless; Type 'B': Open-Flued and Type 'C': Room-Sealed. The letter A, B or C is immediately followed by a number that further defines the type of flue system design, as shown in Table 6.1.

6.3 The flueless appliance takes its air for combustion from within the room in which the appliance is situated and discharges the products of combustion into the same environment.

6.4 Open-flued appliances also take the combustion air from the room but remove the combustion products to the outside air. It should be noted that the appliance often referred to as a 'closed flue', which is discussed in Part 14: Section 14.10, has the same design concept but has no draught diverter but is classed as an open-flue (namely B21 in Table 6.1).

6.5 Room-sealed appliances do not take any air from the room in which the appliance is situated, they take their air direct from outside via a duct. The term 'balanced flue' refers to a system where the air intake is taken from a point adjacent to the flue gas extract and in the same pressure zone, i.e. in balance.

Illustrations of flue/chimney configurations with their relevant classification can be seen in Figure 6.1.

Gas Installation Technology, Third Edition. Andrew S. Burcham, Stephen J. Denney and Roy D. Treloar.
© 2024 John Wiley & Sons Ltd. Published 2024 by John Wiley & Sons Ltd.

Table 6.1 Classification of gas appliance flue systems

Flue type		Flue/chimney design	Natural draught (no fan)	Induced draught (fan downstream of the burner)	Forced draught (fan upstream of the burner)
Flueless	A	Not applicable	A1	A2	A3
Open-flued chimneys	B1	With draught diverter	B11	B12 (A) or B14 (B)	B13
	B2	Without draught diverter	B21	B22	B23
Room-sealed chimneys	C1	Horizontal and 'in balance' to outside	C11	C12	C13
	C2	Inlet and outlet duct connections to multi-stack system (SE duct or U duct)	C21	C22	C23
	C3	Vertical and 'in balance' to outside	C31	C32	C33
	C4	Inlet and outlet connections to leg of 'U' duct system	C41	C42	C43
	C5	Ducted flue and air supply system 'out of balance'	C51	C52	C53
	C6	Appliance purchased without a flue or air supply inlet ducts	C61	C62	C63
	C7	Vertical flue, with draught break in roof void above ducted air intake to appliance, also in roof (e.g. vertex)	C71	C72	C73
	C8	Flue connected to common duct with air supply ducted from outside and therefore 'out of balance'	C81	C82	C83

(A) B12 – Fan downstream of heat exchanger but prior to draught diverter.
(B) B14 – Fan downstream of draught diverter.

Chimney/Flue Material and Specification

Relevant Industry Documents
BS 5440, BS 6644, IGEM/UP/10

6.6 The flueway is generally rectangular or round in shape, depending on the material used. Its cross-sectional area is dependent on the manufacturer's design and the heat input.

The Building Regulations (Approved Document J) give the minimum size of any flue system as the appliance outlet cross-sectional area. However, for a gas fire, a cross-sectional area of 120 cm² (12000 mm²) is needed, which, in effect, needs to be 125 mm in diameter if the flue is round. If the flue is rectangular, such as in a chimney constructed with flue blocks,

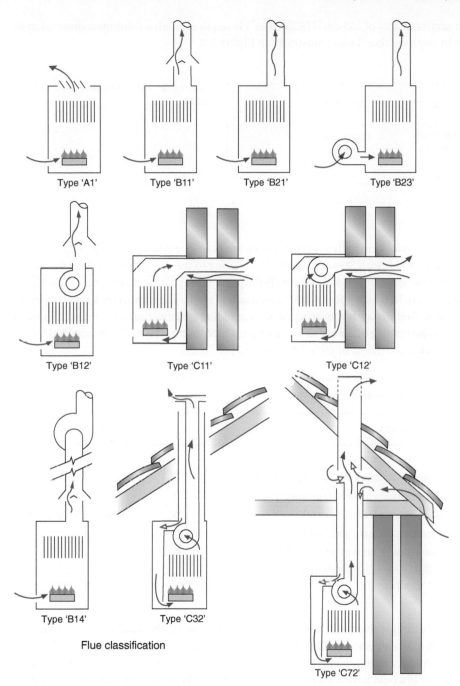

Type 'A1'   Type 'B11'   Type 'B21'   Type 'B23'

Type 'B12'   Type 'C11'   Type 'C12'

Type 'B14'   Type 'C32'

Flue classification

Type 'C72'

Figure 6.1 Illustrations of flue/chimney configurations

a cross-sectional area of 165 cm^2 (16500 mm^2) is required, with a minimum dimension of 90 mm in any direction. This is illustrated in Figure 6.2.

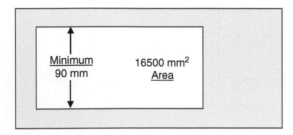

Figure 6.2 Cross-sectional area of flue block

6.7 For many years, operatives have often referred to a flue system as being class 1 or class 2. This was an old British Standard definition relating to the design temperature for which the flue system was made. It was effectively dropped many years ago in favour of specifying the temperature in which a flue system could safely work. The term class 1 refers to those used for solid fuel. Today, however, a new system of flue specification is used, and materials are classified according to the prescribed format, which uses a system of numbers and letters to identify what the flue is suitable for (Table 6.2).

Table 6.2 Defining flue specification details

Defining characteristics		Example
Temperature class (max. working temperature)	T250	
Pressure class: N (negative), P (positive), H (high positive)	P1	
Sootfire resistance class S (with) O (without)	O	
Condensate resistance D (dry) W (wet)	W	T250 P1OW1 R22 C50
Corrosion resistance (class 1 gas; 2 and 3 oils and solid fuel)	1	
Thermal resistance (in units of m^{2}K/W × 100)	R22	
Minimum distance to combustible material	C50	

So, for the example given in Table 6.2, the flue material would be suitable for a system with a maximum flue temperature of 250°C, operating under positive pressure. It would have no resistance to soot fire but would be resistant to condensate with a corrosion resistance of 1, i.e. for gas. It would have a thermal resistance of 2200 W/m^{2}K temperature change and, finally, it should be installed no nearer than 50 mm to any combustible surface.

6.8 Several materials may be used for the passage of flue gases, and it is essential to choose a material that is suitable in accordance with the Building Regulations. Table 6.3 lists some of the available options.

Table 6.3   Typical flue materials

Material	Temperature range (°C)
Acid-resistant brick and insulated brick	≤200
Pre-cast concrete	≤300
Glazed clayware	≤200
Enamelled mild steel	≤450
Stainless steel	≤500
Aluminised steel	≤600
Cast iron	≤500
Aluminium	≤300
Plastic materials (MUPVC)	Variable generally ≤150 °C

Notice Plates

6.9   Where a Type B flue, chimney, hearth or fireplace opening is included in a new building or as part of a refurbishment, the information applicable to the flue system should be indelibly marked on a durable, robust indicator plate fixed at an unobtrusive, but obvious location. This might be positioned

- next to the chimney opening;
- next to the water, gas or electrical supply inlet.

The notice plate, as seen in Figure 6.3, should convey the following information:

- the location of the beginning of the flue system, e.g. fireplace;
- the category of flue and types of appliances that may be fitted to the system;
- the manufacturer's name of the flue material and the type and size;
- the installation date.

Note: The flue designation and installer details are optional.

IMPORTANT SAFETY INFORMATION
Property Address:
164 Beachcroft Road, Anytown, Essex.
The hearth and chimney installed in the front lounge is suitable for:
Any type/design of gas fire
Flue: 175mm id Clay liner supplied by 'All-Flue Ltd'   **Installed:** 10.12.2022
Flue Designation to BS EN 1443: T250 N1 S D 1 R22 C50   Installer: Baytree Construction, Anywhere, Essex.
This label must not be removed or covered.

Figure 6.3   Typical notice plate

Natural Draught Open-Flue Systems

6.10   This design of flue system is the oldest form of flue design and, for many years, was referred to as a conventional flue system. It consists of a flueway through which the products of combustion can pass out to the external environment. The air supply for combustion is drawn into the appliance from the room in which the appliance is installed. Hot air rises up through the flueway by convection. It must be remembered that a flue system may be installed by another person, but the gas installer is responsible for confirming it is adequate.

Convection

6.11   As the air or flue gas molecules are heated, they expand and, in so doing, become less dense. In Figure 6.4, it can be seen that there are only four large hot air molecules in the left-hand basket of the scales, whereas there are sixteen cooler, more dense air molecules on the right-hand side. The weight of each molecule has not increased, only its size. Therefore, the basket with the greater number of molecules bears down with a greater force and causes the lighter volume to lift. By applying this principle to the heated gases in a flueway, it will be seen that the cooler air surrounding the burner forces the lighter gases upwards. As seen in Figure 6.5, to achieve good convection, it is essential that air is free to enter the room in which the appliance is situated, otherwise the supply of heavy, dense cold air would soon be depleted, convection would slow down, and the appliance would spill flue gases into the room. For appliances with a heat input of ≤7 kW, adventitious ventilation (air that comes in through openings in windows and doors) is usually all that is needed, but as the heat input increases, so does the need for a fixed air supply.

Hot gas in which the molecules have expanded

Figure 6.4   How heat affects flue gases

Cooler more dense gas

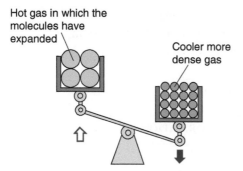

(Heavier more dense gas forces the lighter expanded gas upwards)

6.12   Natural draught open-flued appliances incorporate a draught diverter, see Figure 6.6, which must be installed within the same room as the appliance to assist in the movement of the combustion products. It does this by maintaining a lower pressure within the flueway. In addition, the draught diverter:

- assists in diluting the flue gases and thereby reducing the CO_2 content;
- breaks the pull of the secondary flue;
- prevents down draught blowing back down into the appliance.

Figure 6.5   Concept of convection

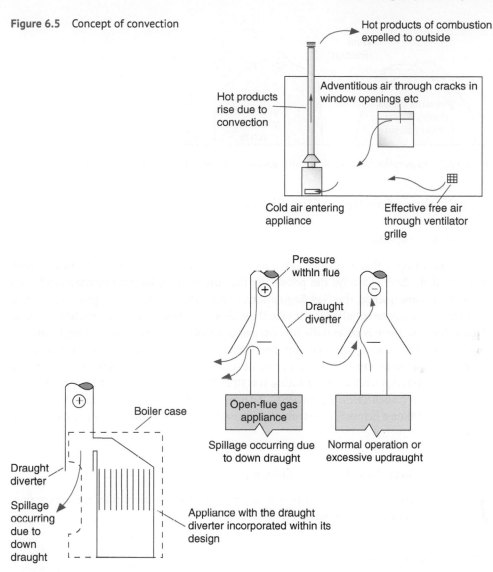

Hot products of combustion expelled to outside

Hot products rise due to convection

Adventitious air through cracks in window openings etc

Cold air entering appliance

Effective free air through ventilator grille

Pressure within flue

Draught diverter

Open-flue gas appliance

Spillage occurring due to down draught

Normal operation or excessive updraught

Boiler case

Draught diverter

Spillage occurring due to down draught

Appliance with the draught diverter incorporated within its design

Figure 6.6   Draught diverter

6.13   When wind blows, it creates positive and negative pressure zones along and around the surfaces of the building; see Figure 6.7. The side on which the wind blows has positive pressure. In designing a good flue system, the aim would be to position the flue terminal just outside the influence of these pressure zones. Unfortunately, wind speeds are not constant and as the wind speed increases, so does the size of the pressure zone. Wind gusts continually vary and so the terminal is occasionally contained within the pressure zone and a negative or positive draught will blow through the flue route. Wind from a down draught will hit the baffle and blow into the room; conversely, where an excessive updraught pulls on the products, the air is drawn in at the draught diverter rather than through the restricted route of the heat exchanger. Both cases help prevent the loss of flame stability.

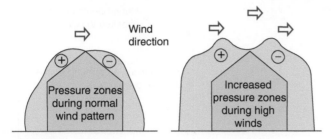

Figure 6.7   Illustration showing why some excessive down draft & updraught conditions occur

Installation of an Open-Flue Chimney

Restricted Locations

6.14   As an open-flue appliance takes its air for combustion from the room in which it is situated, there is always the possibility that the air may become vitiated (lack oxygen). As a consequence, the Gas Regulations restrict the use of such appliances, stating that they must not be installed in, or take their air supply from, a bathroom/shower room. This is because there is potentially less oxygen available in a warm, steamy environment. In addition to this, an open-flued appliance installed in a room that is used for sleeping, i.e. bedroom/bed-sit, must not have a gross heat input greater than 14 kW. It is possible to install an appliance with an input rating less than this; however, it must have some form of safety control, such as a vitiation sensing device (covered in Part 3: Section 3.57), to shut down the appliance before there is any significant build-up of combustion products in the room.

Component Parts of an Open-Flue Chimney

6.15   An open-flue system consists of four parts, namely: primary flue, secondary flue, draught break or diverter and terminal.

Primary and Secondary Flues

The primary flue is the section of pipe from an appliance to a draught break, where fitted, and the pressure within this section is that experienced in the appliance itself. From this point the flue is run to the external environment by what is known as a secondary flue. In order to facilitate the disconnection of an appliance from a flue system, a disconnecting joint will often be found just above the appliance, as shown in Figure 6.8.

Draught Break

This is a component that allows pressure fluctuations to be alleviated within the flue system. There are two different types of draught break, see Figure 6.9: the draught diverter

(described in Item 6.2) and keeps the joint airtight. The flange is made of a
material to compensate for movement, with a hinged door that can be detached
to allow the duct to be taken apart to enable sampling the flue.

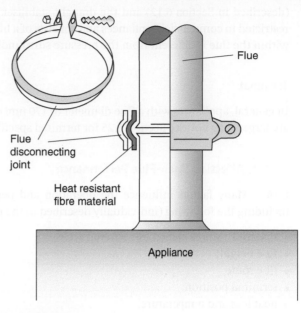

Figure 6.8 Flue system disconnecting joint

Flue

Flue disconnecting joint

Heat resistant fibre material

Appliance

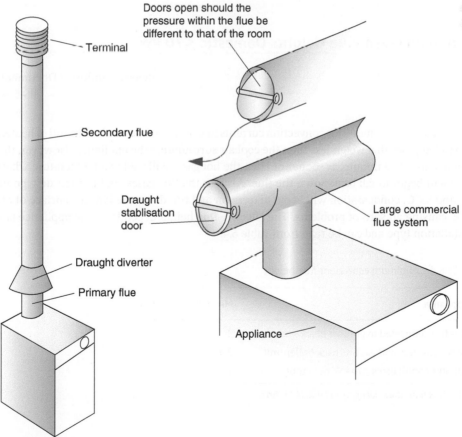

Terminal

Secondary flue

Doors open should the pressure within the flue be different to that of the room

Draught stablisation door

Large commercial flue system

Draught diverter

Primary flue

Appliance

Figure 6.9 Draught break in flue

(described in Section 6.12) and the draught stabiliser. The draught stabiliser is generally restricted to commercial appliances and consists of a hinged door that opens if the pressure within the flue is different from the pressure surrounding the flue.

Terminal

In general, appliances with a flue diameter of 170 mm or less require a terminal; those that are larger may not. See Section 6.25 for terminal specification.

Factors Affecting Open-Flue Performance

6.16   Many factors influence the operation and performance of an open-flue system, including the following (individually described in the previous and following sections):

- materials: shape and cross-sectional area;
- flue height;
- flue route;
- terminal position;
- heat loss and temperature.

Minimum Open-Flue Heights Domestic ≤70 kW

Relevant Industry Documents
BS 5440

6.17   Height is required for convection currents to work. The flue needs to be full of heated, expanded gases that weigh less than the cool air surrounding the appliance. However, this height cannot be infinite, as sooner or later, the hot gases will cool to a temperature where they will begin to fall back down the flue. Also, as the flue gases cool to their dew point, around 55°C, condensation will begin to form on and run down the internal surface of the flue, causing all sorts of problems. The effective flue height depends on the appliance and installation type and can be read from Table 6.4.

Table 6.4   Minimum equivalent flue height

Appliance	Minimum height (m)
Gas fire connected to pre-cast block system	2
Other gas fires and gas fire/back boiler unit	2.4
All other appliances ≤70 kW net input	1

Note: This table does not apply to DFE/ILFE fires.

Flue Route

6.18 The route that the flue takes as it passes up through the building should ideally be as straight as possible, travelling vertically throughout its entire length. A certain flue height is required, as discussed in Section 6.16, to create the draught, but not a particular length - every time there is a change in direction, there is frictional resistance, which decreases the velocity of the flue gas, see Figure 6.10. Bends should be avoided, and where the system is to rely on natural draughts caused by convection, they should be restricted to angles of not less than 135°. The distance from the appliance draught diverter to the first bend needs to be at least 600 mm in order to get a good start to the updraught unless the manufacturer's instructions state otherwise.

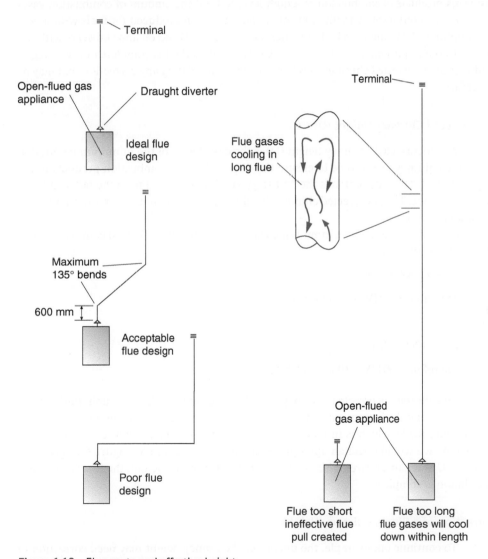

Figure 6.10 Flue route and effective height

Minimum Open-Flue Heights Commercial >70 kW

Relevant Industry Documents
IGEM/UP/10

Chimney Height

6.19   The method to establish the required chimney height and termination location for commercial installations, i.e. >70 kW net, differs from domestic, although many of the principles of operation are the same. They are designed to achieve an exit velocity that safely disperses products of combustion to a high level to limit the amount of combustion gases present at ground level. A method of calculation has been produced for fuels with a very low sulphur (VLS) content, which includes natural gas, LPG, gas oil and kerosene, with reference to the requirements of The Clean Air Act. The method is completed in two stages: finding the uncorrected chimney height before then calculating any correction that may be necessary.

Uncorrected Chimney Height (U)

6.20   An uncorrected chimney height (U) is the calculation of the minimum height that a specific chimney needs to be to ensure the dispersal of the combustion products to the required level. This calculation is the first stage and is determined from the net heat input of the appliance. The uncorrected chimney height (U) can be obtained from the graphs seen in Figures 6.11–6.13.

You will notice that the heat input is measured in Megawatts (MW). To convert between MW and kW, the sums are:

$$MW \times 1000 = kW$$

Therefore, $0.5\,MW \times 1000 = 500\,kW$

Or

$$kW \div 1000 = MW$$

Therefore, $500\,kW \div 1000 = 0.5\,MW$

6.21   To gain the uncorrected chimney height, we select the relevant graph, find our net heat input along the bottom axis, move vertically up to the plotted line and then across to find a figure for U, which is in metres. So, for example, if we had plant equipment rated at 1.5 MW, we would select an appropriate graph; for this example, Figure 6.12 gives the value for U at about 2 metres. This is our uncorrected chimney height; the first stage of our calculation is complete.

Correction of a Single Chimney Height

6.22   To continue our example, the uncorrected chimney height may need correction to increase in length due to the effect that the buildings, or ones nearby, may have on the

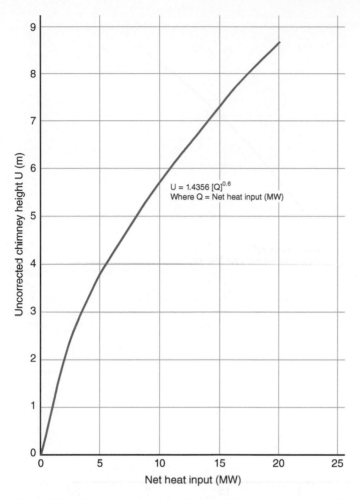

Figure 6.11 Uncorrected chimney height (*U*) for gas appliances ≤20 MW. Source: From IGEM/UP/10

performance of the flue. If the value of U is <2.5 times the height of a building it is serving, or if there are other buildings within a distance of 5 × *U*, a further calculation will be necessary. For **Example A**, Figure 6.14 shows the proposed installation of the flue, and as our value of U, calculated in Section 6.21 (2 m), is <2.5 times the height of the building, a further calculation is required to find the corrected chimney height. The formula for this is:

$$H_B + (0.6 \times U) = Hc.$$

- H_B being the height of the building that the chimney serves.
- *U* being the uncorrected chimney height.
- *Hc* being the corrected height of the chimney.

Therefore, the calculation for **Example A** is:

$$12\,\text{m} + (0.6 \times 2\,\text{m}) = 13.2\,\text{m}$$

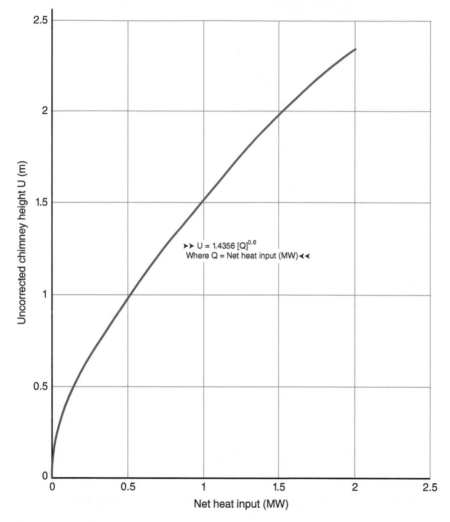

Figure 6.12 Uncorrected chimney height (*U*) VLS fuels for small plant ≤2 MW. Source: From IGEM/UP/10

This figure is to be rounded up to the nearest metre, which gives a corrected chimney height of 14 m.

Note: The calculation for nearby buildings will be the same but will use the height of the tallest building in the calculation.

Correction of Multiple Chimney Heights

6.23   Where, for example, two chimneys are attached to a building, as can be seen in Figure 6.15 **Example B**, the calculation is extended to include taking account of the distance between the two chimneys. If the chimneys are within 5 x *U* of the largest appliance of each other, the aggregate heat input of both is used to determine the value of U for the calculation. We will now calculate the installation in Figure 6.15.

The plant using chimney 1 has 500 kW net heat input. Referencing Figure 6.12, U (the uncorrected chimney height) is 1 m.

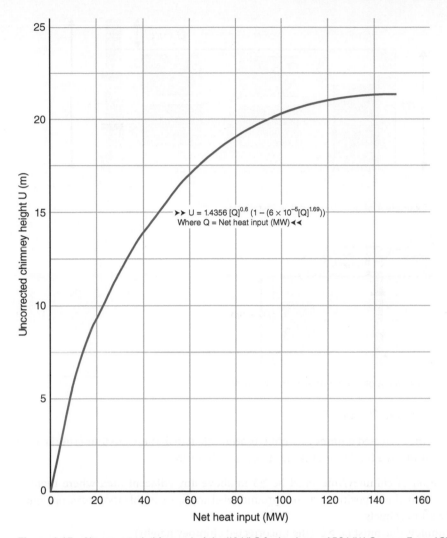

Figure 6.13 Uncorrected chimney height (*U*) VLS fuels plant ≤150 MW. Source: From IGEM/UP/10

The plant using chimney 2 has 1500 kW net heat input. Referencing Figure 6.12, U (the uncorrected chimney height) is 2 m

Chimney 2 has the largest value for U; therefore, we calculate 2 m × 5 = 10 m. The distance between chimney 1 and chimney 2 is just 7 metres, so we now must treat the installations as a group and work out a combined value for U. Now the calculation of U will be using 2000 kW and referencing Figure 6.12, the value of U is now 2.4 m. So, the calculation to correct both chimneys is as follows:

$$H_B + (0.6 \times U) = Hc$$

$$9\,m + (0.6 \times 2.4) = 10.44\,m$$

This will be rounded up to 11 m for both chimneys.

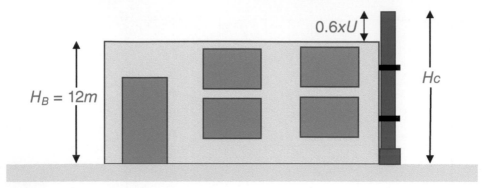

Figure 6.14   Example A of building and flue installation

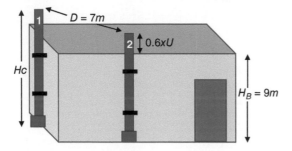

Figure 6.15   Example B of two chimneys on a building

Review of Corrected Chimney Heights

6.24   Once the corrected chimney height has been calculated, there are four criteria to be checked regarding the final flue/chimney position. These are:

1. Termination of chimney/flue must be ≥ 3 m above any adjacent area where there is access. This includes positions such as ground-level areas, accessible roof areas, openable windows, or air inlets.
2. The chimney/flue must be $\geq U$ (the uncorrected chimney height)
3. If there are any buildings within $5 \times U$ of the chimney/flue concerned, the chimney must be higher than the building.
4. If there is another chimney within $5 \times U$ of the chimney being calculated, the aggregate of both chimneys must be used to establish the value of U.

Open-Flue Terminal Design and Location

Relevant Industry Documents
BS 5440, BS 6644 and IGEM/UP/10

6.25   Flues with a diameter ≤ 170 mm require the use of a terminal; for larger flues, a terminal may inhibit the flow and correct dispersal of flue products; this flow is referred to as the efflux velocity. For natural draught appliances, flues should be designed with a target efflux velocity of ≥6 m/s.

The terminal serves many functions, including:

- preventing the entry of rain, snow or debris and even small birds, etc.;
- assisting in the discharge of flue gases;
- minimising up and down draughts.

Where a terminal is incorporated, its outlet opening needs to be such that it maintains an area twice that of the cross-sectional area of the flueway it serves. This outlet should also be of a design that will admit a 6 mm diameter ball but will not allow the entry of a 16 mm diameter ball, see Figure 6.16.

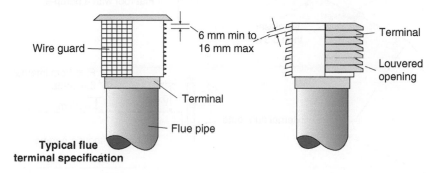

Figure 6.16 Typical flue terminal specification

Open-flue terminal design is the same whether it is a domestic or non-domestic installation, where it differs is the discharge location.

Flue Terminal Discharge Location Above Roof ≤70 kW

6.26   The flue discharge or terminal, where fitted, should not be located where it is likely to cause a nuisance. It also needs to be outside any pressure zone that may affect its performance. Table 6.5 gives an indication of the minimum height to be maintained. Note: Where three or more flues terminate at the same height, unless the outlets are more than 300 mm apart, they should terminate at the same level.

Table 6.5   Flue discharge position for appliances ≤70 kW (See also Figure 6.17)

		Pitched roof	At ridge, or 600 mm above, or at least 1.5 m measured horizontally to the roof line
≤70 kW Net input	Flat roof	With parapet or external flue route	600 mm above the roof line
		Without parapet and providing internal flue route	250 mm above the roof line

Note: For 'fan draught' open-flue terminal position, see Table 6.12.

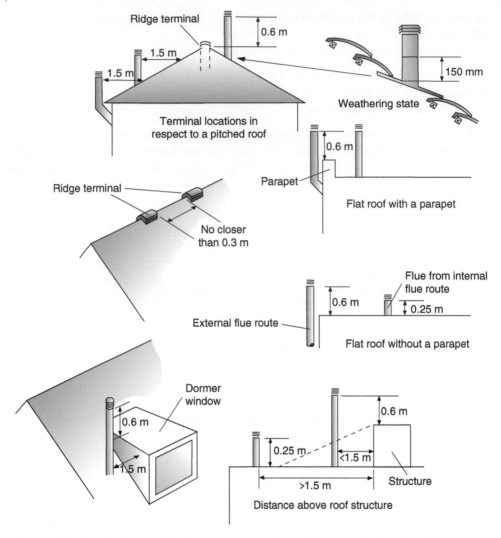

Figure 6.17 Termination height above a roof for appliances less than 70 kW net heat input

Flue Terminal Discharge Location Above Roof >70 kW

6.27 Where the flue discharge is adjacent to a location of general access to the public, it will need to be raised to a height of 3 m above the level of access. Furthermore, to avoid the adverse effect of pressure zones acting upon the terminal, it must be raised to a sufficient height in relation to a roof. Table 6.6 provides the necessary formulas to calculate the minimum height required.

Table 6.6 Minimum height (X) for flue terminal located on a roof for appliances with heat input ≤333 kW

Appliance flue type referencing distance x from Figure 6.18a,b	Formula
Minimum distance X for natural draught flues	$X = 1.5225 \text{ x (net heat input kW)} + 493.43$
Minimum distance X for fan draught flues	$X = 2.6644 \text{ x (net heat input kW)} + 113.49$

Note: In compliance with the Clean Air Act, single or groups of appliances >333 kW net heat input must terminate above roof level.
Source: Data from IGEM/UP/10

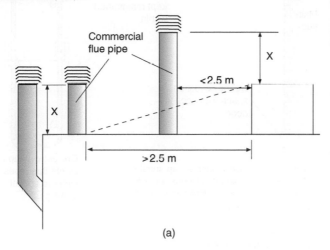

(a)

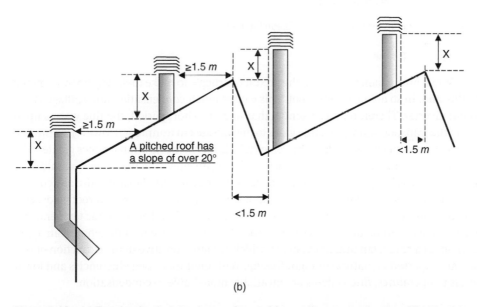

(b)

Figure 6.18 (a) Termination height above a flat roof for appliances greater than 70 kW net heat input. (b) Termination height above a pitched roof for appliances >70 kW net heat input.
Source: Data from IGEM/UP/10

Condensation within Open-Flue Systems

6.28   Prevention of condensation within any flue system needs to be a major consideration at the design stage. Figure 6.19 illustrates some methods of design to combat condensation issues. Under full load conditions, a correctly designed flue system should not give rise to any condensation problems.

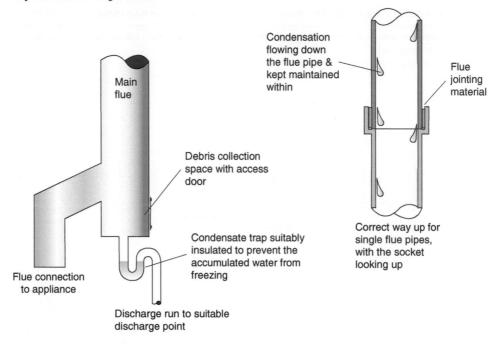

Figure 6.19   Condensation within open-flues/chimneys

Heat Loss and Temperature

6.29   When an appliance is first lit, the products struggle to make their way up through the cold flue and, in so doing, rapidly cool. This causes much condensation and spillage at the draught diverter. Therefore, it is essential that within a short period, the flue warms up to assist in the transportation of the flue gases to the outside environment. However, as the flue gas temperature increases, so does the velocity. This increased flow rate creates greater frictional resistance and increased turbulence, which again slows the flue flow to an ultimate maximum flow velocity. Heat loss from the flue must be prevented to minimise any cooling effects and the associated problems. Pipes that run externally and through roof voids, etc., where they are subject to cooling, must be of a twin wall design. The high flue temperatures needed to make natural draught systems work effectively reduce the efficiency of an appliance, and, as a result, fan draught systems, which are more positive in the extraction of flue gases, are superseding natural draught flueing. With their increased efficiencies and lower flue gas temperatures, flue systems are invariably more liable to condensation.

Condensation Problems

6.30   In a large commercial application, where several appliances are connected to a modular flue system, there are periods where the flue is oversized. This is due to times of low demand where not all of the appliances are required to operate, and this can result in continuous condensation problems. To prevent the condensation from flowing back down into the appliance, it is sometimes possible to fit a minimum 22 mm condensate pipe at the lowest point, which can disperse the liquid to a suitable drain or soakaway. Because Building Regulations do not permit any openings into the flue system, apart from a draught diverter or stabiliser, a trap should be fitted to this pipe, suitably insulated, thus ensuring that no flue products can exit via this route. The material used for any condensate pipe should be non-corrodible. Copper and copper alloys should not be selected due to the acidic nature of the water produced unless an inline neutraliser is fitted to protect against corrosion. With unavoidable sustained condensation, the flue wall or lining and all jointing materials used must be non-permeable, with joints sealed to prevent the condensation from running out of them. Where condensation occurs in an existing masonry chimney that was previously used for an oil or solid fuel appliance, it can mix with the soot deposits and sometimes cause staining to the inner walls of the brickwork.

When designing any flue system, consideration needs to be given to the formation of condensation. Tables 6.7 and 6.8 give the approximate known maximum condensate-free length of flues installed on appliances up to 70 kW net heat input. No tables are available for flues over 70 kW net for which IGEM/UP/10 should be used.

Table 6.7 Approximate maximum condensate free length for individual gas fire

Flue material	Internal	External
225 × 225 brickwork and standard pre-cast flue blocks	12 m	10 m
125 mm single-walled flue pipe (not insulated)	20 m	N/A
125 mm twin-walled flue pipe	33 m	28 m

Table 6.8 Approximate maximum condensate free length for other appliances up to 88% net efficiency

Net heat input (kW)	225 × 225 brickwork and standard pre-cast flue blocks (m)		125 mm twin-walled flue pipe (m)	
	Internal	External	Internal	External
5	4	2	17.5	5
10	6	4	22.5	15
15	10	6	26	19
20	13	9	29	23
25	15	11	32	26
30	17	14	34.5	29
35	18	15.5	37	32
40	18.5	16.5	39	34
45	19	17.5	41.5	37.5
50	20	18	43	40
55	21	19	45	42.5

Brick Chimneys

6.31 Brick has been the traditional material for flues since the earliest times. However, in most instances, it is quite porous, as is the jointing mortar used, and it is, therefore, generally unsuitable for use. Since 1966, brick chimneys have been lined with one of the following:

- clay liners to BS EN 1457;
- concrete liners to BS EN 1857;
- metallic liners to BS 715.

Lining the flue allows gas appliances to be connected without fear of the internal fabrication deteriorating and being subjected to condensation problems. A lined brick chimney offers a very stable and effective flue system into which most appliances can happily discharge their products; more major problems are encountered where a flue is oversized for the installation in question. All new chimneys, including those for the installation of a gas fire, must have a minimum cross-sectional area of 120 cm^2 (12000 mm^2). Should the chimney pot be removed with a gas fire installation and a slab provided with an opening at opposite side outlets, they must still maintain the minimum 12000 mm^2 either side, as seen in Figure 6.20a.

Where condensation problems are likely, it is sometimes possible to line an existing brick chimney with poured/pumped concrete certificated by an accredited test house. This method is not permitted on new masonry chimneys, which must be constructed in line with current building regulations.

Connection to a Brick Chimney

6.32 The traditional construction of a chimney in a domestic dwelling consisted of a 225×225 mm cross-sectional void through which products could pass. Commonly used

Figure 6.20 (a) Slabbed over chimney

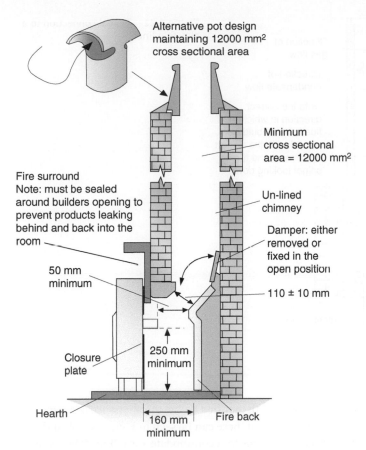

Alternative pot design maintaining 12000 mm² cross sectional area

Minimum cross sectional area = 12000 mm²

Fire surround
Note: must be sealed around builders opening to prevent products leaking behind and back into the room

Un-lined chimney

Damper: either removed or fixed in the open position

50 mm minimum

110 ± 10 mm

Closure plate

250 mm minimum

Hearth

160 mm minimum

Fire back

Figure 6.20 (b) Typical gas fire connection to brick a chimney

clay flue liners have an internal diameter of 175 mm. Both these sizes meet the minimum dimensions required. Where it is necessary to connect to the chimney, the first considera- tion is to ascertain its age and whether it has been lined. If it has been used for other fuels, it will certainly first require sweeping. Any dampers or restrictor plates, seen in Figure 6.20b, will need to be removed. It is possible to secure a damper in the open position where removal is difficult. For a simple gas fire, where the heat input is relatively small, it is permissible to install the appliance to an unlined chimney but, in general, for all other appliances flue lining is essential. Connections to clay flue liners can be achieved, as shown in Figure 6.21, by putting the pipe into the liner at a minimum distance of 150 mm. Alternatively, a flexible flue liner could be considered, as shown in Figure 6.23a–d.

Debris Collection Space

6.33 At the base of open-flued chimneys or large flues where several appliances may join a main flue, a debris collection space (void) should be provided with an access door or similar to allow for regular maintenance to inspect for potential flue blockage. Table 6.9

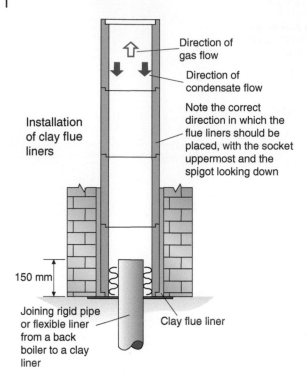

Figure 6.21 Typical connection to a clay flue liner

Installation of clay flue liners

Direction of gas flow

Direction of condensate flow

Note the correct direction in which the flue liners should be placed, with the socket uppermost and the spigot looking down

150 mm

Joining rigid pipe or flexible liner from a back boiler to a clay liner

Clay flue liner

specifies the volume of this void and the depth required below the appliance connection in relation to the type of installation. Examples of these can be seen in Figures 6.20b and 6.22. Ultimately, the void must be of sufficient volume to accommodate any falling debris, such as birds or falling mortar. It must be noted that adherence to relevant appliance manufacturer's instructions is key to also ensuring that the void is not oversized, as this can cause issues with the operation of the flue.

Table 6.9 Minimum void dimensions

Installation Details	Depth Below Appliance Connection	Volume
An appliance fitted to an unlined chimney	250 mm	0.012 m³ (12 l)
An appliance fitted to a lined brick chimney which is new or unused or has just been used with gas	75 mm	0.002 m³ (2 l)
An appliance fitted to a lined brick chimney which has previously been used with solid fuel or oil	250 mm	0.012 m³ (12 l)
An appliance fitted to a flue block chimney or a metal chimney which is new or unused or has just been used with gas	75 mm	0.002 m³ (2 l)
An appliance fitted to a flue block chimney or a metal chimney which has previously been used with solid fuel or oil	250 mm	0.012 m³ (12 l)

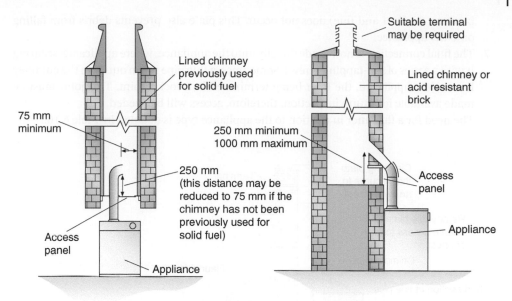

Figure 6.22 Connection to brick chimneys

Flexible Stainless Steel Flue Liners

6.34   These are specially designed, corrugated construction flue liners, which can be passed down through an existing chimney to provide a sound flue system of the correct diameter. These liners must not protrude from the flueway, and where the appliance is external to the chimney, a flue pipe is required to extend to the liner. There should be no joints throughout the length of the flue liner because it would not be possible to inspect them. As a result, care needs to be taken in measuring the length needed, as the liner should be in one continuous piece. The installation is carried out as follows:

1. The existing chimney pot, where applicable, is removed. It may be left if an approved pot/liner plate is used; see Figure 6.23a.
2. The chimney needs to be thoroughly swept to remove any previously accumulated soot deposits and to ensure that it is clear.
3. A rope is dropped down the flue system. This is achieved by tying a weight to one end and passing it through the flue, from above, Figure 6.23b.
4. With an end plug located on the liner end and tied to the rope, the liner is drawn up or down through the flue.
5. The top end is now secured with a clamping plate; a terminal is fitted and the whole area is suitably flaunched to drain the water away from the liner, see Figure 6.23c. Note: Three or more terminals in close proximity to each other must discharge at the same height.
6. A second clamping or debris plate is affixed at the base of the chimney, securing the liner firmly, see Figure 6.23d. The annulus space at the base must be effectively sealed, using mineral wool to ensure secondary flueing (combustion products passing up

between the liner and flue) does not occur. This plate also prevents debris from falling onto the appliance.

7. The final connection is now made directly onto the appliance, where applicable, securing it with a series of self-tapping screws. Sometimes, a flue pipe is run out from the chimney liner to the appliance, the liner being terminated in a socket joint. This joint must be made available for future inspection; therefore, access will be needed.

The need for a flue liner in relation to the appliance type is detailed in Table 6.10.

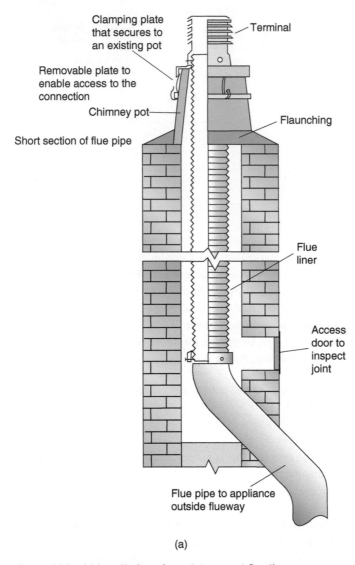

(a)

Figure 6.23 (a) Installation of a stainless steel flue liner

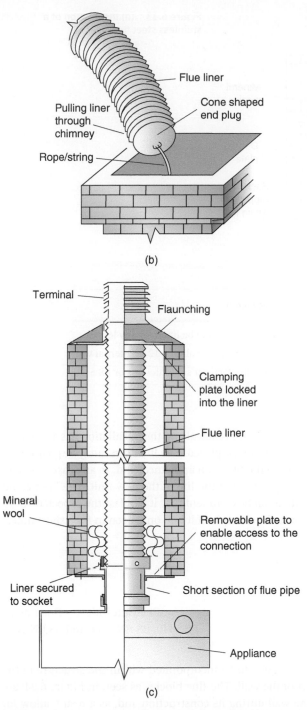

Flue liner

Pulling liner
through
chimney

Cone shaped
end plug

Rope/string

(b)

Terminal

Flaunching

Clamping
plate locked
into the liner

Flue liner

Mineral
wool

Removable plate to
enable access to the
connection

Liner secured
to socket

Short section of flue pipe

Appliance

(c)

Figure 6.23 (b) Installation of a stainless steel flue liner. (c) Installation of a stainless steel flue liner

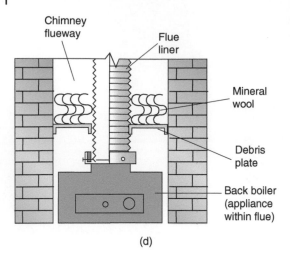

Chimney
flueway

Flue
liner

Mineral
wool

Debris
plate

Back boiler
(appliance
within flue)

(d)

Figure 6.23 (d) Installation of a stainless steel flue liner

Table 6.10   Need for a flue liner

Appliance type	Flue length
Back boiler and gas fire	The flue needs to be lined for any length
Gas fire (with or without a circulator)	>10 m external wall or >12 m internal wall
Other appliances	See manufacturer's instructions or BS 5440

Existing Flexible Liners

6.35   Generally, where an appliance is due for renewal and the installation has an existing stainless steel flue liner, it should preferably be replaced. The liner can rarely be guaranteed to last in good condition for longer than the life of an appliance, and, in any case, it would be unlikely to last for sufficiently long to see out the life of the new unit. The liner could be left, provided that the gas operative can be confident that it will last the expected 10–15 years that the new appliance may last, for example, if it had only been in for a short time.

Pre-cast Flue Blocks

Relevant Industry Documents
BS 5440, BS EN 1806 and BS EN 1858

6.36   The pre-cast flue block is designed to be incorporated within the structure of the building, forming an integral part of the wall. The flue blocks, as seen in Figure 6.24, are bonded into the block work of the wall during its construction and, as a result, allow for a greater volume within the living area. The flue is constructed from the base, where it is necessary to begin with the manufacturer's starter blocks, as shown. These are designed to provide the location for the appliance and support the chimney flue. Raking blocks are

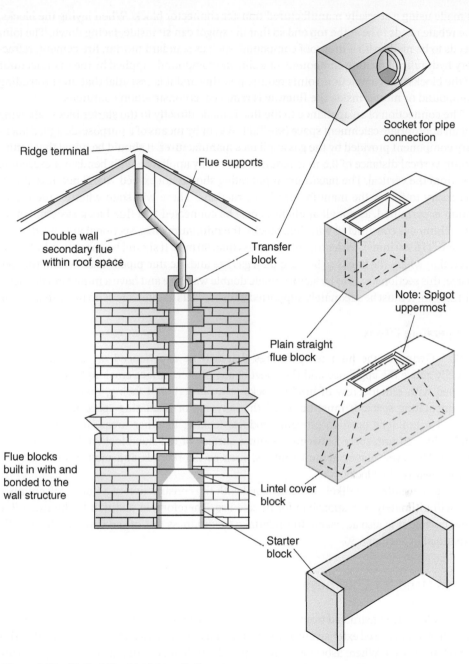

Figure 6.24 Typical flue block installation

available where a vertical rise is not achievable, however, they should be avoided where possible, and in all cases, the manufacturer's instructions should be followed. Invariably, when the flue blocks reach the roof space, the last section is run through this void using a twin wall flue pipe, terminating at the ridge with a suitable terminal; the connection to the blocks is made using a specially manufactured transfer connector block. When laying the blocks, the rebate needs to be at the top end so that the spigot can sit inside, facing down. The joint needs to be made with a fireproof compound, such as standard mortar, fire cement, refractory hydraulically setting compound, or a silicone compound supplied by the manufacturer of the blocks. On completion, joints require pointing and it is essential that any protruding compound or mortar inside the flueway is removed as construction continues.

The connection of an appliance to the flue is made directly to the starter block, allowing for an appropriate catchment space (see Table 6.9) or by means of a purpose-designed ancillary component provided by the gas appliance manufacturer. It should be noted that a minimum vertical distance of 0.6 m is required above any appliance with bends not exceeding 45° from the vertical. The maximum input rating should not exceed 70 kW net. It is always necessary to follow the manufacturer's instructions. These will state whether the installation is permitted, as not all appliances may be connected to a flue block system. For all new chimneys constructed with flue blocks, the minimum cross-sectional area needs to be 165 cm^2 (16500 mm^2), with no dimension less than 90 mm. It should be borne in mind, however, that where the flue is to be used for a gas fire and the flue pipe is used within the roof space, this section needs to be factory-made double wall flue and have a minimum diameter of 125 mm. It must be adequately supported with brackets at intervals of no more than 1.8 m.

Temperature Effects

6.37   Owing to the high temperatures that will be experienced on the wall surface directly above the appliance and the effect of any plaster surface cracking, the flue should be clad with either a row of bricks/blocks or a plasterboard facing, maintaining an air gap/insulating space. It must be noted that any gap or insulating space must be sealed around a builder's open to prevent products leaking to other areas. It is desirable to maintain a 50 mm space between the inner surface of the flue block and any structural timber. However, non-structural timbers, such as floorboards and picture rails, may be placed against the blocks.

The gas installer is unlikely to be involved with the construction of a flue block chimney but will ultimately be responsible for its safe use. Therefore, in addition to the usual flue flow testing, it is also advisable to undertake a soundness test of the system. This is fully explained in Section 6.50.

Pipes Used for Chimneys

6.38   Pipes are constructed from a variety of metallic and non-metallic materials. Asbestos cement was once used extensively until several safer alternatives were found, and today, this is no longer used. Where asbestos is encountered within an existing installation, great care needs to be observed because of its hazardous nature, and information should be sought from the Health and Safety Executive website when removal or disturbance is to be considered. Metallic flue pipes are manufactured from a variety of metals, including stainless steel, cast iron, enamelled pressed steel and aluminium, and are supplied as single or double (twin) walled. Metal pipes should be supported at intervals no greater than 1.8 m and at changes of direction.

6.39 Single-walled pipes should no longer be used for external applications; the space within the roof void is also regarded as an external environment. However, an uppermost protrusion through the roofline is permitted, provided the distance is kept to a minimum. When making connections to single-walled flue systems, the method of jointing is often by means of a spigot and socketed joint, see Figure 6.25. Should this be the case, it is essential that the socket is installed to the top end and that the spigot sits inside, facing down into the fitting. This is to ensure that any jointing medium used to assist in making the joint does not fall out and that any condensation flows back down the pipe internally.

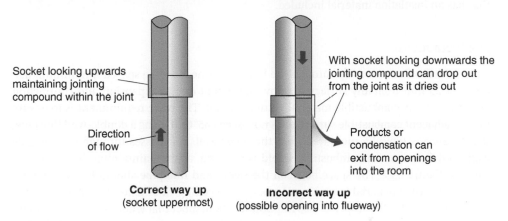

Figure 6.25   Direction of flow for single-walled flue pipe

6.40 Double-walled flue pipes consist of two pipes, one secured inside the other, trapping an air space between the concentric void, see Figure 6.26. This increases its thermal

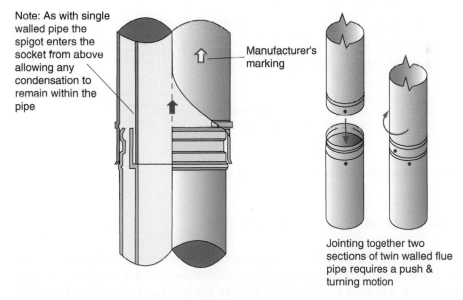

Figure 6.26   Direction of flow for twin-walled flue pipe

ability to keep the heat within and so overcomes many of the condensation problems associated with single-walled flues. It is essential that this design of flue pipe be assembled in accordance with the manufacturer's fixing instructions, ensuring it is installed the correct way up. Connection is usually made by inserting the male coupling into the female coupling of the next component and giving the pipe a slight twist, thus interlocking the two sections. These flues may be used externally. However, the distance is restricted to 3 m unless additional insulation is provided to overcome the problems associated with excessive flue cooling. It is possible to purchase a double-walled flue pipe for external applications that has an insulation material included.

Fire Precautions

6.41 As the flue passes up through the building, it needs to be spaced a minimum distance of 25 mm from any surface to allow movement due to expansion and contraction and prevent any combustible material getting too hot. The basic requirement is to ensure that an adjacent combustible surface does not exceed 65°C. If using a double-wall flue pipe, the distance is typically measured from the inner wall. If it passes through combustible walls and floors, a non-combustible shield is required, again maintaining the 25 mm air gap, see Figure 6.27. The space between the sleeve and flue pipe should be filled with a non-combustible material (e.g. mineral wool) and a ceiling/floor plate should be provided to ensure a suitable fire stop. It is also important to remember that where a chimney system, such as a flue pipe, passes through a different dwelling from the one that it serves, it must be enclosed within a non-combustible material.

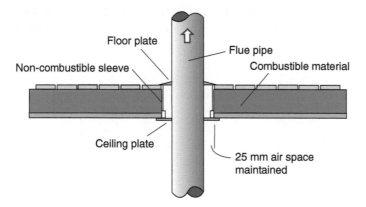

Figure 6.27 Passing flue pipes through combustible surfaces

Fan Draught Open-Flue Chimney Systems

6.42 Unlike the natural draught open-flue systems that work on convection currents, fan draught open-flue chimney systems have the additional motive force of a fan to assist in the extraction of flue gases from the building. The fan may be located either before the appliance (forced draught) or after it (induced draught); see Figure 6.28. Where a fan has

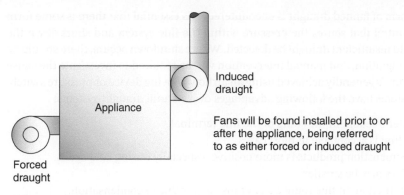

Figure 6.28 Fan location

been located within the flue system away from the appliance, it is invariably because there was some difficulty in designing the flue system to an acceptable standard and possibly the flue route fails to meet the required design criteria. The mechanical assistance of a fan can help overcome horizontal runs of flue and various restrictions such as flue diameter and an excessive number of bends. To cut down on undue draughts, the air supply from outside should ideally be located next to the appliance. Figure 6.29 shows the various British Standard flue classifications in relation to the positioning of a fan on an open-flue system.

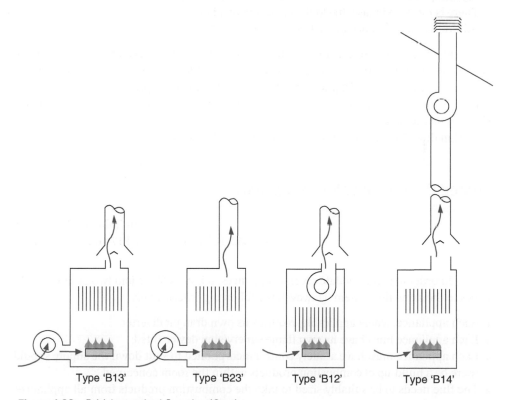

Figure 6.29 British standard flue classification

Where any form of fanned draught is encountered, it is essential that there is some form of automatic control that senses the pressure within the flue system and shuts down the appliance should insufficient draught be detected. Where shutdown occurs, there should be no automatic re-ignition, and manual intervention will be required to investigate the cause of fan failure. This is generally achieved using an airflow-proving device or pressure switch.

Fan-flued systems have the following advantages over natural draught systems:

- There is greater freedom in the siting of the flue terminal.
- Wall termination is acceptable in most cases.
- Removal of combustion products is more positive, especially during cold spells.
- Flue outlet sizes may be smaller.
- There is no restriction on flue route except those applicable to condensation.
- The appliance design can allow for more efficiency, with the possibility of plastic flues being used.
- Greater flue dilution can be achieved.

The disadvantages include the following:

- There are more things to go wrong, and additional safety features are required.
- It can be more expensive.
- They tend to depressurise the room, causing products and smells to be drawn in from adjoining rooms.
- There is increased noise due to the operation of a fan.
- Additional maintenance to the fan unit is required.

Where a fan is to be selected for inclusion into a flue system, several factors need to be considered, including the volume of flow to be allowed for, as well as the dilution air and the likely static pressure drop within the system, about which specialist advice should be sought. Some thought is also needed about the location of the fan so that it is easily accessible for maintenance.

For fan draught open-flue terminal locations, see Section 6.60.

Shared Open-Flue Chimney Systems

Relevant Industry Documents
BS 5440, BS 6644, IGEM/UP17 and IGEM/UP/10

6.43   In order to accommodate several appliances, it is possible to join two or more appliances to the same flue system. However, the following criteria apply:

- Each appliance, where applicable, requires its own draught diverter.
- Each appliance must have its own flame supervision device fitted.
- Each appliance must have a safety control incorporated to shut down the supply should there be a build-up of combustion products within the room concerned.
- The flue needs to be suitably sized to take the combustion products from all appliances connected to the main flue.

- Access needs to be provided to the main flue system.
- Advice on installing several appliances to the same fluc system should always be sought from the manufacturer.

Common Flue System: Appliances within the Same Room (Modular Flue)

6.44   With this system design, all appliances are installed within the same room and have the same burner system. Either all are atmospheric or all forced draught, see Figure 6.30. Gas-burning appliances should not be discharged into flue systems used for solid fuel; however, it is possible that a mix of gas and oil-burning appliances are permitted together in the same system, provided that this is in agreement with the manufacturer's instructions. Where one appliance is to operate for longer periods than the others, it should be positioned nearest to the main flue. If the flue is to operate under natural draught, the maximum number of appliances fed into the same horizontal branch of a flue should be restricted to six, and the minimum height above each draught diverter needs to be 500 mm. However, if the appliances are fed into a vertical chimney/flue, it is advised to limit the number to eight. If more than eight are being considered, it would be advised to ensure the flue is fan-assisted.

Figure 6.30   Common flue systems
(modular flue)

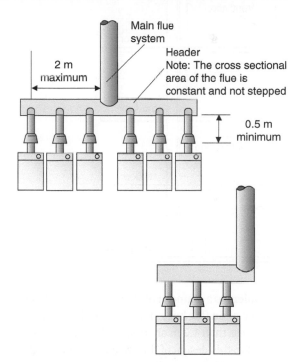

Main flue system
Header
Note: The cross sectional area of tho flue is constant and not stepped
2 m maximum
0.5 m minimum

Branch Flue System: Appliances Installed on Different Floors (Shunt Duct)

6.45   This is a design of shared flue primarily restrictcd to natural draught appliances with limited output, such as domestic boilers, water heaters or gas fires, see Figure 6.31. The main flue must run up through the internal structure of the building, not form part of the external wall and should have a minimum cross-sectional area of 400 cm^2 (40000 mm^2), which is

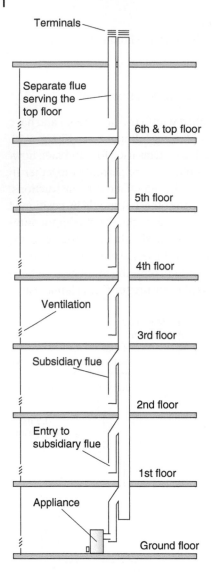

Terminals

Figure 6.31 Branch flue system (shunt duct)

Separate flue
serving the
top floor

6th & top floor

5th floor

4th floor

Ventilation

3rd floor

Subsidiary flue

2nd floor

Entry to
subsidiary flue

1st floor

Appliance

Ground floor

equivalent to a diameter of about 230 mm. Individual appliances must not discharge directly into the main flue but join via a subsidiary flue with a minimum height above the appliance of not less than 1.2 m or, for a gas fire, this distance is increased to 3 m. All appliances connected to this design of flue system must be of the same type (e.g. all gas fires) as indicated in Table 6.11. Any appliances connected to such a system need to be suitably labelled, as seen in Figure 6.32, to indicate that they are installed into a shared flue system; where a replacement appliance is to be installed, it should be of a similar size.

6.46 The responsibility for the main flue system lies with the landlord/owner of the building who must ensure that the complete shared flue system is inspected and tested at intervals of no greater than 12 months.

Table 6.11 Appliances discharging into the subsidiary flue of a branch flue system

| Appliance type | Minimum cross-sectional area of main flue | | | |
| | >400 cm² and < 620 cm² | | ≥620 cm² | |
	Total input (kW)	Maximum appliances	Total input (kW)	Maximum appliances
Fire	30	5	45	7
Instantaneous water heater	300	10	450	10
Storage water heater, boiler or air heater	120	10	180	10

Figure 6.32 Typical label affixed to an appliance warning that it forms part of a shared flue system

Checking and Testing Open-Flue Systems

Relevant Industry Document
BS 5440

6.47   The responsibility that the gas engineer takes for connections to any flue system features highly in the Gas Regulations. Consequently, any operative who fails to recognise fully the dangers of a poor flue system may find himself or herself in contravention of the law, but more importantly, they may install appliances that are potentially dangerous to the occupants of the building. The checks and tests listed below need to be considered every time you work on an appliance and, where necessary, completed in full.

They include visual inspection, flue flow tests, soundness tests, and spillage tests.

Visual Inspection

6.48   A good visual inspection of a flue system invariably identifies potential problems before they are experienced. To assist in the completion of this task a checklist may be followed, an example is shown in Part 8: Figure 8.2. The main points to look for include:

- that it is constructed of the correct materials and complete throughout its entire length, with the flue taking a suitable route;
- internal inspection should reveal no dampers, unless fixed open, register plates or other potential blockages;

- that the flue serves only one location unless it is a shared flue system and the rules in Section 6.43 have been followed;
- that the flue outlet is correctly sited and a terminal fitted if applicable; and
- that there is suitable ventilation.

Flue Flow Test (Integrity Test)

6.49   This test confirms the integrity, in particular, that there is no leakage throughout the route of the flue and that there is only one opening at the bottom, where products enter and one at the top, where the products disperse. It is achieved using a smoke pellet that can produce a minimum volume of 5 m³ of smoke in 30 seconds, see Figure 6.33, where necessary, more than one pellet can be used. The method to adopt for this test is as follows:

1. Close all windows/doors to the room in which the flue is to be checked.
2. Check with an ordinary match to see if there is an evident updraught, if not, introduce heat into the flue by means of a blowlamp until an updraught is present.
3. Ignite a smoke pellet and place it in the base of the flue. In the case of a gas fire, if a closure plate is supplied, this should be placed in situ.
4. The test can be deemed satisfactory if the smoke is seen to rise up into the flue, with no significant escape into the room, and can be seen clearly discharging only from the correct terminal outlet. In addition, during the discharge of smoke all accessible joints, including those in roof voids and cupboards, etc., should be examined for possible leakage.

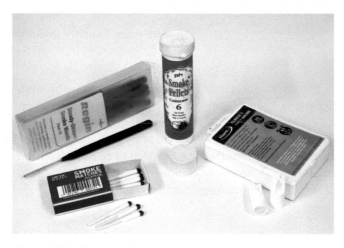

Figure 6.33   Smoke pellets and matches

Soundness Test

6.50   This test will certainly not be required on every occasion, and for many flue systems, their integrity may be without question. However, should you have any doubt as to the condition of a flue then this is a test that must be undertaken as follows:

1. With the flue open at the top and bottom and the flue warmed to assist in the creation of an updraught, insert a smoke pellet/s at the base.
2. When smoke is seen to discharge from the top while the pellet is still issuing smoke, temporarily close off both the base and top.
3. Any joints from which smoke can be seen to issue can now be identified as unsatisfactory.

Spillage Test

6.51 The spillage test is designed to check if the products of combustion are being safely carried to the outside under the worst possible conditions. It is carried out as follows:

1. Close all windows/doors to the room in which the appliance is installed.
2. Switch on any fans within the room, setting these to maximum. If the room contains any other open-flue appliances, including solid fuel and oil, these should also be ignited as they also draw air from the room.
3. After the appliance has been alight for some 5–10 minutes, and with the appliance in operation, test for spillage in accordance with manufacturer's instructions. Where these are unavailable, simply hold a lighted smoke match just inside the draught diverter and run it along its entire lower edge, see Figure 6.34. All smoke, apart from an odd wisp, should be kept within the flue.
4. Should there be any fans in an adjoining room, the test should now be repeated with the adjoining door open and the fans running.

Figure 6.34 Spillage test to a gas fire

Where spillage occurs, it may be possible to leave the appliance running for a further period to see if the problem rectifies as the flue heats up. It may be that the extractor fan is creating a lack of pull. This can be determined by turning off the fan, or opening a window slightly, to see if the flue now clears the smoke, in which case it suggests that more ventilation is needed.

Do not leave an appliance that fails the spillage test in operation. Treat the situation as immediately dangerous.

Room-Sealed Flue Chimneys

6.52   The room-sealed system takes the air required for combustion from outside the room in which the appliance is situated. The room-sealed appliance may or may not be in balance. The term balanced flue refers to the position of the flue terminal in relation to the air supply inlet. For example, in Figure 6.35, the room-sealed appliance is in balance, however, the vertex flued appliance covered in Section 6.59 is room-sealed, but it is not in balance because the air inlet is not directly adjacent to the flue outlet.

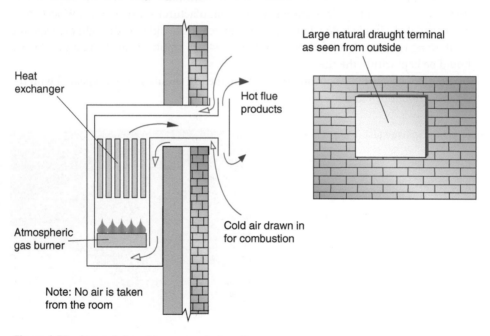

Figure 6.35   Natural draught room-sealed appliance

Natural Draught Appliances

6.53   The terminal outside a building is located in a positive or negative pressure zone, depending on the wind direction. The air inlet is also in this zone, with the pressure differential between each as zero. Thus, the only motive force required to expel the flue gases from the appliance to the outside is that obtained via the heat rising due to convection currents within the heat exchanger. Natural draught systems have been used for many years because of the simplicity of their design. However, with the need for improved efficiencies and the wish to site appliances away from the external wall led to the incorporation of a fan to produce a more positive motive force and have made natural draught units less desirable. Also, where a natural draught appliance is used, the size of the flue terminal is much larger.

Balanced Compartments

6.54   This is a special arrangement in which an open-flued appliance has been installed within a small room or compartment that is completely sealed from any adjoining room. All air openings to the compartment are taken from a point outside, adjacent and within 150 mm below the base of the flue terminal. The air intake is at a high level only for appliances greater than 70 kW net input, see Figure 6.36. For appliances up to 70 kW, the air supply can be at high level only, low level only or at high and low level. Where low-level air is used, it should be ducted to within 300 mm of the floor. Where the ceiling height of the balanced compartment is greater than 300 mm above the base of the skirt of the appliance draught diverter, a high-level opening should also be provided. The opening can be by means of a Y or T piece inserted into the air supply duct. The cross-sectional area of the inserted piece should be the same as the low-level duct, see Figure 6.37. For vent sizes, see Part 7: Ventilation. Access to the compartment is made via a secured panel or locked door, designed so that if it is opened, the appliance will cease to operate. It would incorporate a switch and have a self-closing attachment. Draught excluders would also be fitted around the sides of the door, including the base, to seal further the space within. On the outside of the access door should be affixed a notice stating that the door is to be kept closed at all times. These compartments are generally large enough for the gas operative to work in

Figure 6.36   Balanced compartment high-level ventilation

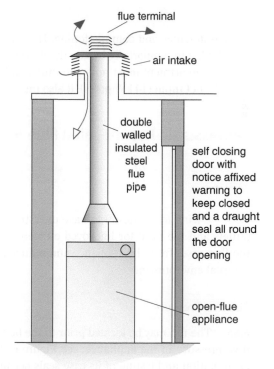

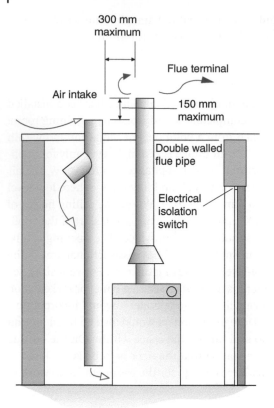

Figure 6.37 Balanced compartment option for appliance ≤70 kW net

for maintenance and commissioning. However, if insufficient room is available, then it is permissible to temporarily bypass the door switch to complete the work. The access door to the compartment must not open into a bath/shower room, and if the appliance is over 12.7 kW net input (14 kW gross), it also must not open onto sleeping accommodation.

Fan-Assisted Room-Sealed Flue Appliance

Relevant Industry Document
BS 5440 and BS 7967

6.55   Owing to the desire to have unrestricted siting of appliances, smaller termination outlets and the need for improved efficiencies, appliances need to rely on a fan to force the flue gases from the appliance and induce the air required for combustion to/from the external environment.

Positive Pressure

6.56   The fan may be located prior to the burner, giving a forced draught, creating a positive pressure in the appliance; see Figure 6.38a and b. This appliance requires rigorous examination and testing of its case seals because any leakage has the potential for fumes to escape into the living space. The procedure to check the integrity of the case seals on a positive-pressure boiler is as follows:

- Prior to replacing the case, examine the boiler, check for water leaks, and ensure the backplate is in sound condition, that it is not distorted, and there is no corrosion. A prescribed method to test a corroded area is to use a screwdriver with sharp knocks to see if it perforates. If there are signs of corrosion, but it is sound, advise the gas user of the potential risk.
- Check case seals, grommets and gaskets are intact and that nothing is likely to get trapped between the seal and the case when it is replaced; see Figure 6.38c.
- Look for signs of discolouration around the appliance.
- Replace the case, ensuring nothing is fouling the seal, screws and fastenings, then locate and tighten sufficiently.
- Light the appliance on its highest setting.
- Run your hand around the casing and backplate to feel for any escaping fan pressure.
- With a flue gas analyser, perform a sweep test by slowly sampling all around the outside of the appliance at a distance of approximately 100 mm. The test should last at least two minutes and include at least two passes around the appliance.

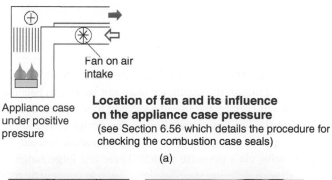

Fan on air intake

Appliance case under positive pressure

Location of fan and its influence on the appliance case pressure
(see Section 6.56 which details the procedure for checking the combustion case seals)

(a)

(b)   (c)

Figure 6.38 (a) Fan located prior to the burner. (b) Positive pressure boiler with the front case removed. (c) Grommets must be intact and in place

- Check any disturbed gas joints with leak detection fluid to confirm that there are no gas escapes and then light an ordinary match or taper and take it around the casing, looking for disturbance of the flame.
- If leakage of products is found, repair as necessary and retest. If unable to remedy, follow the Gas Industry Unsafe Situations Procedure.

Negative Pressure

6.57 Conversely, the fan may be placed after the heat exchanger, causing an induced draught through the appliance, resulting in a negative pressure inside the casing, see Figure 6.39.

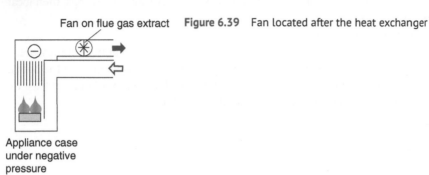

Fan on flue gas extract

Figure 6.39 Fan located after the heat exchanger

Appliance case under negative pressure

6.58 Whenever a fan is incorporated with the system it is imperative that, should it malfunction or fail to create the necessary draught, the appliance must not be able to become established and burn the fuel. This is achieved by the use of a sensing tube, which detects a positive or negative pressure within the flueway, depending on the design of the appliance. This pressure is used to hold open a gas valve, where a zero governor is used, or allows electricity to flow to the gas valve via a pressure switch. There is a large range of designs in fan-assisted appliances, all with differing specifications. Such as concentric rear or side exit horizontal flues, see Figure 6.40a, and concentric or twin pipe vertical flues, see Figure 6.40b. Ultimately, in all cases, the maximum flue distances and the total number of flue bends permitted will need to be referenced from the appliance manufacturer's instructions.

Vertex Flued System

6.59 This type of flue system falls within the category of C7 (see Table 6.1 for flue classification). It is a strange mix in that the appliance and flue, up to the draught break is room-sealed and after the draught break in the roof void, it converts to open-flue to exit through the roof, see Figure 6.41. Accordingly, it is necessary for a spillage test to be performed at the opening in the draught break to test the open flue.

Air is supplied to the roof void with a minimum free air size, as indicated by the manufacturer of the appliance. This air is then drawn into the flue system via an opening in the flueway; it can also enter a duct at this point and pass down via a concentric flue (air passing down the outside void) to the appliance in a room below. Thus, no air is drawn from

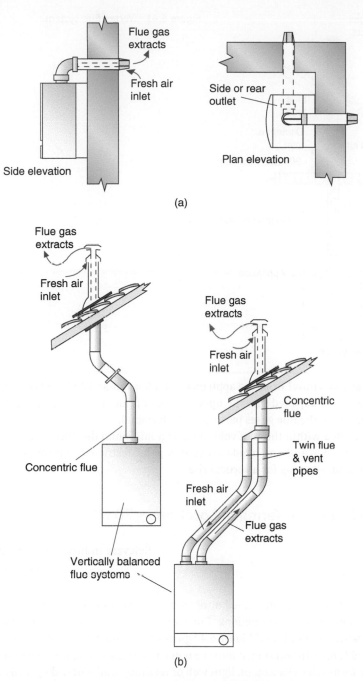

Figure 6.40 (a) Horizontal flue systems. (b) Vertical flue systems

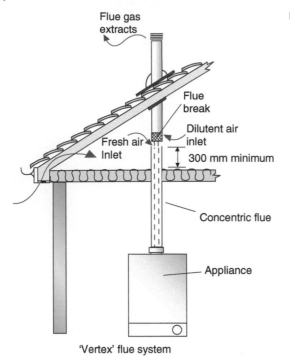

Figure 6.41   Vertex flue system

'Vertex' flue system

the room itself. The hot gases travel from the appliance, which is fan-assisted, and pass up through the central flue to be expelled into the upper section of flue. Diluent air is drawn in at this point and from here the flue gases pass up to the terminal.

The draught break located within the roof void must be a minimum distance above the insulation material of 300 mm,and any bends above this point restricted to a minimum distance of at least 600 mm, allowing for a vertical rise.

Room-Sealed and Fan Flue Terminal Locations ≤70 kW

Relevant Industry Document
BS 5440

6.60   The terminal location for a room-sealed or open 'fan' flued appliance should be in accordance with the manufacturer's instructions. Table 6.12 gives sufficient information to enable you to comply with BS 5440-1, and Figure 6.42 gives a visual illustration of the measurements shown in Table 6.12. In addition, it must be noted that flues should not terminate into an enclosed space formed by a basement, light well or retaining wall where dispersal of combustion products may be problematic. Where this is unavoidable, the terminal must be no more than 1 *m* below the top level of the basement area so that combustion products can disperse freely in the open air. Where a basement area is created by a single retaining wall, which forms an uncovered passageway open at both ends, termination may be acceptable, providing the passageway is at least 1.5 *m* wide.

Table 6.12   Terminal locations for room sealed and open flued fan draught appliances

	Location	Heat input (net)	Room-sealed Nat. draught	Room-sealed Fan draught	Open-flued Fan draught
A	Directly below an opening (e.g. window or air brick)	0–7 kW	300 mm	300 mm	300mm
		7–14 kW	600 mm		
		14–32 kW	1500 mm		
		32–70 kW	2000 mm		
B	Directly above an opening (e.g. window or air brick)	0–32 kW	300 mm	300 mm	300 mm
		32–70 kW	600 mm		
C	Horizontally to an opening (e.g. window or air brick)	0–7 kW	300 mm	300 mm	300 mm
		7–14 kW	400 mm		
		14–70 kW	600 mm		
D	Below gutters and pipes	0–70 kW	300 mm	75 mm	75 mm
E	Below eaves	0–70 kW	300 mm	200 mm	200 mm
F	Below balconies or car ports	0–70 kW	600 mm	20 0mm	200 mm
G	Horizontally to vertical pipes	0–5 kW		75 mm	
		5–70 kW	300 mm	150 mm	150 mm
H	Horizontally to corners	0–70 kW	600 mm	300 mm	200 mm
I	Above ground or flat surface	0–70 kW	300 mm	300 mm	300 mm
J	Surface facing a terminal	0–70 kW	600 mm	600 mm	600 mm
K	Terminal facing terminal	0–70 kW	600 mm	1200 mm	1200 mm
L	Car port opening to inside	0–70 kW	1200 mm	1200 mm	1200 mm
M	Vertical to 2nd terminal on wall	0–70 kW	1500 mm	1500 mm	1500 mm
N	Horizontal to 2nd terminal	0–70 kW	300 mm	300 mm	300 mm
O	Vertically above the roof line	0–70 kW	N/A	300 mm	150 mm
P	Horizontal from upright structure on the roof	0–70 kW	300 mm	300 mm	300 mm
Q	Horizontal from MVHR intake	0–70 kW	1000 mm	1000 mm	N/A
R	Diagonally from an opening in a building on a different wall	0–70 kW	600 mm	600 mm	N/A
S	Distance between two vertical terminals	0–70 kW	600 mm	600 mm	N/A
T	Vertical terminal to an opening	0–70 kW	1500 mm	1500 mm	N/A
U	Vertical terminal from a wall	0–70 kW	500 mm	500 mm	N/A
V	Terminal from adjacent boundary	0–70 kW	300 mm	300 mm	300 mm
W	Terminal facing boundary	0–70 kW	600 mm	600 mm	600 mm
X	Terminal to opening in facing building	0–70 kW	2000 mm	2000 mm	2000 mm
Y	Below Velux window	0–70 kW	2000 mm	2000 mm	2000 mm
Z	Beside a Velux window	0–70 kW	600 mm	600 mm	600 mm

Note: In addition to the table above, no fan draught terminal should be closer than 150mm to any opening in the building fabric, such as a non-opening window. This distance remains at 300mm for natural draught terminals.

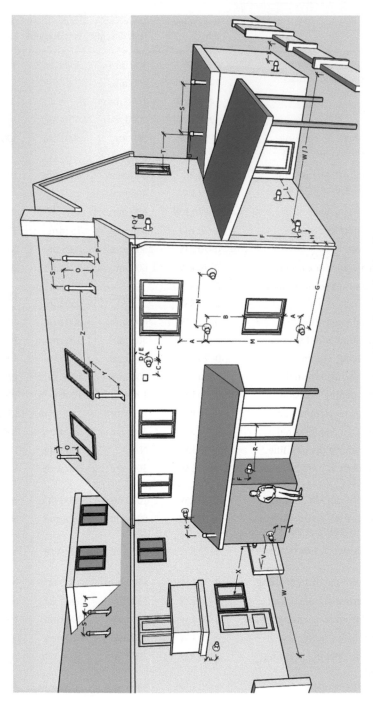

Figure 6.42 Minimum fan-assisted open-flue and room sealed terminal positions. Credit: Geoffrey Eaton.

Where a flue terminates into an inner courtyard area, such as that surrounded by neighbouring flats, consideration needs to be given to the potential accumulation of products of combustion. For new installations or complete building refurbishments new appliances must be connected to a communal flue system (see Section 6.71). However, where an appliance is installed in an existing building the flue can terminate into an inner courtyard providing the height of the structure surrounding the courtyard does not exceed the dimension across the narrowest part of the courtyard. Where the height of the surrounding structure is greater, the appliance shall be connected to a communal flue system. An exception to this could be where the installation is on the upper floor, and it can be determined that the terminal will be no more than 1 *m* below the top level of the structure.

If you encounter an existing installation on a service visit, for example, provided the appliance passes your checks in compliance with Regulation 26:9, it can be left operational.

Siting of Terminals in Relation to Boundaries

6.61   Any terminal outlet should be at least 600 mm from any boundary it is facing. This distance can be reduced to 300 mm, where the terminal is not facing the boundary but running parallel to it. The combustion products should not cause a nuisance to adjoining or adjacent properties and it is recommended that any terminal from a fan-flued appliance is not allowed to discharge combustion products across adjoining boundaries and, where it does occur, a distance of 2 m is maintained from any opening directly opposite. Permission from the local authority may be required if the terminal is to project into an area of public right of way.

Special caution needs to be considered in relation to possible future extensions to adjoining or adjacent properties. For example, if a neighbour extends their property to the boundary line it may adversely affect the performance of your flue system. Thus, the principle that needs to be adopted when selecting a suitable terminal position is to imagine that the neighbouring property has already been extended and therefore if it affects your minimum distance, as shown in Table 6.12, the location should not be chosen.

Terminal Guards and Protection of Combustible Materials

6.62   Care needs to be taken to ensure that no heat damage will be caused. This would include locating a terminal guard around a room-sealed natural draught terminal that is within 2.1 m of any accessible surface. For a room-sealed fan-assisted terminal, the measurement is 2 m; see Figure 6.43a. A minimum space of 50 mm between the guard and terminal needs to be maintained, and they must have no opening bigger than a 16 mm ball. If the terminal is located close to any combustible surface, such as a gutter, soffit or fascia board, a heat shield with a minimum 1 m length will be required to protect the material from damage, see Figure 6.44. Sometimes, the terminal is located where the external surface construction itself is of combustible material; if this is the case, it will be necessary to protect the surface by using a non-combustible plate behind the terminal for a minimum distance of 25 mm beyond the external edges of the terminal.

Terminal guard required if less than:

- 2 m for a room-sealed fan draught

- 2.1 m for a room-sealed natural draught

No part of the guard to be nearer than 50 mm to any part of the flue

(a)

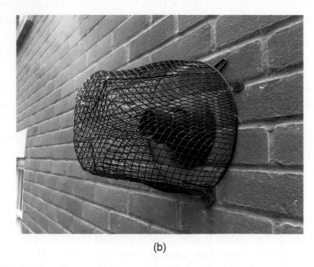

(b)

Figure 6.43 (a) Use of a terminal guard. (b) Photo of terminal guard

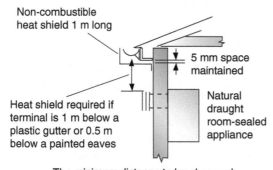

Non-combustible heat shield 1 m long

5 mm space maintained

Heat shield required if terminal is 1 m below a plastic gutter or 0.5 m below a painted eaves

Natural draught room-sealed appliance

The minimum distance to be observed before a heat shield is required

Figure 6.44 Use of a heat shield

Room-Sealed and Open-Flue Terminal Locations >70 kW

Relevant Industry Document
IGEM/UP/10

To an Opening in a Building

6.63   As with domestic appliance installations, for non-domestic, there are specific clearances required for a flue terminal to any opening into a building. However, in this case, to find the correct clearance, a calculation is performed related to the appliance net heat input. Table 6.13 provides the formulas, and Figure 6.45 illustrates where the formulas are used.

Table 6.13   Minimum distance between open-flued/room-sealed terminal and an opening into a building for appliances with heat input >70 kW

Ref. from Figure 6.45	Appliance type	Formula
V	Open-flued (natural or fanned draught)	$9.5156 \times (\text{net heat input kW}) + 833.91$
W	Room-sealed – natural draught	$1.9031 \times (\text{net heat input kW}) + 1866.8$
Y	Room-sealed – fanned draught	$7.232 \times (\text{net heat input kW}) + 93.708$

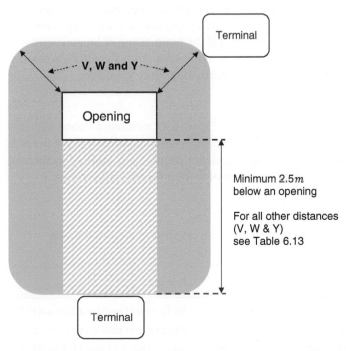

Figure 6.45   Minimum distance of a terminal to any vent or opening into a building

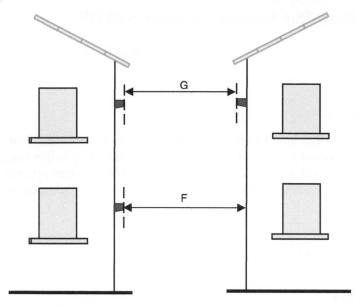

Figure 6.46 Minimum spacing for room-sealed fan flued terminals from facing surfaces and other terminals. Source: Data from IGEM/UP/10

Terminal Clearances From a Facing Wall, a Facing Terminal and Side Walls When in a Recess

6.64 If a terminal position has a surface or another terminal opposite, as illustrated in Figure 6.46, there are minimum clearances that must be obtained. Table 6.14 gives relevant clearances with references G and F. For example, if the input rating of an appliance was 150 kW net and there was a surface facing the flue termination, the calculation to confirm the minimum distance from the end of the terminal to the surface would be:

$$23.126 \times (150) - 618.84 = 2850.06 \text{ mm}$$

If a terminal is to be located within a recess, see Figure 6.47, minimum clearances must be calculated using the relevant formula. For example, if a terminal from a 180 kW net heat input appliance is in a recess, the following calculation would be completed to find the minimum clearance from either side of the terminal to the adjacent wall surfaces:

$$7.232 \times (180) + 93.708 = 1395.468 \text{ mm}$$

Table 6.14 Minimum spacing for room-sealed fan draught terminals to surfaces and other terminals >70 kW

Reference	Flue position	Formula
G	Opposing Terminal (Figure 6.46)	$19.32 \times (\text{net heat input kW}) + 647.59$
F	Opposing Flat Surface (Figure 6.46)	$23.126 \times (\text{net heat input kW}) - 618.84$
Z	Recessed between two or more vertical walls (Figure 6.47)	$7.232 \times (\text{net heat input kW}) + 93.708$

Source: Data from IGEM/UP/10

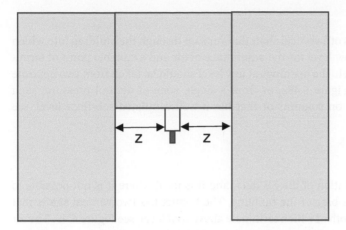

Figure 6.47 Horizontal flue terminals in a recess between two or more vertical walls

It must be noted that even where the correct clearances have been achieved, if there is any doubt about the clearance of products due to factors such as depth of recess, height of building or windows within, etc., plume management kits or vertical flue options must be considered as an alternative to ensure the safe dispersion of products.

Shared Room-Sealed Flue Systems

Relevant Industry Document
BS 5440 and IGEM/UP/17

6.65   Shared room-sealed flue systems within buildings have evolved; originally, there were two specific designs of shared flue that could be encountered: the 'SE' duct and the 'U' duct. They were first developed to allow several room-sealed appliances to share a single flue system within a multi-storey building. Standard efficiency boilers, water heaters and gas fires, designed especially for this type of flue, were installed. The flue system works via the principle of the flue flow effect. Due to the high temperature of the products from the standard efficiency appliances, the flue gases were drawn to the centre of the common flue and rose in a central column. Combustion air is then naturally drawn up the duct wall. This explains why the appliance flue was designed to terminate centrally in the shared flue duct to disperse its hot products and the cool combustion air intake was flush with the inner wall of the duct to receive an air supply.

The introduction of condensing appliances with a much lower flue product temperature renders this type of flue system inadequate because it inhibits the flue flow effect and has the potential to contaminate the outer air supply with combustion products. More recently, however, a new type of shared flue, designed for modern condensing boilers as well as standard efficiency, has entered the market and is now being installed in new multi-occupancy buildings. These systems are called communal flue systems (CFS). Although, the old systems have now been superseded by CFS, it is possible for an operative to encounter appliances connected to the older 'SE' duct and 'U' ducts, therefore, it is necessary for operatives to be able to identify these types of systems; Sections 6.66 to 6.70 give an overview of these. CFS are covered from Sections 6.71 to 6.75.

SE-duct System

6.66 The SE-duct consists of a vertical shaft that runs up through the building into which openings are provided at low level for the admittance of air and a suitable point of termination located at a high level. The openings at low level should be taken from two opposite sides of the building, see Figure 6.48a, or from a single zone of neutral pressure, such as in a building supported on columns or that has a well-ventilated sub-floor level, see Figure 6.48b.

U-duct System

6.67 This design is a variation of the SE-duct, and it is ideal where it is not possible to obtain suitable air from the base of the building. The U-duct has two vertical shafts that run through the building, both of which terminate above roof level, see Figure 6.49. The air for combustion is drawn down through one leg into which it is essential that no appliances are connected.

Flue Design

6.68 The terminal should be positioned away from all other structures on the roof and be at a suitable height as previously specified in Section 6.25 for open-flue terminals. The size of the duct is dependent on the appliances to be installed into the flue system.

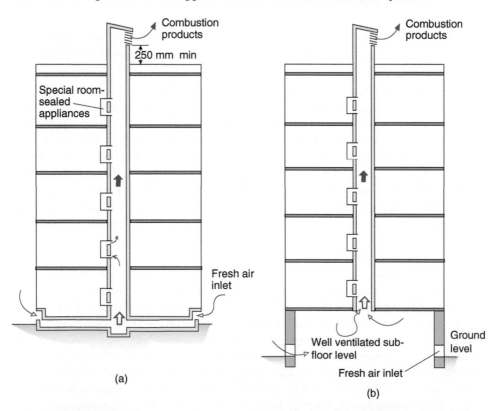

Figure 6.48 (a) Se-duct with openings at low level taken from two opposite sides of the building. (b) Se-duct with a single zone of neutral pressure in a well-ventilated sub-floor level

Figure 6.49 U-duct shared room-sealed system

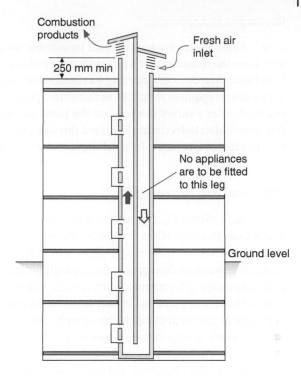

Combustion products ►

Fresh air inlet

250 mm min

No appliances are to be fitted to this leg

Ground level

Inspection and Maintenance

6.69 It is the responsibility of the building owner or landlord to ensure that the flue system remains in a safe condition, and annual inspection and testing should be carried out in order to comply with the current Gas Regulations. Appliances connected to these flue systems will also require annual maintenance checks, as inadequate installation may affect the safe operation of other appliances. Note: All appliances and ventilation systems must be suitably labelled, stating how they are part of a shared system; see Figure 6.50. If you are undertaking a routine service, you do not need to examine the entire flue system.

Figure 6.50 Typical labels on a shared flue system

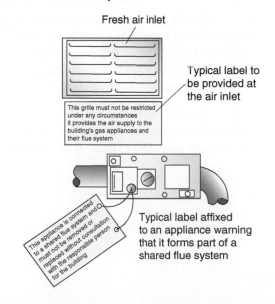

Fresh air inlet

Typical label to be provided at the air inlet

This grille must not be restricted under any circumstances it provides the air supply to the building's gas appliances and their flue system

This appliance is connected to a shared flue system and must not be removed or replaced without consultation with the responsible person for the building

Typical label affixed to an appliance warning that it forms part of a shared flue system

Replacement Appliances

6.70 Replacement appliances must be suitable for connection to this design of flue system, have the manufacturer's permission for installation, and they must not be of a greater heat input than the appliance that has been removed. Any existing holes into the duct that are no longer required should be sealed with a plate made from a suitable non-combustible material. After a survey to eliminate the possibility that asbestos is present, any necessary flue inlet/outlet holes should be drilled through both the sealing plate and duct wall, taking care to prevent any rubble from falling into the main flueway.

Communal Flue Systems (CFS)

6.71 In Section 6.65, it is explained how the older versions of shared flue operated and utilised the flue-flow effect to separate the hot flue gases from the combustion air supply. With the advent of condensing boilers, with a lower flue gas temperature, a new type of shared flue has been developed to enable multiple boilers, condensing and non-condensing, to be connected to a communal flue system. Rather than utilising the flue-flow effect, the systems have a flue and an air duct to maintain separation. A concentric CFS is installed vertically in the multi-storey building with appliances connecting with either a twin flue or a concentric configuration, Figure 6.51 illustrates a twin flue connection. There is a terminal at the top where products are dispersed, and fresh air is admitted into the air duct. At the base of the CFS, there will be a notice plate documenting the specifications for the system installed, including the total number of appliances connected on each floor, their kW rating and type. There are adjustable legs for support, an inspection cover, and a condensate collection unit for the main flue system, typically with a waste trap fitted. It should be noted that each individual appliance still requires its own condensate drain and that it is not acceptable to connect this drain to the CFS, unless the manufacturer states otherwise. There are different configurations of CFS such as, the concentric types, naturally ventilated (CFS(NV)) and positive pressure (CFS(PP)), or an exhaust-only type (CFS/EO). These are further described in Sections 6.72–6.74.

6.72 **CFS(NV)** systems consist of a concentric flue duct under a negative pressure. It can be identified as negative pressure by removing the inspection cover at the base of the CFS, where the inner exhaust flue duct will be an open pipe. BS 5440 gives tables to assist in the design of new CFS(NV) systems based on total heat input of all the appliances and the number of floors the CFS passes through.

6.73 **CFS(PP)** systems can most likely be identified by the base of the flue duct being sealed with a condensate drain trap. If this is the case, each boiler connected to it must have a non-return valve (NRV) fitted in the exhaust flue outlet to prevent reverse circulation of

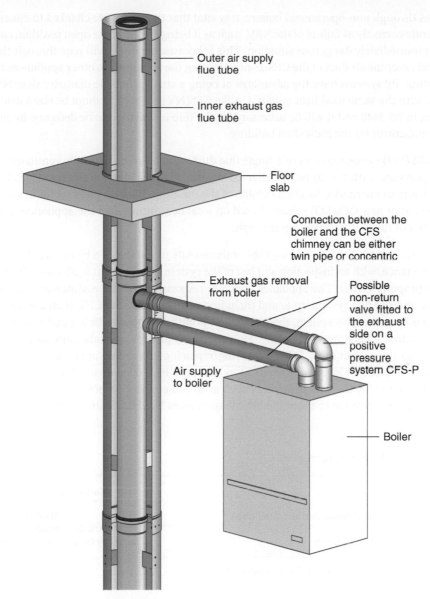

Outer air supply
flue tube

Inner exhaust gas
flue tube

Floor
slab

Connection between the
boiler and the CFS
chimney can be either
twin pipe or concentric

Exhaust gas removal
from boiler

Possible
non-return
valve fitted to
the exhaust
side on a
positive
pressure
system CFS-P

Air supply
to boiler

Boiler

Appliance (boiler) connection to CFS Chimney

*Note: CFS installations can incorporate a combination of concentric style flues and twin flues
    where required.*

Figure 6.51   Gas appliance connection to CFS. Source: From IGEM/UP/17

flue gases through non-operational boilers. It is vital that the valves are checked to ensure they operate correctly as failure of the NRV, such as it being seized in the open position, can cause an immediately dangerous situation. This is because products will pass through the boiler and enter the air duct of the CFS, contaminating the air supply to other appliances in the building. PP systems have the advantage of being a smaller outside diameter than NV systems, with the same total heat output. Unlike CFS(NV), CFS(PP) cannot be sized using the tables in BS 5440 and it will be necessary for the flue installation to be designed by the CFS manufacturer for the individual building.

6.74   CFS(EO) systems consist of a single flue duct, which can be naturally ventilated or positive pressure, with the air being delivered separately to the appliances via a dedicated air duct from an alternative location outside of the building. BS 5440 gives tables to assist in the design of new CFS(EO) systems based on total heat input of all the appliances and the number of floors the CFS passes through.

6.75   The responsibility for ensuring the continued safe use of the CFS by means of ongoing maintenance, with an inspection and test of the system performed at least annually, lies with the property owner. The appliance owner is responsible for the maintenance of the appliance and the flue system up to and including connection to the CFS. If an appliance is removed from the CFS system, the connection points will need to be capped to ensure products cannot escape into a dwelling. Manufacturers provide specialist fittings for this purpose with the correct seals. When removing/replacing an existing appliance the room should be monitored for levels of CO and CO_2 whilst the CFS is open to the property. It is not permissible to modify the CFS in any way such as to add connection points for future use. Table 6.15 gives reference to the suitability of appliances by classification to the different types of CFS.

Table 6.15   Suitability of appliances for CFS

Type classification and description		Suitability for CFS	
Classification	Description	Concentric CFS chimney Duct	Single CFS/EO combustion products only
C_{42} or C_{42} or $C_{(10)2}$	Fan-assisted; fan downstream of the heat exchanger	Yes	No
C_{43} or C_{43p} or $C_{(10)3}$	Fan-assisted; fan upstream of the heat exchanger	Yes	No
C_{82} or C_{82p}	Fan-assisted; fan downstream of the heat exchanger	No	Yes
C_{83} or C_{83p}	Fan-assisted; fan upstream of the heat exchanger	No	Yes

Note: p specifies the inclusion of a non-return valve
Source: Data from IGEM/UP/17

Fan Dilution System

Relevant ACS Qualification
CIGA1

Relevant Industry Document
BS 6644 and IGE/UP/10

6.76   The fan-diluted flue system was developed to overcome the problem of discharging large volumes of flue products at a low level. The principle behind its design concept is that if sufficient fresh air could be mixed with the flue products, it would dilute the final discharge down to an acceptable level. The combustion product level should not exceed the following values:

$$1\% \ CO_2 \le 50 \ \text{ppm} \ CO \le 5 \ \text{ppm} \ NO_x$$

Note: When the total heat input of the appliances exceeds 333 kW net approval must be sought from local authority.

Flue Discharge and Dilution Air Inlet

6.77   From Figure 6.52, one can see that the open-flued appliances are connected to a horizontal flue, positioned above the appliances. This flue has an outlet discharge directly through the wall, with the outlet louvres diverted to direct the products away from the ground upwards at an angle of about 30° from the horizontal. The minimum height for this outlet should be 2 m to the lower edge of the grille for appliances up to 0.9 MW, with this distance increased to 3 m where the gross heat input exceeds 0.9 MW. Flue discharge grilles should not exhaust into areas such as courtyards where the products cannot readily disperse. The dilution air inlet should ideally be taken from the same wall as the flue extract; however, it is possible to take this from another side wall, but problems may be experienced from external wind conditions affecting the operation of the dilution fan. It is also possible, under exceptional circumstances, to take the diluent air directly from the plant room; guidance from IGE/UP/10 should be sought here. The fan unit will be fitted within the flue system. This should be wired so that no appliance may be operated unless the fan has been proven to be operational. The fan and ductwork should be designed to provide a flow velocity of between 6 m/s and 8 m/s and have a flow volume that meets the combustion levels stated in Section 6.76. The volume per second can be determined from the following calculation:

$$\text{Factor} \times \text{Net heat input} \div 3600 = m^3/s$$

where Factor = 10.8 for natural gas and 12.8 for LPG.

Example: Find the minimum volume flow rate where the net input of several natural gas appliances connected to a fan dilution system is 254 kW.

$$10.8 \times 254 \div 3600 = \underline{0.762 \ m^3/s}$$

Within the inlet duct there should also be fitted a lockable damper that cannot be fully closed. It should be adjusted during the commissioning stage to a position that provides a balanced flow to give the desired combustion level, i.e it controls the amount of diluent air drawn in directly from outside. Any flue from an individual appliance should have fitted either a draught diverter or a stabiliser.

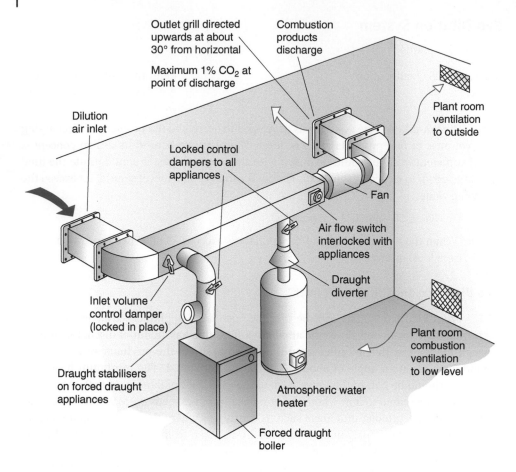

Figure 6.52 Typical arrangement of a fan dilution flue system

Plant-room Air Supply

6.78 Ventilation should be sized in accordance with the appropriate tables in Part 7 and ideally sited on a different wall from that of the flue discharge.

Concealed Flues

Relevant Industry Document
Technical Bulletin 008

Flues Concealed in Voids

6.79 Fan-assisted flues on modern boilers have enabled manufacturers to improve the efficiency of the appliances they develop and with smaller flues, the positioning of the appliance has become more adaptable. However, despite these advantages, an issue has arisen

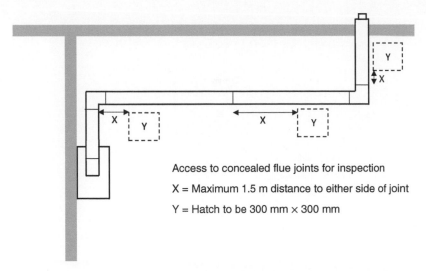

Access to concealed flue joints for inspection

X = Maximum 1.5 m distance to either side of joint

Y = Hatch to be 300 mm × 300 mm

Figure 6.53 Example of a room-sealed fan flue installation in a ceiling void

with respect to ensuring the continued integrity of the flue joints. It is extremely important to follow manufacturer's instructions when installing a flue and, in many cases, screws will be required to hold the joints together. Sadly, deaths have occurred due to faulty joint seals and joints actually falling apart and this has led to a requirement to provide a means of inspection of any room-sealed fan-assisted flue joint. Where you encounter an installation with concealed flue joints without a suitable means of inspection, it will be considered At Risk and recommendations to install inspection hatches must be raised with the responsible person. Figure 6.53 illustrates an example of a room-sealed fan flue installation in a ceiling void. As can be seen, hatches of 300 mm × 300 mm are required at a maximum of 1500 mm from either side of the flue fitting. This is to enable a complete inspection of the joints during a service visit.

1. The first and preferred option to offer your customer would be the installation of suitable inspection hatches, as described in Section 6.79, which would bring the installation up to current standards.
2. If this is not possible, it may be permissible to install a carbon monoxide safety shut-off void monitoring system (COSSVM). This will be interlocked with the gas supply and shut down the boiler if CO is detected in the void. This, along with the other necessary safety checks required by the Gas Regulations, will allow for the continued safe operation of the appliance. This method is not permitted for a new or replacement installation where a suitable means of physical inspection is required. Further advice can be sought from Gas Safe Technical Bulletin 008.

7

Ventilation

Need for Ventilation

> Relevant Industry Documents
> BS 5440, BS 6230 and BS 6644, IGEM/UP/10 and Building Regulations

7.1 Ventilation is required for several reasons, including:

- to provide fresh air to breathe;
- to refresh the condition of the air, e.g. removing smells;
- to provide a means of removing high levels of water vapour;
- for combustion of fuel;
- for cooling down an environment, e.g. compartment or plant room.

The gas operative is particularly interested in the last two points listed above – air for combustion and cooling purposes.

Adventitious Air

7.2 A great deal of the air supply to a building enters through natural or adventitious means. This air enters the building through cracks in floorboards, in window frames and door openings. It is generally accepted that in older properties (built up to 2008) most rooms will provide enough adventitious ventilation for an open-flued appliance of up to 7 kW net heat input. This is equivalent to an air vent with a free area of 35 cm^2. However, this allowance cannot be taken for granted as double glazing, draught-proofing, cavity insulation, extractor fans and various construction methods can have an impact on this natural ventilation. Where double glazing, draught-proofing, cavity insulation, etc., has been added to a property, it is essential to ensure that the ventilation remains effective by undertaking a spillage test. This applies even to installations where the appliance rating is ≤7 kW.

Note: Dwellings built after 2008 will likely have evidence of the building's air-tightness and where the air permeability is confirmed to be lower than 5.0 m^3/h/m^2 at 50 Pa an allowance for adventitious air would not be appropriate. If in any doubt, assume the air permeability is lower than 5.0 m^3/h/m^2 and fit suitable ventilation. In addition, the allowance for adventitious air cannot be applied to flueless appliances or compartment ventilation.

Gas Installation Technology, Third Edition. Andrew S. Burcham, Stephen J. Denney and Roy D. Treloar.
© 2024 John Wiley & Sons Ltd. Published 2024 by John Wiley & Sons Ltd.

Combustion Air

7.3 This is needed for combustion to take place without using up all the oxygen within a room and causing the air to become what is referred to as vitiated (lacking oxygen), resulting in dangerous gases being produced. The amount of air needed depends on the appliance type, location and size. For example, open-flued appliances with a rated net heat input not exceeding 70 kW installed in room or internal space will require permanent ventilation with a minimum free area of 5 cm^2 per kW of the appliance maximum rated input in excess of 7 kW.

The effectiveness of an air supply is something that must be considered on every occasion, not just when installing an appliance. If there is insufficient air available, the consequences may be fatal.

Air for Cooling Purposes

7.4 To prevent the risk of fire and avoid unacceptable temperature conditions, it is essential to monitor the temperature levels within a plant room, enclosure or compartment and ensure that they do not exceed the following:

- 40 °C air temperature within 100 mm of the ceiling level;
- 32 °C air temperature at mid-height (1.5 m above floor level);
- 25 °C air temperature within 100 mm of the floor level.

Where air for cooling is required, two grilles are needed, one at a low level and one at high level. These grilles should be as high and as low as practicable to encourage good convection currents through the area, see Figure 7.1a and b. Both vents must communicate either with the same internal space or to outside air through the same wall, depending on the circumstances. Where the compartment ventilation for an open-flued appliance is taken from a room or internal space, that room or internal space must itself be suitably ventilated to outside air.

Note: For open-flued appliances >70 kW, it is not permitted to take air from an internal space; see Tables 7.8 and 7.13.

For open-flued appliances, the lower vent is twice as big as the high-level vent. This is because some of the air is used in the combustion process and, as a result, passes up into the flue system. Appliance compartments must be labelled to warn against blocking the air vents and using the compartment for storage.

Where louvre doors provide ventilation for a compartment or plant room, the total free area of the slots within the door should not be less than the total high and low-level requirements, and the door should be as tall as practicable to encourage effective airflow, see Figure 7.1c.

Natural and Mechanical Ventilation

7.5 The method of ventilation selected would depend on the ability to provide air by natural means. It may be possible to use some form of mechanical ventilation, however, this would normally be interlocked with the gas supply so that if the fan failed no gas would

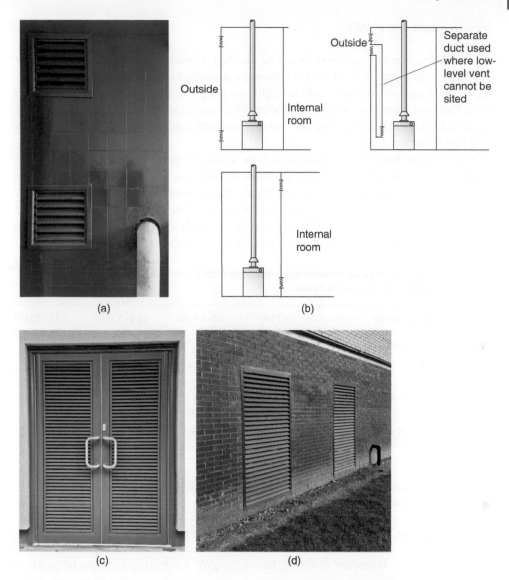

Figure 7.1 (a) High and low-level plant room ventilation, (b) examples of high and low-level compartment ventilation, (c) louvre doors used for plant room ventilation, and (d) large full height commercial plant room ventilation grilles

be allowed to flow to the appliance. Where mechanical extraction is chosen, it is essential to ensure that the air intake exceeds the volume of the extracted air, otherwise, a negative zone may be created, causing open-flued appliances to spill their products back into the room. Catering establishments use the concept of a negative pressure environment to prevent smell accumulation, but this is a special case, and no open-flued appliances would be located within this negative zone. See Ventilation/Extraction in Commercial Kitchens in Part 12.

Ventilation Location

7.6   The location of an air vent should be carefully selected to ensure that a suitable air supply is maintained. Vents should not allow contaminated air or fumes to enter the building or cause nuisance draughts, which might encourage occupants to block or obstruct the vent. Positions where leaves, snow or similar debris may block the vent should be avoided, as should areas liable to flooding. Air vents are not permitted to penetrate a protected area (see Part 4: Section 4.21).

Where an air vent is to be located in close proximity to a flue terminal ≤70 kW net heat input, the minimum distances identified in Table 7.1 should be observed (see also Figure 7.3). Appliances with a net heat input >70 kW must comply with IGEM/UP/10. This standard provides graphs and formulae to calculate minimum distances from a flue terminal to a vent or any other opening into a building. For appliances >70 kW, the minimum distances shown in Figure 7.2 and Table 7.2 should be maintained.

Table 7.1   Minimum distance between open-flued/room-sealed terminal and air vent for appliances of net heat input up to 70 kW

Appliance net heat input	<7 kW	>7–14 kW	>14–32 kW	>32–70 kW
Fan draught appliance				
In any direction	300 mm	300 mm	300 mm	300 mm
Natural draught appliance				
Above terminal	300 mm	600 mm	1500 mm	2000 mm
Horizontally to terminal	300 mm	400 mm	600 mm	600 mm
Below a terminal	300 mm	300 mm	300 mm	600 mm

Table 7.2   Minimum distance between open-flued/room-sealed terminal and air vent for appliances with heat input greater than 70 kW

Ref. from Figure 7.2	Appliance type	Formula
V	Open-flued (natural or fanned draught)	$9.5156 \times (net\ heat\ input\ kW) + 833.91$
W	Room-sealed – natural draught	$1.9031 \times (net\ heat\ input\ kW) + 1866.8$
Y	Room-sealed – fanned draught	$7.232 \times (net\ heat\ input\ kW) + 93.708$

Source: Data from IGEM/UP/10

7.7   Where an air vent into a particular room takes combustion air directly from outside the building, air may enter the room at any position, either at a high level or at a low level, and ideally close to the appliance to reduce the amount of draught. Ducted ventilation from above should be treated with caution as pressures within the room can restrict a good flow of air.

An appliance may need ventilation to be supplied either directly from outside or, in some cases, it may be permitted to be taken from another room. However, care should be taken as

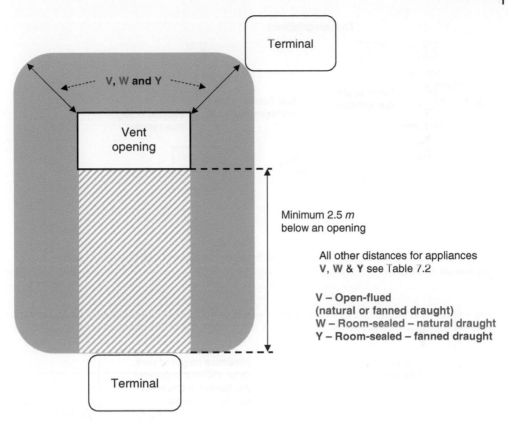

Figure 7.2 Minimum distance of a terminal to any vent or opening into a building >70 kW.
Source: Data from IGEM/UP/10

in many situations – such as with flueless appliances or commercial plant rooms – certain restrictions apply. The ventilation supply must then be obtained either directly from outside or ducted from outside, rather than taking the air from an adjoining space. Where it is acceptable to take a supply of air from another room, the other room must itself have been supplied with fresh air. The communicating grille in such circumstances between each room must be at low level and at a distance no greater than 450 mm above floor level. This is to prevent the spread of smoke in the event of a fire, as shown in Figure 7.4.

Should the air supply need to pass through more than one vent, consideration needs to be given to the effect of the additional resistance. Where the air is to pass through the external air vent and one internal wall or partition vent, the additional resistance is negligible. However, where there is more than one internal air vent, as shown in Figure 7.5, each internal vent will need to be 50% larger than the outside grille in order to take account of the flow resistance (see also Example 7.8 in Section 7.31).

Taking a supply of air from a roof void or an under-floor space may be permitted in certain circumstances, provided that the roof or floor void itself is adequately ventilated and that this space does not communicate with an adjoining property. However, new air vents are not permitted to communicate with a roof space. Underfloor vents to gas space heaters must not be positioned directly below the appliance, and for flueless space heaters must be ducted directly to outside air.

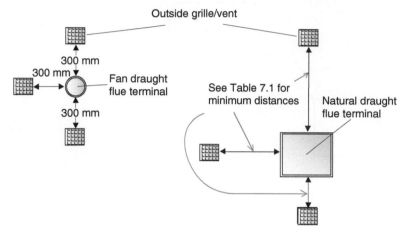

Figure 7.3 Minimum distances to be observed between the air vent and flue terminal for appliances ≤70 kW

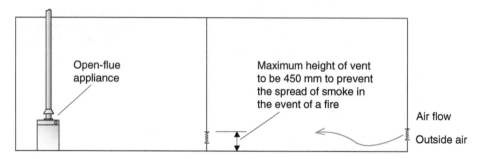

Figure 7.4 Location of air vents when taking the air supply from an adjoining room

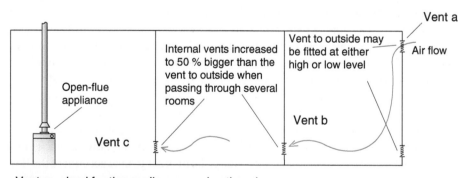

Vent a - sized for the appliance combustion air
Vent b - 50 % larger than vent a (x 1.5)
Vent c - 50 % larger than vent a (x 1.5)

Figure 7.5 Vents in series – increasing in size when passing through several rooms

Special Precautions Where Radon Gas Is a Problem

7.8   Radon is a colourless, odourless and radioactive gas that is found where uranium or radium is present, such as in certain parts of southwest England. Where radon is identified as a problem, below-floor ventilation should not be used. It may be possible to duct the air from outside but the ducted air must be isolated from the underfloor space.

Effective Ventilation

7.9   The size of a ventilation grille for a specific purpose can be calculated by referring to Sections 7.24–7.30 and Tables 7.3–7.16. However, when calculating the vent size, it is essential to know what is meant in terms of 'effective free area'. This is the actual size of the opening through which air can pass and air vents which are commercially marked with their free area should always be used where possible. The size of the actual grille has no bearing on the amount of air that might pass through it. For example, the terracotta air brick shown in Figure 7.7 would not allow the same throughput of air as a plastic or pressed metal grille of the same overall dimensions.

The size of an air grille is usually specified in cm^2, therefore, a randomly chosen grille or one found installed within a wall may physically measure 12×28 cm and, therefore, take up $336\,cm^2$ of wall space. However, this would not be the effective free area of airflow that could pass into the building. This area is determined by calculating the size of one hole and then multiplying this figure by the number of holes in the grille.

Example 7.A   In the diagram of the grille shown in Figure 7.6a, there are 15 holes. Each hole measures 7 mm × 80 mm. Therefore, the effective size would be:

$$7 \times 80 \times 15 = 8400\,mm^2$$

To convert mm^2 to cm^2, simply divide by 100:

$$\therefore 8400 \div 100 = \underline{84\,cm^2}.$$

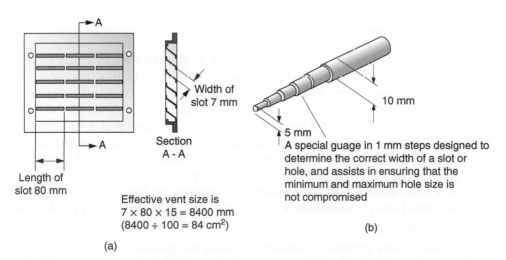

Figure 7.6   (a) Determining effective ventilation and (b) example of a vent sizing gauge

Example 7.B The terracotta air brick shown in Figure 7.7 has 18 holes. To correctly determine the free area of the air brick the cross-sectional area of each hole must first be calculated from the rear of the brick. This is because the aperture decreases from front to back, as illustrated in Figure 7.8.

The area of each hole is 0.8 cm × 0.8 cm. Therefore, the effective free area of the brick would be:

$$0.8 \times 0.8 \times 18 = 11.52 \text{ cm}^2$$

Figure 7.7 Terracotta brick

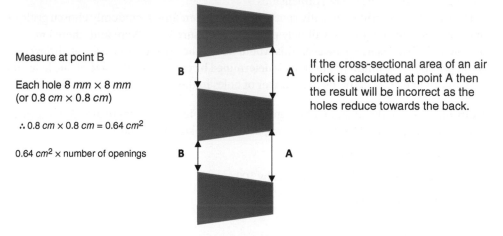

Measure at point B

Each hole 8 *mm* × 8 *mm*
(or 0.8 *cm* × 0.8 *cm*)

∴ 0.8 *cm* × 0.8 *cm* = 0.64 *cm*²

0.64 *cm*² × number of openings

If the cross-sectional area of an air brick is calculated at point A then the result will be incorrect as the holes reduce towards the back.

Figure 7.8 Cross-sectional area of a terracotta brick

7.10 For a vent to be effective, each individual hole needs to be small enough to stop vermin from getting through but not so small that it becomes blocked by dust, flies and general lint. British Standards specify a minimum aperture size of 5 mm for ventilation grilles in all installations, domestic and commercial. While domestic grilles should have an aperture of no smaller than 5 mm but less than 10 mm when communicating with outside air, commercial installations may use larger apertures at times. To prevent entry of birds and vermin into these larger commercial grilles while still allowing adequate airflow, a protective screen is often installed behind the grille. However, no aperture should be less than the 5 mm minimum specified in the standards. In addition, no gauze or fly screen should be incorporated into any ventilation grille, as these can easily become blocked and restrict the free area of the grille.

7.11 Other specific points in air vent design include the following:

- Vents should be non-adjustable, and vents with manual closing devices should not be used.

- No gauze or fly screen; the vent must maintain a minimum aperture size of 5 mm.
- The air vent, when used in a cavity wall, would need to be ducted fully across the cavity void. In order to cut down on draughts and noise transmission from the outside, it is possible to use special ducts that divert the airflow around a series of baffles as shown in Figure 7.9. However, where an air duct incorporates a draught reducing design, tests have shown that the equivalent free area is often restricted by 25–50% over an unrestricted air duct. The equivalent free area of such a duct should be confirmed by the manufacturer.
- Any duct used must not have a cross-sectional area that is less than the effective free area of the ventilation grille or louvre fitted to each end of it.
- Ducts over 3 m in length should be avoided without increasing in size (increase cross-sectional area by 50% for each 3 m section in excess of the first 3 m run).
- The number of 90° bends on a duct should be restricted to a maximum of two.
- Ducts intended to convey the airflow downwards should be avoided.
- Vents, grilles, louvres and ducts should be corrosion-resistant.
- Vents supplying air to open-flued appliances must not communicate with a bedroom or any room containing a bath or shower.
- Must not penetrate a protected area or compromise the fire safety of a building.
- Should be marked as follows: 'IMPORTANT DO NOT BLOCK THIS VENT'.

Note: Trickle vents should not be classed as permanent ventilation.

For examples, see Figures 7.9–7.11.

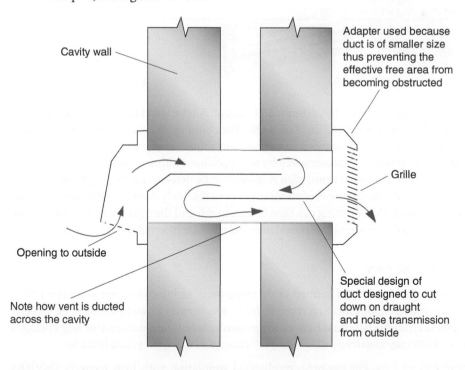

Cavity wall

Adapter used because duct is of smaller size thus preventing the effective free area from becoming obstructed

Grille

Opening to outside

Special design of duct designed to cut down on draught and noise transmission from outside

Note how vent is ducted across the cavity

Figure 7.9 Vent ducted across cavity and designed to cut down on draughts

7.12 The overall effectiveness of ventilation for an open-flued appliance is confirmed by a spillage test. Where a building is altered, such as where double glazing or a fan is installed,

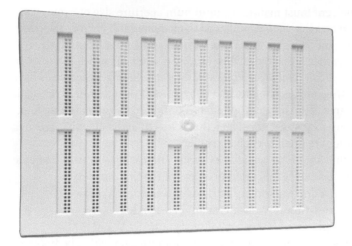

Figure 7.10   An inappropriate grille for ventilation – closable and includes a fly screen

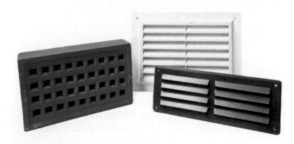

Figure 7.11   Domestic-sized ventilation grilles

a spillage test should always be performed. If spillage is detected at the appliance, the installation should be inspected and any faults with the chimney/flue or ventilation rectified. If spillage is still occurring, the ventilation provision should be increased until the products of combustion are effectively evacuated by the appliance chimney/flue.

The amount of additional ventilation required can be determined by progressively opening a window until the products of combustion are effectively drawn up the chimney/flue. The area of the window opening can then be measured, and the equivalent amount of additional fixed ventilation should be provided to the room or internal space.

Effect of Extract Fans

7.13   Where an extract fan is fitted in the same room or internal space as an open-flued appliance or in a communicating room or inner space, extra thought needs to be given to any adverse effect the fan could have on the performance of the chimney/flue (e.g. spillage).

Fans which may negatively affect the performance of a chimney/flue include:

- Room extract fans. For example, mechanical ventilation with heat recovery (MVHR), kitchen, bathroom, and utility room, etc.
- Cooker hoods.
- Externally vented tumble dryers.

- Ceiling fans (paddle fans).
- Circulating fans (including warm air heating systems).
- Fans in the flues of open-flued appliances.

This list should not be considered exhaustive, as any piece of equipment that has a fan which is designed to remove or circulate air can adversely affect the safe operation of an open-flued appliance and cause spillage. Therefore, where such equipment is encountered, a spillage test, as described in Part 6: Section 6.51, must always be performed. Where the fan has different settings, the test should be repeated at all modes of operation.

Where a spillage test confirms that the fan is having a depressurising affect, additional air vents may be required. This can be determined by following the procedure described in Section 7.12. However, an additional air vent with a free area of 50 cm^2 will be sufficient in most circumstances. If additional ventilation is added, the spillage test should be repeated to confirm satisfactory operation.

Intumescent Vents

7.14   These are special ventilator grilles that are designed to close and prevent the spread of smoke in the event of a fire. They generally employ a lattice arrangement that expands and closes off the holes in the event of extreme heat. It is important to inform the end user that if an intumescent vent has been activated by high temperatures, such as a fire, it must be replaced before any gas appliance reliant on the vent is operated. These should be fitted no higher than 450 mm above the floor level.

Purge Ventilation (Rapid Ventilation)

7.15   Flueless appliances require a ventilation opening direct to outside air. BS 5440-2 stipulates that this should be an openable window or an equivalent; see note to Table 7.5. This is in addition to that which might be required as permanent. This opening direct to outside is referred to as purge ventilation. It is designed to rapidly aid the removal of high concentrations of pollutants such as smells and water vapour released by such activities as cooking, painting, etc.

Purge ventilation is generally achieved by the use of natural openings (windows or doors) or a mechanical extract ventilation system. However, several alternative approaches have been adopted in the past where an opening window was not available. These include:

- an adjustable louvre vent to outside;
- a hinged panel direct to outside;
- an extract fan;
- an open-flue 125 mm in diameter or larger;
- passive stack ventilation.

It should be noted that, at the time of writing, the Building Regulations Approved Document F – Ventilation (2021 edition) only recommends the use of natural openings (e.g. windows or doors) or a mechanical extract ventilation system as a suitable means of purge ventilation.

Internal Rooms and Kitchens

7.16   All newly built kitchens require extract purge ventilation. For intermittent extractors, a minimum extract rate of 30 l/s is required where a cooker hood is used to extract directly to outside or 60 l/s if the fan is located elsewhere in the kitchen. For continuous extraction, a minimum extraction rate of 13 l/s is sufficient. Where a kitchen is refurbished, it is important to ensure that any building work does not make the building any less compliant with the ventilation requirements of the Building Regulations than it was before the work was carried out. Therefore, where an extract fan was already fitted, this should be retained or replaced.

On occasion, you may come across an internal habitable room or kitchen with a flueless appliance that does not have a suitable means of purge ventilation (e.g. where an extension or conservatory has been added to a building, affecting the original purge ventilation). However, the installation must still comply with the Gas Regulations and Building Regulations. Gas Safe Technical Bulletin 005 has traditionally been used in these circumstances, but at the time of writing, this guidance document has been withdrawn and is in the process of being revised. In the meantime, if in any doubt as to whether the installation complies with current regulations, seek guidance from the Local Building Control Body or the technical helpline of the Gas Safe Register. Alternatively, apply the Gas Industry Unsafe Situation Procedure as appropriate.

Passive Stack Ventilation

7.17   In some existing properties, Passive Stack Ventilation (PSV) may be encountered. PSV provides a natural energy-free alternative to mechanical extractor fans, providing a positive pressure to circulate air through the building. It is driven by the natural stack effect in which warm air rises – resulting from internal and external temperature differences and wind-induced pressures – taking with it moisture-laden air directly from 'wet' areas such as kitchens and bathrooms. The air is simply replaced by fresh air through inlet vents situated in the walls or window frames of habitable rooms.

Where PSV has been provided within an existing building as an alternative to the opening window for a flueless gas appliance, ensure the PSV terminal is situated below any open-flue terminal. This is to prevent spillage from the open-flued appliance (Figure 7.12).

Ventilation Sizing and Tables

Relevant Industry Documents
BS 5440, BS 6230, BS 6644 and BS 6896,
IGEM/UP/10 and Building Regulations

7.18   This chapter deals with ventilation sizing within permanent buildings for both natural gas and LPG. However, reference also needs to be made to Part 14, which deals with non-permanent dwellings and to Part 12 for commercial catering, where specific requirements are described. See Table 7.17 for a domestic ready reckoner that can be used for open-flued appliances and compartment ventilation. See also Part 2: Section 2.26, which explains why natural gas and LPG systems can have the same sized air grille.

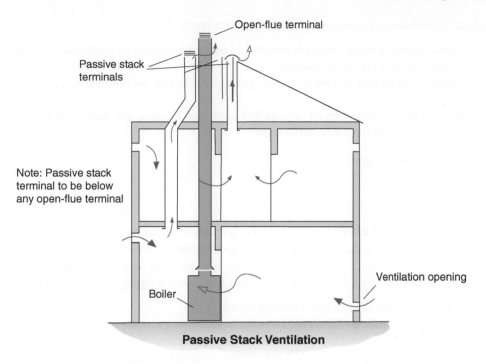

Passive Stack Ventilation

Figure 7.12 Passive stack ventilation

Size of Air Grille Needed

7.19   The size of an air vent into a building for combustion or cooling purposes can be calculated and depends on several factors, including appliance type, size, location and fuel. There may be more than one appliance in a room, and these may be of different designs and burn different fuels. The air supply requirement for a particular appliance is calculated from specific data that can be obtained from various industry documents and standards. This data has been collated in the tables starting in Section 7.24.

Caution: Changes to the Standards

7.20   Prior to the current British Standards, the ventilation calculation requirements were based on 'gross heat inputs'. Today, standards refer to 'net heat input', and therefore, some caution needs to be observed, and a check needs to be made against the manufacturer's data to confirm whether they are quoting net or gross appliance heat input.

It is, therefore, very important to note, when calculating the ventilation requirements for a particular appliance, whether the net or gross heat input is quoted. Where existing appliance manufacturers do not make reference to either gross or net heat input, always assume net, this will ensure your ventilation calculation is not undersized.

- To convert gross into a net heat input, divide by 1.1.
- To convert net into gross, multiply by 1.1.

Example: 60 kW gross input = 60 ÷ 1.1 = <u>54.5 kW net input.</u>

Rooms Containing Appliances Burning Other Fuel Types

7.21   Occasionally a gas appliance is installed in the same space as an appliance burning another fuel type (e.g. oil, wood, solid fuel, etc.). Where this is the case, the other fuel burning appliances are treated as though they are gas appliances and the ventilation calculated accordingly.

If the rated heat output of an oil or solid fuel appliance is stated, the heat input can be calculated by means of the following formula:

$$\frac{\text{Output} \times 10}{6} = \text{Input}$$

Multiple Appliance Installations

7.22   Sometimes, an installation will be encountered in which several appliances are installed within the same location. In this case, a judgement will need to be made to ensure that sufficient ventilation is provided. The following guide from BS 5440-2, which covers domestic ventilation, suggests selecting a vent size based on the <u>largest</u> of the following:

1. the total maximum flueless space heating input;
2. the total maximum open-flued space heating input; or
3. the greatest individual-rated input from any other appliance.
    Note 1: 'Space heating' refers to a gas fire, central heating boiler or air heater/convector.
    Note 2: For decorative fuel effect appliances, see additional notes to Table 7.3.
    Note 3: Where permanent ventilation is required for multiple appliance installations, the vent should be provided between the two appliances wherever practicable.

The exception to number 2 above is where the interconnecting wall between two similar-sized rooms has been removed, and the resultant room has two chimneys, each fitted with a similar type of gas fire with a rated input of less than 7 kW. In this case, an air vent may not be needed.

Servicing and Maintenance

7.23   Whenever carrying out servicing or maintenance work on a gas appliance, the provision of air for combustion or cooling purposes must be verified as adequate. All grilles or louvres should be visually inspected for obstructions or blockages. Wherever possible, insert a screwdriver into the apertures to check for clear plastic film or flyscreens.

Ensure that the air supply is fully ducted across the cavity and, if cavity wall insulation has recently been added, that this has not impeded the ventilation. Ensure that the grilles or louvres are not closable or adjustable. Finally, confirm that the vents are adequately sized for the appliance/s.

Where any defects are identified, these should be rectified before the appliance is used again. Where this is not possible, the installation should be classified in accordance with IGEM/G/11 'Gas industry unsafe situations procedure' and brought to the attention of the responsible person (see Part 8 Installer Responsibility).

Domestic Ventilation Tables

7.24   Notes for Domestic Ventilation Tables

Tables 7.3–7.16 cover a whole range of gas installations from domestic to commercial natural gas applications and include LPG in permanent buildings, and at first glance, they may seem a little daunting. In order to find the required size of the air grille, the following procedure needs to be carried out.

1. Select the correct table based on the room location and flue type.
2. Choose the correct row based on the net input or appliance type.
3. Undertake the calculation or select a size as defined by the table.

The calculations provide the suggested minimum size of ventilation needed for a particular appliance but it must be remembered that where spillage occurs the only cure may be to use a larger vent size.

You may need a little practice to use and understand the tables. However, in Sections 7.31 and 7.32 you will find several worked examples for both domestic and commercial premises, which may prove helpful.

All tables are based on net heat input and identify the 'minimum' ventilation requirement.

$$[\leq \ = \text{less than or equal to;} \ < \ = \text{less than;} \ > \ = \text{greater than}]$$

To convert mm^2 to cm^2, divide by 100
To convert m^2 to cm^2, multiply by 10 000

Table 7.3   Combustion ventilation for open-flued appliances

Appliance type (net input)	Ventilation requirements
Open-flued $\leq$7 kW	No additional ventilation is required
Open-flued >7 kW to $\leq$70 kW	5 cm^2 per kW in excess of 7 kW
Room-sealed appliances	No additional ventilation is required
Decorative fuel effect fires	100 cm^2 for appliances $\leq$20 kW. Where two existing DFEs are installed in the same room, allow 235 cm^2 or, where greater, as stated in the manufacturer's instructions

Notes regarding decorative fuel effect fires (DFE):
Note 1: Air vents shall not be installed in the fireplace recess or builder's opening. All existing air supplies in these positions shall be sealed up. Floor vents can be used as described in Section 7.7 if located outside of the hearth area.
Note 2: Multiple appliance installations.
The 100 cm^2 for a single DFE takes into account the acknowledged 35 cm^2 of adventitious air. Therefore, when another DFE is installed in the same space, 35 cm^2 needs to be added to the overall requirement. Where the second appliance is not a DFE, the principles described in Section 7.22 can be followed, but remember that the adventitious air has already been accounted for (i.e. calculate 5 cm^2 per kW rated heat input for other appliance types).
Note 3: Ventilation requirements, as stated in manufacturer's instructions shall always be followed.

Table 7.4   Compartment ventilation

Appliance type (net input)	Low-level opening		High-level opening	
	To outside	Into room	To outside	Into room
Open-flued ≤70 kW	10 cm² per kW	20 cm² per kW	Half that of low level	Half that of low level
Room-sealed ≤70 kW	5 cm² per kW	10 cm² per kW	Same size as low level	Same size as low level

Table 7.5   Flueless appliance ventilation (direct to outside air)

Appliance type	Net max heat input	Room volume (m³)	Permanent vent size (cm²)	Openable window
Domestic oven, hob or grill	No maximum	<5 m³	100	Yes
		5–10	50 (If door direct to outside 0)	Yes
		>10	0	Yes
Instantaneous water heater	11 kW	<5	Not permitted to be installed	Yes
		5–10	100	Yes
		>10–20	50	Yes
		>20	0	Yes
Tumble dryer	N/A	<3.7 m³/kW of appliance-rated heat input	100	Yes
	N/A	≥3.7 m³/kW of appliance-rated heat input	0	Yes
Refrigerator	N/A		0	No
Natural gas space heater in a room	45 W/m³		55 cm² for every kW in excess of 2.7 kW **plus** 100 cm²	Yes
Natural gas space heater in an internal space	90 W/m³		27.5 cm² for every kW in excess of 5.4 kW **plus** 100 cm²	Yes
LPG space heater to EN449 in a room	50 W/m³	>15	25 cm² for every kW with a minimum of 50 cm² at high and low level	Yes
LPG space heater to EN449 in an internal space	100 W/m³	>15	25 cm² for every kW with a minimum of 50 cm² at high and low level	Yes

Note: Acceptable alternatives to an openable window include adjustable grilles/louvres or hinged panels.

Table 7.6 'Balanced compartment' ventilation (see also Part 6: Section 6.54)

Appliance type/input (net)	Ventilation size	Method of ventilation
≤70 kW	7.5 cm² per kW	Ducted to low level
	12.5 cm² per kW	High level only

Note 1: Low-level ventilation should be ducted to within 300 mm of the compartment floor level.
Note 2: Where the ceiling height of the balanced compartment is greater than 300 mm above the base of the skirt of the appliance draught diverter, a high-level opening should also be provided. The opening can be by means of a Y or T piece inserted into the air supply duct. The cross-sectional area of the inserted piece should be the same as the low-level duct.

Commercial Ventilation Tables

7.25 Notes for Commercial Ventilation Tables

Appliance types:

Type A	–	Flueless appliance
Type B	–	Open-flued appliance
Type B1	–	Open-flued appliance with a draught divertor (natural draught)
Type B2	–	Open-flued appliance without a draught divertor (forced draught)
Type C	–	Room-sealed appliance

An enclosure is considered to be a room not large enough to enter and perform work other than maintenance.

7.26

Gas-Fired Air Heaters

Relevant Industry Documents
BS 6230, IGEM/UP/10

Table 7.7 Air heaters installed in a heated space – natural ventilation

Appliance type Net input	Ventilation requirements	
Indirect-fired air heater >70 kW to 1.8 MW Types B1 and B2	No additional ventilation is required if the air change rate is >0.5 per hour	Where additional ventilation is required, it would need to be 2 cm² per kW provided at low level below the flue connection
Indirect-fired air heater >1.8 MW Types B1 and B2	2 cm² per kW provided at low level below the flue connection	
Direct-fired air heater Type A	The environment into which the flue gas discharges needs to be monitored at various locations where the products may be inhaled to ensure that the carbon dioxide (CO_2) levels do not exceed 0.28% (equivalent to 2800 ppm)	
Room-sealed Type C	No additional ventilation is required	

Table 7.8 Type B and C indirect air heaters installed in a plant room or enclosure – natural ventilation

Appliance type Net input >70 kW	Low-level opening		High-level opening	
	To outside	Into room	To outside	Into room
Open-flued Type B1 and B2 In a plant room	4 cm^2 per kW	Not permitted to be installed	Half that of low level	Not permitted to be installed
Open-flued Type B1 and B2 In an enclosure	10 cm^2 per kW	Not permitted to be installed	Half that of low level	Not permitted to be installed
Room-sealed appliance Type C In a plant room or enclosure	5 cm^2 per kW	10 cm^2 per kW	Same size as low level	Same size as low level

Table 7.9 Type B air heaters installed in a plant room or enclosure – mechanical ventilation

Appliance type	Minimum low-level air inlet m^3/h	Extract high-level air rate m^3/h
Indirect-fired warm air unit with draught diverter Type B1	4.3 per kW	5–10% lower than the inlet rate or via high-level natural ventilation, as identified in Table 7.8
Indirect-fired warm air unit without draught diverter Type B2	4.14 per kW	

Note 1: Automatic control is required to shut down the appliance in the event of fan failure.
Note 2: Air intake must always exceed air extract to ensure no negative zone is created.
Note 3: Systems that use natural air inlet and mechanical extraction are not permitted.

Table 7.10 Direct-fired air heaters (Type A) installed in a plant room or enclosure – natural ventilation

Direct fired heater Net input	Low-level opening	High-level opening
≤70 kW	5 cm^2 per kW	Same size as low level
>70 kW	2.5 cm^2 per kW in excess of 70 kW **plus** 350 cm^2	Same size as low level

Note 1: The environment into which the flue gas discharges needs to be monitored at various locations where the products may be inhaled to ensure that the carbon dioxide (CO_2) levels do not exceed 0.28% (equivalent to 2800 ppm).
Note 2: Both high and low-level vents should be on the same wall and communicate directly to outside air.

Table 7.11 Direct-fired heaters (Type A) installed in a plant room or enclosure – mechanical ventilation

Appliance type	Minimum low-level air inlet m³/h	Extract high-level air rate m³/h
Direct-fired warm air heater Type A	Air ducted to the heater at a rate of 2.4 m³/h per kW	5–10% lower than the inlet rate or via high-level natural ventilation, as identified in Table 7.10

Note 1: Automatic control is required to shut down appliance in the event of fan failure.
Note 2: If the appliance manufacturer states that air to the heater is not to be ducted, ventilation shall be provided in accordance with the manufacturer's installation instructions.
Note 3: The environment into which the flue gas discharges needs to be monitored at various locations where the products may be inhaled to ensure that the carbon dioxide (CO_2) levels do not exceed 0.28% (equivalent to 2800 ppm).
Note 4: Both high and low-level vents should be on the same wall and communicate directly to outside air.

7.27

Overhead Heaters – Type B

Relevant Industry Documents
BS 6896, IGEM/UP/10

Table 7.12 Type B overhead heaters installed in a heated space – natural ventilation

Appliance type Net input	Ventilation requirements	
Type B1 and B2 heaters	No additional ventilation is required if the air change rate is >0.5 per hour	Where additional ventilation is required, it would need to be 2 cm² per kW provided at low level below the flue connection
Special case for Type B1 heaters with a ducted combustion air supply	In the special case where the combustion air is ducted directly to the burner from outside air, then low-level ventilation can be provided, which is no less than 50% of that stated above, i.e. 1 cm² per kW	
Special case for Type B2 heaters with a ducted combustion air supply	In the special case where the combustion air is ducted directly to the burner from outside air, then no additional ventilation provision is required for combustion air or for combustion product dilution	
Type C heater	No additional ventilation is required	

Mechanical Ventilation

7.28 Where mechanical ventilation is employed, it shall be designed to ensure that the building has an air change rate of at least 0.5 per hour. In addition, the following requirements shall be met:

- mechanical inlet (low level) with either mechanical or natural extraction (high level);
- systems that use natural air inlet and mechanical extraction are not permitted;
- where there is mechanical extraction, and there is a flued heater in the same space, the extraction rate shall be 5–10% lower than the inlet rate;
- automatic control required to shut down appliance in the event of fan failure.

7.29

Overhead Heaters – Type A

Relevant Industry Documents
BS EN 13410

The ventilation requirements of Type A flueless overhead heaters may be provided by one of the following methods:

- thermal evacuation;
- mechanical evacuation;
- natural air changes.

In all cases, the ventilation shall be designed in accordance with BS EN 13410. However, thermal or mechanical evacuation may not be required providing:

- the building has natural air changes >1.5 volumes per hour and
- the combined appliance operating heat input is no greater than $5\,W/m^3$.

The environment into which the flue gas discharges also needs to be monitored at various locations, where the products may be inhaled by occupants, to ensure that the carbon dioxide (CO_2) levels do not exceed 0.28% (equivalent to 2800 ppm).

For methods of calculating thermal and mechanical evacuation, see Part 10: Sections 10.31–10.33.

7.30

Gas Fired Hot Water Boilers

Relevant Industry Documents
BS 6644, IGEM/UP/10

Table 7.13   Boilers installed in a boiler/plant room or enclosure – natural ventilation

Appliance type net input >70 kW	Low-level opening		High-level opening	
	To outside	Into room	To outside	Into room
Open-flued Type B In a boiler/plant room	$4\,cm^2$ per kW	Not permitted to be installed	Half that of low level	Not permitted to be installed
Open-flued Type B In an enclosure	$10\,cm^2$ per kW	Not permitted to be installed	Half that of low level	Not permitted to be installed
Room-sealed appliance Type C In an enclosure	$5\,cm^2$ per kW	$10\,cm^2$ per kW	Same size as low level	Same size as low level
Room-sealed appliance Type C In a boiler/plant room	Suitable ventilation to maintain a max temperature of 25 °C at 100 mm from floor level; 32 °C at mid position and 40 °C within 100 mm of the ceiling with a minimum size of at least $2\,cm^2$ per kW at high and low level to outside			

Table 7.14 Mechanical ventilation for boiler/plant rooms

Appliance Type B	Minimum low-level air inlet m³/h	Extract high-level air rate m³/h
Boilers with draught diverter (natural draught)	2.8 per kW	0.73 per kW (±0.18)
Boilers without diverter (forced draught), with or without a draught stabiliser	2.6 per kW	1.25 per kW (±0.18)

Note 1: Automatic control is required to shut down appliance in the event of fan failure.
Note 2: Air intake must always exceed air extract to ensure no negative zone is created.
Note 3: Where mechanical inlet at low level is used with natural extraction at high level, the high-level grilles shall be sized at 2 cm²/kW as detailed in Table 7.13.

Table 7.15 Boilers installed in a heated space – natural ventilation

Appliance type Net input >70 kW	Ventilation requirements	
Open-flued Type B boilers	No additional ventilation is required if the air change rate is >0.5 per hour	Where additional ventilation is required, it would need to be 2 cm² per kW provided at low level (below the flue connection)
Room-sealed Type C boilers MOP ≤ 100 mbar	No additional ventilation is required if the air change rate is >0.5 per hour	Where the air change rate is <0.5 per hour, 2 cm² per kW at high and low level is required

Note: The additional ventilation for Type C boilers is required for pipework safety by the Dangerous Substances and Explosive Atmosphere Regulations.

Table 7.16 Balanced compartment – natural ventilation

Appliance type/ input (net)	Ventilation size	Method of ventilation
>70 kW to ≤500 kW	10 cm² per kW	Permanent opening at high level only
>500 kW	8 cm² per kW in excess of 500 kW **plus** 5000 cm²	

Note: There shall be no other openings into the compartment except in the case of heavier than air gases where an opening of at least 60 cm² shall be provided at low level direct to outside air. In this case, the designer shall verify that the low-level opening does not adversely affect the operation of the compartment.

7.31

Worked Examples for Domestic Premises

Example 7.1 A 22 kW net input open-flued warm air unit has been installed within a compartment, and its ventilation has been taken directly from outside. What is the required minimum size of grille?

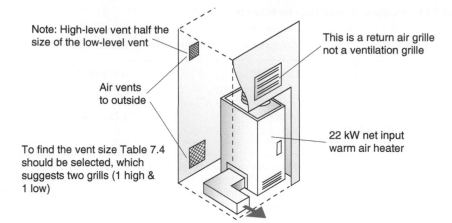

Note: High-level vent half the size of the low-level vent

This is a return air grille not a ventilation grille

Air vents to outside

To find the vent size Table 7.4 should be selected, which suggests two grills (1 high & 1 low)

22 kW net input warm air heater

Step 1: Review the situation and note that the appliance is within a compartment and less than 70 kW, therefore, Table 7.4 should be selected.

Step 2: The appliance is open-flued, therefore, the first row is selected.

Step 3: The table suggests two grilles: one at high level and one at low level, the high-level grille being half the size of the low-level grille.

Thus, the calculation is as follows:

For the low-level grille $= 10 \times 22 = \underline{220 \text{ cm}^2}$; and

for the high-level grille $= 220 \div 2 = \underline{110 \text{ cm}^2}$.

Example 7.2 A 24.2 kW gross input open-flued boiler in a garage has been found, previously installed, with no ventilation grille fitted. What size should the grille be?

First, the gross input needs to be converted to a net input:

Net input $= 24.2 \div 1.1 = 22$ kW

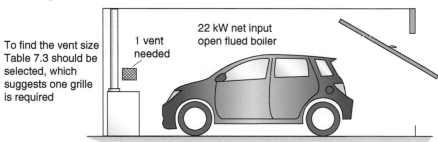

To find the vent size Table 7.3 should be selected, which suggests one grille is required

1 vent needed

22 kW net input open flued boiler

Step 1: With this example, the appliance is installed in a garage. Table 7.3 should be selected because the appliance is not within a compartment.

Step 2: From the table, it can be seen that the second row needs to be selected because the appliance is open-flued, operating at 22 kW and clearly falls within the range >7 kW to ≤70 kW.

Step 3: The table suggests one grille of the following size: 5 cm² per kW in excess of 7 kW. Therefore first, the 7 kW is subtracted from the 22 kW appliance, then the following calculation is made:

$$(22-7) \times 5 = \underline{75\,\text{cm}^2}$$

The grille may be located at any suitable position into the room.

Example 7.3 A domestic cooker of 17.2 kW is to be installed within a kitchen that measures 3 m × 3 m × 2.8 m. There is an outside window. What ventilation is needed?

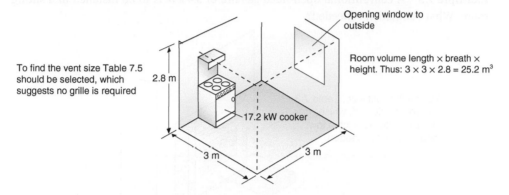

To find the vent size Table 7.5 should be selected, which suggests no grille is required

2.8 m

Opening window to outside

Room volume length × breath × height. Thus: 3 × 3 × 2.8 = 25.2 m³

17.2 kW cooker

3 m   3 m

Step 1: Table 7.5 is selected as the oven is flueless and of a domestic type.
Step 2: The first row is selected; the heat input is irrelevant.
Step 3: The room volume determines the size of the vent, which is calculated as 25.2 m³.

Therefore, it is >10 m³ so no vent is required. However, an opening window is required and that is in place.

Example 7.4 A 14 kW net open-flued water heater has been installed in a kitchen in which there is also a cooker. The room volume is 8 m³. There is not a door opening to outside. What ventilation is needed?

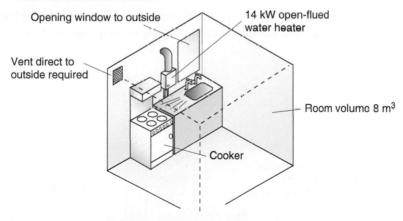

Opening window to outside

14 kW open-flued water heater

Vent direct to outside required

Room volume 8 m³

Cooker

Step 1: Tables 7.3 and 7.5 are both applicable. Because it is a multi-appliance installation, the size of the vent selected should be based on the largest individual requirement (see Multiple Appliance Installations, Section 7.22). So, initially, the requirements for both appliances need to be worked out.
Step 2: This is completed for each appliance, as with previous examples.

Step 3: With the two calculations completed, it is found that the water heater needs $(14 - 7) \times 5 = 35 \, cm^2$ and, from Table 7.5, the cooker requires $50 \, cm^2$.

As the cooker has the larger requirement, a $\underline{50 \, cm^2}$ vent is fitted, plus an openable window.

Example 7.5 A conventional open-flued gas fire of 5.4 kW is to be installed in a sitting room. What ventilation is needed?

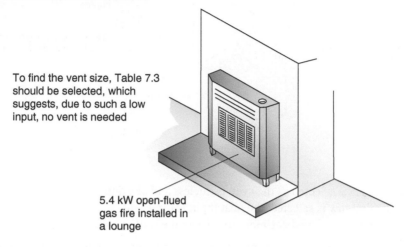

To find the vent size, Table 7.3 should be selected, which suggests, due to such a low input, no vent is needed

5.4 kW open-flued gas fire installed in a lounge

Step 1: Table 7.3 is selected.
Step 2: The first row is chosen as the appliance is less than 7 kW.
Step 3: Because of the low input, sufficient adventitious air can enter the building to support combustion, therefore, <u>no vent is needed</u> (see also Section 7.2).

Example 7.6 A gas fire of 5 kW net has been installed in conjunction with a back boiler that has a maximum net heat input of 17 kW, yet is only set to run at 14 kW. No ventilation has been provided.

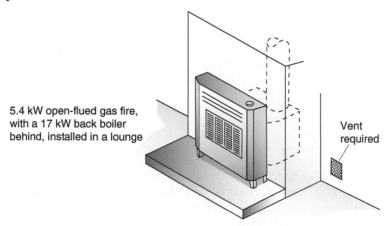

5.4 kW open-flued gas fire, with a 17 kW back boiler behind, installed in a lounge

Vent required

As with Example 7.4, this is a case where two appliances are installed, namely a boiler and a fire, within the same room. Because these are both flued space heating appliances, the heat inputs must be added together. The fact that the boiler is set to run at a lower heat input is irrelevant and its ventilation requirements need to be based on the maximum heat input. Thus, the fire is 5 kW and the boiler 17 kW, therefore, the required ventilation is based on a total input of 22 kW.

Following the steps previously described, Table 7.3 is selected. A vent is required of size:

$$(22 - 7) \times 5 = \underline{75 \text{ cm}^2}$$

Example 7.7 A 22 kW net input open-flued boiler has been installed within a compartment, and its ventilation has been taken directly from the room.

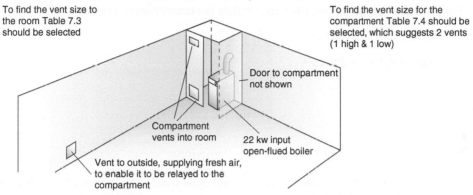

To find the vent size to the room Table 7.3 should be selected

To find the vent size for the compartment Table 7.4 should be selected, which suggests 2 vents (1 high & 1 low)

Door to compartment not shown

Compartment vents into room

22 kw input open-flued boiler

Vent to outside, supplying fresh air, to enable it to be relayed to the compartment

In this domestic situation, it will be seen that two separate tables need to be selected; first, Table 7.3 needs to be consulted in order to find the size of the combustion ventilation grille needed to serve the room in which the appliance is situated. Then, Table 7.4 is used to determine the vent sizes for the compartment itself. Thus, the vent sizes would be:

Room ventilation: $(22 - 7) \times 5 = \underline{75 \text{ cm}^2}$

Compartment low level: $20 \times 22 = \underline{440 \text{ cm}^2}$ and high level $440 \div 2 = \underline{220 \text{ cm}^2}$

Example 7.8 In this final domestic ventilation scenario, a 40 kW net input open-flued boiler has been installed within a compartment. The air for combustion has to pass through an additional two vents before it reaches the room in which the boiler compartment is located. To calculate the correct vent sizes, Tables 7.3 and 7.4 need to be consulted along with the information found in Section 7.7.

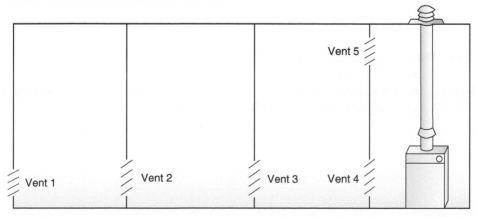

Vent 5

Vent 1

Vent 2

Vent 3

Vent 4

Firstly, by using Table 7.3, the combustion ventilation grille size can be determined (vent 1):

Vent 1 $(40 - 7) \times 5 = 165\,\text{cm}^2$

From the information in Section 7.7, we find that where the air needs to pass through more than one internal air vent, each internal vent will need to be 50% larger than the outside air vent in order to take into account airflow resistance. Therefore, vents 2 and 3 are sized as follows:

Vent 2 (50% larger than vent 1) $165 \times 1.5 = 247.5\,\text{cm}^2$

Vent 3 (50% larger than vent 1) $165 \times 1.5 = 247.5\,\text{cm}^2$

Finally, Table 7.4 can be consulted to calculate the compartment ventilation:

Vent 4 $20 \times 40 = 800\,\text{cm}^2$

Vent 5 $800 \div 2 = 400\,\text{cm}^2$

7.32

Worked Examples for Commercial Premises

Example 7.9   A 92 kW net input forced draught open-flued (B2) warm air heater is to be installed within a workshop. What ventilation would be required?

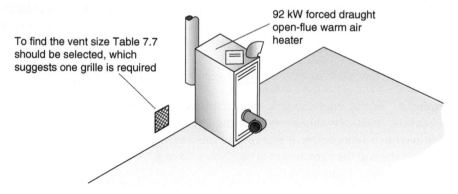

To find the vent size Table 7.7 should be selected, which suggests one grille is required

92 kW forced draught open-flue warm air heater

Step 1: With this example, the appliance is installed in a workshop (a heated space). Therefore, Table 7.7 should be selected.

Step 2: From this table and the example, it can be seen that the appliance is operating at 92 kW and clearly falls within the band >70 kW to 1.8 MW.

Step 3: The table gives options and suggests that a grille is required only if the air change rate is <0.5 room volumes/h (in this example, the air change rate is unknown); therefore, additional ventilation is required.

The vent size can be calculated as $2\,\text{cm}^2$ per kW.
Therefore $2 \times 92 = \underline{184\,\text{cm}^2}$ of effective free air will be needed.

Example 7.10 A 300 kW net input forced draught open-flued (B2) warm air unit is to be installed within a workshop. The room has a known air change rate of 2 volumes of air per hour. What ventilation will be required?

This example follows the same format as the previous example; therefore, steps 1 and 2 are as before. When tackling step 3 to undertake the calculation, it will be seen that because the room volume has an air change rate exceeding 0.5, no ventilation is, in fact, required for this installation, and sufficient air is already available.

Example 7.11 A 92 kW net input forced draught open-flued boiler is to be installed within an compartment (or enclosure). What ventilation is required?

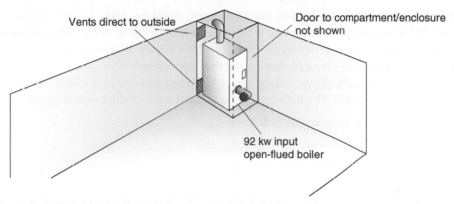

Step 1: With the appliance located in an enclosure, Table 7.13 should be selected.
Step 2: The appliance is open-flued so the second row is selected.
Step 3: The table suggests two grilles: one at high level and one at low level, both taking their air from outside. The high-level grille is half the size of the low-level grille.

Thus, the calculation is as follows:

Low-level grille $= 10 \times 92 = \underline{920\,\text{cm}^2}$; and

high-level grille $= 920 \div 2 = \underline{460\,\text{cm}^2}$.

If the same appliance had been installed within a plant room, the first row of Table 7.13 would need to be selected, giving the ventilation grille sizes as follows:

Low-level grille $= 92 \times 4 = \underline{368\,\text{cm}^2}$; and

high-level grille $= 368 \div 2 = \underline{184\,\text{cm}^2}$.

Thus, as can be seen, smaller grilles are required. This is because a plant room is larger than a small compartment or enclosure and not subject to such a large heat gain.

Example 7.12 Two 230 kW net input forced draught open-flued boilers are installed within a plant room. What ventilation would be required?

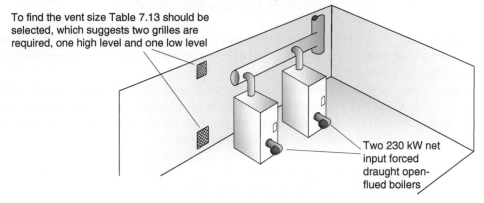

To find the vent size Table 7.13 should be selected, which suggests two grilles are required, one high level and one low level

Two 230 kW net input forced draught open-flued boilers

Step 1: The appliances are located in a plant room, therefore, Table 7.13 is selected. They are of the same type, therefore, it is best to take the sum total of the two appliances in determining the ventilation needs. Thus, $2 \times 230 = 460$ kW is allowed for.

Step 2: The appliances are open-flued so the first row is selected.

Step 3: The table suggests two grilles: one at high level and one at low level, both taking their air from outside. The high-level grille is half the size of the low-level grille.

Thus, the calculation would be as follows:

Low-level grille $= 460 \times 4 = \underline{1840\ cm^2}$; and

high-level grille $= 1840 \div 2 = \underline{920\ cm^2}$.

Example 7.13 A 105 kW net input natural draught open-flued boiler is installed within a balanced compartment.

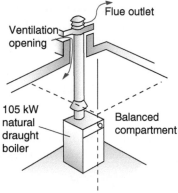

Flue outlet

Ventilation opening

105 kW natural draught boiler

Balanced compartment

The size of the ventilation for this will be found in Table 7.16. The first row is selected, as the appliance is <500 kW, suggesting a permanent vent at high level of 10 cm^2 per kW.

Therefore: $10 \times 105 = \underline{1050\ cm^2}$.

If the boiler were above 500 kW, say 750 kW net heat input, using the second row of Table 7.16, the ventilation would be calculated as follows:

8 cm^2 for every kW in excess of 500 kW plus 5000 cm^2

$$\therefore (750 - 500) \times 8 \text{ cm}^2 = 2000 \text{ cm}^2$$

Plus 5000 cm^2

Total high-level ventilation 2000 + 5000 = 7000 cm^2

Example 7.14   A 250 kW direct-fired air heater is installed in a plant room, which discharges its heated air and products of combustion into a warehouse via ducting. State the ventilation requirements for the plant room, where this is to be provided by natural means.

For this example, the second row of Table 7.10 would need to be checked, which suggests a low and a high-level grille are required, both having a free air intake of at least 2.5 cm^2 per kW in excess of 70 kW plus an allowance of 350 cm^2.

Therefore 2.5 × (250–70) + 350 = 800 cm^2.

This size of vent would be required directly to the outside at both high and low levels within the plant room.

In addition to the calculated air requirement, it is essential that the warehouse environment is monitored to ensure that the CO$_2$ levels do not exceed 0.28%.

Example 7.15   A natural draught open-flued boiler is installed within a plant room. Mechanical ventilation is to be provided. The maximum net heat input of the appliance is 154 kW. State the required mechanical ventilation rate.

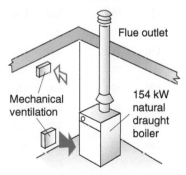

Table 7.14 is selected for this and because the boiler has a draught diverter, row one is chosen. This suggests the following:

Minimum low-level 'air intake' to be 2.8 m^3/h per kW.

Therefore 2.8 × 154 = 431.2 m^3/h.

Minimum high-level extract rate to be 0.73 m^3/h per kW.

Therefore 0.73 × 154 = 112.42 m^3/h.

7.33

Domestic Ventilation Ready Reckoner

Table 7.17   Open flued appliance combustion ventilation and compartment ventilation

Net heat input kW	Open-flued appliance combustion ventilation (air supply to room)	Compartment ventilation					
		Open-flued				Room-sealed	
		Vented to outside		Vented into room		Vent to outside high and low	Vent into room high and low
		Low	High	Low	High		
1	0	10	5	20	10	5	10
2	0	20	10	40	20	10	20
3	0	30	15	60	30	15	30
4	0	40	20	80	40	20	40
5	0	50	25	100	50	25	50
6	0	60	30	120	60	30	60
7	0	70	35	140	70	35	70
8	5	80	40	160	80	40	80
9	10	90	45	180	90	45	90
10	15	100	50	200	100	50	100
11	20	110	55	220	110	55	110
12	25	120	60	240	120	60	120
13	30	130	65	260	130	65	130
14	35	140	70	280	140	70	140
15	40	150	75	300	150	75	150
16	45	160	80	320	160	80	160
17	50	170	85	340	170	85	170
18	55	180	90	360	180	90	180
19	60	190	95	380	190	95	190
20	65	200	100	400	200	100	200
21	70	210	105	420	210	105	210
22	75	220	110	440	220	110	220
23	80	230	115	460	230	115	230
24	85	240	120	480	240	120	240
25	90	250	125	500	250	125	250
26	95	260	130	520	260	130	260
27	100	270	135	540	270	135	270
28	105	280	140	560	280	140	280
29	110	290	145	580	290	145	290

Table 7.17 (Continued)

Net heat input kW	Open-flued appliance combustion ventilation (air supply to room)	Compartment ventilation					
		Open-flued				Room-sealed	
		Vented to outside		Vented into room		Vent to outside high and low	Vent into room high and low
		Low	High	Low	High		
30	115	300	150	600	300	150	300
31	120	310	155	620	310	155	310
32	125	320	160	640	320	160	320
33	130	330	165	660	330	165	330
34	135	340	170	680	340	170	340
35	140	350	175	700	350	175	350
36	145	360	180	720	360	180	360
37	150	370	185	740	370	185	370
38	155	380	190	760	380	190	380
39	160	390	195	780	390	195	390
40	165	400	200	800	400	200	400
41	170	410	205	820	410	205	410
42	175	420	210	840	420	210	420
43	180	430	215	860	430	215	430
44	185	440	220	880	440	220	440
45	190	450	225	900	450	225	450
46	195	460	230	920	460	230	460
47	200	470	235	940	470	235	470
48	205	480	240	960	480	240	480
49	210	490	245	980	490	245	490
50	215	500	250	1000	500	250	500
51	220	510	255	1020	510	255	510
52	225	520	260	1040	520	260	520
53	230	530	265	1060	530	265	530
54	235	540	270	1080	540	270	540
55	240	550	275	1100	550	275	550
56	245	560	280	1120	560	280	560
57	250	570	285	1140	570	285	570
58	255	580	290	1160	580	290	580
59	260	590	295	1180	590	295	590
60	265	600	300	1200	600	300	600
61	270	610	305	1220	610	305	610

(Continued)

Table 7.17 (Continued)

Net heat input kW	Open-flued appliance combustion ventilation (air supply to room)	Compartment ventilation					
		Open-flued				Room-sealed	
		Vented to outside		Vented into room		Vent to outside high and low	Vent into room high and low
		Low	High	Low	High		
62	275	620	310	1240	620	310	620
63	280	630	315	1260	630	315	630
64	285	640	320	1280	640	320	640
65	290	650	325	1300	650	325	650
66	295	660	330	1320	660	330	660
67	300	670	335	1340	670	335	670
68	305	680	340	1360	680	340	680
69	310	690	345	1380	690	345	690
70	315	700	350	1400	700	350	700

Note: Where gross heat input is given by appliance manufacturer divide by 1.1 to calculate net heat input kW figure

8

Gas Installer Responsibility

Working on Gas Installations and Appliances

8.1   All new gas installation work must be carried out in accordance with the Gas Safety (Installation and Use) Regulations, Building Regulations, relevant British Standards and, where appropriate, the appliance manufacturer's instructions. In addition, any new appliances installed must conform to the Gas Appliance (Safety) Regulations.

Definition of 'Working' on a Gas Installation or Appliance

8.2   Many of the Gas Regulations imply that various checks and tests need to be undertaken. However, it would not be possible to trade within the industry if you took ownership and responsibility for every part of a gas installation that you worked on. So, what are you responsible for? Well, first, you have a duty of care. This means that if you know or suspect that a system or appliance is operating incorrectly, you must take reasonable steps to make the installation safe. This includes all your observations on the system, not only the part on which you have been working. This is discussed later in the book (Gas Industry Unsafe Situations, Sections 8.20–8.26). What is deemed as working on an installation or appliance? Regulation 2 of the Gas Regulations clearly defines 'work' and includes any of the following activities:

- installing, or re-connecting;
- repairing, maintaining or servicing;
- disconnecting;
- purging (when simply reinstating the gas supply, purging through an appliance is not deemed as work unless it affects safety).

Further study of the regulations identifies the section of the system that you will need to take responsibility for, and ensure that it is gas-tight. This includes all parts of the installation from nearest valves upstream and downstream of the section on which you are working. For example, if you visit a property to service a boiler in a garage, all the work could be undertaken without entering the property, as the local isolation valve will be adjacent to the appliance. In addition, most of the work could be undertaken at the boiler; the only other gas pipework you may expect to see would be at or around the gas meter.

Gas Installation Technology, Third Edition. Andrew S. Burcham, Stephen J. Denney and Roy D. Treloar.
© 2024 John Wiley & Sons Ltd. Published 2024 by John Wiley & Sons Ltd.

Thus, should there subsequently be an incident in relation to a gas appliance in the house, you could not be expected to have known about it and, therefore, are not responsible for it. However, gas operatives do have additional responsibilities if an appliance is encountered whilst working on another appliance, when tightness testing or purging, for example. For further details, see Section 8.4.

Mandatory Checks When Working on a Gas Appliance

8.3 After undertaking any work on a gas appliance, gas engineers are required to carry out certain checks and tests to ensure that the appliance and any associated flue are operating safely. Therefore, you must be familiar with Regulation 26: Gas Appliances – Safety Precautions, and in particular, paragraph 9. This regulation details the minimum essential tasks that must be undertaken whenever working on a gas appliance. These checks must include:

a) the effectiveness of any flue;
b) the combustion air supply;
c) the operating pressure or heat input, or where necessary both (if it is not reasonably practicable to test the operating pressure or heat input, or where necessary both, then the combustion performance must be checked) and
d) the safe functioning of the appliance.

Appendix 3 of the Gas Regulations provides details of the appropriate tests, checks and examinations that would satisfy the requirements of Regulation 26 (9), these would include confirmation of the following:

- the appliance is suitable and correctly set for the gas type being supplied;
- the appliance is suitable for the room in which it is installed and has adequate ventilation provision for combustion and cooling;
- there is suitable and adequate provision for the removal of combustion products to outside air. This must include a visual inspection of the flue and any necessary checks, e.g. flue flow and spillage tests;
- where these can be checked, confirm that the operating pressure and heat input match the settings provided by the manufacturer. Where it is not possible to check the operating pressure or heat input, the combustion performance of the appliance must be checked (see Section 8.53);
- the safe functioning of an appliance, e.g. confirm the correct operation of safety devices, case seals and fail-safes, that the appliance is stable and secure, appliance gas tightness and that it has the correct flame picture.

In addition, reference should be made to the manufacturer's instructions for any appliance-specific tests.

Visual Risk Assessment of Gas Appliances

8.4 As described in Section 8.2, there will be occasions when a gas operative needs to visually inspect other gas appliances within a property. For example, if a boiler is being serviced in a kitchen and a cooker is installed in the same room, you would be required to visually inspect the cooker for any defects. Table 8.1 gives guidance on the minimum visual checks that should be performed to ensure that no appliance is left in an unsafe situation. These visual checks are the minimum recommended where no other gas work has been undertaken on the particular appliance. If work has been undertaken on the appliance, then in addition to the checks outlined in Table 8.1, a number of further mandatory checks must be performed as described in Section 8.3. There is currently no requirement to record the details of a visual risk assessment, but it would be wise to at least record that the assessment has taken place.

If, during a visual inspection, an unsafe situation is identified or suspected, further investigation will be required, and recommendations for remedial action given to the responsible person. Where it is not possible or feasible to rectify any unsafe situation, the current Gas Industry Unsafe Situations Procedure (GIUSP) shall be followed – see Section 8.20.

Table 8.1 Minimum required checks when conducting a visual risk assessment of gas appliances

Minimum required visual checks	Scenario		
	A tightness test is performed on an installation, and the appliance formed part of the test	Where work has been undertaken elsewhere on a system and an appliance is purged of air	An appliance is encountered whilst working on another appliance
Location	Yes	Yes	Yes
Secure and stable	Yes	Yes	Yes
Signs of distress	Yes	Yes	Yes
Flue	Yes	Yes	Yes
Ventilation	Yes	Yes	Yes
Flame picture	Considered best practice	Yes	

Notes: Details of checks required:
Location: Confirm the appliance has been installed in a suitable location that meets the requirements of the Gas Regulations. For example, non-room-sealed appliances must not be installed in a room that contains a bath or shower.
Secure and stable: Check that the appliance is fixed securely and that it is stable. For example, check freestanding cooker chain/bracket.
Signs of distress: Signs of distress could include discolouration or heat damage on the appliance or surrounding building material.
Flue: Confirm the appliance is suitably flued and has provision for the removal of products of combustion to outside air.
Ventilation: Where necessary, is there suitable and sufficient ventilation for combustion and cooling purposes to allow the appliance to operate safely?
Flame picture: Light all burners on the appliance and check the flame picture. Is it correct for the appliance type?
Source: Data from IGEM/G/11

Commissioning

8.5   When a gas appliance, or whole system of pipework, has been installed, the installation/appliance needs to be commissioned. Commissioning is the checking of the installation to make sure that it is working safely; where this is not completed, the appliance **must** be disconnected from the gas supply and sealed with an appropriate fitting. A label should also be attached to the appliance stating that the appliance should not be reconnected to the gas supply unless it can be fully commissioned. Commissioning a system will identify any operational defects, and if this turns out to be, for example, a poor flame picture, ineffective flue or inadequate gas pressure, then it may be an installation defect rather than a fault in the appliance itself. However, if an appliance is faulty, stripping it down to find the cause of the problem may well invalidate any warranty or guarantee, therefore, the manufacturer should be contacted.

Commissioning is not restricted to a new appliance and following any work on an old appliance, 're-commissioning' would be needed. This is because it is the last person to have worked on an appliance who deems whether the appliance is safe to use.

Gas Work Notification

8.6 The Building Regulations (England and Wales) require that the installation or replacement of any heat-producing gas appliance be notified to the Local Authority Building Control Department. This applies to both domestic and non-domestic buildings and includes any associated heating or hot water services connected to the gas appliance.

Gas operatives have a legal obligation to self-certify that certain gas work they carry out has been commissioned in accordance with the manufacturer's instructions and complies with Building Regulations. They can fulfil this legal obligation by notifying any relevant gas work through the Gas Safe Register who will, for a nominal fee, send the building owner a certificate of compliance and notify the relevant Local Authority that the work has been suitably completed.

There is currently no mandatory requirement to notify gas work in Scotland or Northern Ireland. However, gas operatives can voluntarily submit details of gas work undertaken in these regions to the Gas Safe Register who will in turn provide the customer with a Declaration of Safety Certificate.

The 'Benchmark' Scheme

8.7 Installers of heating and hot water appliances are legally required to complete a commissioning checklist to confirm that the appliance has been installed in accordance with the manufacturer's instructions and in line with current Building Regulations (England and Wales). 'Benchmark' is a nationally recognised scheme for appliance manufacturers and installers to ensure that the installation, commissioning and servicing of heating and hot water appliances are carried out by a competent person and meet appropriate Building Regulations.

Benchmark Scheme members, including all boiler manufacturers in the United Kingdom, currently include a Benchmark Commissioning Checklist and Service Record on the inside back pages of their product installation instructions. *Note*: For boiler installations paper-based records are being phased out in favour of a digitised version of the checklist via the Benchmark app.

Installers should complete the commissioning documentation in full and leave this with the customer. The customer should also sign the document to confirm that they have received a full and clear explanation of the appliance operation instructions. In addition, scheme members have agreed that the completion of the Benchmark Commissioning Checklist is a condition of the manufacturer's warranty.

Record of Work Undertaken

8.8   To ensure that nothing is left to chance, the gas engineer may well follow a checklist, such as the one shown in Figure 8.1b, filling in the details that are applicable to the work undertaken. These records of work are not just a tool to assist the engineer to complete the work, they also provide evidence of the actual work completed. The record of work checklist illustrated is generic, covering a wide spectrum, and is one of many found throughout this book. Other records relate to specific appliances (for example, see 'Commissioning and Servicing Records' for domestic appliances in Part 9). These records should be copied or tailored, if required, to suit your needs, inserting your company details and logo, and then used every time that a system or appliance is worked on. Alternatively, a number of commercial organisations produce pre-designed duplicate forms for purchase.

An explanation of the tasks included in the checklist shown in Figure 8.1b is given in Figure 8.1a. This contains a rationale explaining why a test is undertaken and how each task is executed along with the section reference for further study. *Note*: The checklist shown allows for only one appliance, therefore, where several appliances are to be commissioned, several forms will be needed.

A commissioning checklist can be indefinite in its design, and the version shown only illustrates the advantage of using some form of guide to ensure that you have undertaken all that is possible to make sure that the gas appliance/s are working safely. Without a checklist, it is easy to miss something. The manufacturer's instruction booklet will list other specific items to be checked but may not include all tasks listed on the checklist (see also Figure 8.2).

Commissioning Gas Installation Work Undertaken by Others

8.9   Remember, for properties to which the Gas Regulations apply, only a Gas Safe registered engineer can carry out gas work. Therefore, it is not permissible for a registered engineer to commission and 'sign off' work carried out by a non-registered individual. If this were to happen, both the registered and non-registered parties could face prosecution. For example, while a non-registered individual can install the water side of a heating system, a Gas Safe registered engineer must be engaged to install the boiler, flue, gas pipework and even make the final water connections to the boiler.

Knowingly commissioning gas installation work undertaken by a non-registered person is illegal and also supports these activities. If a registered business is considering commissioning work that they have not undertaken they need to consider this very seriously. Regulation 33 makes it clear that it is the commissioning engineer who takes full responsibility for the installation. This would include confirming that the appliance or installation meets the regulations in full and the requirements of the manufacturer's installation instructions. If Gas Safe were to subsequently carry out an inspection and find defects, it would be the commissioning engineer who would be held responsible and would be required to undertake any remedial works.

	Rationale for the checklist shown in Figure 8.1b and when to undertake checks/test	Section Reference
1.	**Appliance data badge details:** These should always be entered as they indicate the appliance on which you have been working and, therefore, allow your findings to be cross-referenced with the appliance data. Also, check that the appliance is suitable for gas use.	–
2.	**Visual inspection of pipework:** This must be undertaken on all occasions. Inspect pipework within the section on which you are working.	Part 4
3.	**System tightness test:** This would be undertaken on all new installations and any other circumstance where the work performed might affect the gas tightness or has caused the pressure to be lost.	Part 5
4.	**Operating pressure of meter regulator:** This check is to verify that operating pressure at the outlet of the meter is within the correct range for the safe operation of the installed gas appliances. It is often undertaken to investigate a problem where insufficient gas is available at the appliance inlet. It should also be verified that the meter regulator is adequately sealed.	2.56
5.	**Standing/lock-up pressure:** When undertaking the testing of a new installation, or at times when the manometer has been connected to the gas meter, confirm the regulator locks up at a pressure not exceeding 30 m bar for natural gas and for propane or butane within the tolerances indicated in Part 3: Table 3.2.	2.56
6.	**Purging** When supplying gas for the first time or reinstating the supply that has been depressurised.	Part 5
7.	**Appliance complying with the manufacturer's instructions:** This is clearly one of the most important checks that needs to be undertaken and should be seen as a major requirement to ensure safe operation. The manufacturer also identifies specific checks that may need to be undertaken.	–
8.	**Appliance level and secure:** No appliance can be expected to be safe where inappropriate movement is allowed.	–
9.	**Clearance from combustible materials:** The appliance and flue system should be installed so as to prevent heat distress or ignition to any part of the building material.	6.41, 12.8
10.	**Preliminary electrical checks:** Five required electrical checks have been included here and must only be undertaken by competent operatives. Where gas installers do not possess the knowledge and skills of this essential electrical work, they should have this work inspected by a competent electrician.	15.13
11.	**Appliance operating pressure/inlet pressure:** Where appropriate, the burner pressure should be checked to ensure that it is in accordance with the manufacturer's data. This is a mandatory check that should be undertaken every time you work on such an appliance. On occasions, it may not be possible to check the appliance burner pressure, e.g., appliances with an air/gas ratio valve. In these instances, the inlet pressure should be confirmed as adequate, and the heat input should be checked using the index on the meter. Where a meter is not fitted, and it is not possible to check either the operating pressure or heat input, the combustion performance must be checked.	2.56
12.	**Heat input and Gas rate:** A mandatory check that will confirm that the correct injector has been installed in relation to the pressure and that the correct amount of gas is being consumed over a defined period of time.	2.66–2.73
13.	**Main burner flame picture good** A check that allows you to identify combustion problems generally a stable blue flame is sought.	2.32

Figure 8.1a Rationale to commissioning checklist shown in Figure 8.1b

	Rationale for the checklist shown in Figure 8.1b and when to undertake checks/test	Reference
14.	**Pilot flame correct:** For appliances with a permanent pilot flame, this check looks to see that the flame is playing on the correct location, allowing for smooth ignition. It also confirms that it is not unnecessarily large, so wasting fuel and prematurely shortening the life of the thermocouple.	–
15.	**Flame supervision device operational:** Another mandatory check confirming that the appliance would shut down in the event of a flame failure.	3.46
16.	**Thermostat operational:** Again another mandatory check to confirm that the appliance goes off as the temperature reaches its designed safe limit.	3.24, 3.47
17.	**Ignition devices:** Spark ignition devices etc., should be checked to ensure trouble-free operation; this includes electrodes and leads.	3.52
18.	**Appliance tightness test:** A test, often overlooked, on the gas pipework and controls downstream of the appliance isolation valve. The test is achieved by use of leak detection fluid, or a gas detector when the gas is flowing to the burners. There is also a specific 'Appliance Connection Test', which is applied in a commercial setting.	5.44
19.	**Ventilation:** Where appropriate, the size and location need to be confirmed. Included are checks for combustion and cooling air.	Part 7
20.	**Visual inspection of flue system:** Many items need to be reviewed here including the materials used, flue route, termination location, fire hazards, cleaning access into the flue system etc. The list is so great that it warrants its own checklist, and an example is shown in Figure 8.2.	Figure 8.2
21.	**Flue flow test** A must on every occasion where an open-flued appliance is in use. The flue flow test gives a good indication that the flue draught is sufficient to meet the needs of the appliance. It also helps identify any potential leaks. However, it should not be confused with a flue soundness test, where the smoke is trapped in the flue system.	6.49
22.	**Case seals intact and in good condition:** An essential check to confirm the safe operation of an appliance and that no products of combustion are able to escape into the room.	6.65
23.	**Spillage test:** Again, a necessary test when working on an open-flued appliance. The test is designed to confirm that the appliance products of combustion are being effectively evacuated to outside air by the flue system under the worst possible conditions.	6.51
24.	**Combustion performance analysis:** These checks may not be necessary for all types of gas appliances, but it is a test that could be undertaken to confirm the safe operation of a flue system. It must be understood, however, that the reading is only a window event, and in a poor installation, the reading may alter for the worse in the fullness of time as the air within the room becomes vitiated. Note: Since 1 April 2014, it has been a requirement to record the *CO* and combustion ratio on the Benchmark Commissioning Checklist when commissioning a condensing boiler incorporating an air/gas ratio control valve.	8.53–8.62
25.	**Appliance efficiency:** For the domestic appliance this is not an essential test. For the commercial appliance, it may form part of the commissioning procedure to ensure the correct appliance setup. Often, part of the readings provided by a flue gas analyser include the appliance efficiency.	
26.	**Working pressure drops across system:** Often undertaken when installing a new appliance to confirm that the pipework is of adequate size.	2.46, 2.52, 2.53

Note: The numbering in this figure relates to the numbers given to specific tasks in Figure 8.1b

Figure 8.1a *(Continued)*

Gas Installation Commissioning Checklist				
Gas Installer Details	**Client Details**	1.	**Appliance Date Badge Details**	
Name:	Name:		Model/Serial No:	
Gas Safe. Reg. No:	Address:		Gas Type: Natural ☐ LPG ☐	
Address:			Heat Input: max .....kWmin... ..kW	
			Burner Pressure Range:.....-..... mbar	
			Gas Council No:.........................	
2.	**Preliminary System Checks:** Visual inspection of pipework		PASS ☐ FAIL ☐	
3.	Tightness test, to include let-by		PASS ☐ FAIL ☐ N/A ☐	
4.	Operating pressure of meter regulator (......mbar)		PASS ☐ FAIL ☐ N/A ☐	
5.	Standing/lock-up pressure of system (Lock-up ......mbar)		PASS ☐ FAIL ☐ N/A ☐	
6.	Appliance/system purged of air		PASS ☐ FAIL ☐ N/A ☐	
7.	Appliance complies with manufacturer's instructions		PASS ☐ FAIL ☐	
8.	Appliance level and secure		PASS ☐ FAIL ☐	
9.	Clearance from combustible materials		PASS ☐ FAIL ☐	
10.	**Preliminary Electrical Checks:** Conductors secure		PASS ☐ FAIL ☐ N/A ☐	
	Earth continuity and bonding are maintained		PASS ☐ FAIL ☐ N/A ☐	
	Polarity correct		PASS ☐ FAIL ☐ N/A ☐	
	Insulation resistance (> 1 M Ohm)		PASS ☐ FAIL ☐ N/A ☐	
	Fuse rating: ......amp		PASS ☐ FAIL ☐	
11.	**Gas Utilisation Checks:** Appliance operating/inlet pressure		mbar ...... PASS ☐ FAIL ☐	
12.	Heat input........kW Gas rate.........m^3		PASS ☐ FAIL ☐ N/A ☐	
13.	Main burner flame picture is good		PASS ☐ FAIL ☐ N/A ☐	
14.	Pilot flame correct		PASS ☐ FAIL ☐ N/A ☐	
15.	Flame supervision device operational		PASS ☐ FAIL ☐ N/A ☐	
16.	Thermostat operational		PASS ☐ FAIL ☐ N/A ☐	
17.	Ignition devices		PASS ☐ FAIL ☐ N/A ☐	
18.	Appliance tightness test		PASS ☐ FAIL ☐	
19.	**Flue and Ventilation Checks:** Combustion ventilation grille size		cm^2......PASS ☐ FAIL ☐ N/A ☐	
19.	Cooling ventilation grille sizes: High......cm^2; Low...... cm^2		PASS ☐ FAIL ☐ N/A ☐	
20.	Visual inspection of a flue system		PASS ☐ FAIL ☐ N/A ☐	
21.	Flue flow test		PASS ☐ FAIL ☐ N/A ☐	
22.	Case seals are intact and in good condition		PASS ☐ FAIL ☐ N/A ☐	
23.	Spillage test		PASS ☐ FAIL ☐ N/A ☐	
24.	Combustion performance analysis CO.......ppm CO^2......% ratio...........CO/CO_2		PASS ☐ FAIL ☐ N/A ☐	
25.	Appliance efficiency		%...........PASS ☐ FAIL ☐ N/A ☐	
26.	**Post System Checks:** System design pressure loss ......mbar (max. 1 mbar NG/2 mbar LPG)		PASS ☐ FAIL ☐ N/A ☐	
	Safe operation of appliance explained to customer		YES ☐ NO ☐	
Recommendations and/or Warning Notice issued: YES ☐ NO ☐				
Benchmark Commissioning Checklist completed: YES ☐ N/A ☐				
Appliance Safe to Use YES ☐ NO ☐				
Date:			Service Due Date:	
Installer's Signature:		Customer's Signature:		

Note: For an explanation of the number references see Figure 8.1a

Figure 8.1b   Example of a gas commissioning checklist

Chimney/Flue Inspection and Testing Checklist			
Chimney notice plate correctly located	Yes ☐ No ☐ N/A ☐		
Catchment space correct size and clear of debris	Yes ☐ No ☐ N/A ☐		
Fireplace surround sealed	Yes ☐ No ☐ N/A ☐		
Closure plate sealed	Yes ☐ No ☐ N/A ☐		
Pre-cast flue: Correctly constructed Catchment space correct (usually 3 starter blocks) Catchment space sealed	Yes ☐ No ☐ N/A ☐ Yes ☐ No ☐ N/A ☐ Yes ☐ No ☐ N/A ☐		
Hearth construction	Suitable ☐ Unsuitable ☐ N/A ☐		
Materials used and jointing method	Suitable ☐ Unsuitable ☐		
Flexible liner correctly sealed at base and terminal (Including debris plate and annular space sealed)	Yes ☐ No ☐ N/A ☐		
Is the flue route suitable and continuous?	Yes ☐ No ☐		
Flue route clear of obstructions	Yes ☐ No ☐		
Bends (for natural draught no greater than 45°)	Suitable ☐ Unsuitable ☐ N/A ☐		
Cleaning access	Suitable ☐ Unsuitable ☐ N/A ☐		
Openings into chimney correctly sealed (e.g., pipework and annular space)	Yes ☐ No ☐ N/A ☐		
Free from damage, breaks and corrosion	Yes ☐ No ☐		
Does the flue only serve one appliance?	Yes ☐ No ☐		
Draught divertor correctly installed and in good condition	Yes ☐ No ☐ N/A ☐		
Minimum of 600 mm vertical flue above draught divertor	Yes ☐ No ☐ N/A ☐		
Flue supports	Suitable ☐ Unsuitable ☐		
Terminal position	Suitable ☐ Unsuitable ☐		
Terminal	Suitable ☐ Unsuitable ☐ N/A ☐		
Terminal guard required	Yes ☐ No ☐		
Terminal guard clearance from flue adequate (minimum 50 mm)	Yes ☐ No ☐ N/A ☐		
Ridge terminal sealed and secure	Yes ☐ No ☐ N/A ☐		
Signs of spillage	Yes ☐ No ☐		
Clearance from combustible materials (minimum 25 mm)	Suitable ☐ Unsuitable ☐		
Fire stop maintained	Yes ☐ No ☐ N/A ☐		
Fan flow proving device	Suitable ☐ Unsuitable ☐ N/A ☐		
Flue soundness test	Pass ☐ Fail ☐ N/A ☐		
Flue flow test	Pass ☐ Fail ☐ N/A ☐		
Spillage test	Pass ☐ Fail ☐ N/A ☐		
Concealed room sealed flue accessible with suitably sized and located inspection hatches (300 x 300 mm & no more than 1.5 m from a joint)	Yes ☐ No ☐ N/A ☐		
Room sealed flue integrity test $O_2 \geq 20.6\%$ $CO_2 \leq 0.2\%$	Pass ☐ Fail ☐ N/A ☐ Yes ☐ No ☐ Yes ☐ No ☐		
Combustion performance analysis CO.......ppm CO^2....... % CO/CO_2 ratio.......	Pass ☐ Fail ☐ N/A ☐		

Figure 8.2 Chimney/flue inspection and testing checklist

Servicing Gas Appliances/Installations

Servicing

8.10   In general, the maintenance or servicing of any gas appliance should be undertaken at 12-monthly intervals. However, it is essential to observe the manufacturer's instructions, which may give further guidance. Consideration should also be given to the environment into which the appliance has been installed: for example, how often is the appliance used? What is the state of the surrounding environment? Are there vulnerable people within close proximity? Is there a large volume of dust, etc., produced within the environment or is there a presence of 'black dust'? (see Section 8.12). Factors such as these may well lead you to recommend more frequent inspection and maintenance.

Servicing of an appliance can be described as examining the working of the components and ensuring that it is operating as the manufacturer intended by appropriate cleaning, re-greasing, replacing and exchanging of parts not fit for use. Invariably the manufacturer's instructions will give specific details of the servicing for an appliance. The checklist shown in Figure 8.3 may act as a guide to the many tasks to review. The service concludes with the re-commissioning of the appliance as previously described.

Unfortunately, the servicing of many appliances is sometimes neglected, and often the customer only calls out the gas engineer when a fault/problem occurs. With this in mind, always try to ascertain the general performance of the appliance prior to any work and, above all, check its full operation before starting to identify if any faults already exist. This simple act can save later misunderstandings with the customer.

Why Service an Appliance?

8.11   There are many reasons for undertaking a service including maintaining maximum efficiency, ensuring continued trouble-free operation and ensuring continued safe operation. Air currents, whether caused by natural convection or forced draught systems, may cause dust, animal fur, lint, etc., to partially block the passages through which the air, gas and flue products pass. If these spaces are restricted, the way in which the gas and products flow is affected. For example, if the primary airway to a burner becomes blocked, insufficient air will be drawn in to support complete combustion. If the gas injector is blocked, the gas rate will be reduced, and primary air intake will be increased, again affecting combustion.

Black Dust

8.12   Where high levels of hydrogen oxide, which is found naturally in gas, are present in the gas supply, it can attack the metal pipework. Where copper is attacked, a black film of copper sulphide is produced on its surface. Where this black dust, as it is commonly called, flakes off from the surface, it can clog injectors and valves and, in extremely bad cases, it can cause a blockage, reducing the appliance heat input.

Gas Service Checklist					
Gas Installer Details	**Client Details**		**Appliance Date Badge Details**		
Name:	Name:		Model/Serial No:		
Gas Safe Reg. No:	Address:		Gas Type: Natural ☐ LPG ☐		
Address:			Heat Input: max...... *kW* min...... *kW*		
			Burner Pressure Range: ..... – ..... *mbar*		
			Gas Council No:		
Component checked, dismantled and/or cleaned			**Remedial Work Required/Notes**		
	Yes	**No**	**N/A**		
Is the appliance operational	☐	☐			
General condition of pipework/appliance	☐	☐	☐		
Air intake/lint arrestor free from dust etc.	☐	☐	☐		
Primary air port clear	☐	☐	☐		
Main injectors correct size and undamaged	☐	☐	☐		
Pilot injectors correct size and undamaged	☐	☐	☐		
Burner surfaces free from blockage	☐	☐	☐		
Burner surfaces free from damage	☐	☐	☐		
Control taps and valves working freely	☐	☐	☐		
Ignition devices, electrodes and leads	☐	☐	☐		
Filters	☐	☐	☐		
Fan louvres free from lint	☐	☐	☐		
Fan motor, air tubes	☐	☐	☐		
Pressure switches and connections	☐	☐	☐		
Combustion chamber and seals effective	☐	☐	☐		
Heat exchanger	☐	☐	☐		
Case seals	☐	☐	☐		
Refractory plaques	☐	☐	☐		
Appliance and flue seals	☐	☐	☐		
Flue inspection/testing check completed	☐	☐	☐		
Debris collection space	☐	☐	☐		
Fan flow and proving switches	☐	☐	☐		
Draught stabiliser	☐	☐	☐		
Thermostats	☐	☐	☐		
Flame supervision device	☐	☐	☐		
Oxygen depletion system or ASD	☐	☐	☐		
Remaining appliance controls	☐	☐	☐		
Other/external controls	☐	☐	☐		
Electrical wiring	☐	☐	☐		
Combustion and cooling ventilation suitable	☐	☐	☐		
Combustion performance analysis	☐	☐	☐		
CO.......*ppm* CO^2....... % CO/CO_2 ratio..............					
Date:	**Next Service Due Date:**				
Installer's Signature:			**Customer's Signature:**		

Figure 8.3 Gas service checklist

Installation of Second-hand Appliances

8.13   The Gas Appliances Regulations, as described in Part 1: Section 1.21, are not applicable to second-hand gas appliances. There are applicable legislation, the Gas Cooking Appliance (Safety) Regulations and the Heating Appliance (Fireguards) Regulations, but no such Regulations are as important as the Gas Safety (Installation and Use) Regulations. These place an obligation on the installer to ensure that the appliance is **safe to use**.

The supply and installation of second-hand appliances frequently pose a problem because it is often difficult to assess the condition of an appliance until the commissioning stage. The manufacturer's instructions are possibly the best guide to the installation of any appliance. However, these may be out of date in terms of compliance with current Gas Regulations and Building Regulations.

The Gas Regulations require the operative to check the physical condition of a previously used gas appliance before it is installed. This is because it may not be possible to check it completely once installed. In addition to the installation of second-hand appliances, this regulation applies even when an appliance is moved from one location to another within the same room.

An appliance that is found disconnected from the gas supply should be treated as second-hand and requires the operative to verify that it is safe for continued use before reinstating it. However, an appliance disconnected for servicing or repair is not considered to be second-hand.

Points to consider concerning second-hand appliances include:

- Are the manufacturer's instructions available?
- Is the data badge intact and legible?
- Does it meet current standards (e.g. oxygen depletion system or atmospheric sensing device incorporated)?
- Have any modifications been made and is it complete?
- Are spare parts available?
- Is the appliance suitable for the gas being used?
- Is its general condition good?
- Has it been tested before and after installation?
- Will it operate safely?

If in any doubt, do not fit the appliance!

Landlords' Gas Safety Responsibilities

8.14   Under Part F of the Gas Regulations it is a requirement that landlords ensure that all the gas appliances and flues under their control are maintained in a safe condition. Furthermore, suitable records must be provided and kept for a minimum period of two years.

In this context, a landlord is anyone who rents out a property under a lease or a licence that is shorter than seven years. Landlords' gas safety duties extend to a wide range of property types and includes but is not limited to:

- leased accommodation;
- rented accommodation, both private sector and local authorities;
- part of a building, such as where a room is let out in private households;
- hotels, bed and breakfast, hostels and bed-sits;
- holiday cottages, chalets, flats and caravans;
- hired out narrow boats used on inland waterways.

Landlords must arrange for a gas safety check to be carried out at least every 12 months on all gas appliances and relevant chimneys/flues that they own. For newly installed gas appliances the safety check must be performed within 12 months of installation.

Gas safety checks also apply to additional appliances such as water heaters and central heating boilers serving a relevant property listed above, even though the appliance itself may not be located in the tenant's accommodation.

8.15   It is an offence for a landlord to have a tenanted property without a valid gas safety record. Landlords can arrange for the gas safety checks to be carried out any time between 10 and 12 calendar months after the previous inspection and still retain the original expiry date. For example, if a Gas Safety Record expires on 31 December 2024, the landlord can arrange for the gas safety checks to be carried out anytime between 31 October and 31 December 2024. The new Gas Safety Record would be given an expiry date of 31 December 2025. To benefit from this flexibility and retain the deadline date, landlords must keep the record until a further two gas safety checks have been carried out as evidence that they have complied with the law. If the gas safety checks are carried out *within* 10 months of the previous record, then this will have the effect of 'resetting the clock' and the new Gas Safety Record will expire 12 months after this new record.

8.16   Although the gas installation pipework is not covered by the annual gas safety check, the Health and Safety Executive (HSE) recommends that landlords request a registered gas engineer to visually inspect and test the installation for gas tightness during the annual gas safety check. This is because landlords are required to ensure that all gas installation pipework is maintained in a safe condition, which will normally involve a programme of regular inspection along with any necessary repairs.

8.17   The landlord has no legal duty to inspect appliances 'owned' by the tenant, such as a cooker or gas fire, or the flue to which such a gas fire is solely connected. The tenant is responsible for the gas safety of appliances that they own and the flue to which they are connected if that flue only serves their appliance. However, as landlords have a duty of care under the Health and Safety at Work Act, once again, the HSE strongly recommends that all flues connected to gas appliances in the property are included in the annual gas safety check, even when they only serve appliances that the landlord does not own. In addition, if a registered gas engineer encounters an appliance which is owned by the tenant, it is recommended that a visual inspection of the appliance is carried out. The results of this inspection should be recorded on the Landlords Gas Safety Record. For details of a visual inspection, see Section 8.4.

8.18   A Landlord's Gas Safety Record form may be purchased from commercial organisations or can be produced by oneself; an example is shown in Figure 8.4. As a

Landlord Gas Safety Record				
Gas Installer Details		**Tenant Details**	**Landlord/Agent Details**	
Company Name:		Name:	Name:	
Gas Safe Reg. No:		Address:	Address:	
Address:				
Tel:		Tel:	Tel:	

Details	Appliance 1	Appliance 2	Appliance 3	Appliance 4
Type				
Location				
Make				
Model				
Flue design	OF☐ RS☐ FL☐	OF☐RS☐ FL☐	OF☐ RS☐ FL☐	OF☐RS☐ FL☐
Owner	Tenant☐ Landlord☐	Tenant☐ Landlord☐	Tenant☐ Landlord☐	Tenant☐ Landlord☐
Full inspection (for all landlord-owned appliances/flues)	Yes☐ N/A☐	Yes☐ N/A☐	Yes☐ N/A☐	Yes☐ N/A☐
Visual inspection (minimum required for tenant-owned appliances)	Yes☐ N/A☐	Yes☐ N/A☐	Yes☐ N/A☐	Yes☐ N/A☐
Operating pressure	......$mbar$	......$mbar$	......$mbar$	......$mbar$
Heat input if applicable	......kW	......kW	......kW	......kW
FSD operational	Yes☐ No☐ N/A☐	Yes☐ No☐ N/A☐	Yes☐ No☐ N/A☐	Yes☐ No☐ N/A☐
Thermostat operational	Yes☐ No☐ N/A☐	Yes☐ No☐ N/A☐	Yes☐ No☐ N/A☐	Yes☐ No☐ N/A☐
Ventilation satisfactory	Yes☐ No☐ N/A☐	Yes☐ No☐ N/A☐	Yes☐ No☐ N/A☐	Yes☐ No☐ N/A☐
General condition	Satisfactory Yes☐ No☐	Satisfactory Yes☐ No☐	Satisfactory Yes☐ No☐	Satisfactory Yes☐ No☐
Visual flue condition	Pass☐ Fail☐ N/A☐	Pass☐ Fail☐ N/A☐	Pass☐ Fail☐ N/A☐	Pass☐ Fail☐ N/A☐
Flue flow	Pass☐ Fail☐ N/A☐	Pass☐ Fail☐ N/A☐	Pass☐ Fail☐ N/A☐	Pass☐ Fail☐ N/A☐
Spillage	Pass☐ Fail☐ N/A☐	Pass☐ Fail☐ N/A☐	Pass☐ Fail☐ N/A☐	Pass☐ Fail☐ N/A☐
Termination	Pass☐ Fail☐ N/A☐	Pass☐ Fail☐ N/A☐	Pass☐ Fail☐ N/A☐	Pass☐ Fail☐ N/A☐
Combustion performance satisfactory	Pass☐ Fail☐ N/A☐ CO.....ppm CO_2.......% CO/CO_2 Ratio.........	Pass☐ Fail☐ N/A☐ CO.....ppm CO_2.......% CO/CO_2 Ratio.........	Pass☐ Fail☐ N/A☐ CO.....ppm CO_2.......% CO/CO_2 Ratio.........	Pass☐ Fail☐ N/A☐ CO.....ppm CO_2.......% CO/CO_2 Ratio.........
Service undertaken	Yes☐ No☐	Yes☐ No☐	Yes☐ No☐	Yes☐ No☐
Approved CO alarm fitted (BS 50291)	Yes☐ No☐ N/A☐	Yes☐ No☐ N/A☐	Yes☐ No☐ N/A☐	Yes☐ No☐ N/A☐
Appliance safe to use*	Yes☐ No☐	Yes☐ No☐	Yes☐ No☐	Yes☐ No☐
Warning notice issued	Yes☐ No☐	Yes☐ No☐	Yes☐ No☐	Yes☐ No☐
Defects and remedial work undertaken				

Installation pipework satisfactory Yes☐ No☐   Tightness Test Satisfactory Yes☐ No☐ N/A☐
Notes/Recommendations:

IMPORTANT NOTE: For appliances **not** owned by the landlord, the 'Appliance safe for use' response is based on a visual check for obvious defects only.

Issue Number:	Issue Date:		

NEXT SAFETY INSPECTION DUE WITHIN 12 MONTHS OF DATE OF ISSUE

Issued by:	**ID No:**	**Received by:**
Name: **Signature:**		**Name:** **Signature:**

Figure 8.4 Example of a Landlord's Gas Safety Record

minimum, it must contain the following information and be numbered so that it can be tracked:

1. Date of the appliance/flue inspection.
2. Address of the property.
3. Name and address of landlord.
4. Description and location of each appliance/flue checked.
5. Any identified defects and remedial action taken.
6. Confirmation that the flue and air supply are effective, the operating pressure and/or heat input are correct and that the appliance is operating safely.
7. The name, signature and Gas Safe registration number of the individual undertaking the checks.

8.19   If a room that contains a gas appliance has been converted into sleeping accommodation, any existing gas appliance needs to be inspected to ensure that it complies with current standards and regulations in force. Where this is not the case, the landlord should ensure the appliance is disconnected or replaced. However, this does not apply if the accommodation was used for such purposes prior to the current Gas Safety Regulations.

When a tenant vacates a property, the landlord needs to ensure that all gas pipework, appliances and flues are safe before re-letting. Occupants may have removed appliances unsafely or left their own appliances in place.

Finally, in addition to the Landlord's Gas Safety Record, landlords have a separate legal duty to provide ongoing maintenance of all gas appliances that they own in a tenanted property. The frequency and scope of maintenance and servicing of a gas appliance will be determined by the appliance manufacturer.

Gas Industry Unsafe Situations

Relevant Industry Document
IGEM/G/11

8.20   *Note*: The procedures outlined in this section cover both domestic and non-domestic premises supplied by Natural Gas or Liquified Petroleum Gas (LPG). Although the Gas Regulations do not generally apply to agricultural premises, sewage works, factories, mines and quarries, the general principles of the information contained in this section may be applied to these premises.

8.21   Regulation 34 of the Gas Regulations stipulates that the responsible person for a premises must not allow the use of a gas appliance if they know, or suspect, that it cannot be used without constituting a danger to any person. In this context, the responsible person can be the occupier/owner of the premises or any person with authority at that point in time to take appropriate action in relation to the property or gas fitting.

Often, the responsible person will become aware of an unsafe situation after being informed by a registered gas engineer. Gas operatives have a legal duty to take all reasonable steps to inform the responsible person for the premises – in writing – of any unsafe or

Note: Minimum recommended size 105 *mm x 74 mm*

Figure 8.5   An example of a 'DANGER DO NOT USE' label

dangerous situations they encounter. The written statement should also make it clear that its continued use is an offence and a contravention of the law. In fact, the Gas Regulations make it an offence for any person to use a gas installation/appliance once they have been informed that it is unsafe (see Figures 8.5 and 8.6).

8.22   When a gas operative encounters an unsafe situation within the scope of their competence, their priority should be to give appropriate advice and take suitable action to protect life and property. Therefore, it is essential that they are able to identify any unsafe gas installation/appliance that presents an immediate or potential danger and take the correct action to prevent this danger from occurring. To assist gas operatives in this endeavour, the gas industry has produced an unsafe situation procedure. The procedure is published by the Institution of Gas Engineers and Managers and entitled *IGEM/G/11 The Gas Industry Unsafe Situation Procedure* (GIUSP). This document is readily accessible to all Gas Safe registered businesses/engineers.

Dealing with Unsafe Situations

8.23   Gas operatives that identify an unsafe gas installation/appliance shall either classify it as:

- **Immediately Dangerous (ID)**

  An ID situation is one that poses an immediate danger to life and/or property, such as a gas leak or combustion products spilling from a flue.

  *Note: S*ome immediately dangerous situations will also be RIDDOR reportable, see Section 8.32.

  Or;

<table>
<tr><td colspan="9" align="center">**Warning Notice**</td></tr>
<tr>
<td colspan="3">**Engineer Details**
Name:
Gas Safe Registration No:
Address:</td>
<td colspan="6">**Customer Details**
Name:
Address:

Telephone No:</td>
</tr>
</table>

	Appliance Details		Defects (tick)					
	Gas Appliance/ Installation Type	**Gas Appliance/ Installation Location**	**Gas Escape**	**Meter**	**Pipework**	**Chimney/ Flue**	**Ventilation**	**Other (specify)**
1								
2								
3								
4								
5								

	Classification (tick)		
	Immediately Dangerous (ID) With permission from the responsible person the installation has been disconnected, sealed from the gas supply and labelled **"Danger Do Not Use"**	**At Risk** With permission from the responsible person the installation has been turned off and labelled **"Danger Do Not Use"**	**At Risk** The installation has been classified At Risk but turning off will not remove the risk **Please follow advice below**
1			
2			
3			
4			
5			

Has permission has been refused?   **YES/NO**   If YES, Gas Emergency Contact Centre Ref. No:

	Faults Found	**Remedial Action Required**
1		
2		
3		
4		
5		

Details of RIDDOR situations identified and reported (where applicable)

I acknowledge receipt of this warning notice and understand that I must not use or permit the use of any unsafe appliance until the defects have been rectified. To do so would put me in breach of the **Gas Safety (Installation and Use) Regulations 1998.**

Received By (Print Name):	Issued By (print name):
Signature:	Signature:
Date:	Date:
	Gas Safe Licence no:

Figure 8.6 Example of a Warning Notice

- **At Risk (AR)**

An AR situation is one that has the potential to pose a danger to life and/or property in the future. For example, an inadequately supported flue or insufficient ventilation but without products of combustion entering the room.

The GIUSP contains many common, and some less common, examples of unsafe situations that may be encountered and gives guidance on the classification of risk and actions to be taken. The GIUSP cannot be expected to contain every conceivable unsafe situation that may exist, so gas operatives are often required to exercise their professional engineering judgement, within the scope of their competency, to determine the risk. If in any doubt, they should seek further expert advice such as that provided by the Gas Safe Register technical helpline; see Part 1: Section 1.2 for contact details.

8.24   If a gas operative identifies an unsafe situation at non-domestic premises which fall outside the scope of the Gas Regulations, the principles outlined in the GIUSP can be used to classify the level of risk and determine the appropriate action to take. However, engineers must consult with the site's responsible person before taking any action (unless in extreme circumstances where a delay in taking action would result in an immediate danger to life and/or property). This is because some industrial process equipment can present an immediate health and safety risk if not shut down in a controlled manner. The responsible person will then use their professional judgement and knowledge of the equipment, along with a risk assessment, to determine the safest course of action. This may at times, deviate from the advice contained in the GIUSP. Whatever the outcome, it is essential that the gas operative keeps full and accurate records of all checks and tests that they perform. Advice regarding the preparation of risk assessments can be obtained from IGEM/UP/16.

Unsafe Situation Procedures

8.25   When a gas operative encounters an unsafe situation, they must follow a prescribed procedure to ensure safety. In most cases, this will involve informing the responsible person of the danger, that the installation/appliance should not be used, and either disconnecting and sealing the gas supply or turning the gas supply off, depending on the level of risk. However, in a limited number of cases, turning off the installation will not remove the risk. This will often be related to faults upstream of the emergency control valve or associated with LPG bulk storage tanks. In these situations, the gas operative should advise the responsible person that the installation is unsafe and who they should contact for further investigation; this will normally be the gas transporter or LPG supplier (see Figure 8.7b).

The principal objective if you have a concern for safety is to protect life and property and make the situation safe. Figure 8.7a and b provide the prescribed course of action that gas operatives should take to meet this legal duty. If you are ever in any doubt, always take the course of action that will ensure safety.

8.26   In some existing installations, you may encounter a situation where the gas installation or appliance does not meet the requirements of current industry standards but is not unsafe. In these cases, engineers will need to use their judgement around what advice to give the gas user or responsible person and the feasibility of its implementation. This advice can be given on a job report or service record (see Figure 8.3).

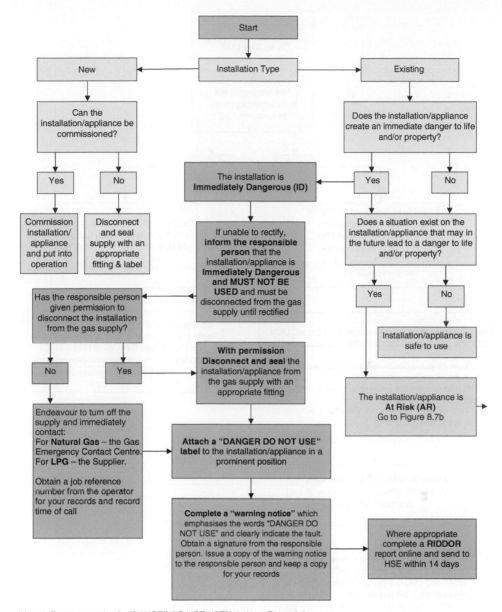

Start

Installation Type
New ← **Existing**

New
Can the installation/appliance be commissioned?
Yes — No

Yes:
Commission installation/appliance and put into operation

No:
Disconnect and seal supply with an appropriate fitting & label

The installation is **Immediately Dangerous (ID)**

If unable to rectify, **inform the responsible person** that the installation/appliance is **Immediately Dangerous and MUST NOT BE USED** and must be disconnected from the gas supply until rectified

Has the responsible person given permission to disconnect the installation from the gas supply?
No — Yes

No:
Endeavour to turn off the supply and immediately contact:
For **Natural Gas** – the Gas Emergency Contact Centre.
For **LPG** – the Supplier.

Obtain a job reference number from the operator for your records and record time of call

With permission Disconnect and seal the installation/appliance from the gas supply with an appropriate fitting

Attach a "DANGER DO NOT USE" label to the installation/appliance in a prominent position

Complete a "warning notice" which emphasises the words "DANGER DO NOT USE" and clearly indicate the fault. Obtain a signature from the responsible person. Issue a copy of the warning notice to the responsible person and keep a copy for your records

Existing
Does the installation/appliance create an immediate danger to life and/or property?
Yes — No

Yes → The installation is Immediately Dangerous (ID)

No:
Does a situation exist on the installation/appliance that may in the future lead to a danger to life and/or property?
Yes — No

No:
Installation/appliance is safe to use

Yes:
The installation/appliance is **At Risk (AR)**
Go to Figure 8.7b

Where appropriate complete a **RIDDOR** report online and send to HSE within 14 days

Note 1: For an example of a "DANGER DO NOT USE" label see Figure 8.5
Note 2: For an example of a warning notice see Figure 8.6

Figure 8.7a   Dealing with unsafe situations

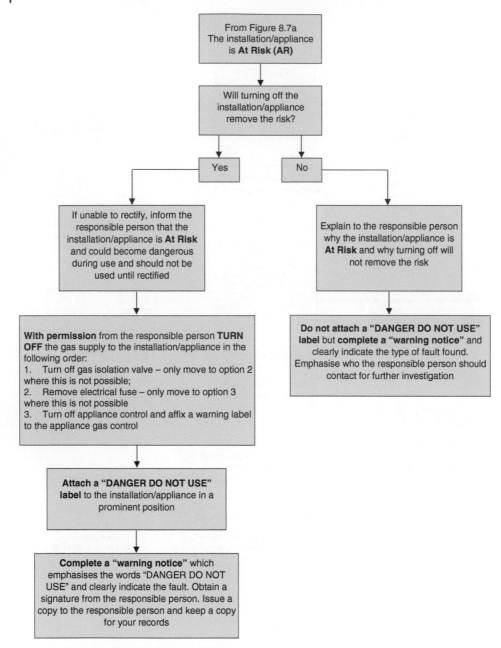

Note 1: For an example of a "DANGER DO NOT USE" label see Figure 8.5
Note 2: For an example of a warning notice see Figure 8.6

Figure 8.7b Dealing with unsafe situations

Gas Escape Procedures

Responsibilities of the Gas User (Responsible Person for the Premises)

8.27　Where the responsible person for the premises knows or suspects that there is a gas leak or fumes escaping into the building, they must act immediately by turning off the appliance and contacting a registered gas engineer. Where the suspected gas leak is on the system, they should endeavour to turn off the gas supply at the emergency control valve. If the smell of gas persists after shutting off at the emergency control valve, then they must report the gas escape immediately to the LPG supplier or to one of the National Gas Emergency call centres listed in Section 8.30. The gas supply must not be reinstated until remedial action has been taken. This information will usually be displayed on the primary meter or, in some circumstances, adjacent to the emergency control valve.

Responsibilities of the Gas Operative (OFF-SITE)

8.28　When a gas operative receives a report of a gas escape or fumes and they are not on-site, the following advice needs to be given to the responsible person:

- **Turn off the gas if it is safe to do so**
  For natural gas:
  Turn off the emergency control valve at the meter. If the emergency control is in a cellar or basement and there is a smell of gas in the cellar or basement, do not enter but immediately evacuate the building.
  For LPG:
  Bulk supply – turn off the isolation valve outside the building;
  Metered supply – turn off at the meter;
  Cylinder supply – turn off all cylinder valves.
  Turn off all sources of ignition, do not operate any light switches and do not smoke.
- Ventilate the building by opening doors and windows and get into fresh air.
- Call the relevant Gas Emergency number, giving the address and location of the gas emergency and contact details.
- Ensure access is made available to the premises for the gas operative and/or emergency service provider.

Responsibilities of the Gas Operative (ON-SITE)

8.29　Where the gas operative is on site, it is clearly possible to instigate the above actions with the responsible person's permission. It must be fully understood, however, that the gas operative, although fully responsible for their own actions, may not have permission to shut off the gas supply. This is usual in a commercial situation where interrupting the gas supply to an industrial process may itself lead to a dangerous situation, possibly resulting in a third-party claim against the gas operative for losses and damage. Where permission is not given to make the situation safe, the gas operative must treat the situation as Immediately Dangerous and follow the procedure previously described under 'Unsafe Situation Procedures' in Section 8.25. See also Figure 8.8.

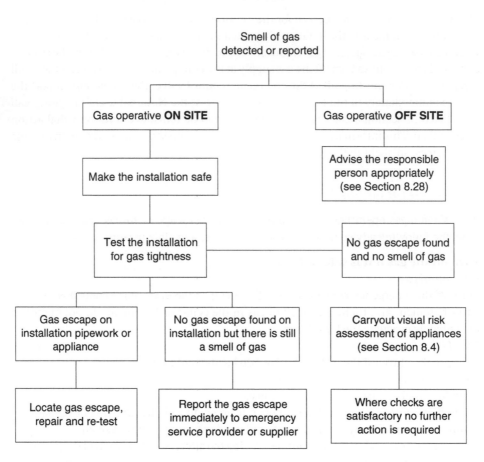

Figure 8.8 Gas escape procedure

When reporting a gas escape to the emergency service provider/LPG supplier, be prepared to provide the following information:

- Address of the property and where the smell is most noticeable.
- Address and telephone number of the gas user.
- Address and telephone number of the person reporting the gas escape.
- Time the gas escape was first discovered.
- Whether the emergency control is turned off or not. If not, the reason why. Also, whether the gas can still be smelt and if it is inside or outside.
- Any special circumstances such as access problems or vulnerable residents.

On reporting the gas escape, the service provider will give you the following information, which you should keep for reference:

- the job number;
- date and time of the report;
- the name of the person to whom you reported the incident.

National Emergency Gas Service Provider and Gas Supplier Contact Details

8.30 Natural Gas

- England, Scotland and Wales: 0800 111 999.
- Northern Ireland: 0800 002 001.
- Isle of Man: 0808 1624 444.
- Guernsey: 01481 749000.
- Jersey: 01534 755555

8.31 LPG

- Bulk storage: see telephone number on the bulk storage vessel or at the meter.
- Cylinder supply: the supplier is often identified on the cylinder. Then, search for the gas supplier details online or in the local telephone directory.

Reporting of Injuries, Diseases and Dangerous Occurrences Regulations (RIDDOR)

8.32 RIDDOR is applicable to all aspects of health and safety. The notes here focus on those areas that affect gas safety. When a situation is encountered that involves the discovery of a dangerous gas installation, you will have to decide if the situation needs to be notified to the Health and Safety Executive (HSE). If the situation meets the following criteria, then it must be notified under RIDDOR:

1. The situation is a result of poor workmanship or design and
2. It is so dangerous that it has caused, or could cause, death, unconsciousness or the taking to hospital of a person.

Most RIDDOR reportable situations will be classified as Immediately Dangerous as defined in the current Gas Industry Unsafe Situation Procedure.

Examples of RIDDOR Reportable Situations

- Poor installation work, resulting in a gas escape outside the allowable tolerances of a tightness test.

- Uncapped or open ends connected to a live gas supply.
- Signs of spillage or combustion problems resulting from bad practices.
- Inappropriate materials are being used, such as a garden hose or plastic plumbing fittings.
- Flued appliances are not being flued correctly.
- Appliances are being supplied with the incorrect gas type.
- Instances where a safety device has been made inoperative.
- If an appliance has become dangerous as a result of faulty servicing.

    Note 1: This list should not be considered exhaustive.

    Note 2: Where an engineer corrects a defect upon discovery, the incident still needs to be reported and failure to do so may result in prosecution.

When and What to Report

8.33   Under RIDDOR Regulation 11, there are two gas safety areas that must be reported:

- RIDDOR 11 (1) – Gas Incident

    This applies when a person has died, been rendered unconscious or taken to hospital as a result of carbon monoxide poisoning, an escape of unburnt gas, or a fire or explosion caused by an escape of gas.

    RIDDOR 11 (1) shall be reported to the HSE by the gas transporter/supplier. The HSE must be notified without delay (within two hours of attending the incident), and a report must be sent online within 14 days of the incident.

    Note: Where a gas engineer encounters a *Gas Incident*, they MUST NOT disturb the scene. However, if it is safe to do so, they should make the installation safe. They should immediately contact the emergency service provider for natural gas or the supplier for LPG, informing them of the incident and, where necessary, contact the emergency services. In addition, record all actions taken, as these may be required for any subsequent investigation.

- RIDDOR 11 (2) – Dangerous Gas Fittings

    Under this legislation, gas businesses or engineers have the responsibility to report to the HSE any dangerous gas appliance/installation that has caused or is likely to cause death, unconsciousness or a person being taken to hospital due to poor workmanship or design.

    RIDDOR 11 (2) reports must be sent to the HSE online within 14 days of discovery.

    All RIDDOR reports can be reported online via the HSE's RIDDOR website: www.hse.gov.uk/riddor/report.htm. The website also provides a telephone number, which should be used for reporting RIDDOR 11 (1) incidents **only.**

Dealing with Unsafe Situations That Are NOT RIDDOR Reportable

8.34   Some gas fittings/installations may not be installed in accordance with current standards or legislation. However, unless there is a good reason to believe that they are dangerous, there is no need to report them to the HSE. It is about questioning the competence of the gas operative who completed the installation in the first place and whether they should be allowed to operate unchecked. If gas engineers wish to report these situations to the HSE as a matter of concern for further investigation, they can do this online via the HSE website below. Registered businesses/engineers can also report illegal gas work and competency concerns to Gas Safe for further investigation via the engineer resource hub on the Gas Safe website.

HSE – www.hse.gov.uk/contact/tell-us-about-a-health-and-safety-issue.htm
Gas Safe Register – www.gassaferegister.co.uk/engineer/resource-hub

Carbon Monoxide Detection

Relevant Industry Documents
Building Regulations, BS EN 50291

8.35   Should the level of carbon dioxide (CO_2) in the ambient air rise to around 1.5–2% (15 000–20 000 ppm) where an appliance is operating, the effect will be that the air will become vitiated (lacking in oxygen). This, in turn, will mean the appliance would produce carbon monoxide (CO) as part of the combustion process. Excessive levels of CO_2 in the air, in excess of 1.5%, can be the pre-cursor to a rapid rise in the production of CO. CO is significantly more toxic to humans and animals than CO_2, and therefore, regulations are in place to detect its presence in dwellings to protect occupants.

Carbon Monoxide Alarms

8.36   It is mandatory that all privately rented homes and social housing have a CO alarm installed in any room used as living accommodation that has a fixed combustion appliance (excluding gas cookers). Building Regulations (England) also require that a CO alarm is fitted in all dwellings whenever a new or replacement fixed gas appliance is installed (gas appliances used solely for cooking are excluded from this requirement). *Note*: Similar legislation is in place in all other regions of the United Kingdom.

8.37   For a CO alarm to be effective, it must be installed and tested in accordance with the manufacturer's instructions and comply with BS EN 50291. The installation of a CO alarm is intended as an additional means of protection where a gas appliance is used and is no substitute for correct installation, adequate maintenance and servicing of the gas appliance.

Electric and Battery Operated CO Detector Alarms

8.38   These alarms need to be installed in the same room as the appliance and preferably be fixed to the ceiling at least 300 mm away from any wall and between 1 and 3 m horizontally from the appliance. If the alarm is to be wall mounted, it should be as high as possible and above any doors or windows but no closer than 150 mm to the ceiling. However, always refer to the manufacturer's instructions for the correct location. BS EN 50291 requires any alarm to activate:

- within 120 minutes where there is 30 ppm of CO in the room;
- within 60–90 minutes where there is 50 ppm of CO in the room;
- within 10–40 minutes where there is 100 ppm of CO in the room;
- within 3 minutes where there is 300 ppm of CO in the room.

CO alarms in dwellings should be 'Type A' battery-powered alarms. The battery should be designed to last for the working life of the alarm and be manufactured with a warning device to alert users when the alarm needs replacing. Alternatively, 'Type A' mains powered alarms with fixed wiring (not plug-in types) may be used as long as they incorporate a warning device that will alert the user if the sensor fails.

CO Indicator Cards

8.39 A 'spot' type detector is not an alarm and only provides a visual indicator. It has an orange spot which turns grey or black when CO is detected in the atmosphere in concentrations as low as 100 ppm. Its performance can be impaired by humidity or even neutralised by halogens, nitrous gases or ammonia (these could be produced from cat litter and babies' nappies). Cleaning solvents will also affect their use. They also have a very limited life span of perhaps only three to six months once removed from their packaging. These indicator cards are not a safe means of CO monitoring and detection and do not meet the requirements of the Building Regulations.

Gas Operative's Response to an Activated CO Alarm

Relevant Industry Documents
BS 7967, IGEM/G/11 Supplement 1, BS EN 50291

8.40 If a gas operative receives a report of fumes, smells, spillage or the activation of a CO alarm, their first priority is to protect life and property. If not on-site at the time of the report, the occupant should be advised to immediately:

- turn off the gas supply to all appliances;
- ventilate the property and get into fresh air;

- seek medical attention if feeling unwell;
- notify the emergency service provider/LPG supplier who will attend to make the situation safe.

8.41 Gas operatives may be asked to attend to the report of fumes or the activation of a CO alarm after the attendance of the emergency service provider (ESP) or LPG supplier. All investigations shall only be performed by engineers that have the competencies to work on all types of gas appliances to be encountered at the property and should follow the basic safety principles in the order listed below:

1. protect life and property;
2. locate all fuel-burning appliances;
3. identify the source of any gas escape, spillage, smells or leaks of combustion products;
4. confirm the correct installation and safe operation of all suspect gas appliances;
5. make safe any defective appliance in accordance with the Gas Industry Unsafe Situations Procedure (GIUSP) and advise the responsible person of any remedial work required to make the installation/appliance safe;
6. complete all reports/warning notices and actions as required by the GIUSP.

If there are appliances in the property that burn fuels other than gas (wood, solid fuel, oil, etc.), then the safe operation of those appliances must be confirmed by a suitably competent person, e.g. HETAS or OFTEC qualified.

8.42 You must always assume CO is present until it has been confirmed otherwise. Therefore, the first task is to use a portable electronic gas analyser, or a personal CO alarm, to ensure you are not entering an environment with dangerously high levels of CO. If using a personal CO alarm, ensure that it is calibrated and used according to the manufacturer's instructions. If the atmosphere is to be checked using a gas analyser, once again, the equipment must be within the calibration date, and the air should be sampled at head height for 60 seconds in every room entered. Where this proves to be in excess of 30 ppm, all occupants must be evacuated from the affected area and advised to relocate to a CO-free atmosphere.

8.43 The highest concentration of CO will likely be within the room where the source of CO exists. However, gas operatives must be aware of the potential for CO to migrate between rooms through open doorways or gaps and cracks in the building structure. This movement can even occur between adjacent properties. Therefore, if the source of CO cannot be identified at the property being investigated, neighbouring properties should be suspected, and the ESP or LPG supplier should be contacted for further investigation.

8.44 A report of fumes, smells, spillage or the activation of a CO alarm may be investigated by a Gas Safe engineer who holds the necessary competencies for all appliances installed at the property without the need for further specialist training. However, in the following situations, the investigation should be escalated to an engineer holding CMDDA1:

Situation	Recommended Action
Failure to identify the source of fumes/smell or CO alarm activation	Classify as ID and follow GIUSP Further specialist investigation is required by an engineer holding CMDDA1
No obvious cause of CO alarm activation has been identified, and all appliances are operating satisfactorily	To rule out other sources of CO or ambient CO, escalate for further specialist investigation by an engineer holding CMDDA1
The gas user reports previous fumes/smells or CO alarm activation at the property within the past three months with no identified cause	If off-site report to the ESP/LPG supplier If on site disconnect and seal the gas supply. Escalate for further specialist investigation by an engineer holding CMDDA1

Source: Data from IGEM/G/11 Supplement 1

Sweep Test for Room-Sealed and Open-Flued Appliances

8.45   A 'sweep test' is an initial test to check for CO in the atmosphere around room-sealed and open-flued appliances.

Prepare the analyser for use by switching it on and zeroing in outside air. For open-flued appliances ensure that all windows and doors are closed and any fans are turned on. Begin in the room where the CO alarm was activated or symptoms were reported, and operate one appliance at a time at full rate. Set the analyser to measure CO and slowly move the sampling probe above, below and all around the appliance and flue system being tested.

Keep the probe end approximately 100 mm away from the appliance/flue and test each appliance in turn for a minimum of two minutes. During the two minutes test ensure a minimum of two passes around the appliance/flue are conducted.

Where CO is detected, the suspect appliance will require a detailed examination and the defect rectified or the appliance made safe by disconnection and sealing from the gas way with an appropriate fitting.

Determining the Ambient Level of CO in a Room (CO Build-up Test)

Relevant ACS Qualification
CMDDA1

Relevant Industry Document
BS 7967

8.46   Where it is suspected that CO is being produced by an appliance, it will be necessary to check its operation as identified below, taking samples of ambient air with a portable electronic analyser. Testing the environment is completed in two stages: first, before any appliances have been operated; and second, with the appliance/s in operation.

Stage 1: Testing for CO before the operation of appliances

- Using a suitable gas analyser conforming to BS EN 50379-3, set to measure CO and position an open-ended sampling probe approximately 2 m above floor level in the centre of the room at least 1 m away from any suspect appliance – see Figure 8.9.

Figure 8.9 A portable electronic gas analyser fitted to a tripod stand for a CO build-up test

- Ventilate the room until the inside and outside levels of CO are the same.
- Close all external windows and doors and again record the CO levels in the room over a period of 15 minutes. If the level of CO starts to rise during the test, then it suggests the CO is migrating from another source, and this should be investigated appropriately, seeking expert advice where required. Possible sources include, among other things: cigarette smoke; other fuel-burning appliances in close proximity to the property; vehicles or generators; landfill sites and drains.

Where no CO levels above ambient are recorded, then proceed to stage 2.

Stage 2: Testing for CO with the appliances in operation

- With suspect appliances operating one at a time, determine the levels of CO for each appliance depending on type as described in the following sections, checking for spillage where appropriate, and record the CO levels in the room. The combustion performance of suspect appliances should also be recorded as described in Section 8.53 onwards.
- Repeat the procedure for all appliances separately and then in combination until it is clear what appliance/s, if any, are responsible for the CO.

Rooms with Open-Flued and Room-Sealed Appliances

8.47   Operate the appliance at its full rate until the CO level in the room stabilises or begins to fall, whichever comes first. Determine the action to take as below:

- CO level less than 10 ppm.
  At these levels the source of CO could be caused by a number of common activities, such as cigarette smoke or vehicle exhausts on a busy road. If the increase of CO is caused by the

gas appliance, then this should be investigated further and, where necessary, appropriate remedial action taken. *Note*: Some open-flued gas appliances may occasionally spill products of combustion during adverse weather conditions. Where the CO can be attributed to this normal operation of the appliance, and there is no potential for higher levels of CO, then at these levels, the appliance can be considered to be operating satisfactorily.

- CO level of 10 ppm or above.
  The appliance requires further investigation. Remedial work should be carried out, and the test should be repeated. If the fault cannot be rectified, the appliance should be classified as Immediately Dangerous.

- If CO levels ever exceed 30 ppm, immediately turn off the appliance, ventilate the room and evacuate the property. This appliance must be treated as Immediately Dangerous. Do not re-enter the property until CO levels are below 10 ppm.

Rooms with Flueless Cookers (Including Built-in Units)

8.48   Ensure all fixed ventilation is unobstructed. Operate the appliance with all external windows and doors closed. Place a saucepan with a flat base of between 160 and 220 mm in diameter containing 1 l of water onto each of the largest two hotplate burners and cover with a lid. Light the burners and turn to maximum. As the water boils, turn it down to a simmer setting and top up with water if necessary to avoid the saucepans boiling dry. At the same time place the grill pan to its highest position, light the grill and set to maximum; turn the grill off after 30 minutes. Also, turn on the oven to its mid-range position or gas mark 5. Where not all burners can be operated, only use those burners applicable. Now record the CO levels at one-minute intervals. If the CO reading:

- begins to fall without exceeding 30 ppm or
- does not exceed 30 ppm for more than 20 minutes and begins to fall but never exceeds 90 ppm at any time it indicates a satisfactory test.

However, it may still be appropriate to consider the cooker to be a source of CO. For example, very large utensils may cause flame chilling, as will damaged grill frets.

If the CO level exceeds 90 ppm at any time, the test must be stopped and the room ventilated to a safe level before further investigation. If it is possible to rectify the fault, do so before re-testing. Alternatively, classify the appliance as Immediately Dangerous.

Rooms with Flueless Water Heaters

8.49   Ensure all fixed ventilation is unobstructed and operate the appliance with all external windows and doors closed at its maximum setting for a period of five minutes *(this is the maximum period that a flueless water heater should be operated)*. Record the CO levels at one-minute intervals. Determine the action to take as below:

- CO level less than 10 ppm.
  If the increase of CO is caused by the gas appliance, then this should be investigated further and, where necessary, appropriate remedial action taken followed by a re-test.

- CO level of 10 ppm or above within the five minutes test.
  The test must be stopped and the cause investigated and where possible rectified. Alternatively, classify the appliance as Immediately Dangerous.

Rooms with Flueless Space Heaters (Including LPG Cabinet Heaters)

8.50   Ensure all fixed ventilation is unobstructed with all external windows and doors closed. Operate the appliance at its maximum setting for a period of 30 minutes and record the CO levels at one-minute intervals. Determine the action to take as below:

- CO level less than 10 ppm.
  If the increase of CO is caused by the gas appliance then this should be investigated further and, where necessary, appropriate remedial action taken followed by a re-test.
- CO level of 10 ppm or above within the 30 minutes test.
  The test must be stopped and the cause investigated and where possible rectified. Alternatively, classify the appliance as Immediately Dangerous.

Additional Tests

8.51   Where chimneys and flues pass through other rooms these rooms may also require to be checked as CO can migrate around a property. Voids containing concealed flues should also be checked where access is possible. Where it is not possible to identify that an appliance or flue system is producing CO, turn off all the appliances and ventilate the room until the inside and outside levels of CO are the same and move the test probe around the room until the point of entry is established.

   If it is suspected that the source of CO is from a landfill site or from a drain then the appropriate authority should be informed. Where the CO level is 10 ppm or above and the source of CO cannot be identified then the emergency services should be notified (e.g. the police or fire services).

Ambient Air Testing for CO$_2$ in a Non-Domestic Setting

Relevant Industry Document
BS 7967

8.52   At times it may be necessary to test the levels of CO$_2$ in ambient air to determine if a gas appliance has been installed correctly and is operating in accordance with the manufacturer's instructions. When selecting a portable electronic device designed to detect and measure CO$_2$ concentrations in ambient air ensure it conforms to BS EN 50543 and is suitably calibrated (Figure 8.10).

   After an initial inspection of the gas appliance and any obvious defects have been corrected, the test procedure can be performed as follows:

1) Zero and purge the test equipment in outdoor air and record CO$_2$ level.
   Note: Outdoor background levels of CO$_2$ should be in the region of 400 ppm unless taking readings, for example, by a busy road or on an industrial estate where levels could be significantly higher.

2) With the agreement of the responsible person turn off all fossil fuel burning appliances in the test area and ventilate until the indoor CO_2 level falls to approximately the same as the outdoor level.

3) Close all external doors, windows and customer adjustable ventilation.

4) Position the test equipment sampling probe approximately 2 m above floor level in the centre of the area and at least 1 m away from the gas appliance. Operate the appliance and record the CO_2 level over a minimum 15 minutes period. The level of CO_2 should not exceed 2800 ppm.

5) Where the level of CO_2 exceeds 2800 ppm a risk assessment should determine if the appliance can be safely turned off, the area should be ventilated and occupants evacuated from the affected area. The investigation should not continue until the level of CO_2 has fallen below 2800 ppm.

6) Wherever practicable the affected appliance should be repaired and retested. Where remedial works are not possible, follow guidance in the current Gas Industry Unsafe Situations Procedure.

7) Final gas safety checks should be undertaken and the responsible person informed of all results and actions taken.

Figure 8.10
Example of a portable electronic CO_2 detector conforming BS EN 50543

Note: Where a mechanical extract/ventilation (MEV) system is installed to provide air and evacuate fumes and vapour, such as those found in commercial kitchens, this should be fully operational during ambient air testing and the level of CO_2 inside the MEV canopy should also be confirmed.

Combustion Performance Analysis

Relevant ACS Qualification	Relevant Industry Document
CPA1	BS 7967, BS EN 50379 and Gas Safe Technical Bulletin 143

8.53 CO poisoning has been attributed to many fatalities over the years, plus there have been many non-fatal casualties, and not only with natural gas but also with other fuels. By means of flue gas analysers, we have the ability to monitor the levels of CO production from an appliance and if too high, take the appropriate action.

Note: Checking and recording the CO and combustion performance ratios (CO/CO_2) for all newly installed condensing gas boilers incorporating an air/gas ratio control valve became mandatory from 1 April 2014. If available, manufacturer's guidance must be followed when carrying out these essential commissioning checks. The flowchart in Figure 8.11 is intended as a generic guide and may be referred to when commissioning a newly installed condensing boiler.

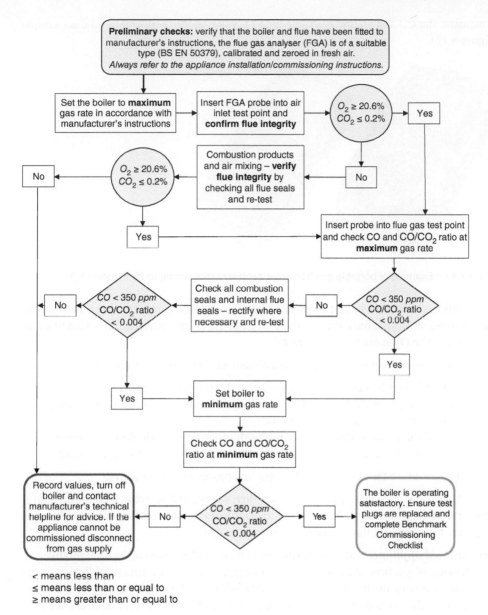

Figure 8.11 Combustion checks when commissioning newly installed condensing boilers

Electronic Combustion Gas Analysers

8.54 There are many makes of combustion analysers with potentially many variables, so when using, always read the operating instructions fully. The analyser must conform to BS EN 50379-3 or BS 7927 and have a current certificate of calibration. Gas analysers conforming to these standards measure the combustion products in different ways depending on the type of sensor fitted. They will either measure the CO and CO_2 and calculate the O_2 value, or measure the CO and O_2 and calculate the CO_2 value in the combustion product sample (Figure 8.12).

Figure 8.12   Example of portable electronic gas analysers conforming to BS EN 50379-3

8.55   All gas analysers should be switched on, zeroed and purged in fresh outdoor air and for those analysers that calculate the CO_2 from the O_2 measurement it is important to confirm that the O_2 in outdoor air is 20.9%.

8.56   The analyser will be supplied with different sampling probes to include:

- An open-ended probe suitable for open-flued and room-sealed appliances.
- A cooker oven/grill multi-hole probe, consisting of seven 1.8 mm diameter holes evenly spaced along the length of a 250 mm probe.
- A bent multi-hole probe designed for use with other types of flueless appliances. This probe has five 1.8 mm diameter holes evenly distributed over a 125 mm sampling probe.

See Figure 8.13a–e for examples of the sampling probes described above.

8.57   If the appliance has a sampling point, then the procedure prescribed in the manufacturer's instructions should be followed to sample the flue gases. If specific guidance is not given, the sample of flue gas products should be taken from:

- ***Open-flued appliances***: at least 200 mm into the secondary flue via the draught diverter. In the case of gas fires, if the probe cannot be positioned in the stream of flue products without removing the fire then it will not be possible to take a reliable sample of flue gas products as the fire is designed to operate. Therefore, this test should not be carried out.
- ***Room-sealed appliances***: into the sampling port provided by the manufacturer or 200 mm inside the exhaust outlet of the flue terminal.
- ***Flueless appliances***: with the holes facing downwards in the stream of combustion products just above the outlet grille using the appropriate multi-holed sampling probe.

Examples are shown in Figure 8.14a–d.

8.58   Unless stated otherwise in the appliance manufacturer's instructions, the appliance should be adjusted to operate at full gas rate and the flue gas sample should have stabilised or begun to decrease within a period of 30 minutes, whichever is the shortest duration. When sampling from any appliance, the probe needs to be adjusted or moved around until

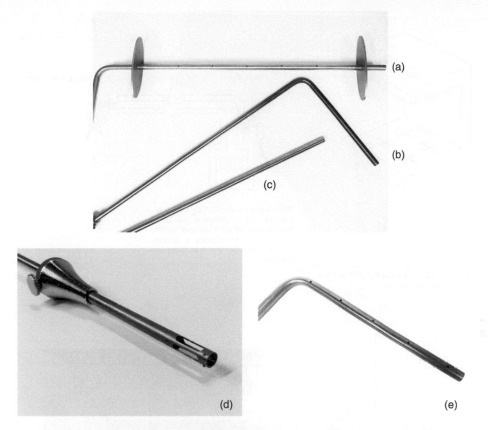

Figure 8.13 Example of flue gas analysis sampling probes. (a) A cooker oven/grill multi-hole probe; (b) a bent multi-hole probe designed for use with other types of flueless appliances; (c) an open-ended probe suitable for open-flued and room-sealed appliances; (d) an open-ended sample probe with adjustable depth stop and (e) close up view of angled sampling probe showing the five evenly spaced 1.8 mm holes over 125 mm

the highest level of CO_2 or the lowest level of O_2 are found, bearing in mind that it may take several seconds for the sample products to pass along the length of tube to the analyser.

Where the CO is excessively high you should immediately remove the probe to prevent damage to the analyser and purge it in fresh air until the CO and CO_2 readings return to zero. Clearly, the appliance would require remedial action and will need to be treated as either At Risk or Immediately Dangerous, depending on the flue type, see Section 8.62.

Safety Checks and Servicing

8.59   Flue gas analysis can be completed as identified above and save the installer undertaking a full strip down service to a gas boiler. Provided that the CO/CO_2 ratio reading is less than 0.004, then there is no requirement to dismantle the appliance to clean it. However, you must still be aware of the Gas Regulations requirement to check the effectiveness of any flue, air supply, operating pressure/heat input and the safe functioning of the appliance, see Section 8.3 for more details. In addition, any manufacturer specific tasks must also be completed.

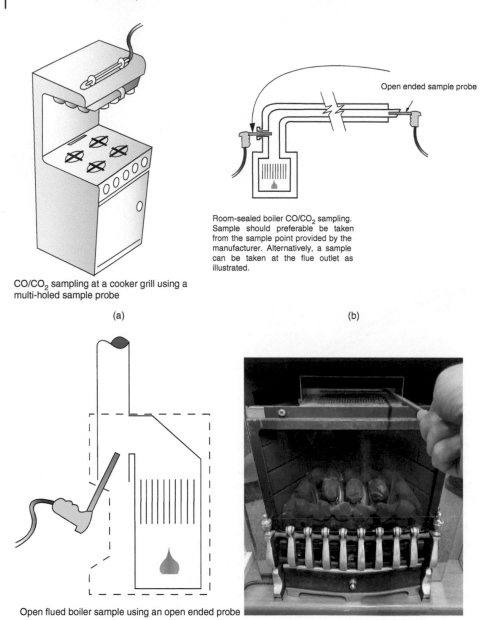

CO/CO$_2$ sampling at a cooker grill using a
multi-holed sample probe

(a)

Open ended sample probe

Room-sealed boiler CO/CO$_2$ sampling.
Sample should preferable be taken
from the sample point provided by the
manufacturer. Alternatively, a sample
can be taken at the flue outlet as
illustrated.

(b)

Open flued boiler sample using an open ended probe

(c)

(d)

Figure 8.14 Sampling for combustion performance. (a) CO/CO$_2$ sampling at a cooker grill using a
multi-holed sample probe; (b) Room-sealed boiler CO/CO$_2$ sampling; (c) open-flued boiler sample
using an open-ended probe and (d) sampling combustion products on a flueless gas fire using an
angled sampling probe

Maximum Combustion CO/CO$_2$ Ratios

8.60   If the appliance manufacturer does not give guidance on the maximum combustion performance ratios (CO/CO$_2$) then the ratio must not exceed those given in Table 8.2. Some appliances may initially fail to be within the correct ratio level because they are unable to burn gas effectively until the appliance is hot. A cooker is a prime example of this as it can produce CO in excess of 10 ppm for short periods of time. But it would be deemed acceptable provided that the cooker conforms to the requirements of BS 7967 and is installed correctly.

8.61   Appliances that have had replacement components fitted may also initially fail until the newness of the part has burned off. If a new component is fitted, it is recommended that the appliance is operated at full rate for a period of at least 10 minutes before the combustion performance test is repeated.

Table 8.2   Combustion performance CO/CO$_2$ action levels for domestic gas appliances

Appliance type		Maximum CO/CO$_2$ ratio	Maximum CO/CO$_2$ ratio for appliances with and air/gas ratio valve
Central heating boiler (including back boiler)		0.0080	0.0040
Back boiler used in combination with a fire		0.020	
Combination boiler		0.0080	0.0040
Circulator		0.010	
Gas Fire – open-flued (Type B)		0.020	
Gas Fire – room-sealed (Type C) live fuel effect		0.020	
Gas Fire – other room-sealed (Type C)		0.0080	
Gas Fire – flueless (Type A)		0.0010	
Flueless LPG cabinet heater		0.0040	
Water heater – flued or flueless		0.020	
Warm air unit		0.0080	0.0040
Flueless cookers	Oven	0.0080	
	Hob	Visual inspection of flame picture	
	Grill (CE marked)	0.010	
	Grill (not CE marked)	0.020	
Flued range ovens		0.020	
Tumble dryers	Flued	0.010	
	Flueless	0.0010	
Refrigerators (LPG)		0.0070	
Gas lights (LPG)		0.020	

8.62   Open-flued or room-sealed appliances that fail to be within the permitted CO/CO$_2$ ratio should be classified as At Risk, whereas if a flueless appliance fails to fall below the maximum tolerance, it must be classified as Immediately Dangerous.

9

Domestic Appliances

Gas Fires and Space Heaters

Relevant ACS Qualification   Relevant Industry Document
HTR1   BS 5871

9.1   There is a large range of gas fires and wall or freestanding heaters. These appliances may be either flueless, open-flued, or room-sealed, each designed to provide a level of comfort and appearance to suit the environment in which they are installed. A range of appliances that fall within this spectrum of the gas industry are shown in Figure 9.1. The term 'space heater' refers to an appliance designed to heat a 'space' or individual room. A space heater provides heat to a room where required, unlike a large central heating system that takes time to warm up many rooms. The heat is primarily distributed from these appliances by convection currents. Radiant heat is also given off, especially from appliances that use radiants, to transmit the infrared heat rays freely from the appliance. Convection is described in Part 6: Section 6.11. Radiation is further identified in Part 10: Section 10.22.

Radiant Heaters

9.2   These appliances are usually mobile; no heat exchanger is incorporated in this design, and all heat is obtained directly from the effect of burning the fuel on a fireclay plaque, which glows red-hot.

Convector Heaters

9.3   These appliances warm the environment by convection currents. Any radiants incorporated are purely there for decorative purposes and are not incorporated to improve efficiency. Apart from the decorative convector heater, they are generally quite plain in appearance, comprising a metal case with louvres at the top through which the hot air passes out into the room. The cold air is drawn into the heater at a low level below the heat exchanger.

Gas Installation Technology, Third Edition. Andrew S. Burcham, Stephen J. Denney and Roy D. Treloar.
© 2024 John Wiley & Sons Ltd. Published 2024 by John Wiley & Sons Ltd.

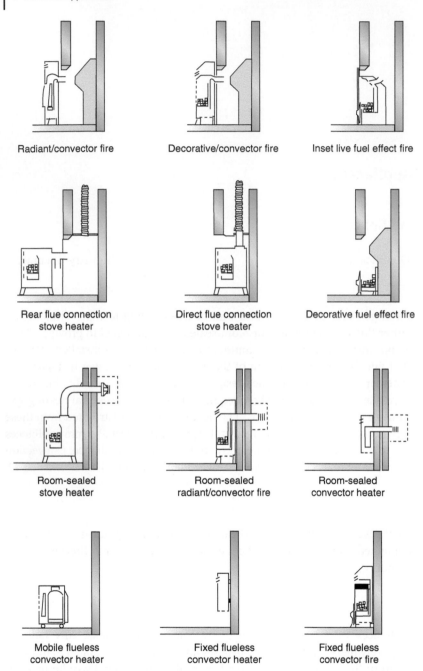

Figure 9.1 Gas fires and space heaters

Radiant Convector Heaters

9.4   These appliances operate as detailed in Section 9.3, but they include radiants as part of their design and, as such, generally prove to be the most efficient form of space heating within this category of appliances.

Decorative Heaters

9.5   There are several designs of decorative heaters, including the inset live fuel effect (ILFE), decorative fuel effect (DFE), and the heating stove, to name a few. These heaters have been developed to create the illusion of a wood or coal-burning appliance without the mess and inconvenience of using solid fuel.

Condensing Heaters

9.6   These gas fires operated with efficiencies of up to 90%, unlike the previous heaters. However, due to the high cost of manufacture, they have fallen out of favour and are no longer produced.

Component Parts of a Gas Fire

9.7   A typical gas fire consists of the following components, as illustrated in Figure 9.2.

Outer Case

This is an enamelled pressed steel surround, possibly with some other material attached to give the desired decorative finish; it also serves to protect the user from hot surfaces. A fireguard or glass panel, designed to prevent anyone from touching the red-hot radiants, is attached to the front. A chromium-plated or stainless steel reflector is often incorporated. This serves two purposes: to enhance the appearance of the fire and to assist in reflecting the heat energy away from the case and floor.

Firebox

This is the metal box in which the combustion process takes place. There are radiants and firebricks in the firebox. The top or sides of the firebox usually form the draught diverter. The firebox also directs the combustion products to the heat exchanger and flue.

Radiants

A radiant is a fireclay, ceramic or volcanic lava rock that is incorporated in order to increase the efficiency and/or improve the appearance of the appliance. When surrounded by the gas flame, the radiant glows to a white or red-hot temperature of around 900°C.

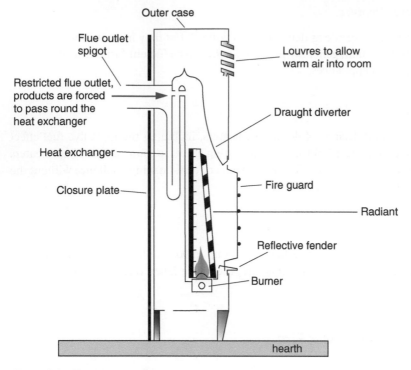

Outer case

Flue outlet spigot

Louvres to allow warm air into room

Restricted flue outlet, products are forced to pass round the heat exchanger

Draught diverter

Heat exchanger

Closure plate

Fire guard

Radiant

Reflective fender

Burner

hearth

Figure 9.2 Component parts of a fire

The radiants need to be located in accordance with the manufacturer's instructions to ensure that incomplete combustion and severe sooting do not occur. The combustion products pass up through the radiants and pass out through the heat exchanger and flue. During maintenance work, it is often good practice to rotate the order of radiants to extend their life. In other words, if there are four radiants, move the more commonly used central ones to the outside and the lesser used outer ones to the centre. This will ensure even usage.

Firebrick

This is a refractory lining, sometimes located behind the radiants or coals. It is designed to prevent the appliance from becoming overheated and cracking. It also prevents heat from being lost into the chimney.

Burners

The general burner design has been described in Part 2: Section 2.40. However, in gas fires, in order to allow a pair of radiants to glow independently from the full set of four, the injectors are sometimes positioned in an arrangement referred to as a duplex design. Therefore, for clarity, if only one injector is used, it is referred to as a simplex burner; where more than one injector is used, the arrangement is called a duplex burner (see Figure 9.3).

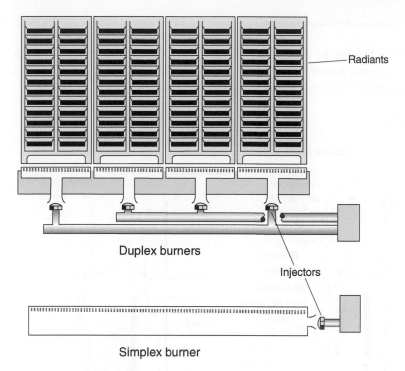

Figure 9.3 Simplex and duplex burner arrangements

Heat Exchanger

This consists of a flat metal enclosure located above the firebox. Cool air enters the base of the fire and passes over the flat surface, quickly heating up before passing through louvres in the top front edge of the casing. It is essential to check the condition of the heat exchanger thoroughly, as they are prone to fatigue cracking as a result of the huge heat transference, as seen in Section 9.41 and Figure 9.25.

Open-Flued Radiant Convector Gas Fires

9.8 This design of fire takes the air for combustion from inside the room. The products of combustion are expelled from the room into a chimney or system of flue pipework of no less than 125 mm diameter, usually by means of convection currents. Owing to the comparatively small amount of fuel consumption, less than 7 kW, no additional ventilation is generally required to support the combustion process. The air in the room is replaced by adventitious means, through cracks in window frames, etc. These fires have typically been found to be around 65% efficient.

The design of the fire has changed over the years to encompass several variations since the traditional design of the 1950s, which typically had four fireclay radiants positioned at the front to give the effect of coal or wood being burnt. Some fires incorporate a glass front, allowing the solid fuel-burning effect to be seen behind the glass. This prevents material from falling onto the exposed flame and can also have the advantage of increasing the overall efficiency.

Cool air enters the base of the fire, passes round through the heat exchanger, and out through a series of louvres located at the top of the fire. At the same time, radiant heat is directed into the room to be heated. This is illustrated in Figure 9.4.

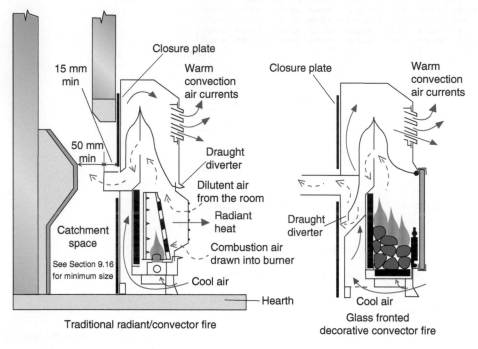

Figure 9.4   Open-flued radiant convector gas fires

Closure Plate

9.9   To enable connection to the builder's opening at the base of a chimney and to allow access to the flueway for servicing purposes, a closure plate is used. The closure plate is just a thin sheet of metal about 0.5 mm thick, supplied with the appliance, in which there is a slot through which the flue spigot of the fire can be located. In addition, the manufacturer incorporates a relief opening into the design of the closure plate. This could be a specific size hole somewhere in the plate, or alternatively, the flue spigot hole is designed to be larger than the spigot. This is illustrated in Figure 9.5a. The size of the relief opening controls the amount of air that can be drawn in from the room without creating too great a draught and also maintains a ventilation rate of about 60–70 m³/h, equivalent to approximately two air changes per hour. The plate must be sufficiently sealed to the opening, along all edges, with heat-proof tape or other suitable sealing material capable of withstanding temperatures up to 100°C, see Figure 9.5b. This must be flexible enough to seal along uneven surfaces, such as stonework. Failure to provide a suitable seal may lead to too much air passing into the flue due to a strong pull, starving the burner of sufficient combustion air and causing the flames to be drawn back downwards away from their intended direction.

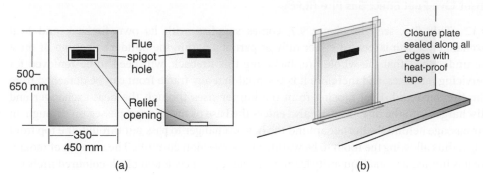

Figure 9.5 (a) Typical Closure Plate. (b) Sealing the closure plate to the opening

Flue Spigot Restrictor

9.10   A typical domestic dwelling has a flue draught of about 0.1 mbar pressure, which is more than sufficient to pull the products of combustion through the heat exchanger without pulling too much air through the appliance. If the flue draught is too great, the additional air pulled through the heat exchanger has the effect of cooling it down, thereby reducing the overall efficiency. To overcome the pull of a high chimney, some manufacturers recommend the use of a spigot restrictor, see Figure 9.6, which creates resistance to flow through the appliance.

Figure 9.6   Typical flue spigot restrictor

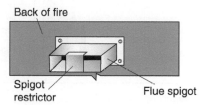

Open-Flued Solid Fuel Effect Fires and Heaters

9.11   These heaters, because they give the illusion of an exposed flame similar to that given off by solid fuel, often have lower efficiencies than the more conventional radiant convector heater. Efficiencies of not much greater than 45% can be expected. The appliances have a firebox on which imitation logs or coals are placed and through which the gas can pass to burn and give the desired flame pattern. The logs/coals must be positioned strictly in accordance with the manufacturer's instructions. Failure to do so may lead to high levels of carbon monoxide (CO) and soot being produced. For the same reason, it is essential that the customer is advised how the appliance operates and is instructed not to throw paper and cigarette ends, etc., onto the burning flame as the accumulating ash will have a detrimental effect. Designs include the inset live fuel effect (ILFE) and heating stove, which connect to the flue system. However, other gas fires, including open-flued radiant convector heaters, decorative fuel effect (DFE) fires, and flueless gas fires, also give a solid fuel effect, see Section 9.29.

Inset Live Fuel Effect Gas Fire (ILFE)

9.12   This fire, seen in Figure 9.7, comes supplied with its own fireback and heat exchanger, which is located either fully or partially within the builder's opening. It must be understood that the whole fire, including this fireback, will need to be removed for servicing purposes, and therefore it is essential that any frame front can be detached from the fire surround. Heat passing from the burner rises through the heat exchanger and discharges into the flue. Cold air also enters the base of the fire and passes up through an air passage behind the burner around the heat exchanger to pass out through the top front edge, thus allowing the room to be warmed by convection currents. The amount of radiant heat is limited to a small quantity from the incandescent coals and black-coloured fireback.

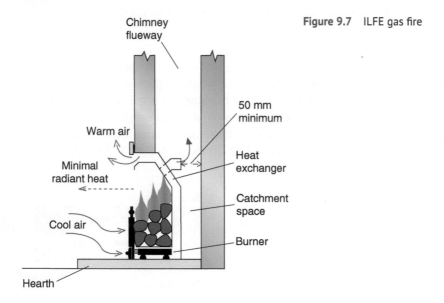

Figure 9.7   ILFE gas fire

The minimum flue size where an ILFE fire is installed is 125 mm in diameter. Additional ventilation may or may not be required, depending on the input rating of the appliance; the manufacturer's instructions should be consulted.

Heating Stove

9.13   Two basic designs of heating stoves are available, as illustrated in Figure 9.8a: those with a vertical flue pipe and those with a rear flue spigot that passes through a closure plate, as with the radiant convector gas fire. For the stove with a flue pipe connection, as seen in Figure 9.8b, the flue will need a minimum diameter equal to the size of the appliance outlet. As with the gas fire, unless the manufacturer specifies differently, a heating stove will need to be positioned on a suitable hearth as described in Section 9.17. It should be noted that when installing a heating stove, it is not acceptable to make the connection to an unlined chimney, and either a clay or metal liner is needed. This would need to be suitably sealed at the base and, where necessary, an inspection hatch provided.

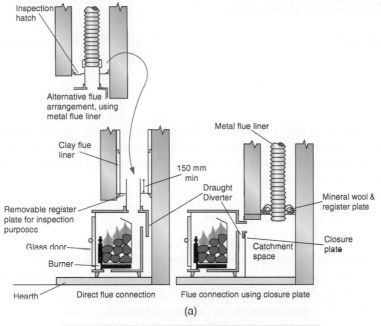

Inspection hatch

Alternative flue arrangement, using metal flue liner

Clay flue liner

Metal flue liner

150 mm min

Draught Diverter

Removable register plate for inspection purposes

Mineral wool & register plate

Glass door

Burner

Catchment space

Closure plate

Hearth

Direct flue connection

Flue connection using closure plate

(a)

(b)

Figure 9.8 (a) Typical designs of heating stove that are available. (b) A heating stove

Installation of Open-Flued Gas Fires

Relevant ACS Qualification
HTR1

Relevant Industry Document
BS 5871

Flue Termination

9.14   It is not normally necessary to fit a terminal to any flue system where the flue is greater than 170 mm in diameter. Where a terminal does become necessary, the design will have to meet the requirements of the manufacturer and follow the recommendations in Part 6: Section 6.25, Flues and Chimneys.

Connecting Fires and Heaters to an Existing Flue

9.15   Generally, in a traditionally designed dwelling, due to the low heat input of a gas fire, there is no requirement to line the chimney unless the flue length exceeds the distance identified in Part 6: Section 6.34, Table 6.10. However, the manufacturer's instructions will need to be consulted for clarification. An existing flue will need to be inspected fully to ensure that it is suitable for the installation and of the correct material. It will be necessary to sweep the flue, especially if it was previously used for appliances burning fuels other than gas. Older chimneys will need to be inspected to ensure that there is no evidence of restriction within the flue, such as a damper plate or an old restrictor. Where these are evident, they will need to be removed. In the case of a damper, securing it in an open position will suffice. It may be possible to leave a redundant solid fuel back boiler within the opening, providing it does not restrict the void (see Section 9.16). However, where the boiler has been drained of water, drilling a small 6 mm hole in its base will prevent any pressure from building up inside and causing it to become unsafe. Any existing ventilation from openings into the builders opening will need to be sealed off, as this may affect the operation of the appliance. The flue will need to be inspected throughout its entire length to assess its condition and to ensure that it only serves one location. A flue flow test will need to be completed successfully to ascertain its condition. Flue flow testing is described in Part 6: Section 6.49. Following a check on the condition of the flue, the catchment space and fireplace opening will need to be inspected to ensure that they are adequate for the purpose.

Catchment Space

9.16   This is the opening located behind an appliance, usually as part of the builder's opening or flue box, in which any debris that falls down the chimney can collect. DFE appliances do not have a catchment space; however, conventional radiant convector fires, ILFE, and heating stoves do. The size of this void is determined by the flue system. For unlined flues, such as illustrated in Figure 9.9a, the space should have a minimum depth of 250 mm below the appliance flue connection and a volume of 0.012 m^3. For a lined flue, as seen in Figure 9.9b. the depth below the flue connection can be reduced to 75 mm provided the flue system has never been used with solid fuel or oil. Table 9.1 provides a reference to the minimum void dimensions for different scenarios.

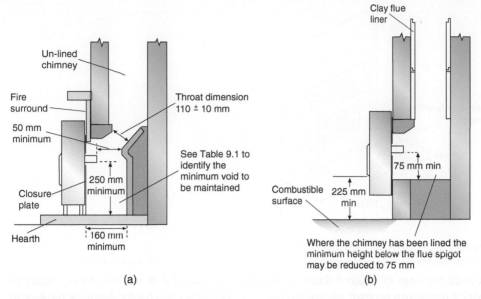

Figure 9.9 (a) Minimum catchment space for unlined chimney. (b) Minimum catchment space for lined chimney

Table 9.1 Minimum void dimensions

Installation details	Depth below appliance connection	Volume
An appliance fitted to an unlined chimney	250 mm	0.012 m³ (12 l)
An appliance fitted to a lined brick chimney which is new or unused or has just been used with gas	75 mm	0.002 m³ (2 l)
An appliance fitted to a lined brick chimney which has previously been used with solid fuel or oil	250 mm	0.012 m³ (12 l)
An appliance fitted to a flue block chimney or a metal chimney which is new or unused or has just been used with gas	75 mm	0.002 m³ (2 l)
An appliance fitted to a flue block chimney or a metal chimney which has previously been used with solid fuel or oil	250 mm	0.012 m³ (12 l)

Hearth Construction

9.17 Where a hearth is required (see Figure 9.10), it must be of fireproof construction and have a minimum thickness of 12 mm. For DFE and ILFE gas fires, an additional up-stand is required along the front and side edges of 50 mm to discourage the placing of rugs on the hearth. This may be achieved by the use of a fender plate. Where a hearth is located at the base of a builder's opening, it will be necessary to extend it forward a minimum distance of 300 mm in front of the fire surround and a distance of 150 mm to each side

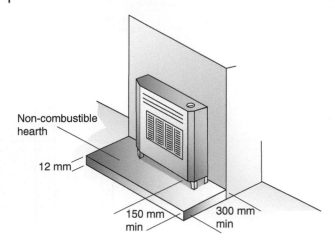

Figure 9.10 Minimum hearth dimensions

beyond the edge of a naked flame or incandescent material. It will also be necessary to extend the hearth beneath any naked flame. See Part 6: Section 6.9 for details of the hearth notice plate.

Fire Protection

9.18   The following will need to be considered with regard to adjacent combustible surfaces.

Rear wall: Where the fire is to be located against a surround, additional protection should be provided to minimise the effects of heat transmission unless the manufacturer states otherwise. No combustible material should be placed within the fireplace or builder's opening.

Side-wall: If the fire is to be located in a corner, it is essential where the construction includes combustible material to ensure that a minimum distance of 500 mm is observed between the wall and the flame or incandescent material.

Shelf: An appliance may only be located below a combustible shelf where the manufacturer's instructions clearly specify that it is permitted.

Wall-mounted fires: Where a fire is mounted above floor level into a wall, no hearth is needed, provided a distance of 225 mm is maintained between any combustible floor covering and the flame or incandescent material; see Figure 9.9b.

Connecting a Radiant or Convector Heater to the Flue Opening

9.19   Using a secured closure plate, as described in Section 9.9, the flue passes through the slot or hole provided. Where an existing fireback or chair brick is in place, it is essential to ensure that a minimum distance of 50 mm is maintained and that the void provides the required catchment space, as detailed in Table 9.1. Sometimes, due to the design of the fire surround, it will be necessary to fit a flue spigot extension piece, as shown in Figure 9.11.

Figure 9.11 Fire incorporating flue spigot extension

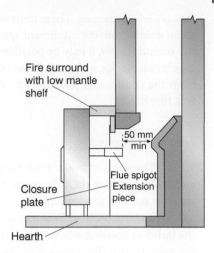

Fire surround
with low mantle
shelf

50 mm
min

Flue spigot
Extension
piece

Closure
plate

Hearth

Connecting a Heater to a Pre-Cast Flue Opening

9.20 An appliance can only be fitted to a pre-cast flue if it is specifically identified as an installation option by the manufacturer. When the flue spigot passes into a pre-cast flue block system via the closure plate, a cooler device may be required, subject to the manufacturer's instructions (see Figure 9.12)

Figure 9.12 Cooler device

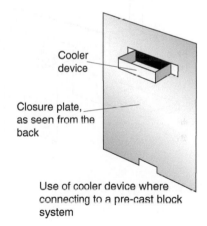

Cooler
device

Closure plate,
as seen from the
back

Use of cooler device where
connecting to a pre-cast block
system

Oversized Openings and Voids

9.21 If a fire or heater is to be installed to an opening that is larger than the appliance closure plate, it will be necessary to reduce the opening size accordingly. This may be achieved by simply bricking up the opening using non-porous bricks; alternatively, it is acceptable to use a panel of non-combustible material. Materials such as wood cannot be covered in cladding and used as the heat within the opening may result in a fire. Whatever method is used, the final opening should be small enough that the closure plate fits over it. However, the opening must not be just the size of the fire-flue spigot, or in other words, what is known as a letterbox opening. There must be a facility to remove the closure plate and be able to clean debris from the catchment space. Should the void itself be too large, as identified by the manufacturer, it may be possible to reduce the size using non-porous bricks or blocks. Insulation blocks, such as 'Celcons', should not be used as they will absorb the water vapour from the combustion products and slowly deteriorate. Alternatively, it may be possible to fit a flue box.

Flue Box or Collector

9.22 These metal boxes have been especially designed for the installation of gas fires and heaters where there is no chimney or the existing chimney is unsuitable. The box must be secured in accordance with the manufacturer's instructions and be positioned on a non-combustible base. When in position, the box can be treated in much the same way as the builders opening; however, you can only install manufacturer-approved appliances into the system with the closure plate secured as necessary. As can be seen in Figure 9.13, the flue box is connected to a lined chimney or where a false chimney and breast are proposed, a rigid flue pipe could be used. Pipes should pass into the box at a low level, and the hole must be sealed with a heat-resistant non-setting sealant.

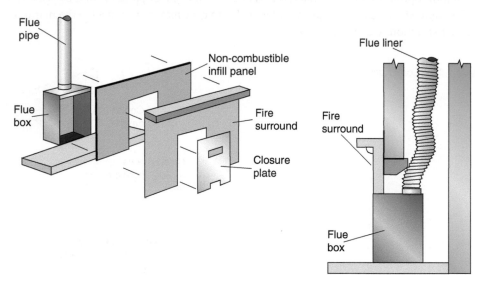

Figure 9.13 Flue box installation

Openings into the Fireplace

9.23 It is absolutely essential that there are no openings into the builders opening except that allowed for in the closure plate. In all cases, any additional air that may be drawn into the fire via this unintended route may lead to poor combustion and adversely affect flue performance. Specific points to look for include:

- poorly fitted proprietary fire surrounds or infill panels;
- gaps in pre-cast flue blocks or where they may adjoin a dry-lined wall;
- unsealed redundant pipes that enter the void;
- poorly sealed and secured closure plates;
- wrongly sized closure plates.

Direct Flue Connection

9.24   In the past few years, fires and heaters with direct flue connections have come onto the market; see Figure 9.14. Where these are to be fitted, it is essential that they are installed to a lined or twin wall flue system and not to an unlined masonry chimney unless the manufacturer states otherwise, such as where the appliance is of the cassette type, using a certified flue kit (as shown in Figure 9.15), and even then the flue must be in good condition. Generally, the appliance should be treated similarly to a boiler connection and not connected simply through a closure plate. They may look like traditional gas fires or heating stoves, but they have their own installation requirements as well as alternative fixing methods.

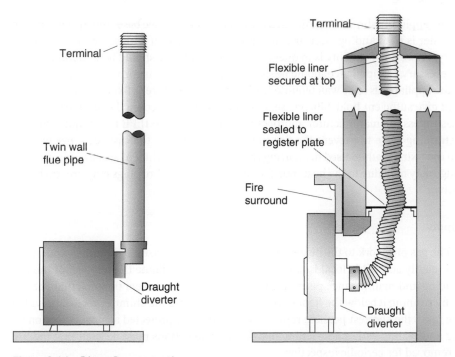

Figure 9.14   Direct flue connections

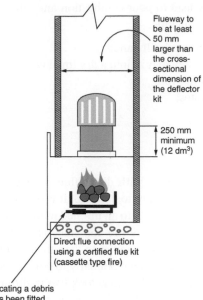

Flueway to be at least 50 mm larger than the cross-sectional dimension of the deflector kit

250 mm minimum (12 dm³)

Direct flue connection using a certified flue kit (cassette type fire)

Label attached indicating a debris deflector flue kit has been fitted

Figure 9.15 Direct flue connection using a certified flue kit (cassette-type fire)

Securing the Fire

9.25 The appliance should be positioned on a sound, level, stable base, and, in general, the fire or heater is freestanding. Securing the appliance to the wall with screws is not always necessary. However, if stability is in question, additional fixing should be undertaken, and where necessary, the appliance should be secured to the wall to ensure no undue movement. Where the appliance is to be wall mounted, all of the manufacturer's fixing points should be used and give sufficient hold. Silicone sealant should, on no account, be used for securing purposes unless the manufacturer approves this method. Some ILFE fires require the use of a restraining cable to prevent the fire tipping; where this is the case, care needs to be given to the fixing method used in securing the cable in the catchment space as plastic-type wall plugs may melt due to the heat transfer along the cable. Fibre plugs may prove a better alternative.

Gas Supply

9.26 The gas pipework to the heater must be a permanently fixed rigid supply and incorporate a readily accessible local isolating control valve to facilitate the removal of the fire for servicing and maintenance purposes. It is permissible to conceal the pipe that runs to the fire by running it below the floor or through the wall into the chimney recess, provided the pipe takes the shortest possible route and is adequately protected against corrosion by sleeving and sealing with a non-setting mastic. All connections must be such that the fire can be removed for periodic inspection.

Ventilation

9.27 All gas appliances need an adequate supply of air for combustion, and the gas fire is no exception. However, adventitious ventilation (air that enters through cracks in doors and window openings, etc.) is sufficient to supply an open-flued radiant/convector or ILFE appliance of up to 7 kW input, provided that the manufacturer's instructions do not state otherwise. For fires in excess of 7 kW, see Part 7: Section 7.24. For decorative fuel effect gas fires (DFE), see Section 9.32.

Restricted Locations Under Current Gas Safety Regulations

9.28   The installation of open-flued and flueless appliances is prohibited in rooms containing *showers and baths*. Appliances installed in *bedrooms and bed-sits*, over 12.7 kW net (14 kW gross) input must be room-sealed. Alternatively, an open-flued appliance of less than 12.7 kW net input may be installed, providing a safety control, such as a vitiation sensing device (see Part 3: Section 3.57) is incorporated. *Basements and cellar* locations have restrictions on the use of LPG appliances.

Decorative Fuel Effect (DFE) Gas Fires

9.29   This heater consists of no more than a firebox and burner. The unit is simply located within the builders/fireplace opening, flue box, or under an independent canopy, and no direct connection is made to the flue. Because there is no heat exchanger, warmed convection currents passing from the appliances are minimal, and the only heat supplied to the room is the radiant heat from the coals. Generally, these appliances are used as a focal point, and the room will have to be heated by some other means. It must be noted, however, that Approved Document L to the Building Regulations states that only one decorative fuel effect gas fire may be installed per 100 m² of an existing dwelling floor area. This is due to the efficiency of such gas fires where a majority of the heat produced exits via the chimney.

Flue Size

9.30   The minimum flue size to be looked for when installing these appliances is 175 mm across the axis. When installing a DFE, the information from the previous sections, dealing with the installation of open-flued fires, should be referred to, as these details also apply. The chair-brick, where fitted, assists in directing the products into the flueway. It is not necessary to alter the size if connecting into an existing chimney, provided the flue system has been proved to work safely. However, it must not be assumed that a DFE appliance positioned below a flue system will work simply because the hot products of combustion would be expected to rise upwards.

The fireplace opening in relation to the flue height needs to be considered to work out a minimum cross-sectional flue dimension, as a diameter greater than 175 mm may be required. For example, looking at Figure 9.16, the height of the fireplace opening is 0.75 m, and the width is 0.6 m, the opening area will be $0.75 \times 0.6 = 0.45$ m². If this is to be installed to a flue of height 6 m, referencing the flue sizing chart in Figure 9.17, it will be seen that a flue diameter of either 225 or 250 mm will be required. Large diameter flues, particularly when they are short, can be prone to downdraught; therefore, the smaller diameter of 225 mm should be selected.

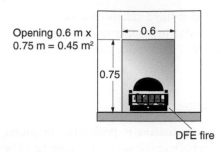

Opening 0.6 m x
0.75 m = 0.45 m²

0.6

0.75

DFE fire

Figure 9.16 Fireplace opening

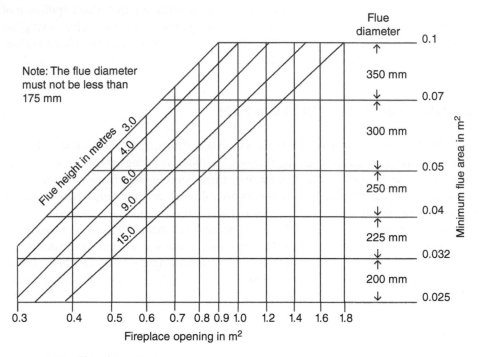

Note: The flue diameter
must not be less than
175 mm

Flue height in metres

3.0
4.0
6.0
9.0
15.0

Flue
diameter

350 mm

300 mm

250 mm

225 mm

200 mm

0.1

0.07

0.05

0.04

0.032

0.025

Minimum flue area in m²

Fireplace opening in m²

0.3   0.4   0.5   0.6   0.7   0.8 0.9 1.0   1.2   1.4   1.6   1.8

Figure 9.17   Flue sizing chart

Installations Below Independent Canopies

9.31   Where an independent canopy is to be used to collect the flue products, it should be sited no more than 400 mm above the fire bed. The outer edges of the canopy should extend by the amounts shown in Figure 9.18. The flue connection should be positioned at the top of the canopy with no further opening into the flue system. The angle of the canopy opening should be no more than 45° from the vertical.

If a DFE fire is freestanding beneath a canopy or similar flue, the hearth will need to be extended 300 mm beyond the fire in all directions, and where such a fire is adjacent to a wall within this 300 mm space, the wall will need to be adequately protected.

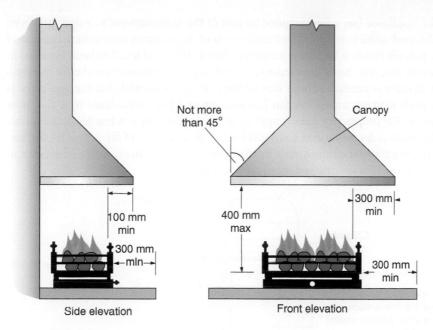

Figure 9.18 DFE fire installed below a canopy

Ventilation

9.32   The ventilation should be no less than 100 cm^2 for appliances of up to 20 kW input unless the manufacturer states otherwise.

Previously, before Approved Document L to the Building Regulations changed, it was acceptable to install two DFE gas fires in a room, regardless of size, but restrictions have now been placed on this; see Section 9.29 for a further explanation. You may, however, encounter existing installations where two DFE fires are installed in the same room, and you will need to confirm that the ventilation is correct. The requirements allow 100 cm^2 for a single DFE, but it needs to be remembered that adventitious air can only be used once. Therefore, for each further DFE, an additional 35 cm^2 will need to be added. The calculation for two DFE gas fires in a room is 100 cm^2 + 100 cm^2 + 35 cm^2 = 235 cm^2. Preferably, the vent should be located between the two appliances. It should be noted that the ventilation should not be located within the hearth opening area, and the ventilation must be obtained via a grille direct to the outside or a ventilated floor void.

Fan Flued and Room-Sealed Space Heaters

Fan Flued Systems

9.33   Fans are sometimes used to assist in the removal of combustion products with a space heater, particularly in the case of installations that incorporate a false chimney breast.

Usually, the appliance has a fan included as part of the manufacturer's design; however, occasionally, such units have a fan fitted at the point of the chimney termination in place of a chimney pot, see Figure 9.19. One example is when a DFE appliance has been installed in a pub or restaurant, and due to the nature of the extracting system and ventilation changes, additional draught is needed. Should this be the case, it is essential that the appliance is prevented from working unless the fan has been proved to be functional by a flue flow sensing device. For proprietary fan draught systems, discharging at a low level, the terminal usually needs to be positioned so that there is a free passage of air across its surface. However, the appliance manufacturer's instructions need to be checked for compliance in terms of maximum flue lengths and terminal positions.

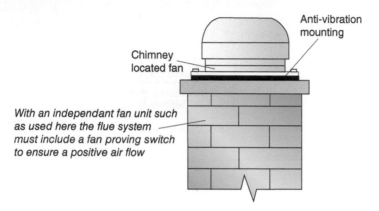

Figure 9.19 Typical chimney-mounted exhaust fan

Many heaters, such as the one illustrated in Figure 9.20, have a fan fitted as part of the manufacturer's design; these are often wrongly seen as being room-sealed, but they take the air from the room in which the fire is positioned. The Gas Regulations permit the installation of pipework to a fanned draught living flame effect fire to be installed in a cavity wall. However, as seen in Part 4: Section 4.38, the pipe must be enclosed within a gas-tight sleeve and take the shortest possible route. The sleeve will need to be sealed at the point at which it enters the fire.

Room-Sealed Heaters

9.34   There are two distinct designs of room-sealed units: those that have a glass-fronted panel designed to provide a radiant element and/or characteristic solid fuel effect and those that are just a unit from which heat is distributed by convection, see Figure 9.21. The room-sealed heater needs to be installed with reference to the manufacturer's instructions, in particular the termination requirements and location. See Part 6: Table 6.12 for suitable terminal positions. It should be noted that with the room-sealed heater, no ventilation is taken from the room in which it is situated, and therefore, no additional ventilation to the room will be required. The heat-resistant glass panel fitted to these units is normally removable for servicing, etc., in which case it is essential that the condition of the seal used is maintained, thus preventing any combustion products from entering the room.

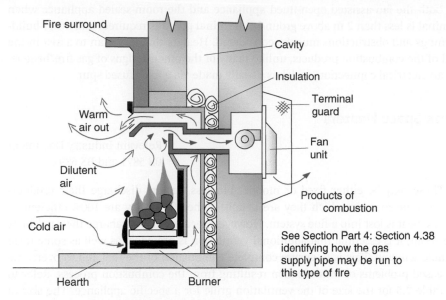

Figure 9.20 Fan-flued fire designed to fit within a cavity wall

Labels in figure:
- Fire surround
- Cavity
- Insulation
- Terminal guard
- Warm air out
- Fan unit
- Dilutent air
- Products of combustion
- Cold air
- See Section Part 4: Section 4.38 identifying how the gas supply pipe may be run to this type of fire
- Hearth
- Burner

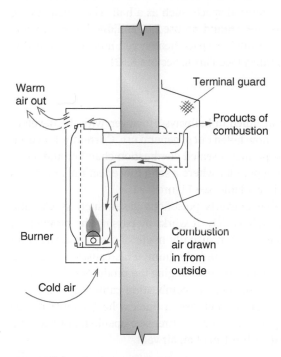

Figure 9.21 Room-sealed heater

Labels in figure:
- Warm air out
- Terminal guard
- Products of combustion
- Burner
- Combustion air drawn in from outside
- Cold air

With both the fan-assisted open-flued appliance and the room-sealed appliance when the terminal is less than 2 m above ground, a terminal guard is required, and other building openings and obstructions must be considered. Heaters that use a fan to assist in the removal of the combustion products, unlike many of the other designs of gas fire/heaters, require an electrical connection, which is usually made via a fixed fused spur.

Flueless Space Heaters

Relevant Industry Document
BS 5871 and BS 6891

9.35 These may be either fixed or mobile. Flueless heaters discharge their products directly into the room in which they are installed; therefore, they are 100% efficient as none of the heat is lost into a flue system. However, because they discharge their products into the room, it is essential that additional ventilation is provided, as well as some form of openable window. This is to ensure complete combustion of the fuel and to overcome the increased problems of condensation resulting from the combustion process. Refer to Part 7: Table 7.5 for the size of the ventilation grille for a specific appliance. The size of the room would also need to be considered to accommodate such an appliance. There is a maximum appliance-rated input of $45\,\text{W/m}^3$ if the heater is fitted in a room, such as a living room, and $90\,\text{W/m}^3$ if fitted in an internal space, such as a hall. These heaters are generally quite low in output and, therefore, should be used in conjunction with some other form of heating. When working on a flueless space heater, you must carry out flue gas analysis on every occasion to ensure safety (see Part 8: Section 8.53).

Fixed Flueless Heaters

9.36 Traditionally, fixed heaters were installed in hallways. However, with the development of the room-sealed heater, they have now fallen out of general use. Currently, however, there is a revival with a design that incorporates a sealed combustion chamber and a catalytic converter. These heaters are proving popular where a solid fuel effect fire or stove is wanted, and no chimney or flue system is available, see Figure 9.22.

The catalytic converter is a device that converts poisonous gases, such as CO and aldehydes, into less harmful emissions, such as CO_2. It works by passing the products of combustion through an inner honeycomb section, usually made of a ceramic structure coated in a metal such as platinum, palladium and rhodium. This causes a chemical reaction in the flue gases and speeds up the process of changing the products to less harmful ones. Due to the potential problems of incomplete combustion caused by too much air movement, unless the manufacturer states otherwise, a flueless heater must not be installed in an opening under an existing flue, in front of a fireplace opening, or near to an existing flue. It must also not be installed within 1 m of an air vent.

Mobile Flueless Heaters

9.37 Flueless heaters are not restricted to fixed appliances, and there are mobile heaters that use a butane gas cylinder housed within the case, see Figure 9.23. Propane must not

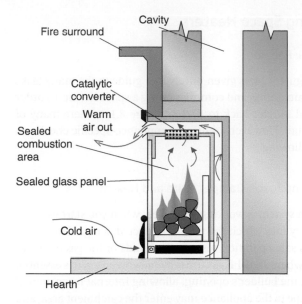

Figure 9.22 Fixed flueless heater

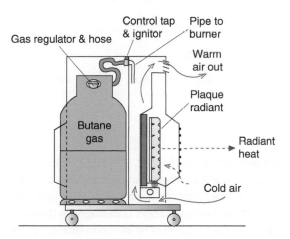

Figure 9.23 Mobile heater

be used with these mobile heaters because of the higher cylinder pressures. When using a mobile cabinet heater that uses a series of three radiant plaques, allowing one, two or all three burners to be lit, it should be noted that the ventilation requirements alter for each input setting, and having the appliance on full may be permitted in one room but not in another, where only one radiant may be lit, due to room size and available ventilation. The size of vent required is given in Part 7: Table 7.5. These heaters generally have a label inside the case, usually where the cylinder connects, warning of the requirements. As these heaters run on LPG, they are restricted from use in cellars and basements where any escape of gas could prove disastrous.

Commissioning and Servicing Space Heaters

Commissioning and Servicing Work Record

9.38   The Work Record, seen in Figure 9.24, is given purely as a guide to the many tasks to be undertaken when servicing/maintaining and commissioning a space heater. In order to complete the form, you may need to refer back to Part 8: Figure 8.1a, where many of the tasks are explained in greater detail. However, additional appliance-specific checks on space heaters include some of the following.

Visual Inspection of Flue, Catchment Space, Damper Plate and Hearth

9.39   You cannot assume that a flue system is working effectively when you start work on a gas fire. The flue system or chimney into which the fire is connected is regarded as part of the installation, and therefore, its safe operation rests in the hands of the gas engineer working on the fire. So, for example, where an ILFE gas fire is being worked on, it is essential that the whole unit is removed from the builder's opening, allowing internal inspection of the flue and catchment space. The pipe to the appliance may enter the catchment area, and special care needs to be given at inspection for corrosion problems and its effective seal into the chamber. The Flue Inspection and Testing Checklist in Part 8: Figure 8.2 may prove useful in completing this task.

Flue/Hearth Notice Plate Located and Correctly Filled In

9.40   A plate should be in place that gives details of the flue, location and hearth (see Part 6: Figure 6.3).

Metal Fatigue of the Heat Exchanger

9.41   Owing to the intensive heat generated within the firebox, sometimes cracks will appear behind the radiant and firebrick, see Figure 9.25; these can only be seen by completely removing them from the appliance. These cracks are the result of the continued expansion and contraction of the metal. Where the heat exchanger is damaged in this way, the fire needs to be condemned as the combustion products may be drawn in through these cracks to be discharged around the room, giving rise to the production of CO.

Fireguard in Place

9.42   A fireguard may be required in order to comply with the Heating Appliances (Fireguards) (Safety) Regulations; this is particularly important where vulnerable persons may be put at risk from the open flames or hot surfaces. The fireguard needs to be secured to prevent its removal.

Certificate/Record of Space Heater Service/Commission		
Gas Installer Details	**Client Details**	**Appliance Date Badge Details**
Name:	Name:	Model/Serial No:
Gas Safe Reg. No:	Address:	Gas Type: Natural ☐ LPG ☐
Address:		Heat Input: max...... kW min...... kW
		Burner Pressure Range: ...... –...... mbar
		Gas Council No:
Date:	**Appliance Location:** **Install/Commission ☐ Service/Commission ☐**	
Preliminary System Checks:	Compliance with manufacturer's instructions	PASS ☐ FAIL ☐
	General visual inspection of pipework	PASS ☐ FAIL ☐
	Clearance from combustible materials	PASS ☐ FAIL ☐
	Visual inspection of flue, catchment space, damper plate and hearth	PASS ☐ FAIL ☐ N/A ☐
	Flue/hearth notice plate located and correctly filled in	PASS ☐ FAIL ☐ N/A ☐
	Closure plate sealed along all edges	PASS ☐ FAIL ☐ N/A ☐
	Flue flow test	PASS ☐ FAIL ☐ N/A ☐
	Appliance level and secure	PASS ☐ FAIL ☐
	Electrical connections	PASS ☐ FAIL ☐ N/A ☐
	Bonding maintained	PASS ☐ FAIL ☐ N/A ☐
	Fuse rating: .... amps	PASS ☐ FAIL ☐ N/A ☐
	System tightness test; to include let -by	PASS ☐ FAIL ☐ N/A ☐
	Standing pressure of system (lock-up........mbar)	PASS ☐ FAIL ☐ N/A ☐
	Appliance/System is purged of air	PASS ☐
Service/Commission Checks:	Clean primary air ports and lint arrestor	PASS ☐ FAIL ☐ N/A ☐
	Clean/Check condition of injectors, burners and radiants/coals	PASS ☐ FAIL ☐ N/A ☐
	Check condition of heat exchanger, examining for metal fatigue	PASS ☐ FAIL ☐
	Radiants/Coals correctly located and aligned	PASS ☐ FAIL ☐ N/A ☐
	Check for easy operation and grease, if necessary, control taps	PASS ☐ FAIL ☐ N/A ☐
	Ignition devices effective including condition of electrodes, leads and battery	PASS ☐ FAIL ☐ N/A ☐
	Clean and check operation of fans	PASS ☐ FAIL ☐ N/A ☐
	Pilot flame correct	PASS ☐ FAIL ☐ N/A ☐
	Inlet Operating pressure: .... mbar	PASS ☐ FAIL ☐
	Burner pressure: .... mbar	PASS ☐ FAIL ☐
	Heat input........ kW Gas rate.........m^3	PASS ☐ FAIL ☐ N/A ☐
	Flame picture good	PASS ☐ FAIL ☐ N/A ☐
	Flame supervision device operational	PASS ☐ FAIL ☐ N/A ☐
	Condition of frame and combustion seals effective	PASS ☐ FAIL ☐ N/A ☐
	Appliance tightness check	PASS ☐ FAIL ☐ N/A ☐
	Flue flow test	PASS ☐ FAIL ☐ N/A ☐
	Spillage test	PASS ☐ FAIL ☐ N/A ☐
	Operating thermostat correct	PASS ☐ FAIL ☐ N/A ☐
	Flue guard fitted to low level terminals	PASS ☐ FAIL ☐ N/A ☐
	Additional ventilation grille, where necessary: .... cm^2	PASS ☐ FAIL ☐ N/A ☐
	Fireguard in place in presence of vulnerable persons	YES ☐ NO ☐
Post System Checks:	Meter operating pressure (............ mbar)	PASS ☐ FAIL ☐ N/A ☐
	System design pressure loss........mbar (max. 1 mbar NG / 2 mbar LPG)	PASS ☐ FAIL ☐ N/A ☐
	Combustion performance analysis CO.....ppm CO^2......% CO/CO_2 ratio...........	PASS ☐ FAIL ☐ N/A ☐
	Safe operation of appliance explained to customer	YES ☐ NO ☐
Recommendations and/or Warning Notice issued: YES ☐ NO ☐ **Appliance Safe to Use YES ☐ NO ☐ Next Service Due:**		
Installer's Signature:	**Customer's Signature:**	

Figure 9.24 Commissioning and servicing record for space heaters

Figure 9.25 Radiants removed revealing crack formed in heat exchanger behind (See Section 9.41)

Specific Notes Applicable to Mobile Cabinet Heaters

9.43   Because the supply of gas is via a butane gas bottle, a check needs to be made on the condition of the hose connections and regulator to make sure they are in sound working order. It is generally recommended that all flexible hoses are replaced every ten years; however, any hose that shows signs of fatigue needs replacing. A good test is to bend the hose back on itself, forming a very tight bend; this will expose any hair-line cracks that may be developing. The radiant plaques should be inspected, and it is recommended that damaged or cracked plaques be replaced. When re-bedding the plaque, the fire cement should be allowed to dry for at least 24 h before lighting the fire. Where a catalytic panel has been incorporated with the heater this should be inspected for wear, which is often seen as bald patches or holes. This may need replacing, and care needs to be observed as catalytic heaters made before 1983 may include asbestos. Where the heater is mounted on rollers, the casters should be generally examined to ensure smooth running and the control tap must operate freely.

Fault Finding

9.44   The typical faults with gas fires are also common to any burner or control device in a gas appliance. However, the list in Figure 9.26 may help to identify specific problems.

Fault	Possible cause
Pilot will not light	• Gas supply was turned off • Air in pipe • Pilot injector blocked or incorrect size • Incorrect spark gap • No spark or lead not connected properly
Poor pilot flame and will not stay alight	• Thermocouple connection loose • Pilot flame too small or pilot tube blocked • Faulty thermocouple or thermo-electric valve • Inadequate ventilation • ODS defective or blocked
Poor heat output and lack of warm air in the room	• Poorly sealed closure plate, allowing heat to be lost up the flue • Inadequate gas pressure • Blocked gas injectors • Flame reversal, resulting from incorrect or poorly sealed closure plate • Is a spigot restrictor needed? • Faulty thermostat, where fitted • High curb or up-stand at the lower edge of the fire preventing cool air from circulating through the heat exchanger
Staining to an outer case or wall areas surrounding the fire	• Inadequate catchment space, causing spillage • Inadequate ventilation, causing spillage • Blocked or restricted flue, causing spillage
Poor combustion, e.g. yellow flames and soot	• Burner and primary air intake blocked • Incorrectly positioned radiants or coals
Uneven flame pattern or ghosting where the flames begin to emerge from the top of the radiant	• Faulty or linted burner • Insufficient air • Damaged radiants • Blocked injector, where duplex injectors are used • A thick rug restricting airflow

Figure 9.26 Gas fire fault diagnosis chart

Domestic Gas Cookers

Relevant ACS Qualification
CKR1

Relevant Industry Document
BS 6172 and IGEM/G/5

9.45 The domestic gas cooker consists of three separate components, namely the grill, oven, and hob/hotplate. These may be purchased separately or as one complete unit and are designed to be installed with the associated kitchen cupboards or as a freestanding appliance.

The Grill

9.46   The grill is used for toasting, grilling or browning previously cooked food. It works by directing radiant heat that has been produced on the surface of a red-hot fret, mesh or pressed steel onto the food (pictured in Figure 9.27).

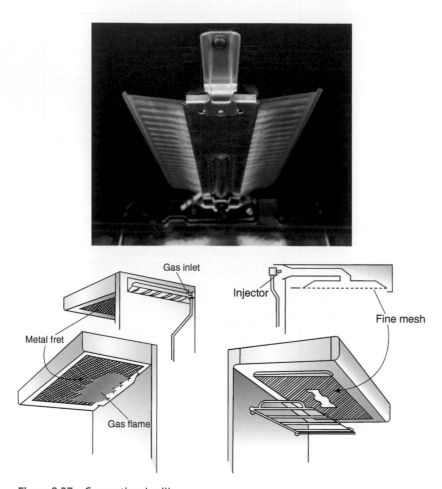

Figure 9.27   Conventional grill

The Oven

9.47   The oven is used for warming, roasting and baking. It works by surrounding the food with hot convection currents. The oven temperature is adjusted by a thermostat that varies the amount of heat surrounding the food to be cooked. The grading ranges from a simmer or economy setting of around 100°C through to gas mark 9, which generates temperatures of around 245°C. Many domestic ovens found in the UK work by natural convection currents, see Figure 9.28a. The burner located at the rear of the base creates a circulation of hot gases that eventually discharge from the rear of the oven by means of a flue. Owing to the relatively slow circulatory motion, temperature zones invariable develop in the oven, with the hottest region at the top.

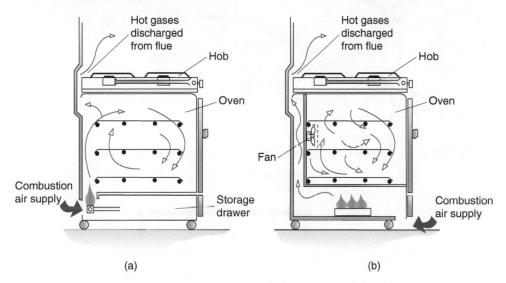

(a)　　　　　　　　　　　　　　　　(b)

Figure 9.28　(a) Directly heated convection oven. (b) Indirectly heated fan-assisted oven

In another design of oven, illustrated in Figure 9.28b, the gas burner is located outside the food compartment, and hot air is allowed to enter via various ports, so producing a more even spread of heat temperature throughout the oven. This design is often used with an additional fan to give improved efficiency.

The Hob or Hotplate

9.48　The hob, illustrated in Figure 9.29a, is used for boiling, frying, steaming, simmering and braising. It primarily works by the conduction of heat from the flame through the pan surface. However, a certain amount of radiant heat is also generated. A large volume of primary air is provided so that the pan can be located as close as possible to the burner head and so gain the heat from the flame. The flame, however, should not be allowed to lick up the side of the pan, as heat will be wasted. The hotplate, as seen in Figure 9.29b, is usually a solid steel flat plate or ceramic surface on which the pan sits, or it can take the form of a griddle, which is used for dry-frying foods such as eggs, bacon, hamburgers, etc. The flame is located beneath the metal and heats it to the desired temperature.

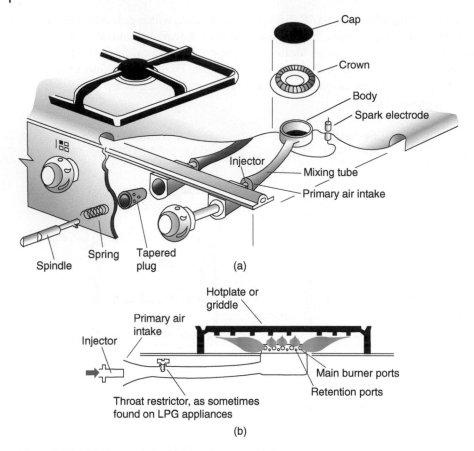

Figure 9.29 (a) The gas hob. (b) A hotplate or griddle

The heat setting of a hob or hotplate often has a grading, such as simmer, medium and high. The flame does not adjust to the heat requirements of the food being cooked, and if set too high, the contents of the pan may boil over or burn. However, having said that some appliances have thermostatically controlled burners. They work using a contact sensor that is filled with a volatile fluid that, when heated, expands, forcing the gas valve to close down to a by-pass rate.

Hobs are often found with a glass drop-down lid that is designed to provide a smoother, pleasing appearance. Where these are incorporated, it is essential to understand the additional safety feature, a safety cut-off device, which is incorporated to prevent gas passing to the burners when the lid is closed. This works in a number of ways, one of which, illustrated in Figure 9.30a, allows a pin to be pushed in to open the valve when the lid is raised. Figure 9.30b and c show photographs of the device in the open and closed positions. *Note:* When tightness testing, this valve must be open; otherwise, several joints will remain unchecked; therefore, the lid must be in the upright position.

With freestanding cookers, all control taps and the oven thermostat are located on a pipe named the float rail, which is located behind the control knob fascia. During any

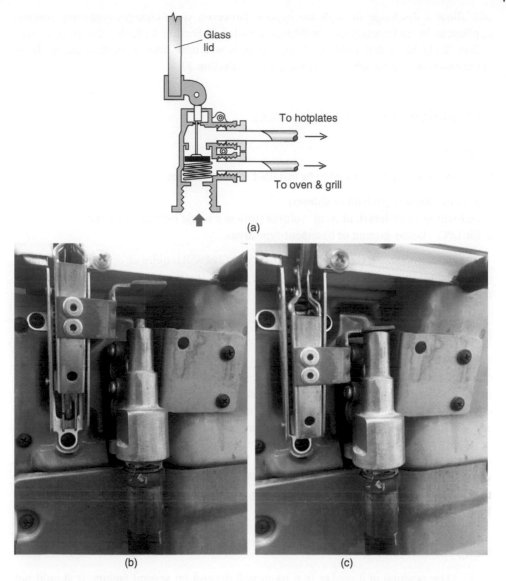

Figure 9.30 (a) Cooker safety cut-off valve. (b) Safety cut-off device lid open. (c) Safety cut-off device

maintenance work, all these gas connections should be sprayed with leak detection fluid to ensure they are gas-tight.

Historically, gas burners used with the associated parts of the cooker were generally unprotected in terms of flame failure. This meant that should the flame go out, gas would be allowed to discharge freely into the room. The oven would have a flame supervision device fitted, usually of the liquid vapour type. With this type of device, on initial light up, there will be a low flame via the bypass for a period of 20–30 s before the main gas comes up. If flame failure occurs, the device will shut down the main flow of gas but would still allow a discharge through the bypass. However, since changes regarding flueless appliances in multi-occupancy buildings, detailed in Section 9.50, this design of cooker is less likely to be encountered. Newer models will now have a thermoelectric flame supervision device, which is described in Part 3: Section 3.42.

Installation of Cooking Appliances

Location

9.49   A cooker is not allowed to be installed in the following areas.

- a room containing a bath or shower;
- bed-sitting room less than 20 m^3 volume (unless a single hotplate burner);
- for LPG – below-ground or basement-type areas.

The room into which the appliance is to be installed will require an openable window or similar adjustable opening. In addition, if the room is small, additional ventilation may be required, as identified in Part 7: Table 7.5.

Multi-Occupancy Buildings

9.50   The requirements regarding flame supervision in multi-occupancy buildings changed in 2008. All new cookers in these locations now require a method of flame failure protection on each burner. Subsequently, manufacturers began to provide thermo-electric devices as standard on all cookers to all burners, including the oven, replacing the liquid expansion method previously used. As a result, most new cookers purchased today will be of this design regardless of the location it is to be fitted. It must be noted that the rule only applies to flats, maisonettes and other buildings divided into individual dwellings when a new hob/cooker is being installed. Where a new or replacement hose is fitted to a cooker in a multi-occupancy building, the hose shall meet the requirements of BS EN 14800.

Siting Requirements

9.51   The position of a cooker in a room will depend on several factors. It should not be affected by draughts from windows, nor must it affect adjacent appliances such as a refrigerator. It may be necessary to have an adjacent electrical point, and a cooker will generally need to be near other appliances, worktops, etc., for convenience. However, wherever it is sited, care will need to be taken to ensure that materials in close proximity will not be in danger of catching fire. The appliance manufacturer generally gives clear guidance as to the correct siting; however, the dimensions given in Figure 9.31 can be taken as suitable minimum recommendations.

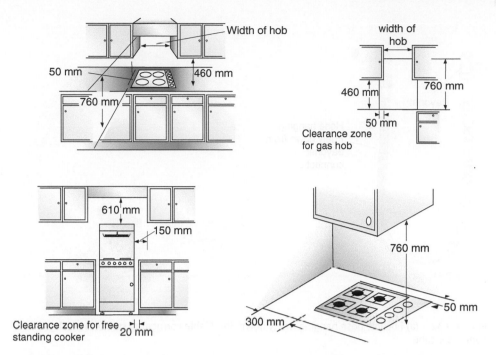

Figure 9.31 Minimum clearances to be maintained between appliance and combustible materials

Gas Connections

9.52 This may be by either a rigid connection with an accessible isolation valve, as found in ranges or where a gas hob is installed directly into a worktop, or, as is often the case, by a flexible connector and self-sealing bayonet valve. The valve must be accessible for disconnection purposes, and the flexible hose should hang freely down, thus avoiding any undue stress to the rubber (see Figure 9.32a). Historically, hoses to BS 669-1 had two separate versions, one for natural gas and one marked with a red line along its length to identify it as suitable for LPG. The LPG version is no longer in production. Therefore, hoses to BS EN 14800, limited to environments with a temperature not exceeding 60°C, which are suitable for both LPG and natural gas, can be used instead. Natural gas hoses to BS 669-1 can be in environments of up to 95°C.

Connections to built-in ovens and hobs shall be by means of a rigid connection or if approved by the appliance manufacturer:

1. Pliable corrugated stainless steel tube (PCSST) to BS 7838 or BS EN 15266. The length of unsupported PCSST to the appliance shall be not more than 500 mm.
2. A flexible connection to BS 669-1 or BS EN 14800 with a self-sealing plug.

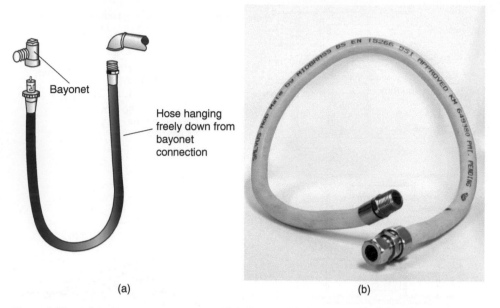

Bayonet

Hose hanging freely down from bayonet connection

(a)

(b)

Figure 9.32 (a) Flexible hose hanging freely down. (b) Pliable corrugated stainless steel gas hob connector tube

Note: PCSST manufactured to the requirements of BS EN 15266 has an environmental temperature limitation of 60°C. PCSST manufactured to BS 7838 can be in environments of up to 95°C.

Cooker Stability

9.53 With a freestanding cooker, it is possible that when the oven door is opened and a minimal weight applied, the cooker might tip forward. This could have disastrous effects, especially if a pan of boiling water is on the hob. For this reason, a stability bracket or chain must be installed, as shown in Figure 9.33.

Domestic Flued Cooking Range

Relevant ACS Qualification
CKHB1 or CENWAT with CKR1

9.54 This is a cooker that is based on the traditional cast iron design of a solid fuel or oil-burning range. There are two designs: those with atmospheric burners and those that use a forced draught burner. These appliances invariably have more than one function, such as heating water for central heating and domestic purposes, and so require a flue system, which may be either of open-flue or balanced flue design.

Those ranges designed with an atmospheric burner often include a maintained flame to keep the appliance at a constant temperature, thus mirroring the traditional design.

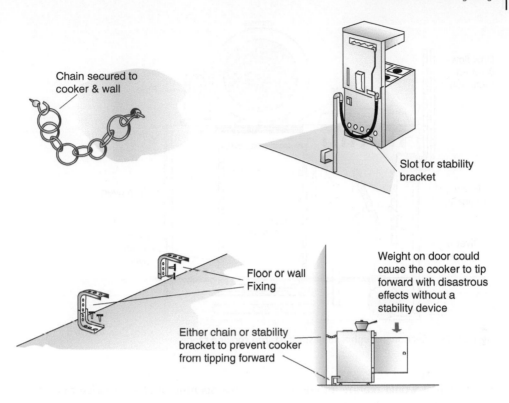

Figure 9.33 Free-standing cooker stability

Invariably, two separate ovens and two large hotplates are incorporated, operating at differing temperatures with graduated zones, see Figure 9.34.

Some designs of this type of cooker are too heavy to be supplied as a complete unit, and it is the specialist installers responsibility to ensure that all the internal parts are correctly located to ensure safe and adequate heat transfer. Owing to the weight, which may be several hundred kilograms, a suitable non-combustible hearth with a minimum thickness of 12 mm needs to be constructed.

The gas connection to a range needs to be of the rigid type because it is a flued appliance, and so a flexible connection, as used for the normal freestanding cooker, must not be used. Following the isolating valve, a disconnecting joint, such as a union connector, is required for servicing purposes.

Flueing

9.55 The flue design depends on the appliance. However, when an open flued appliance is encountered, care should be taken to follow manufacturer's instructions and ensure that the flue is not less than the appliance spigot size. Existing chimneys need to be lined with a double skin liner, certified as being suitable for use with solid fuel, and the void between the liner and chimney filled with a suitable insulating material such as vermiculite.

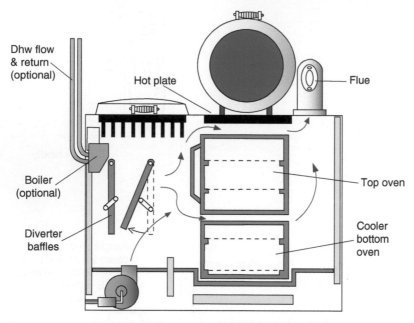

Figure 9.34 Section through a domestic cast iron cooking range

The terminal would require an effective free area opening of at least twice that of the cross-sectional area of the flue system used.

Ventilation

9.56 Ventilation requirements for range cookers will be different from those for a more traditional cooker, which, of course, is flueless. Generally, a range would be treated the same as any open-flued appliance, as described in Part 7: Ventilation. However, these appliances would also require an openable window.

Commissioning and Servicing Cookers

Work Record for Cookers

9.57 The Work Record in Figure 9.35 is purely given as a guide to the many tasks to be undertaken when servicing/maintaining and commissioning gas cookers. In order to complete the form, you may need to refer back to Part 8, Figure 8.1a, where many of the tasks were explained in further detail. However, the additional appliance-specific activities for cookers include some of the following.

Certificate/Record of Cooker Heater Service/Commission		
Gas Installer Details	**Client Details**	**Appliance Date Badge Details**
Name:	Name:	Model/Serial No:
Gas Safe Reg. No:	Address:	Gas Type: Natural ☐ LPG ☐
Address:		Heat Input: max...... kW min...... kW
		Operating Pressure........ mbar
		Gas Council No:
Date:	**Appliance Location:** Install/Commission ☐ Service/Commission ☐	
Preliminary System Checks:	Compliance with manufacturer's instructions	PASS ☐ FAIL ☐
	General visual inspection of pipework	PASS ☐ FAIL ☐
	Clearance from combustible materials	PASS ☐ FAIL ☐
	Gas connection e.g. bayonet and cooker hose in sound condition	PASS ☐ FAIL ☐ N/A ☐
	Stability bracket/chain effective	PASS ☐ FAIL ☐ N/A ☐
	Appliance level and secure	PASS ☐ FAIL ☐
	Electrical connections	PASS ☐ FAIL ☐ N/A ☐
	Bonding maintained	PASS ☐ FAIL ☐ N/A ☐
	Fuse rating: .... amps	PASS ☐ FAIL ☐ N/A ☐
	System tightness test; to include let-by	PASS ☐ FAIL ☐ N/A ☐
	Standing pressure of system	PASS ☐ FAIL ☐ N/A ☐
	Appliance/system is purged of air	PASS ☐
	Service/Commission Checks: Clean primary air ports	PASS ☐ FAIL ☐ N/A ☐
	Clean injectors; burners; burner rings and grill frets	PASS ☐ FAIL ☐ N/A ☐
	Check, ease and grease, if necessary, control taps	PASS ☐ FAIL ☐ N/A ☐
Ignition	devices effective e.g. condition of electrodes, leads and battery	PASS ☐ FAIL ☐ N/A ☐
	Pilot flame correct	PASS ☐ FAIL ☐ N/A ☐
	Inlet operating pressure: .... mbar	PASS ☐ FAIL ☐
	Total heat input: .... kW	PASS ☐ FAIL ☐ N/A ☐
	Flame picture good	PASS ☐ FAIL ☐ N/A ☐
	Simmer settings to all burners and oven by-pass	PASS ☐ FAIL ☐
	Flame supervision devices operational	PASS ☐ FAIL ☐ N/A ☐
	Oven flueway is clear	PASS ☐ FAIL ☐ N/A ☐
	Appliance tightness check	PASS ☐ FAIL ☐ N/A ☐
	Lid safety cut-off device is effective	PASS ☐ FAIL ☐ N/A ☐
	Oven and door seals are effective	PASS ☐ FAIL ☐ N/A ☐
	Oven thermostat correct	PASS ☐ FAIL ☐ N/A ☐
	Ancillary equipment (timers, oven lights, fans, etc.)	PASS ☐ FAIL ☐ N/A ☐
	Openable window or equivalent	PASS ☐ FAIL ☐
	Additional ventilation grille, where necessary .... cm²	PASS ☐ FAIL ☐ N/A ☐
	Post System Checks: Meter operating pressure (............ mbar)	PASS ☐ FAIL ☐ N/A ☐
	System design pressure loss........mbar (max. 1 mbar NG / 2 mbar LPG)	PASS ☐ FAIL ☐ N/A ☐
	Safe operation of appliance explained to customer	YES ☐ NO ☐
Recommendations and/or Urgent Notification		
Appliance Safe to Use YES ☐ NO ☐	**Next Service Due:**	
Installer's Signature:	**Customer's Signature:**	

Figure 9.35 Commissioning and servicing record for cookers

Cleaning Grill Frets

9.58 When servicing a cooker, particular attention should be given to the inspection of the grill frets for damage, such as buckling or splits that might result in flame impingement, etc., leading to the production of high levels of carbon monoxide (CO). This is often overlooked by the inexperienced gas engineer, and it has been responsible for a number of CO poisoning incidents.

Checking the Simmer Settings on the Hob

9.59   Turn the gas tap to its lowest setting. If the flame goes out, the smallest hole in the valve is blocked, possibly with grease.

Checking the Appliance for Gas Leaks

9.60   All exposed pipework and fittings should be sprayed with a leak detection solution when the gas is flowing to the burners. Particular attention should be paid to the pipe leading up to the grill, where a compression joint is often found. When a gas tap has been reassembled following a repair, the valve should also be sprayed.

Lid Safety Cut-Off Device

9.61   To check the correct operation of this safety device, with the hot plate flames burning, lower the glass lid and check that the flame is extinguished.

Checking the Door Seals

9.62   This is a simple test carried out by trapping a piece of paper (0.25 mm thick) in the top and both sides of the door and pulling it out. A resistance should be felt. Where there is little resistance, heat could possibly escape, so cooling the oven and, invariably, drying out the control taps. The base of the door does not need to be checked as this is where the air is drawn into the oven.

Testing the Oven Thermostat and By-Pass

9.63   This is achieved using a thermometer, such as the one pictured in Figure 9.36, to compare gas settings against those of the manufacturer. *Note:* It is generally not possible to re-calibrate the oven thermostat, and therefore, it may need replacing if the temperature of

Figure 9.36   Typical oven thermometer

Table 9.2 Typical oven thermostat temperature settings

Gas mark	Approx. (°C)	Oven heat
S/E	105–120	Very cool
1	135–140	Cool
2	150	Cool
3	160	Warm
4	175	Moderate
5	190	Fairly hot
6	202	Hot
7	220	Hot
8	230	Very hot
9	250	Very hot

the oven is incorrect. Table 9.2 may be used as a general guide to oven cooking temperatures. An oven thermostat is usually tested at gas mark 5 (190°C).

Set the oven thermostat to gas mark 5 (190°C). Place an oven thermometer centrally within the oven and close the door. Allow the oven temperature to rise for 10–15 minutes. The temperature should eventually be satisfied, and the thermostat allows the flame to drop to the by-pass rate. Check that the thermometer reading at by-pass rate is approximately 190°C. If there is a variance to this temperature, the thermostat may be faulty and require replacement. Turn the temperature down to gas mark 1, causing the thermostat to firmly close and observe the flame. If the flame goes out, the by-pass is blocked and the screw to the side of the thermostat will need to be removed to dislodge any excess grease, etc.

Fault Finding

9.64   Many of the faults to be found with cookers are the result of spillage of food, blocking burner ports, injectors and ignition electrodes. Faults with hobs and grills are usually self-evident. Oven defects such as uneven cooking are more difficult to diagnose and may be the result of misplaced linings or distorted shelves or the cooker itself being out of level. The chart in Figure 9.37 lists some of the many faults that may be encountered.

Instantaneous Gas Water Heaters

Relevant ACS Qualification
WAT1, CENWAT

Relevant Industry Document
BS 5546, BS 8558 and BS EN 806

9.65   There are two types of instantaneous water heaters: single-point and multipoint heaters. A combination boiler is a form of multipoint, but it also functions as part of the

Fault	Possible cause
Incorrect flame picture or insufficient heat	• Incorrect gas pressure • Blocked injector or wrong size • Blocked or damaged burner head or frets • Aeration port blocked or setting incorrect • Insufficient ventilation in a room • Poor door seals • Faulty oven thermostat • Combustion outlet blocked
No or poor ignition	• Damaged, dirty or wrongly positioned electrodes • Faulty batteries or electric ignition system • Incorrect gas pressure • Faulty flame supervision device • Aeration port blocked or setting incorrect • Damaged flash tube
Uneven cooking	• Appliance not level • Blocked or damaged burner head or grill frets • Blocked injector or wrong size • Poor door seals • Faulty oven thermostat or incorrectly positioned probe • Faulty or misplaced oven linings • Combustion outlet blocked • Distorted shelves
Hot control taps	• Poor oven door seals

Figure 9.37 Fault diagnosis chart for cookers

central heating system and will therefore be dealt with later in Section 9.88. The general principle of the instantaneous heater is that when cold water passes into the unit it flows up around the combustion chamber and through the heat exchanger, where it is rapidly heated to the required temperature at the hot draw-off point. The water flow rate will, therefore, determine the outlet temperature. As the water flows through the unit, the movement will be detected by some form of differential valve. This, in turn, brings on the gas supply to the main burners, warming the water. When the supply is turned off, the static no-flow state is identified, and the gas supply is cut off. The remaining cold water in the heater now cools the unit and prevents the water from boiling. The water supply may be either mains fed or fed via a feed cistern. Early designs of water heaters used a bi-metallic strip as the flame supervision device; however, apart from in older models, a thermal-electric flame supervision device that uses a thermocouple is usually used. One of the biggest problems for these heaters is the issue of scale build-up in hard water areas. The condition manifests itself by a low rate of water flow through the heater, with squealing and kettling noises. Furthermore, it can cause damage to the heat exchanger or combustion chamber, where the heat has been restricted from escaping.

The Single Point

9.66 These heaters are used in close proximity to their point of use and are usually used to serve only one or two sanitary appliances. They may have a swivel spout or be plumbed in fully to a system of pipework, running to the various outlets. A typical heater has less

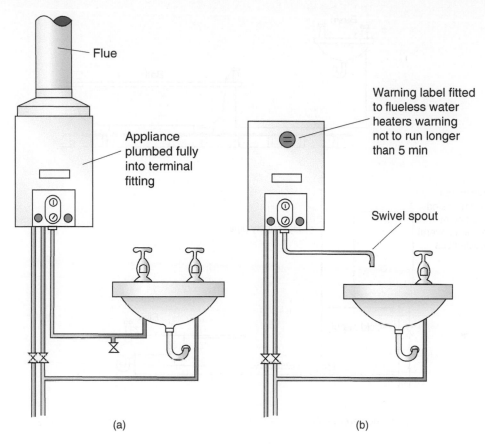

Figure 9.38 (a) Flued single point. (b) Flueless single point

than 11 kW net input, with a water flow rate of around 2.5 l/min, allowing a temperature rise of about 35°C.

This type of unit may be either flued (Figure 9.38a) or flueless (Figure 9.38b). If a flueless heater is used, a warning sticker needs to be fixed in a prominent position on the front of the heater warning of a maximum running period of five minutes; this is designed as a safeguard against raising the levels of combustion products, in particular carbon monoxide, within the room. A flueless heater must be installed in the room where it is to be used and not in an adjoining room, where it would need to be flued.

The Multi-point

9.67 This heater, illustrated in Figure 9.39, is designed to serve several outlets and typical heat inputs of around 30 kW net, giving water flow rates of around 6.5 l/min, and a 35°C temperature rise can be expected. Today, all models are of the room-sealed type. A combination boiler serves effectively as a multi-point yet has the additional function of providing central heating.

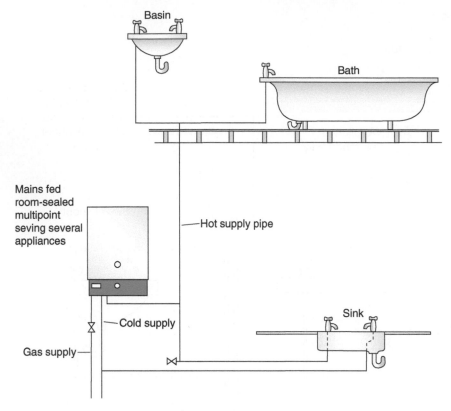

Figure 9.39 Multipoint water heater

Instantaneous Water Heater Operation

Traditional Water Heater Operation

9.68 The following sections will look at the more traditional style of water heater and their method of operation, Section 9.73 looks at how instantaneous water heating has evolved.

For the instantaneous water heater to function, the water and gas controls work in conjunction with each other and are usually specifically designed for the appliance. There are many variations for individual heaters; however, the components and operation identified here are of a typical design. Sections 9.69–9.72 describe the operation of various controls that enable the appliance to operate. For reference, Figure 9.40a shows the appliance in the off position, with no water flowing. Figure 9.40b shows the appliance operating with the water and gas flowing through the appliance.

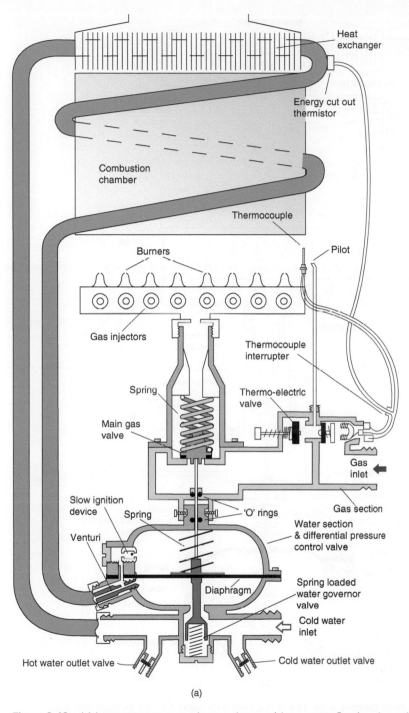

(a)

Figure 9.40   (a) Instantaneous water heater shown with no water flowing through unit

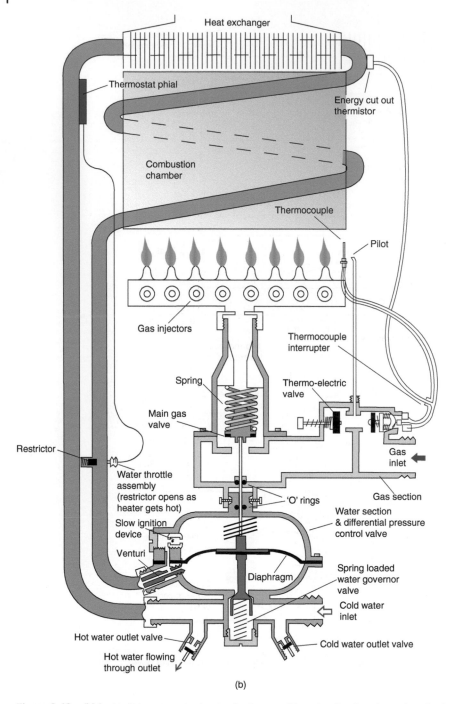

Heat exchanger

Thermostat phial

Energy cut out thermistor

Combustion chamber

Thermocouple

Pilot

Gas injectors

Thermocouple interrupter

Spring

Thermo-electric valve

Main gas valve

Restrictor

Gas inlet

Water throttle assembly (restrictor opens as heater gets hot)

Gas section

'O' rings

Slow ignition device

Water section & differential pressure control valve

Venturi

Diaphragm

Spring loaded water governor valve

Cold water inlet

Hot water outlet valve

Hot water flowing through outlet

Cold water outlet valve

(b)

Figure 9.40 (b) Instantaneous water heater is shown with water flowing through unit plus additional thermostat

Components and their Operation

Differential Pressure Control Valve

9.69   This valve, seen in Figure 9.41, has the function of automatically opening/closing the gas supply to the main burner when water is flowing through the heater. It also ensures that sufficient volume flows before bringing on the gas to prevent the water temperature from rising above 55°C, which would lead to scaling up and overheating. The valve works by passing the water through a venturi, which has the effect of reducing the pressure from above a diaphragm, causing it to lift in response to the negative pressure.

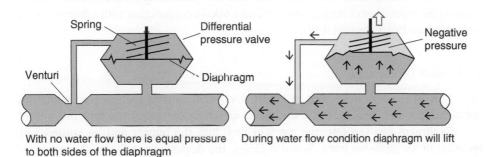

With no water flow there is equal pressure to both sides of the diaphragm

During water flow condition diaphragm will lift

Figure 9.41   Valve lift due to the venturi

The venturi works on the theory that reducing the bore of a pipe causes the water flow to increase in velocity, just like putting your finger over the end of a hosepipe. The increased velocity is an energy force, and the energy is obtained by giving up some other form of energy, in this case, pressure.

As the diaphragm lifts, it forces the gas valve open, allowing gas to the injector manifold to be ignited by a previously established pilot flame. When there is limited flow or no water flow through the valve, the pressure differential between each side of the diaphragm is minimal, and the spring will cause the gas valve to close.

Water Governor and Throttle

9.70   For cistern-fed water heaters where the pressure is low, adjustment of the water flow is usually by means of a water throttle, which is no more than a screw-in restrictor. For mains-fed appliances, however, variable water flows are experienced and the use of a water governor is required. The governor usually forms part of the appliance and consists of a spring-loaded valve acting on the underside of the diaphragm. As water flows into the heater it passes through and around this valve and on through the venturi, which allows the diaphragm to lift. If the water pressure is too high, the governor valve assumes its highest position, reducing the inlet flow through the seating. However, where low pressure is experienced, the strong spring acting on the gas valve tries to re-close the diaphragm and, in so doing, allows the governor valve to open further, enabling increased water flow.

Temperature Control

9.71 The temperature of the water flowing out at the draw-off points is determined by the volume and speed at which the water flows through the unit. With some units, this adjustment is made by a temperature selector located on the front, thus altering the volume of water flow through the selector and enabling the user to select variable temperatures for the outlet point. Some models have a water throttle in line, which restricts the water flow. As the appliance heats up, a thermostat operates, and a volatile fluid is forced into the bellows chamber. This forces open the throttle, allowing a greater volume of water to flow and so the temperature is maintained. To prevent the unit from overheating an energy cut-off device is incorporated with most modern units. This is simply a thermistor, which is located on the top section of the heating unit and works in conjunction with a thermocouple interrupter (see Part 3: Section 3.59).

Slow Ignition Device

9.72 This is a device, illustrated in Figure 9.42, that controls the speed at which the diaphragm lifts and, consequently, the speed at which the gas flows into the burner. If the diaphragm lifts too fast, the gas will flow into the burner and be ignited with something of an explosive force. When it closes, the device responds rapidly to prevent the appliance overheating. There are several designs of slow ignition devices. The one described here is positioned in the low-pressure duct from the venturi. As the water flows from the space above the diaphragm, due to the action of the venturi, its velocity is restricted by a small ball bearing, blocking a central port hole, so the valve can only lift slowly as the water is drawn out. Conversely, when the cold supply is closed, the water flowing back to this space above the diaphragm can pass quite rapidly through the central hole as well as through the restricted space. The amount/rate of diaphragm lift can be adjusted by screwing the device further in or out of the housing until smooth ignition is achieved.

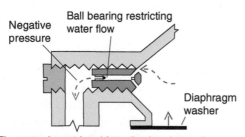

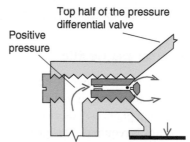

The water is restricted from flowing due to the position of the ball bearing, thus allowing for a slow lifting of the diaphragm and consequently a slow opening of the gas valve

Water can flow back quickly as the ball bearing no longer restricts the through way, thus the gas valve quickly closes

Action as water flows through the appliance

Action as water stops flowing into the appliance

Figure 9.42 Operation of the slow ignition device

Modern Instantaneous Water Heaters

9.73   As described in the previous sections, older design water heaters operate by utilising water pressure to lift a diaphragm, mechanically open a gas valve and turn on a burner, which is ignited by a permanent pilot. The flow of water is fixed at a speed that will allow it to be sufficiently heated. Whereas modern instantaneous water heaters, as seen in Figure 9.43, typically operate with the use of electronics. The water flow is detected by a control system which activates an electronic ignition system, without a permanent pilot, to ignite the burner. The water temperature is monitored by various sensors that, in conjunction with the control system, modulate the burner accordingly. The hot water temperature can be programmed to a set temperature and can be altered by the user. Some manufacturers have developed system designs for the commercial market that involve linking various water heaters together mechanically and electrically, see Part 10: Section 10.21.

Figure 9.43   Modern Instantaneous water heater. Credit: Rinnai UK

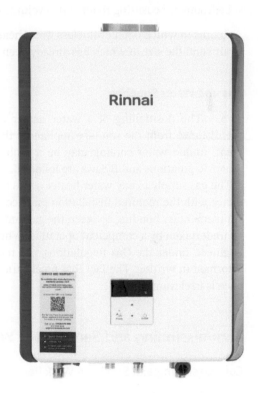

Installation of Water Heaters

Location

9.74   If the appliance is to be room-sealed, which will more likely be the case, it may be installed in any location within reason, with the exception of an LPG appliance, which must not be located below ground level or in a low-lying area. Care needs to be observed to ensure that any compartment is prevented from getting too hot and clearly the terminal would have to be located in accordance with the manufacturer's instructions. If the appliance is fan-assisted, there are special requirements if installed in a bath or shower room to ensure that all electrical components are in accordance with BS 7671.

Historically, open-flued and flueless water heaters were installed, but this does not form part of modern practice. However, it is worth noting the restrictions applicable to such installations. Regulation 30 of the Gas Regulations states that it is not permissible to install an open-flued or flueless appliance in any of the following locations:

- bath or shower room;
- bedroom or bed-sitting room where the appliance is more than 14 kW gross input; where it is less than 14 kW, it would need to incorporate some form of atmospheric sensing device;
- for LPG – below ground level or basement-type areas;
- bedroom or bedsitting room with a volume of less than 20 m^3.

Rooms in which open or flueless water heaters are installed will also require some form of air vent, the size of which has already been discussed in Part 7, Ventilation.

Gas and Water Supplies

9.75 The positioning of a water heater depends on several factors, not least being the distance from the sanitary appliances that it is to serve. Should the distance be too great, undue water cooling may be a problem. Therefore, it is important that current Water Regulations and Bylaws are followed.

The gas supply to any water heater needs to be of fixed pipework and installed in accordance with the required installation practices. Where a water heater is a new installation, supplementary bonding between the gas and water pipework may be required and should be undertaken by a competent operative. Where this work cannot be completed by the gas engineer, under the Gas Regulations, the responsible person for the property should be informed in writing. The fact that no electrical connections have been made to the appliance is irrelevant.

Commissioning and Servicing of Water Heaters

Commissioning and Servicing Work Record

9.76 The Work Record, shown in Figure 9.44, is purely given as a guide to the many tasks to be undertaken when servicing/maintaining and commissioning a water heater. The form can be used for both instantaneous and storage water heaters. In order to complete the form, you may need to refer back to Part 8: Figure 8.1a, where many of the tasks are explained in more detail. However, additional appliance-specific checks on these appliances may include some of the following activities.

Certificate/Record of Water Heater Service/Commission		
Gas Installer Details	**Client Details**	**Appliance Date Badge Details**
Name:	Name:	Model/Serial No:
Gas Safe Reg. No:	Address:	Gas Type: Natural □ LPG □
Address:		Heat Input: max…… kW min…… kW
		Operating Pressure Range: …… –…… mbar
		Gas Council No:
Date:	**Appliance Location:** Install/Commission □ Service/Commission □	
Preliminary System Checks: Compliance with manufacturer's instructions		PASS □ FAIL □
General visual inspection of pipework		PASS □ FAIL □
Clearance from combustible materials		PASS □ FAIL □
Flue notice plate located and correctly filled in		PASS □ FAIL □ N/A □
Visual inspection of flue and flue flow performance test		PASS □ FAIL □ N/A □
Appliance level and secure		PASS □ FAIL □
Electrical connections		PASS □ FAIL □ N/A □
Bonding maintained		PASS □ FAIL □ N/A □
Fuse rating: ………… amps		PASS □ FAIL □ N/A □
System tightness test; to include let-by		PASS □ FAIL □ N/A □
Standing pressure of system		PASS □ FAIL □ N/A □
Appliance/system is purged of air		PASS □
Service/Commission Checks: Clean primary air ports and lint arrestor		PASS □ FAIL □ N/A □
Clean/check condition of injectors; burners		PASS □ FAIL □ N/A □
Condition of heat exchanger for scale build-up and sacrificial anode		PASS □ FAIL □
Clean and free movement of water governor, slow ignition device and venturi		PASS □ FAIL □ N/A □
Check, ease and grease, if necessary, control taps		PASS □ FAIL □ N/A □
Ignition devices effective including condition of electrodes, leads		PASS □ FAIL □ N/A □
Clean and check operation of fans		PASS □ FAIL □ N/A □
Pilot flame correct		PASS □ FAIL □ N/A □
Inlet operating pressure: ………. mbar Burner pressure: ………… mbar		PASS □ FAIL □
Total heat input: ………… kW		PASS □ FAIL □ N/A □
Flame picture good		PASS □ FAIL □ N/A □
Flame supervision device operational		PASS □ FAIL □ N/A □
Atmosphere sensing device operational		PASS □ FAIL □ N/A □
Flow rate: ……L/s Temperature rise: ……°C		PASS □ FAIL □
Condition of frame and combustion seals effective		PASS □ FAIL □ N/A □
Appliance tightness check		PASS □ FAIL □ N/A □
Spillage tests		PASS □ FAIL □ N/A □
Operating thermostat correct		PASS □ FAIL □ N/A □
Flue guard fitted to low-level terminals		PASS □ FAIL □ N/A □
Additional direct ventilation grille required: ………… cm^2		PASS □ FAIL □ N/A □
High-level compartment ventilation where required: ………… cm^2		PASS □ FAIL □ N/A □
Low-level compartment ventilation where required: ………… cm^2		PASS □ FAIL □ N/A □
Post System Checks: Meter operating pressure (………… mbar)		PASS □ FAIL □ N/A □
System design pressure loss…..…mbar (max. 1 mbar NG / 2 mbar LPG)…		PASS □ FAIL □ N/A □
Combustion performance analysis CO…...ppm CO^2……% CO / CO_2 ratio………..		PASS □ FAIL □ N/A □
Safe operation of appliance explained to customer		YES □ NO □
Recommendations and/or Urgent Notification		
Appliance Safe to Use YES □ NO □ Next Service Due:		
Installer's Signature:	**Customer's Signature:**	

Figure 9.44 Commissioning and servicing record for water heaters

Clean and Free Movement of Water Governor and Slow Ignition Device (Older Instantaneous Water Heaters Only)

9.77 This is simply undertaken when the water is isolated by removing the appropriate housing screw in the water section. In general, the working parts should move freely, and where appropriate, water industry approved silicone based grease can be used. Other working parts in the water section should also be checked, including the free movement of the diaphragm and gas valve spindles. Care should be taken when replacing the slow ignition device to ensure that the ignition sequence is a smooth, quiet operation, adjusting in or out as necessary. Note: With some models, certain components are not removable.

Checking the Flow Rate and Temperature Rise

9.78 On the older style water heaters, it is critical to set the flow rate in order to achieve and maintain a set temperature rise, as indicated by the appliance manufacturer. The flow rate and temperature rise are measured using a flow cup and thermometer, as shown in Figure 9.45. The water volume is read from the scale as the water flows through the cup. The temperature rise should be taken first with the water running cold, then again at its allocated temperature, any adjustment in temperature rise being made in accordance with the manufacturer's instructions. *Note:* On modern units, typically, the heaters automatically adjust gas and water flow to achieve and maintain a set temperature.

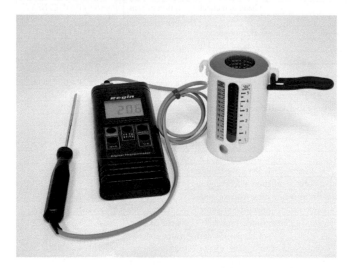

Figure 9.45   Flow cup and digital thermometer

Checking the Operation of an Atmosphere Sensing Device (ASD)

9.79 The manufacturer of the appliance may give instructions as to the method of testing the ASD if one is fitted to an appliance.

Scale Build-Up

9.80   Scale can be a major problem with all direct water heaters, in particular instantaneous heaters, where the water tubes can become blocked or reduced in size. Scale is the result of calcium carbonate deposits that are carried in suspension in hard water districts. Should the water be heated to a temperature in excess of 60°C the limescale, as it is called, is given up and deposited in the vessel. It is possible to descale the heat exchanger of an instantaneous water heater by slowly passing a proprietary descalent solution through the heater unit. Scale accumulation within a heater is generally identified by reduced water flow rates, excessive temperature rise or a noisy appliance.

Fault Finding

9.81   A useful instantaneous water heater fault diagnosis chart can be seen in Figure 9.46.

Fault	Possible cause
Pilot will not light	• Gas supply was turned off • Air in pipe • Pilot injector blocked or incorrect size • Incorrect spark gap • No spark or lead not connected properly
Poor pilot flame and will not stay alight	• Thermocouple connection loose • Energy cut-out connection loose • Pilot flame too small or pilot tube blocked • Faulty thermocouple to thermo-electric valve • Faulty thermal switch or energy cut-out device • Inadequate ventilation
Main burner will not light	• Low water flow rate, e.g. blocked filter • Low gas pressure • Faulty diaphragm • Gas valve push rod jammed • Slow ignition device incorrectly set or stuck
Poor water flow rate	• Blocked filter • Scaled heat exchanger • Poor inlet water supply • Water governor sticking
High water flow rate	• Faulty diaphragm or water governor sticking • Gas valve push rod sticking
Low water temperature	• Gas pressure too low • Faulty diaphragm or water governor sticking • Gas valve push rod sticking • Slow ignition device incorrectly set
Noisy heater	• Scaled heat exchanger • Incorrectly set slow ignition device • Burner ports blocked
Smells	• Faulty case or flue seals • Flueless heater and failure to open window • Newness of appliance

Figure 9.46   Instantaneous water heater fault diagnosis chart

Domestic Gas Boilers

Relevant ACS Qualification
CENWAT

Relevant Industry Document
BS 6798

9.82   There are two types of hot water boilers: the open-flued and the room-sealed appliance. Since the introduction of the Boiler Efficiency Regulations of 1995 and their subsequent amendments, the natural draught open-flued boiler failed to meet the requirements as laid down, and these are no longer installed. Open and room-sealed appliances are discussed in more detail in Part 6: Flues and Chimneys.

Energy-related Product Directive (ErP)

9.83   The Energy-related Product (ErP) Directive replaced SEDBUK as the vehicle to set standards for the efficiency and performance of heating systems in September 2015.

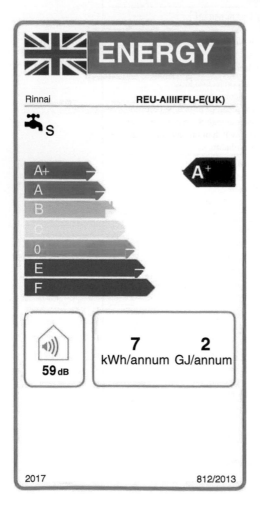

Figure 9.47   ErP label

SEDBUK, which is an acronym for Seasonal Efficiency of Domestic Boilers in the UK, is still used for Standard Assessment Procedure (SAP) modelling. SAP is the method used to calculate the energy performance of dwellings.

It is incumbent on the installer to ensure that the ErP procedure is followed. A new boiler will come with an ErP label that is required to be supplied with the appliance; see Figure 9.47. If the boiler is being installed along with multiple products, such as a new cylinder, an additional ErP label will need to be produced to reflect the package that is being installed. The label can either be produced by the merchant supplying the items as a package or the installer will purchase the goods separately and produce their own label using an online tool. The online tool is available on boiler manufacturers and government websites.

With the government's target in relation to 'Net Zero', there will be further developments to the standards regarding the heating of our homes and other buildings. Therefore, it is the heating installers responsibility to keep up to date and ensure that they are following the current legislation and guidance.

Boiler Type and System Design

9.84 The system illustrated in Figure 9.48 shows an example of a standard central heating design. The water in the system may be open to the atmosphere and fed via a feed and expansion vessel, called an open system. Alternatively, the boiler may be fed directly from a mains supply via a temporary filling loop and incorporate a sealed expansion vessel, in which case the system is called a closed or sealed system.

The boiler serving the system can be one of several designs:

- a heat-only condensing boiler;
- a condensing combination boiler;
- a condensing system boiler.

Presently, when upgrading/replacing an existing boiler, in order to comply with the Building Regulations, the system design must be fully pumped, with thermostatic radiator valves (TRVs) fitted to all radiators except those located within the same room as the room thermostat. It may also be necessary to replace the domestic hot water cylinder with one that is more efficient. The system must be fully interlocked so that the boiler will not fire unless heat is called for. This is achieved by means of a room thermostat and a cylinder stat as appropriate. For new or fully refurbished systems, there is a requirement that the heating and hot water system is designed to operate with a maximum flow temperature of 55°C. This is to future-proof the system ready for a transition to low-carbon forms of heating, see Part 1: Section 1.35. There are numerous other factors to consider in relation to system design and many more than this book can adequately describe. For more information, further reading is required. Ultimately, Part L of the Building Regulations must be fully complied with, and it is important to keep up to date with any changes. At the time of writing, the Future Homes Standard is planned to be introduced by 2025 and will require future-proofing of new build homes with low-carbon heating. So, it is vital to keep abreast of developments and adapt accordingly.

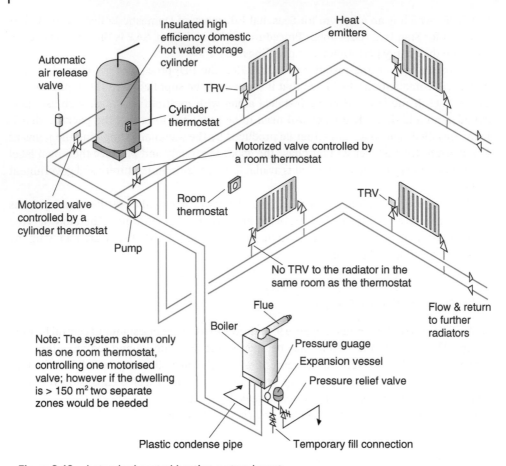

Automatic air release valve

Insulated high efficiency domestic hot water storage cylinder

Heat emitters

TRV

Cylinder thermostat

Motorized valve controlled by a room thermostat

Motorized valve controlled by a cylinder thermostat

Room thermostat

TRV

Pump

No TRV to the radiator in the same room as the thermostat

Flow & return to further radiators

Flue

Boiler

Note: The system shown only has one room thermostat, controlling one motorised valve; however if the dwelling is > 150 m² two separate zones would be needed

Pressure guage

Expansion vessel

Pressure relief valve

Plastic condense pipe

Temporary fill connection

Figure 9.48 A standard central heating system layout

Conventional or Traditional Gas Boilers

9.85   The older type of traditional gas boiler consists of a multi-functional gas valve, a burner/firebox, a heat exchanger and the combustion products collection point/connection to the flue system. Some models incorporate a fan to assist in the expulsion of the combustion products, and the unit may be either open-flued or room-sealed, depending on the age of the appliance. The boiler may be freestanding, as seen in Figure 9.49, wall mounted or a back boiler unit, as illustrated in Figure 9.50. A back boiler unit will be incorporated with a gas fire and installed within the a room located within the chimney catchment space, hidden behind the fire. All domestic boilers installed today are of the fan-assisted room-sealed design and incorporate a pressure sensing device to only allow gas to flow to the burner when the fan has been proved to be functioning.

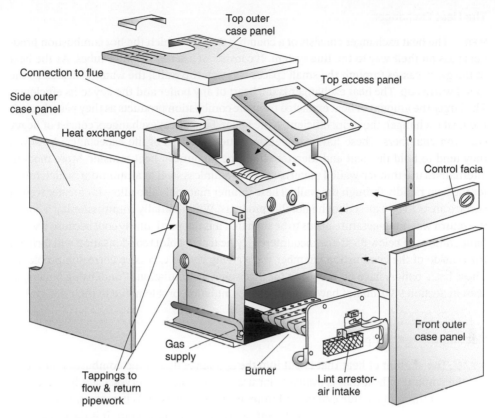

Top outer case panel

Connection to flue

Top access panel

Side outer case panel

Heat exchanger

Control facia

Gas supply

Tappings to flow & return pipework

Burner

Lint arrestor-air intake

Front outer case panel

Figure 9.49   Free-standing open-flued boiler

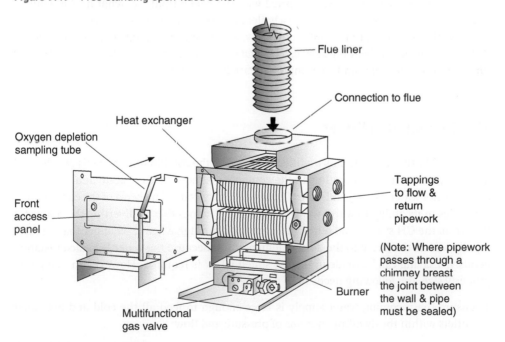

Flue liner

Connection to flue

Heat exchanger

Oxygen depletion sampling tube

Tappings to flow & return pipework

Front access panel

(Note: Where pipework passes through a chimney breast the joint between the wall & pipe must be sealed)

Multifunctional gas valve

Burner

Figure 9.50   Back boiler unit

The Heat Exchanger

9.86   The heat exchanger consists of a chamber through which the hot combustion products pass on their way to the flue system. It consists of a series of water tubes. As the heat of the gases passes through the small spaces between the walls, the water is heated as the metal warms up. The heat exchanger is the heart of any boiler and the key to its efficiency. The larger the amount of heat extracted from the combustion products as they pass through the heat exchanger, the more efficient is the boiler. Some heat exchangers consist of heavy cast iron chambers. These have excellent qualities in terms of life expectancy; however, they tend to hold the heat and therefore limit the amount of heat transfer. More modern alternatives use thinner-walled materials such as stainless steel or aluminium, which transfer the heat rapidly through the walls. These thinner materials also allow for greater wetted surface areas to be exposed to the hot flue products. When initially commissioning a boiler, the return water temperature needs to be such that it is not particularly cool because if water temperatures of below 55°C are encountered (typical dew point) condensation will form on the outside of the combustion chamber walls giving rise to excessive corrosion problems. These basic boilers have no provision for this water accumulation. Condensing boilers, as seen in Section 9.90, do not have this water accumulation problem due to their design.

Heat Input/Heat Output

9.87   The amount of heat that is put into the appliance due to the combustion of fuel is not the amount of heat that is available for use by the system it serves. Clearly, some heat will be lost through the flue system and from the appliance itself. When commissioning an appliance, the gas engineer is primarily concerned with the heat input. It is the designer or central heating installer who is concerned with the heat output, i.e. what heat is actually available for use. Thus, when selecting a boiler for a particular purpose, it is the heat output available that should be important. Conversely, when commissioning an appliance, the gas installer confirms the heat input by calculating the gas consumption used over a period of time. See also 'Gas Rates and Heat Input' in Part 2: Section 2.66.

The Combination Boiler

9.88   Unlike the conventional/traditional gas boiler, the combination boiler is usually sold as a complete unit, incorporating all the components required to operate and control the domestic hot water (DHW) and central heating (CH) systems. The heart of the combination boiler lies in its ability to warm the DHW instantly. It does this by diverting the hot water flow from the CH system and temporarily passes it through a heat exchanger that, in turn, rapidly warms the DHW to the desired temperature. Because the water is heated instantaneously, a saving can be made in that no stored DHW is required. However, in relation to property size, the following needs to be considered:

1. ensure the incoming water supply is large enough to feed all the cold and hot water outlets within the dwelling in terms of pressure and flow;

2. it must be remembered that while the DHW is being heated, no warming of the CH will take place.

Furthermore, Building Regulations Approved Document L requires a new or replacement combination boiler installed in an existing dwelling to include additional energy efficiency measures. There are four acceptable options to choose from, and at least one of the following should be installed:

- Flue gas heat recovery.
- Weather compensation.
- Load compensation.
- Smart thermostat with automation and optimisation.

Operation of the Combination Boiler

9.89   A schematic section through a combination boiler is shown in Figure 9.51. It operates as follows.

Central Heating Mode

If the time clock and thermostat call for heat, the pump runs. This allows the water to flow through primary pipework within the boiler and eventually out into the CH circuit. As the water flows past the CH flow switch within the boiler, it allows the exhaust fan to operate, which initiates the gas flow. This in turn generates the flame within the combustion chamber. The boiler continues to run until the thermostat is satisfied.

Domestic Hot Water Mode

When water is drawn off from a tap, water flows through the boiler and secondary heat exchanger and, in so doing, activates the DHW flow switch. This brings on the pump and operates the three-way control to allow water to divert and flow through the heat exchanger. With water flowing round the boiler within the primary circuit, the CH flow switch allows the boiler to fire up, as above. Heat flows round within the boiler and, as it passes over the secondary heat exchanger, warms the water instantaneously. Note, as previously stated, the DHW has precedence over the CH and, as a consequence, if the CH and DHW were activated at the same time, the CH would temporarily cease.

The Condensing Boiler

9.90   The condensing boiler is a highly efficient appliance that extracts as much of the heat from the flue gases as possible rather than allow it to be dispersed to the external environment and waste fuel. In order to gain the maximum efficiency from a condensing appliance, it is essential that the system is designed to operate at lower water temperatures. Systems such as those relying on cooler flow and return temperatures prove to be the most effective, such as underfloor heating. Where radiators are used, the return to the boiler should

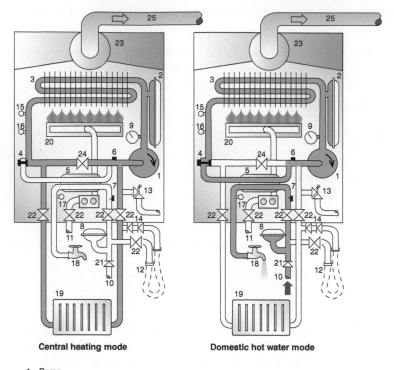

Central heating mode

Domestic hot water mode

1. Pump
2. Primary expansion vessel
3. Primary heat exchanger
4. Three-way control valve
5. Secondary heat exchanger
6. CH flow switch
7. DHW flow switch
8. Secondary expansion vessel
9. Pressure gauge
10. Cold supply main
11. Gas supply
12. Temporary connection
13. Pressure relief valve
14. Double check valve
15. High limit thermostat
16. CH thermostat
17. DHW thermostat
18. Hot water draw off point
19. Heat emitter
20. Burner
21. Non-return valve
22. Isolation valve
23. Exhaust fan
24. By-pass
25. Flue

Figure 9.51 Typical layout of a combination boiler showing hot water and central heating circuits within

have a temperature differential of 20°C lower than the flow. *Note:* A forced draught burner is invariably used to give improved efficiency. Due to the relatively low flue temperatures that can be experienced with these appliances manufacturers often simply supply plastic flue pipes.

Principle of Operation

9.91   When gas is burnt, water vapour (H_2O) is produced as a result of the combustion process (see Part 2: Section 2.21). In the more traditional/conventional appliance, this water vapor is dispelled from the appliance along with the other combustion products. However, the design of the combustion chamber in a condensing boiler consists of a very tight network of closely fitted waterways and baffles, or in some models, two individual heat exchangers, as illustrated in Figure 9.52. The flue gases are cooled extensively down to temperatures typically of around 35–50°C, often allowing for a plastic flue. Water vapour within the flue gases condenses to its liquid form at a temperature of about 55°C, therefore water forms within the appliance and runs down the inside, collecting in a condensate trap in the base of the unit. By referral to the combustion process identified in Part 2, Section 2.22 it will be found that for every 1 m³ of natural gas consumed 2 m³ of water vapour are produced. This water vapour, if cooled, will condense to produce 1.25 l of water. The water produced is full of latent heat and every litre of water produced will give off approximately 2.3 MJ of heat energy. This equates to approximately 0.64 kW, and the energy is therefore reclaimed in a condensing boiler rather than simply being discharged out through the flue system. A typical condensing boiler will produce over 3 l of water an hour.

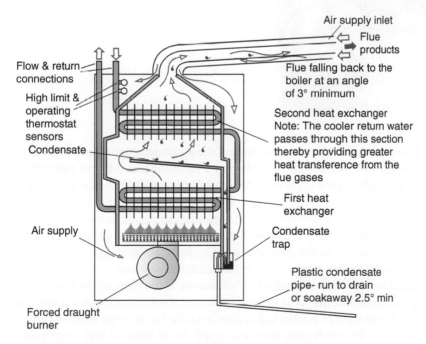

Figure 9.52 Condensing boiler utilising two heat exchangers

Disposal of the Condensate

9.92 Condensate will also form within the flue system and, as a consequence, the flue will need to be routed in an uphill direction so that this water can also drain back to the condensate trap. Figure 9.53 shows a modern condensing boiler and you will notice the condensate trap, slightly left of centre in the base of the unit. Clearly, for such an effective heat exchanger to work, the flue products will need to be expelled from the appliance by the use of a fan draught flue system.

Figure 9.53 Modern condensing boiler

The condensate collected in the trap at the base of the appliance discharges to a drain or soakaway when sufficient volume has been collected. Discharging as a volume rather than a continued dripping effect aims to prevent the water freezing within the discharge pipe. Because of the nature of the condensate (slightly acidic), metallic pipe is not recommended for the condensate drain due to the potential of corrosion, so plastic materials are used. Alternatively, it is permissible, subject to manufacturer's approval, to install an in-line neutralising device which automatically shuts down the boiler if the condensate is no longer neutralised. This would allow a copper waste to be installed downstream.

The drain should have a minimum fall of 2.5°. However, if the route of the condensate requires the pipe to rise before dropping to the termination, a suitable condense pump can be employed. The pump must be interlocked to shut down the boiler in the event of failure.

The preferred option for condensate termination will be an internal foul water point. This method significantly reduces the risk of the condensate freezing. If this option is not available, termination to an external foul water point is acceptable; however, every effort is required to keep at much of the pipe internal as possible before exiting outside. External pipe runs should be restricted to no more than 3 m in length. If the pipe connects under a sink it must be prior to a trap with a minimum 75 mm seal.

Other options for termination include rainwater gullies or hoppers that pass to a combined foul and rainwater drainage system. This termination of condense into a greywater recovery system is not an acceptable method.

Further details for specific condense installation, including combined pressure relief and condense pipe, can be referenced from BS 6798.

Pluming Effects

9.93 Because the water content of the flue gases has condensed or is close to the point where condensation occurs, a white vapor is often seen discharging from the flue terminal; this is referred to as pluming. It can be seen at all times of the year; however it tends to be more pronounced when the weather is cooler. Because of the nuisance factor involved, particularly with neighbours, the site of a flue terminal should be carefully chosen and should be at least 2 m away from any openings into adjacent buildings. Manufacturers' instructions may permit the use of a plume management kit to provide a high-level discharge.

Installation of Domestic Gas Boilers

Restricted Locations

9.94 Open-flued appliances must not be installed in rooms containing baths or showers. For bedrooms and bed-sits no open-flued appliance >14 kW gross heat input is allowed, neither can they be fitted in cupboards that open into these rooms. For appliances ≤14 kW, a vitiation sensing device must be incorporated to turn off the appliance if there is a dangerous build-up of fumes. If the appliance is room-sealed, then no such ruling exists, and there are no prohibited locations. LPG boilers, however, are not allowed to be located in basements and similar positions where any possible gas escapes could allow gas to build up. If a room-sealed boiler is located in a bathroom, the electrical controls must comply with BS 7671.

Compartments

9.95 A very common place to locate the boiler is within a compartment. A compartment is not a specific size; it may be small or large, depending on the size of the appliance. In general, it is a purpose-built rigid structure in which nothing else is used or stored. However, sometimes an airing cupboard is used, see Figure 9.54, and the boiler needs to be separated from the clothes by a perforated partition. This may consist of expanded metal with perforations no larger than 13 mm. An open-flue passing through an airing space should be of the insulated double wall type, this will provide adequate protection. However, you may encounter older installations where a single wall open-flue passes up through an airing space; this must be protected with a minimum 25 mm air gap between the flue pipe and the contents of the cupboard. Ideally, the compartment should

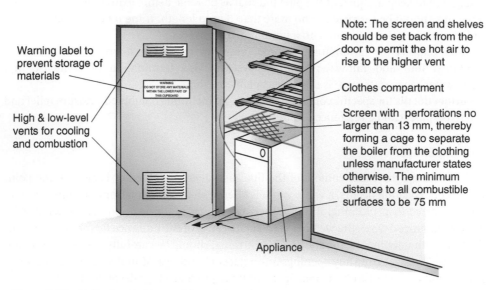

Warning label to prevent storage of materials

High & low-level vents for cooling and combustion

WARNING
DO NOT STORE ANY MATERIALS
WITHIN THE LOWER PART OF
THIS CUPBOARD

Note: The screen and shelves should be set back from the door to permit the hot air to rise to the higher vent

Clothes compartment

Screen with perforations no larger than 13 mm, thereby forming a cage to separate the boiler from the clothing unless manufacturer states otherwise. The minimum distance to all combustible surfaces to be 75 mm

Appliance

Figure 9.54   Boiler located in an airing cupboard

not be located under the stairways of buildings higher than two storeys. However, if there are more than two storeys, the whole cupboard must be lined with a protective material of no less than 30 minutes fire resistance. Some of the main requirements, in the absence of the manufacturer's specific instructions, for a compartment include the following:

- Internal surfaces need to be 75 mm away from the boiler unless non-combustible or adequate means of fire protection have been applied.
- Adequate ventilation must be provided, and open-flued appliances must not communicate with a bedroom or a room containing a bath or shower, see Section 9.94.
- Adequate space must be provided to service the boiler and permit its removal if necessary.
- A notice should be located at a suitable position, warning against storage.

Roof Spaces

9.96   Where a boiler is to be located in a roof void, as illustrated in Figure 9.55, specific additional installation points need to be considered, including the following:

- A permanent means of easy access (e.g. a ladder) must be provided and the area around the loft opening should have a guard fitted.
- A suitably floored route to the boiler must be provided, with sufficient area for servicing, as necessary.
- Items stored in the roof void must not be allowed to be in contact with the boiler.
- Within the loft, fixed lighting must be installed and a means of isolating the boiler from inside and from outside the roof space provided.

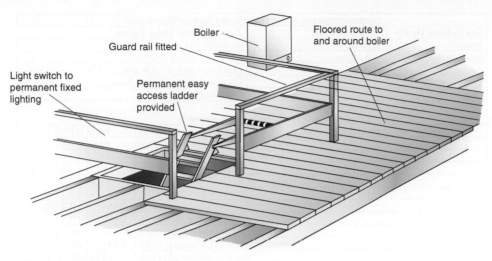

Light switch to permanent fixed lighting

Guard rail fitted

Boiler

Permanent easy access ladder provided

Floored route to and around boiler

Figure 9.55   Boiler located in roof space

Flue Termination

9.97   Domestic gas boilers installed today are of the room-sealed type; therefore the flue requirements are generally quite specific as laid down by the manufacturer of the appliance.

Ventilation

9.98   The ventilation requirements for a boiler depend on its type and location. Where it is installed in a room and is room-sealed, no ventilation is necessary. However, ventilation may be required if it is installed in a compartment; the requirements for this will be dictated by the manufacturer.

Commissioning and Servicing Gas Boilers

Commissioning and Servicing Work Record

9.99   The Work Record (see Figure 9.56) is given purely as a guide to the many tasks to be undertaken when servicing/maintaining and commissioning a boiler. In order to complete the form you may need to refer back to Part 8: Figure 8.1a, where many of the tasks were explained in more detail. However, additional appliance-specific checks on boilers include some of the following activities.

Notice Plate

9.100   Ensure the notice plate, if applicable, is correctly located and filled in; see Part 6: Section 6.9.

Certificate/Record of Hot Water Boiler Service/Commission		
Gas Installer Details	**Client Details**	**Appliance Date Badge Details**
Name:	Name:	Model/Serial No:
Gas Safe Reg. No:	Address:	Gas Type: Natural ☐ LPG ☐
Address:		Heat Input: max...... kW min...... kW
		Operating/Burner Pressure Range: ......–...... mbar
		Gas Council No:
Date:	**Appliance Location** **Install/Commission ☐ Service/Commission ☐**	
Preliminary System Checks: Compliance with manufacturer's instructions		PASS ☐ FAIL ☐
General visual inspection of pipework		PASS ☐ FAIL ☐
Clearance from combustible materials		PASS ☐ FAIL ☐
Visual inspection of flue		PASS ☐ FAIL ☐ N/A ☐
Flue notice plate was located and correctly filled in		PASS ☐ FAIL ☐ N/A ☐
Flue flow performance test		PASS ☐ FAIL ☐ N/A ☐
Electrical connections		PASS ☐ FAIL ☐ N/A ☐
Bonding maintained		PASS ☐ FAIL ☐ N/A ☐
Fuse rating: ...... amps		PASS ☐ FAIL ☐ N/A ☐
System tightness test; to include let-by		PASS ☐ FAIL ☐ N/A ☐
Standing pressure of system		PASS ☐ FAIL ☐ N/A ☐
Appliance/system is purged of air		PASS ☐ FAIL ☐ N/A ☐
Service/Commission Checks: Clean primary airports and lint arrestor		PASS ☐ FAIL ☐ N/A ☐
Clean/check condition of injectors; burners		PASS ☐ FAIL ☐ N/A ☐
Check condition of heat exchanger and sweep through		PASS ☐ FAIL ☐
Ignition devices effective including condition of electrodes, leads		PASS ☐ FAIL ☐ N/A ☐
Clean and check operation of fans		PASS ☐ FAIL ☐ N/A ☐
Pilot and main burner flame correct		PASS ☐ FAIL ☐ N/A ☐
Inlet operating pressure: .... mbar		PASS ☐ FAIL ☐ N/A ☐
Burner pressure: .... mbar		PASS ☐ FAIL ☐ N/A ☐
Heat input........ kW Gas rate.........m^3		PASS ☐ FAIL ☐ N/A ☐
Flame supervision device operational		PASS ☐ FAIL ☐ N/A ☐
Atmosphere sensing device operational		PASS ☐ FAIL ☐ N/A ☐
Condensate trap cleaned and condensate pipe effective		PASS ☐ FAIL ☐ N/A ☐
Combustion seals effective		PASS ☐ FAIL ☐ N/A ☐
Appliance tightness check		PASS ☐ FAIL ☐ N/A ☐
Spillage tests		PASS ☐ FAIL ☐ N/A ☐
Operating thermostat correct		PASS ☐ FAIL ☐ N/A ☐
Flue guard fitted to low-level terminals		PASS ☐ FAIL ☐ N/A ☐
Pressure relief valve effective		PASS ☐ FAIL ☐ N/A ☐
Additional direct ventilation grille required: .... cm^2		PASS ☐ FAIL ☐ N/A ☐
High level compartment ventilation where required: .... cm^2		PASS ☐ FAIL ☐ N/A ☐
Low level compartment ventilation where required: ....cm^2		PASS ☐ FAIL ☐ N/A ☐
Post System Checks: Meter working pressure (............ mbar)		PASS ☐ FAIL ☐ N/A ☐
System design pressure loss.......mbar (max. 1 mbar NG / 2 mbar LPG)		PASS ☐ FAIL ☐ N/A ☐
Combustion performance analysis CO......ppm CO^2......% CO / CO_2 ratio...........		PASS ☐ FAIL ☐ N/A ☐
Safe operation of appliance explained to customer		YES ☐ NO ☐
Central Heating Checklist Completed		YES ☐ NO ☐
Recommendations and/or Urgent Notification Benchmark Logbook completed YES ☐ NO ☐ **Appliance Safe to Use** YES ☐ NO ☐ **Next Service Due:**		
Installer's Signature:	**Customer's Signature:**	

Figure 9.56 Commissioning and servicing record for boilers

Check Condition of the Heat Exchanger and Sweep Through

9.101   On many of the modern boilers that use low water content heat exchangers, greater care is required not to damage the delicate fins, which may require no more than a gentle brushing. With the older, more robust cast iron heat exchangers, a selection of stiff brushes can be passed through the water tubes in the hope of removing any loose rust, etc.

Atmosphere Sensing Device Operational

9.102   There is no laid down procedure for checking the correct operation of an ASD located in an open-flued boiler. Some systems rely on oxygen depletion, whereas others rely on sensing the temperature at the draught diverter (see Part 3: Section 3.59). However, as an installer/service engineer, you can check for accumulation of dust, etc. around the sensing points and check that the components are correctly and securely in place.

Condensate Trap Cleaned and Condensate Pipe Effective

9.103   For the condensing type boiler, the condensate trap may need to be removed in order to give it a thorough clean; debris may have accumulated inside, which may lead to a blockage. The discharge pipe leading to the outside should also be checked to ensure that it is not damaged and still operates effectively.

Condition of Casing and Combustion Seals Effective

9.104   Generally, this requires no more than a good visual inspection of all joints for corrosion or poor sealing/gasket material. However, where the appliance case is subject to positive fan pressure, the appliance seals can be tested by following the procedure described in Part 6: Section 6.56.

Pressure Relief Valve Effective

9.105   For sealed systems, the pressure relief valve test lever should be operated to confirm that the valve opens and discharges the water safely into the discharge pipe and closes effectively on completion. The discharge pipework should also be inspected for damage and correct location.

Central Heating Checklist Completed

9.106   In addition to the boiler service or installation, it is always good practice to use this opportunity to inspect the condition of the central heating system and pipework. As with all tasks, a checklist such as that shown in Figure 9.57 could be completed. BS 7593 maintains that inhibiter levels should be checked, and any inline filter should be cleaned on an annual basis.

Inspection Record of Hot Water Central Heating System		
Gas Installer Details	**Client Details**	**System Details**
Name:	Name:	Open System ☐ Closed System ☐
Gas Safe Reg. N°:	Address:	Fully Pumped ☐ Gravity Primaries ☐
Address:		Conventional Boiler ☐
		Combination Boiler ☐
		Condensing Boiler ☐

Date:	This inspection was completed in conjunction with a check of the gas installation of the boiler: **Yes** ☐ **No** Boiler location:	

Components Inspected	**Notes**
Boiler: location acceptable	PASS ☐ FAIL ☐
Pipework: no leaks, secure and in accordance with good practices	PASS ☐ FAIL ☐
Pump: speed correct: (flow and return temperature differential ....°C)	PASS ☐ FAIL ☐ N/A ☐
TRVs fitted to all radiators except rooms with the thermostats	Yes ☐ No ☐
System balanced	PASS ☐ FAIL ☐ N/A ☐
No pumping-over air drawing in air from F & E	PASS ☐ FAIL ☐ N/A ☐
Pump noise minimal	PASS ☐ FAIL ☐ N/A ☐
Motorised valve operational	PASS ☐ FAIL ☐ N/A ☐
Radiator/heat emitter valves operational	PASS ☐ FAIL ☐ N/A ☐
Sealed system at correct pressure (sealed/closed system)	PASS ☐ FAIL ☐ N/A ☐
Temporary filling loop disconnected (sealed/closed system)	PASS FAIL N/A ☐
Expansion vessel at correct pressure (sealed/closed system)	PASS ☐ FAIL ☐ N/A ☐
Pressure relief operational at designed pressure (sealed/closed system)	PASS ☐ FAIL ☐ N/A ☐
Discharge from pressure relief at a safe location (sealed/closed system)	PASS ☐ FAIL N/A ☐
Automatic air admittance valves work freely	PASS ☐ FAIL N/A ☐
F & E cistern adequately insulated	PASS ☐ FAIL ☐ N/A ☐
F& E water level adjusted and float-operated valve functioning	PASS ☐ FAIL ☐ N/A
Overflow pipe secure and at a visible location	PASS ☐ FAIL ☐ N/A ☐
Condition of water within system showing no major corrosion	PASS ☐ FAIL ☐ N/A ☐
Removed air from system	Yes ☐ No ☐
Cleaned inline filter	Yes ☐ No ☐
System flushed	Yes ☐ No ☐
Corrosion inhibitor added	Yes ☐ No ☐
Inhibitor checked	Yes ☐ No ☐
Programmer and time clock were correctly adjusted.	Yes ☐ No ☐
Room thermostats, cylinder and frost thermostats are operational	Yes ☐ No ☐
All other external controls are operational	Yes ☐ No ☐
Condense pipe correctly installed	PASS ☐ FAIL ☐ N/A ☐

Recommendations and/or Urgent Notification	
Installers Signature:	**Customers Signature:**

Figure 9.57 Inspection record of hot water central heating system

Fault Diagnosis

9.107 Many of the faults are associated with modern fan-assisted appliances using printed circuit boards (PCBs). To assist fault diagnosis, quite extensive fault-finding charts will be found at the back of the manufacturer's installation instructions, taking you through the various options available in diagnosing why a boiler will not work. Some manufacturers have even gone that bit further and have incorporated a point to enable the connection of a computer that tells you exactly what is required for the operation of the boiler. Figure 9.58 covers a few common faults and may assist a newly qualified engineer when they encounter an older style boiler.

Fault	Possible cause
No flame will establish within the boiler, e.g. pilot flame or main burner in the case of electronic ignition	• Gas or electrical supply is turned off • Faulty fuse • Air in pipe • Pilot injector blocked or incorrect size • Incorrect spark gap • No spark or spark lead not connected properly • Faulty fan or tubes to pressure switch • Faulty pressure switch or poor connections
Poor pilot flame and will not stay alight	• Thermocouple connection loose • Pilot flame too small or pilot tube blocked • Faulty thermocouple or thermo-electric valve • Inadequate ventilation • Defective atmosphere sensing device or oxygen depletion tube
Poor heat output and lack of heat from the appliance	• Inadequate gas pressure • Blocked gas injectors • Incorrectly fitted gas injectors • Faulty thermostat
Poor flame picture	• Incorrect gas pressure • Insufficient air intake or lint arrestor blocked • Blocked heat exchanger • Poor or blocked flue system • Incorrect injectors • Damaged burner
Noisy boiler	• Gas pressure too high • Pump speed incorrect • Flames impinging onto heat exchanger • Scale build-up within boiler • By-pass insufficiently open • Loose screws in boiler casing or incorrectly fitted case

Figure 9.58   Basic fault diagnosis chart – boiler

Domestic Ducted Warm Air Heaters

Relevant ACS Qualification
DAH1

Relevant Industry Document
BS 5864

9.108   A warm air unit consists of a burner enclosed in a heat exchanger, around which air is passed, assisted by the draught created by a fan. The warmed air is then circulated around the building through a system of ductwork to be discharged into the various rooms via register grilles. The cooler air returns to the warm air heater simply by being sucked from the room via a second grille in the wall leading back via the passageway or hall. There are three types of fan-assisted warm air units, each illustrated in Figure 9.59:

• the down-flow unit;
• the up-flow unit;
• the horizontal unit.

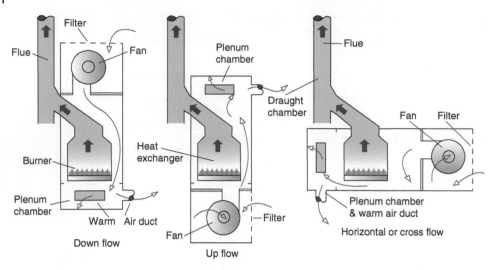

Figure 9.59 Types of warm air heater (open-flued)

Some warm air units, referred to as 'Modairflow' or even temperature (ET), are designed to give variable heat outputs, thus saving fuel. Basically, used in conjunction with a thermistor-type room thermostat, they will bring on the firing of the warm air unit only if needed and will turn off the burner intermittently, simply circulating the warm air at a reduced fan flow rate.

Some units incorporate a circulator to provide a hot water supply system or to be used as additional background heating. Where this is the case, the circulator operates as an individual appliance within the unit, however it shares the same flue system. If a circulator is incorporated with a warm air unit, it is essential that sufficient ventilation is provided to serve both appliances.

Earlier models of warm heaters will predominately be of the open-flued type, however, manufacturers now produce condensing room-sealed heaters with an example shown in Figure 9.60.

Warm Air Duct System

9.109 Various ducting systems are outside the scope of this book. However, Figure 9.61 illustrates a typical installation of ductwork and how the warm air moves around the property in a circuit back to the air heater via the return air duct. As warm air leaves the heat exchanger of the appliance, it collects inside a box-shaped plenum chamber. This plenum is designed to equalise the air pressure inside and distribute it to the various supply ducts. For the down-flow unit, the plenum chamber needs to be strong enough to support the weight, as it stands on the floor on which the heater is placed. If the plenum is located within a timber floor the area must be insulated to withstand temperatures up to 120°C.

Figure 9.60 Condensing room-sealed warm air heater.
Credit: Johnson and Starley Ltd

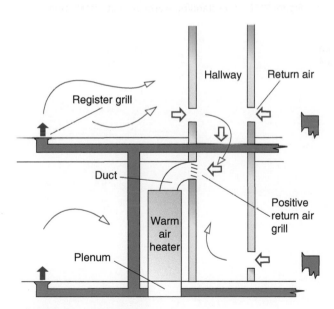

Figure 9.61 Air circulation

Return Air Ducting

9.110   The return air is circulated back to the hall and passes back into the heater, via a filter, for reheating. Where the heater is open-flued and located in a compartment it is essential that the return air is suitably ducted from outside the compartment to the appliance return air inlet to ensure that the operation of the fan, which creates negative pressure, does not adversely affect the safe operation of the flue. This does not apply to a room-sealed unit, although it is still necessary to ensure that the path of the return air to the appliance is not obstructed. Sometimes outside air is introduced into the heater return air duct; where this is the case a minimum flow of 2.2 m^3/h should be provided for every 1 kW heat input.

Installations of Warm Air Heaters

9.111   Warm air units are installed as either freestanding, slot-fix or within a compartment; the latter tends to be the most common. The freestanding model, seen in Figure 9.62, is often made to suit the height of the room by using a top closure set and is designed to be located back to the wall or, ideally in a corner. Slot-fix models, as illustrated in Figure 9.63, are specially designed to be located within a purposely designed area between two surfaces that protect the sides. Only appliances that are designed for slot-fix applications can be fitted in this way.

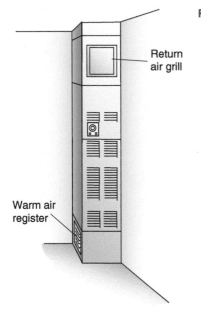

Figure 9.62   Freestanding warm air unit installation

Return air grill

Warm air register

Restricted Locations

9.112   The restricted locations for warm air heaters are the same as those for boilers as described in Section 9.94.

Figure 9.63   Slot-fix warm air unit installation

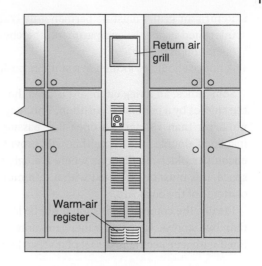

Compartments

9.113   Where the warm air unit is installed within a compartment, as seen in Figure 9.64, the following points need to be considered:

- The internal surfaces should be 75 mm from the heater unless non-combustible.
- Adequate ventilation must be provided, both for cooling and combustion air (see Part 7: Ventilation), and open-flued appliances must not communicate with a bathroom or bedroom, see Restricted Locations for boilers, see Section 9.94.
- Adequate space must be provided to service the heater and permit its removal if necessary.

Figure 9.64   Typical warm air unit compartment installation

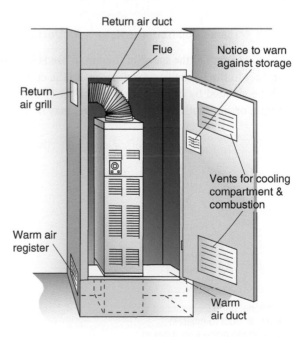

- Return air grilles must be permanently connected via a suitable duct to the return air inlet sited inside the compartment, thereby overcoming possible spillage caused by the operation of the fan.
- A notice should be located at a suitable position, warning against storage.

Sometimes, an airing cupboard is used, however the heater needs to be separated from the clothes by a perforated partition. This may consist of expanded metal with perforations no larger than 13 mm. An open-flue passing through an airing space should be of the insulated double wall type, this will provide adequate protection. However, you may encounter older installations where a single wall open-flue passes up through an airing space, this must be protected with a minimum 25 mm air gap between the flue pipe and contents of the cupboard.

Ideally, the compartment should not be located under the stairways of buildings higher than two storeys. However, if there are more than two storeys, the whole cupboard must be lined with a protective material of no less than 30 minutes fire resistance.

Specific Notes on the Installation of Return Air Grilles

9.114 When installing return air grilles. The following should be noted:

- Grilles should not be located more than 450 mm above the floor to prevent the spread of smoke in the event of a fire.
- Communication between bedrooms should be avoided for the sake of privacy.
- No return air should be taken from kitchens, bath/shower rooms and toilets to avoid smell and moisture transference.

Noise Transmission from the Return Air Ducting

9.115 Owing to the speed of the fan, noise is sometimes transmitted through the return air duct to the habitable area adjacent to the compartment. As shown in Figure 9.65, this can be avoided by incorporating a bend or two in the return air duct, thus increasing its length, or, alternatively, lagging the duct may help.

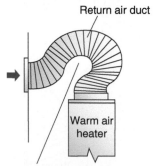

Return air duct

Warm air heater

Additional bend incorporated in return air duct to assist in overcoming noise transmission

Figure 9.65 Overcoming noise transmission in a return air duct

Commissioning and Servicing Warm Air Heaters

Commissioning and Servicing Work Record

9.116 The Work Record, as seen in Figure 9.66, is given purely as a guide to the many tasks to be undertaken when servicing/maintaining and commissioning a warm air unit. In order to complete the form you may need to refer back to Part 8: Figure 8.1a, where many of the tasks are explained in further detail. However, additional appliance-specific checks on these units include some of the following.

Notice plate is located and correctly filled in: See Part 6: Section 6.9.

Check Condition of Heat Exchanger

9.117 Distortion often occurs, resulting in cracking at the welded seams due to metal fatigue. This is due to the method of manufacturing the heat exchanger from pressed steel and the continual expansion and contraction of the metal. Therefore it is essential that a thorough visual inspection is undertaken because if a crack should develop the circulation air would be blown into the heat exchanger, causing flame turbulence and possible spillage. Less likely, but also possible, is that combustion products may get drawn into the air duct system and be discharged around the whole dwelling. The heat exchanger should be checked by shining a powerful torch inside to allow close inspection of the welded joints. A good indicator is that the flame picture will be disrupted when the fan kicks in to blow the warm air through the building. When in doubt, the heat exchanger could be tested as follows:

- Light the appliance and allow it to heat up for some 5–10 minutes.
- Switch off the appliance and insert a smoke pellet, placed on a non-combustible surface, into the heat exchanger towards the rear.
- While it is burning, close all warm air register grilles, except the one nearest the heater.
- Finally, switch on the circulation fan and observe the open register grille for traces of smoke.

Positive Return Air Path

9.118 Since the operation of convection currents within an open-flue system caused by the airflow fan can be affected, the return air must be suitably ducted back to the air inlet within the heater compartment. Where this is not the case in an existing installation, the appliance must be regarded as 'At Risk'. At the point where the return air enters the heater a filter is located. It is essential this is kept clear of lint, as, should the filter become blocked, the heater will simply overheat causing cycling of the limit switch.

Operating Thermostats Correct, Including High Limit Stat and Fan Switch

9.119 The high limit or overheat thermostat should be checked as per manufacturer's instructions. If the air does not circulate through the warm air heater, the heat exchanger

Certificate/Record of Warm Air Unit Service/Commission		
Gas Installer Details	**Client Details**	**Appliance Date Badge Details**
Name:	Name:	Model/Serial No:
Gas Safe Reg. No:	Address:	Gas Type: Natural □ LPG □
Address:		Heat Input: max...... kW min...... kW
		Operating/Burner Pressure Range: ...... –...... mbar
		Gas Council No:
Date:	**Appliance Location:** Install/Commission □ Service/Commission □	
Preliminary System Checks:	Compliance with manufacturer's instructions	PASS □ FAIL □
	General visual inspection of pipework	PASS □ FAIL □
	Clearance from combustible materials and compartment warning notice up	PASS □ FAIL □
	Visual inspection of flue	PASS □ FAIL □ N/A □
	Flue notice plate was located and correctly filled in	PASS □ FAIL □ N/A □
	Flue flow performance test	PASS □ FAIL □ N/A □
	Appliance level and secure	PASS □ FAIL □
	Electrical connections	PASS □ FAIL □ N/A □
	Bonding maintained	PASS □ FAIL □ N/A □
	Fuse rating: .... amps	PASS □ FAIL □ N/A □
	System tightness test; to include let-by	PASS □ FAIL □ N/A □
	Standing pressure of system	PASS □ FAIL □ N/A □
	Appliance/system is purged of air	PASS □
Service/Commission Checks:	Clean primary airports and lint arrestor	PASS □ FAIL □ N/A □
	Clean/check condition of injectors, burners	PASS □ FAIL □ N/A □
	Check condition of the heat exchanger	PASS □ FAIL □
	Check, ease and grease, if necessary, control taps	PASS □ FAIL □ N/A □
	Ignition devices effective including condition of electrodes, leads	PASS □ FAIL □ N/A □
	Clean and check operation of fans and filters	PASS □ FAIL □ N/A □
	Pilot flame correct	PASS □ FAIL □ N/A □
Inlet operating pressure: .... mbar	Burner pressure: .... mbar	PASS □ FAIL □
	Total heat input: .... kW	PASS □ FAIL □ N/A □
	Flame picture good	PASS □ FAIL □ N/A □
	Flame supervision device operational	PASS □ FAIL □ N/A □
	Combustion seals and plenum seals effective	PASS □ FAIL □ N/A □
	Positive return air path	PASS □ FAIL □ N/A □
	Appliance tightness check	PASS □ FAIL □ N/A □
	Operating thermostats correct, including high limit stat and fan switch	PASS □ FAIL □ N/A □
	Temperature differential through unit: ....°C	PASS □ FAIL □
	Spillage tests	PASS □ FAIL □ N/A □
	Distribution grilles and dampers	PASS □ FAIL □ N/A □
	Additional direct ventilation grille required: .... cm^2	PASS □ FAIL □ N/A □
	High-level compartment ventilation where required: .... cm^2	PASS □ FAIL □ N/A □
	Low-level compartment ventilation where required: .... cm^2	PASS □ FAIL □ N/A □
Post System Checks:	Meter operating pressure (............ mbar)	PASS □ FAIL □ N/A □
	System design pressure loss........mbar (max. 1 mbar NG / 2 mbar LPG)	PASS □ FAIL □ N/A □
	Combustion performance analysis CO.....ppm CO^2......% CO/CO_2 ratio..........	PASS □ FAIL □ N/A □
	Safe operation of appliance explained to customer	YES □ NO □
Recommendations and/or Urgent Notification		
Appliance Safe to Use YES □ NO □ Next Service Due:		
Installer's Signature:	**Customer's Signature:**	

Figure 9.66 Commissioning and servicing record for warm air units

will quickly become overheated. One of two controls may overcome this problem: a limit thermostat set at 95°C and an overheat thermostat set at 110°C. Either one or both may be included.

The *limit stat* turns off the gas supply at the upper limit of 95°C and re-lights the appliance when the temperature falls to about 80°C. This automatic resetting causes the heater to cycle on and off every few minutes.

The *overheat stat* differs in that if the temperature reaches the upper limit of 110°C, the appliance will shut down with no automatic re-set facility. The overheat stat is often an additional control in down-flow heaters because of the high heat rise that can be experienced when the fan is slow to switch off.

To test this control, it may be possible to block the filter inlet with a piece of card or dust sheet or, alternatively, run the appliance with the heater alight, but the fan disconnected. The heater should shut down within two or three minutes.

The *fan switch is* a thermally operated switch that prevents cold air blowing into the rooms before the appliance has had time to warm up. The fan begins to operate when a predetermined temperature is reached. Operating temperatures are: fan on 58°C, fan off 38°C.

The *Summer/winter switch* allows the user to switch the control to the summer position to allow unheated air to be blown into the rooms.

Temperature Differential Through the Basic Unit (Not Modairflow)

9.120 This is a check of the temperature rise through the appliance. It is simply carried out by taking a temperature reading at the air intake to the appliance and at the first warm air diffuser grille. The rise should be as indicated by the manufacturer, usually in the region of 50°C ± 5°, which may mean balancing the system or adjusting the fan speed if necessary.

Spillage Test

9.121 This is usually carried out as described in Part 6: Section 6.51. However, in some instances, the draught diverter is totally inaccessible, and an alternative method is used as follows:

1. Pre-heat the appliance for some 5–10 minutes.
2. Turn off the appliance and insert a small smoke pellet on a non-combustible surface inside the combustion chamber, replacing the cover plate. The pellet selected should not be too large, otherwise the volume of smoke emitted would give an unrealistic test.
3. Look for the presence of smoke in the general area of the draught diverter.

When the spillage test has been completed it should then be repeated with the fan running, or summer airflow switch set to 'on'. This aims to ensure the heater has been correctly sealed to the base duct/plenum and does not create a negative pressure, pulling the combustion products into the duct system.

Distribution Grilles and Dampers

9.122 Check to see that the register grilles located in all the rooms are obtaining sufficient heat, adjusting the damper located at the rear of the distribution grille as necessary. A check should also be made to ensure that the diffusers open and close freely.

Figure 9.67 provides a useful guide to fault finding on the more traditional warm air units. As always, the manufacturer's instructions should also be consulted.

Fault	Possible cause
Poor pilot flame and will not stay alight	• Gas supply is turned off, or air in pipe • Thermocouple connection loose • Pilot flame too small or pilot tube blocked • Faulty thermocouple or thermo-electric valve Inadequate ventilation • Defective atmosphere sensing device or oxygen depletion tube
Pilot is established, but the main burner will not ignite	• Electricity supply is turned off, or a faulty fuse • Controls not calling for heat or set to 'summer' setting • Loose electrical connections • Faulty gas solenoid valve • Faulty thermostat (room or limit stat)
Main burner lights, but fan fails to operate after preheat period	• Faulty fan assembly, e.g. electrical connection loose, defective switch or fan belt (if fitted) • Fan setting incorrect • Operating gas pressure too low
Main burner lights intermittently with fan running	• Operating gas pressure too high • Fan speed incorrect • Air filter or return air path restricted • Most outlet diffusers closed
Fan running intermittently with main burner on	• Fan switch setting incorrect • Operating gas pressure too low
Fan runs for long time after main burner shuts off, or intermittently	• Fan switch setting incorrect Poor flame picture • Split or blocked heat exchanger • Insufficient air intake or lint arrestor blocked • Poor or blocked flue system • Incorrect or blocked injectors
Noisy operation	• Gas pressure too high • Noisy fan motor • Fan speed too high
Main burner does not switch off	• Faulty multi-function valve

Figure 9.67   Basic fault diagnosis chart – warm air

Domestic Tumble Dryers

Relevant ACS Qualification
LAU1

Relevant Industry Document
BS 7624

9.123 The domestic gas tumble dryer, pictured in Figure 9.68, looks to all intents and purposes just like the electrical version, which is far more common. It consists of the

Figure 9.68 Domestic gas tumble dryer

same components, including an electronic drive belt, drum and control panel for the user. The only real difference is the method used to heat the air that is warmed to pass through the tumbling clothes. This is achieved by using a gas burner located at the base of the unit that rapidly warms the air drawn in for combustion. A fan pulls the products through the drum and eventually expels them from the rear of the appliance. An exhaust vent, supplied with the unit, disperses the products to outside. The dryer should not operate with the door in the open position; when the door is closed it operates a timer switch allowing the appliance to operate.

Component Parts

Burner

9.124 This is usually of pressed steel construction and is fitted inside a metal tube, with the gas flame burning horizontally in the direction of airflow to the back of the appliance. Because the draught is induced through the combustion chamber, flame detection is usually by means of flame rectification. To provide an added level of safety, two safety shut-off valves are usually incorporated, as shown in Figure 9.69. If the dryer fails to recognise an established flame within approximately 10 s, it will go into lockout and the appliance will need to be turned off and on again to re-set the gas control sequence.

Drum

9.125 This is the heart of the tumble dryer, it consists of a stainless steel cylinder of around 115 l capacity and turns at approximately 50–60 revolutions/min, driven by a belt fixed to the outer circumference and motor. It usually rotates in a clockwise direction, but it stops

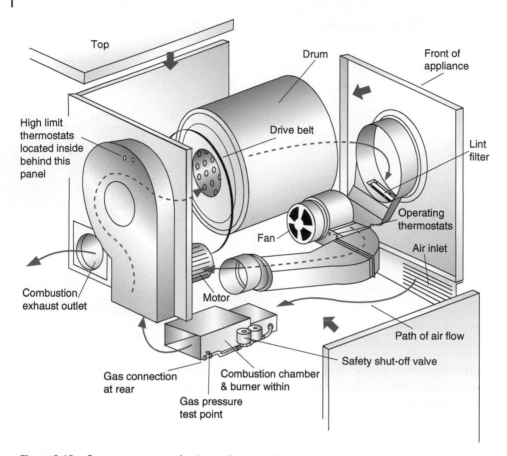

Figure 9.69 Component parts of a domestic gas tumble dryer

and turns anticlockwise for short periods to enable the clothes to untangle so that they are free to move.

Thermostat Control

9.126   Two operating thermostats are generally incorporated to provide drying temperatures of around 50°C and 60°C. These are located in the exhaust duct just after the fan. In order to prevent overheating and damage to the unit or clothes, a thermostat operating at around 110°C is also incorporated to the top rear of the drum housing. This thermostat will cut off the gas supply but allow its re-ignition as the drum cools. A final high limit overheat thermostat, operating at 120°C, is also located at this point and will shut down the appliance if the 110°C thermostat fails.

Lint Filter

9.127   This is located just inside the front door. It collects the vast amount of fluff, etc., generated from the clothes during tumble drying, and it is essential that the customer is instructed how to remove and clean out the filter regularly; failure to do this will result in the appliance short cycling.

Installation of Domestic Tumble Dryers

Restricted Locations

9.128   Tumble dryers must not be installed in bath/shower rooms and can only be installed within a bedroom/bed-sitting room or garage where the manufacturer permits and the room volume is at least 7 m³/kW appliance input. The positioning in a protected stairway, such as in flats over two storeys high, is also restricted. LPG appliances in basements, etc., are also not allowed.

The exhaust vent should not discharge into the confines of a covered alleyway, such as between two adjoining properties, where the combustion products may accumulate. These products may contain carbon monoxide, which may eventually find its way into an inhabited area.

Location and Clearances

9.129   Figure 9.70 illustrates the installation requirements. The tumble dryer will usually fit in a space 600 × 600 mm. Where the dryer is to be positioned under a worktop, a 15 mm minimum space between the top of the dryer and worktop should be maintained to allow for ventilation. A free space to the front of the appliance should be allowed so that air can be drawn into the appliance and to allow the appliance to be pulled right out for maintenance and servicing purposes. It is possible to stack a tumble dryer directly on top of a compatibly sized washing machine, provided that the correct recommended stacking kit is used. In such a case, a restraining device must be used, as pictured in Figure 9.71.

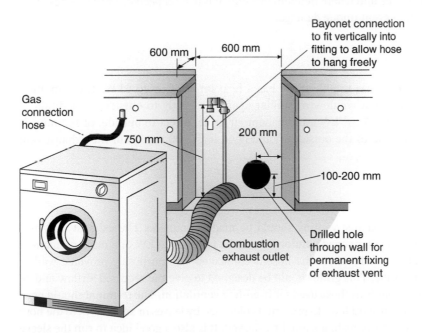

Figure 9.70   Provision for exhaust outlet & installation requirements

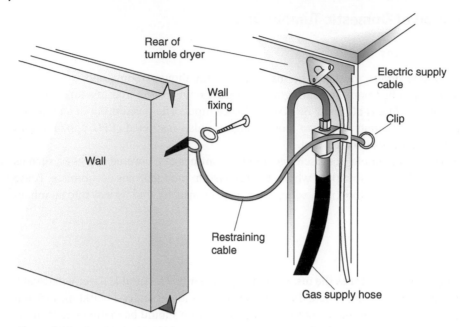

Rear of
tumble dryer

Electric supply
cable

Wall
fixing

Clip

Wall

Restraining
cable

Gas supply hose

Figure 9.71   Restraining device

Gas Connection

9.130   The gas supply is connected via a flexible hose and bayonet connection. This allows for limited movement, and where frequent movement is to be expected, a restraining device should be used to prevent undue damage.

Ventilation

9.131   The domestic gas tumble dryer has a heat input of up to 6 kW. The minimum requirement is to have a window or similar opening directly to outside. In addition, if the room is less than 3.7 m^3/kW a permanent air grille, direct to outside, of 100 mm^2 is required. Any air grille to the outside should be at least 300 mm from the exhaust vent termination.

Exhaust Venting

9.132   In all cases, exhaust venting should be made to outside. The exhaust vent may simply be hung out of an opened window where the input is <3 kW or a more permanent arrangement can be made where the hose is connected to a purposely provided wall or window terminal grille. This grille should be designed to give the required airflow and be up to the same standards as those used for fixed-free ventilation. The terminal should be at least 300 mm above ground level to prevent its blockage by leaves or snow. Where the hose is to pass through a cavity wall, it should be sleeved. It is also a good idea to run the sleeve slightly downwards to the external face to prevent the entry of rain, etc.

Domestic Gas Refrigerators

Relevant ACS Qualification
REFLP2

9.133   The gas fridge often operates as a multi-fuel appliance, particularly when used in caravans or touring vehicles, using either gas or electricity to warm an ammonia-water mix. Gas refrigerators are of the absorption type, illustrated in Figure 9.72, and work as follows:

1. A mix of ammonia and water is heated within the heater compartment of a sealed unit by a small gas flame and, as the liquid heats up, ammonia is given off as a gas which rises to the condenser.
2. Air surrounding the condenser allows the ammonia to cool, and it condenses back into its liquid form and drains by gravity into the evaporator, which is located in the cooling compartment of the refrigerator.
3. Hydrogen gas within the evaporator lowers the pressure and, as a result, the ammonia evaporates. For evaporation to occur the ammonia needs to extract heat from its surroundings, in this case, from the air inside the fridge.
4. The hydrogen-ammonia mix then falls, due to its density, to the absorber. The absorber is a series of small tubes fed with a trickle of water from the heater compartment. This water rapidly dissolves the ammonia, leaving the hydrogen, which is now free to rise back to the evaporator.

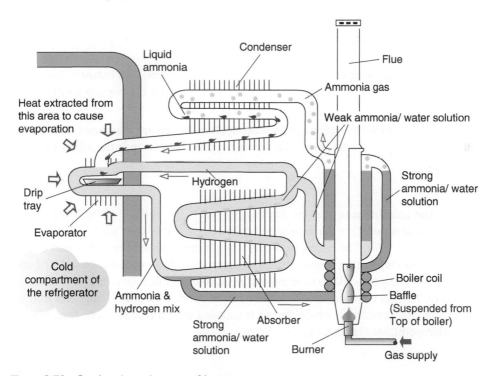

Figure 9.72   Section through a gas refrigerator

5. The newly replenished water-ammonia mix now flows to the lowest point in the system, where there is a coil located around the combustion chamber. As the liquid heats up, bubbles form in the boiling liquid and rise up into the top of the heater, whereupon the ammonia is driven off, and so another cycle begins.

Most of the cooling unit is mounted behind the back of the refrigerator cabinet, with only the evaporator located inside at the top. The liquids contained within the unit, referred to as the refrigerants, cannot be replaced on-site, and where damaged or faulty, the unit will need to be replaced. The actual cabinet usually consists of an inner lining, typically of plastic or aluminium, and an outer skin of mild steel with the space between filled with polyurethane foam or, for insulation near the boiler, glass fibre.

The Burner

9.134   Located at the base of the boiler will be the burner, which consists of an injector and small burner head that generates a very small flame. As heat rises in the combustion chamber it is deflected by a twisted stainless steel baffle that is suspended from the top of the heater and is designed to radiate and deflect the heat towards the boiler coil. The heat then passes up through a small central flue to be discharged either into the room or run externally. The amount of heat generated is controlled by a liquid expansion thermostat with a variable setting to compensate for hot and cold weather.

Leisure Appliances

Relevant ACS Qualification   Relevant Industry Documents
LEI1   BS EN 498 and BS 2977

9.135   The term leisure appliances includes appliances that are associated with outdoor activities and may be situated around a terrace or patio. They include the following:

Barbecue

9.136   The gas barbecue may be freestanding on a flat surface or built into a non-combustible structure. One of the most common designs is that used with LPG as a mobile unit (see Figure 9.73). The barbecue consists of a burner located beneath a grill cooking plate and with long-lasting ceramic briquettes spread out to provide the heat distribution medium. A barbeque should be sited away from all combustible materials and overhanging shrubbery. The gas connection to a barbecue is made via a flexible connection from gas bottles sited underneath or, where the gas is supplied from a piped supply, from a locally fixed micropoint, see Figure 9.74.

Figure 9.73 Gas Barbecue

Figure 9.74 Micropoint or leisure gas connection

Patio Heater

9.137   Freestanding patio heaters operate typically with a heat output from 6 to 14 kW, which allows an area of 25 m² to be warmed. They stand at a height of around 2.3 m. Some models consist of a burner with a round heat emitter positioned at the top of a pedestal, with a reflector above to direct the heat downwards. A whole range of heaters is available, including a tabletop version, operating at around 2.8–6 kW. The heater may be fixed and secured firmly to the ground and connected to the gas supply by a permanent rigid pipe, or it may form part of a mobile unit on wheels and be supplied with an LPG gas bottle. The siting/positioning of these heaters needs some thought, and generally a minimum distance of 1.5 m should be maintained between the heater and any combustible surface. The heater is ignited with either piezo or electronic ignition. An example of one of the many designs can be seen in Figure 9.75.

Figure 9.75   Patio heater

Gas Light and Flambeaux

9.138   There are several designs of gas light. The traditional gas lamp uses a mantle on which the flame burns, and a bright light is emitted. This design was very common in older style caravans, but today it is more commonly found as a single light, located on a pedestal or wall in a courtyard or similar venue using a gas mantle, a selection of these can be seen in Figure 9.76a–c. Another form of gas light does not use a mantle but has a traditional fishtail burner to give an olde-worlde effect with a flickering or vestal flame. The last type of gas light to be described, known as the gas flambeaux, is basically a decorative gas burning torch (see Figure 9.77). These come in a whole range of shapes and designs including wall units, bowls and mounted on pedestals. Their sizes range from a small gas flame of around 6 kW

(a)   (b)   (c)

Figure 9.76   (a) Gaslamp. (b) Gas Street lamp. (c) Gas Street lamp. Credit: Ian Cook

Figure 9.77   Gas flambeaux. Credit: Ian Cook

to a more dramatic flame, outside the scope of leisure appliances, of up to 60 kW input. As with the other appliances listed here, these appliances can run on LPG or natural gas. The modern gas light switches on automatically, possibly using a photo-electric cell to open the gas solenoid to the main burner as the light fades.

Greenhouse Heaters and Gas Pokers

Relevant ACS Qualification
LEI1

Greenhouse Heater

9.139   Gas greenhouse heaters, as seen in Figure 9.78, can be supplied to operate on either LPG or natural gas, usually operating between 2 and 4 kW. A 2 kW heater is large enough to provide frost protection over an area of 11 m². Larger heaters of 4 kW would be sufficient to provide frost protection over an area of 32 m². The heater may be freestanding or it may be permanently installed on a flat non-combustible surface or wall-mounted, depending on the appliance manufacturer's instructions. Where the heater is installed in a greenhouse that forms part of a dwelling, a rigid gas connection must be made. If the greenhouse is independent of the house a flexible connection is used. The heater incorporates a self-extinguishing device in case it is knocked over. The single greenhouse heater is a flueless appliance and as such it is essential that the greenhouse is adequately ventilated. One feature of a greenhouse heater is that it produces a vast quantity of carbon dioxide (CO_2), which many plants thrive on. One of the biggest problems with these heaters is the spiders' webs that build up in them during the summer months when the heater is not being used. These cause all sorts of problems, particularly with the pilot assembly where the injector is almost impossible to clean and replacement is often needed. The best solution is to disconnect and store the heater away during the summer months.

Figure 9.78   Greenhouse heater.
Credit: Lifestyle Appliances

Ventilation

9.140   For greenhouses attached to a dwelling, the minimum ventilation required is an openable window. However, additional ventilation may be required, depending on the manufacturer's instructions. It must also be noted that the size of a greenhouse heater is restricted to 90 W/m^3 of its volume where there is an opening into a dwelling. Where the greenhouse stands alone and has no openings into any buildings and the appliance net input rating does not exceed 2.7 kW, no additional ventilation is required. However, for appliances with inputs greater than this, two fixed free air grilles will be required, one at a low level and one at high level, providing a minimum effective area of 39 cm^2 for every kilowatt in excess of 2.7 kW.

Gas Poker

9.141   The gas poker, although rarely seen today, is pictured in Figure 9.79. It consists of a handle and flattened tube through which gas can pass at various positions. This is attached to a specially designed reinforced flex, suitable for ambient temperatures up to 70°C and 95°C touch temperatures. Basically, the gas poker can be used to establish a coal fire without the need to use paper and wood by bringing the fuel up to ignition temperature.

When using a gas poker, it is essential to ensure that all the holes through which the gas can pass are alight; otherwise, unburnt gas can pass to the fire bed and subsequently up the chimney, resulting in a 'boom'.

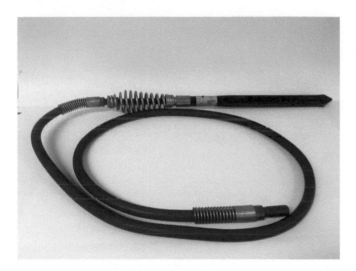

Figure 9.79   Gas poker

10

Commercial Appliances

Commercial Boilers

Relevant ACS Qualification
CIGA1

Relevant Industry Documents
BS 6644 and IGE/UP/10

10.1 Boilers over 70 kW net heat input fall within the category of commercial appliances, but in terms of commercial boiler operation, it would not be uncommon for appliances to operate at inputs of 500 kW to over 5000 kW. The range is quite vast. Today, with improved boiler designs and the need for weight reduction, boilers are much smaller than their predecessors. Sometimes, a series of boilers, as seen in Figure 10.1, are installed in a row and connected together via an arrangement of headers, these are referred to as modules. Boiler design tends to be of one of the following types.

Figure 10.1 Range of modular boilers

Gas Installation Technology, Third Edition. Andrew S. Burcham, Stephen J. Denney and Roy D. Treloar.
© 2024 John Wiley & Sons Ltd. Published 2024 by John Wiley & Sons Ltd.

Sectional Boiler

10.2 This appliance consists of a number of individual cast iron sections that contain the waterways, an illustration of which can be seen in Figure 10.2a with a photograph in Figure 10.2b. Each section is joined to the adjacent section by the flow header at the top and usually two return connections at the bottom. An appropriate gasket material is used between each section, and the whole system is clamped together via long bolts. The number of sections joined together denotes the total heat output that the boiler can provide. The burner is specifically selected to suit the final size of the boiler.

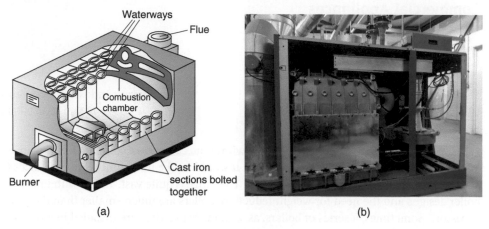

Figure 10.2   (a) Sectional boiler. (b) Sectional boiler opened up

Shell Boiler

10.3   These boilers can generally operate at higher pressures than sectional boilers and are often used where high-temperature hot water or steam generation is required. There are several designs of shell boilers, but they all work on the principle that the flue products are forced through a series of fire tubes that are surrounded by water on their way to the flue system, as illustrated in Figure 10.3a and b. In order to keep down the size of the boiler, the flue products are often turned through 180° to make a pass through another heat exchanger, thereby extracting more heat from the combustion products. Figure 10.3c is a photograph of a shell boiler opened up, and you will notice a few of the spiral agitators have been withdrawn from the flue tubes. These agitators assist to give a more even distribution of the heat.

Modular Boilers

10.4   The modular boiler is basically a series of between two and six smaller boilers installed together and connected by a series of headers. As a result of advances in technology, these boilers are lighter in weight and are suitable for rooftop installation. Mounting the boiler at the rooftop also cuts down on the total weight of the boiler, as it no longer needs to support the additional weight of water in a large system. Each module is usually small enough to pass through doorways and can be transported in a passenger lift. Above all, they are more easily installed and serviced. It is possible to service an individual boiler

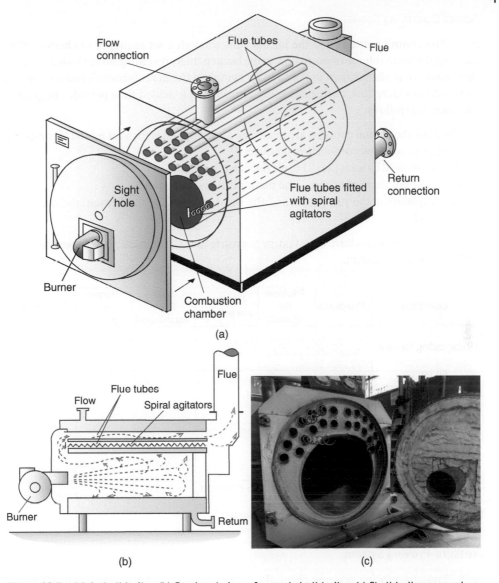

Figure 10.3 (a) A shell boiler. (b) Sectional view of a steel shell boiler. (c) Shell boiler opened up

without shutting down the whole system and where a fault condition occurs, the supply is not totally lost. Modular boilers also have the added advantage of being able to cope with variable heating loads.

Commercial Boiler Gas Control Systems

10.5 The gas supply to a large commercial boiler may consist of a single multi-functional gas valve or it may consist of a 'gas train' comprising of separate gas controls such as a regulator and safety shut-off valves.

Burner Operating Sequence

10.6 The control unit fitted to the burner runs through a set sequence of checks before allowing the main volume of gas to flow to the burner, thus preventing an explosive ignition. This sequence is often shown in the form of a graph in the manufacturer's instructions, as illustrated in Figure 10.4, each stage being displayed as a shaded time period. The general sequence is as follows:

1. When the thermostat contacts are made, the air fan operates for a set period to pre-purge the appliance of any combustion products.
2. This is followed by a set period to establish a pilot flame.
3. With the pilot flame confirmed, the main gas flame commences.
4. On completion of the run period and the satisfaction of the thermostat, the boiler will switch off.

Note: Some burners also have a post purge to ensure that all products are removed from the appliance and flue system.

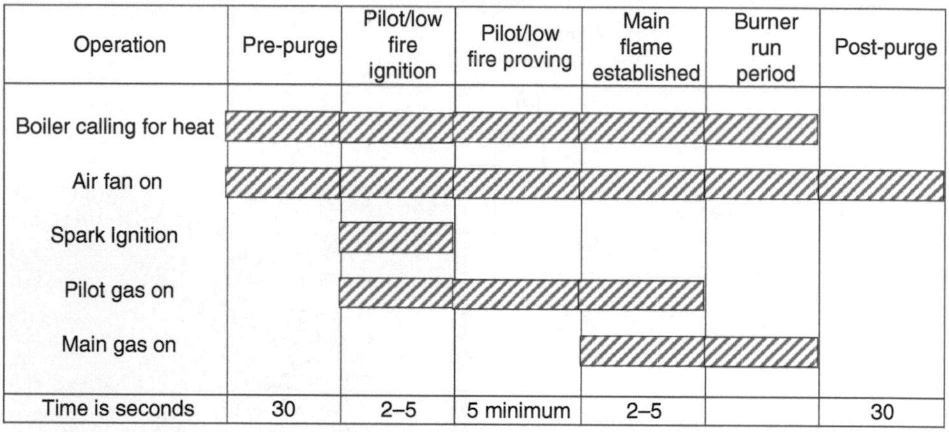

Operation	Pre-purge	Pilot/low fire ignition	Pilot/low fire proving	Main flame established	Burner run period	Post-purge
Boiler calling for heat	/////	/////	/////	/////	/////	
Air fan on	/////	/////	/////	/////	/////	/////
Spark Ignition		/////				
Pilot gas on		/////	/////	/////		
Main gas on				/////	/////	
Time is seconds	30	2–5	5 minimum	2–5		30

Figure 10.4 Typical burner control sequence diagram

Pressure-Proving System

10.7 Pressure-proving systems check the correct operation of the safety shut-off valves before the burner can begin its operating cycle. Where a problem is encountered at any stage of proving, the system will go into lock-out and prevent ignition. There are several proving systems, including the following, which rely on *sequential proving* (see Figure 10.5):

1. With all SSOVs closed, valve 'A' opens for two to three seconds to release the pressure from the enclosed section of pipework and re-closes.
2. The pressure is now monitored for a set period by pressure switch 1 to check that the pressure does not rise above 5 mbar, thus indicating that no let-by is occurring from the valves in either direction.
3. Assuming no pressure rises, the inlet SSOV opens for two to three seconds to charge the section between the valves and re-closes.

4. The pressure is now monitored for a set period by pressure switch 2 to check that the pressure does not drop, indicating a leak through the downstream valves. If proved, the boiler ignition cycle is initiated.

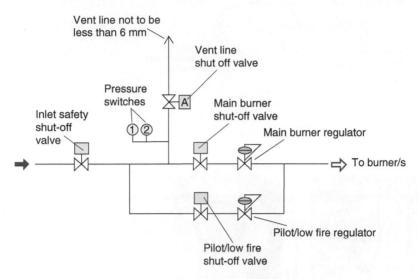

Figure 10.5 Gas train showing the location of safety shut-off valves

When commissioning non-domestic gas appliances, in general, the same commissioning procedures and checks are undertaken as identified in Part 8: Figure 8.1b. However, because of the nature of burning potentially large volumes of gas, the engineer has to undertake a dry run prior to the introduction of gas to confirm that the appliance will shut down to a safe condition. On introducing gas for the first time, the engineer needs to confirm the correct operation of the ignition sequence. For this, it is essential to simulate pilot or start gas failure, checking the flame safeguard goes to lock-out. This is best achieved by simply turning off the pilot or start gas supply.

Commercial Warm Air Heaters

Relevant ACS Qualifications
CIGA1 and CDGA1

Relevant Industry Document
BS 6230

10.8   Warm air heating falls into two categories:

1. direct-fired air heaters (See Figure 10.6a–c);
2. indirect-fired air heaters (see Figure 10.7).

Both designs of air heaters may be floor mounted, secured to a wall or installed suspended from the roof trusses of the structure. High-level heaters are commonly referred to as unit air heaters, and they prove popular where floor space is at a premium. Manufacturer's

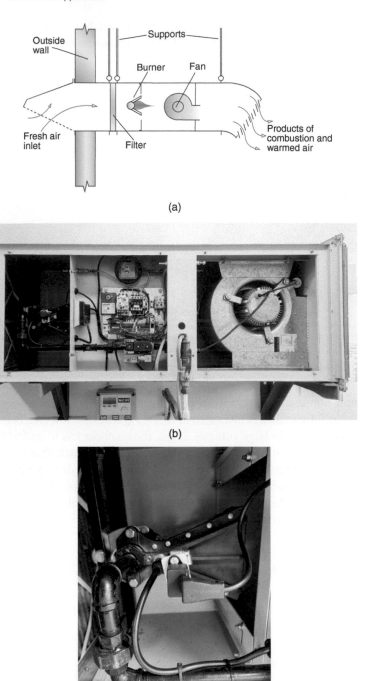

(a)

(b)

(c)

Figure 10.6 (a) Section through a direct-fired heater. (b) Side panels removed to exposed burner, controls and fan. (c) Heater burner

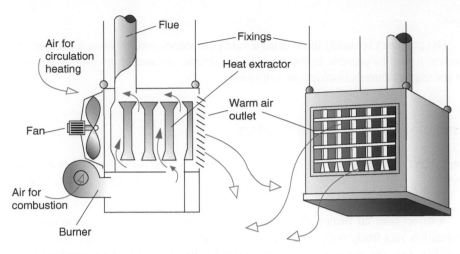

Figure 10.7 Suspended indirect-fired warm air heater

instructions will provide information regarding the necessary clearances required around and below the heaters and, as always, must be followed.

The air heater is fitted with similar controls to those used for domestic warm air heaters and includes a temperature-limiting thermostat and fan control to ensure that the air temperature blown into the room is at an acceptable level. They also have the facility to blow cool draughts in summer. The heaters are independent, serving the area where they are located, however units can be found utilising a series of ductwork, serving several outlets.

Ducted Air Distribution and Return Air

10.9   Generally, any ductwork needs to be as short as possible, giving warmth without the need for an excessive positive draught from the warm air outlet grille. Materials used for ductwork must not be a fire risk, and they need to be of adequate strength and of sufficient durability to withstand any internal and external temperatures and loads under normal operating conditions. If the air heater is installed in a plant room, the return air intake and warm air outlet must be fully ducted to prevent any interference with the safe operation of the flue. The openings into the plant room must also be suitably fire-stopped. Return air ducts and inlet points must not be located in areas where smells, dust or fumes could be drawn into the appliance, affecting its combustion performance and distributing the smells and odours around the entire building.

Hazardous Areas

10.10   Air heaters should not be used to supply warmed air to hazardous areas unless:

- all incoming air is from outside;
- the outlets from register grilles are at least 1.8 m above the floor.

Pipework

10.11   Gas pipework is usually made using a rigid pipe connection, however for suspended appliances, it may be possible to use a semi-rigid flexible connector or metallic gas flex, which the manufacturer's instructions will specify.

Commercial Direct Fired Air Heaters

Relevant ACS Qualifications   Relevant Industry Document
CDGA1   BS 6230 and IGE/UP/10

10.12   A direct-fired air heater, shown in Figure 10.6a–c, is one in which the products of combustion mix freely with the heated air and are passed out, through distribution outlet grilles, into the space to be heated. Generally, the air required for combustion is taken directly from outside, however direct-fired air heaters, such as the one pictured in Figure 10.8, may be positioned where it is dependent on natural ventilation from within the room. Because these heaters discharge their products into the heated environment, it is essential that during commissioning, environmental analysis checks are carried out in the room to ensure that the carbon monoxide (CO) and carbon dioxide (CO_2) levels

Figure 10.8   Suspended direct-fired warm air heater. Credit: Maywick Ltd.

remain acceptable. The maximum exposure limits in any space should not exceed 0.001% CO (10 ppm) and 0.28% CO_2 (2800 ppm). Modern units have fitted, as part of their design, a CO_2 limiting control to monitor the environment continually, shutting down the appliance should the levels rise excessively. Where such a control is incorporated, it needs to be regularly calibrated. It has been estimated that you require 37 m³/h of outside air for each 1 kW of gross heat energy input to keep the CO_2 levels below the 0.28% maximum level. The heater may be independent and a permanent fixture, incorporating various outlet diffusers and connected to a system of ductwork or a mobile air heater may be found. Mobile or transportable heaters must not be controlled by time switches or other remote controls.

The temperature of the air discharged from these heaters should not exceed 60°C; to maintain such temperatures invariably, dilution of the heated air with fresh or room air may be necessary. The efficiency of a direct-fired air heater is very high, typically over 90%, as there is no heat exchanger or fuel wasted into a flue system. A well-designed direct-fired air heater installation tends to slightly pressurise the room in which the heated air passes. This has the effect of preventing cold draughts from entering the room.

Ventilation

10.13   Where a direct-fired heater is located in a large open space, no additional ventilation is generally required, provided that the maximum exposure limits identified in Section 10.12 are not exceeded. Where the heater is installed in a plant room, passing the warmed air into the room to be heated by a system of ductwork, ventilation will be required to keep the relatively small space cool. The sizing of such a ventilation grille was discussed in Part 7: Section 7.19. Where a direct-fired heater is installed in an environment that relies on a closable ventilation system or an extract system, a safety system interlock must be provided to shut down the appliance if the air movement is interrupted.

Commercial Indirect-Fired Air Heaters

Relevant ACS Qualifications   Relevant Industry Document
CIGA1   BS 6230 and IGE/UP/10

10.14   Unlike the direct-fired air heater described in Section 10.12, the products of combustion are kept separate from the heated air that is discharged into the space to be heated. Indirect air heaters have a heat exchanger and flue system through which the combustion products pass, see Figure 10.9.

Indirect heaters are manufactured in a range of sizes from 15 to 140 kW if suspended. This increases up to 590 kW for the freestanding model (see Figure 10.10). A fan is located in the heater to draw cool air into the unit from the room and force it across the heat exchanger, where it is rapidly heated and discharged through the system of ductwork or, as is often the case, an outlet louvre grille forming part of the unit.

Air heaters may have either an atmospheric or forced draught burner installed. However, as the output increases, appliances generally use forced draught burners.

Figure 10.9 Suspended indirect fired warm air heater

Figure 10.10 Freestanding indirect fired warm air heater. Credit: Powrmatic Ltd.

Ventilation

10.15   The standard for ventilation air should follow the same general guidelines as for all open-flued appliances installed in rooms or plant rooms, as described in Part 7, Ventilation. However, the following specific points also need to be considered:

- Air should not be taken from areas where it is likely to be contaminated, for example, by odours and exhaust fumes from vehicles, etc., as these would subsequently be blown out through the warm air distribution system.
- Particular attention needs to be paid to warm air carrying high levels of water vapour, which is released as condensation when the air touches the cold surfaces of steel structures and windows or through natural cooling in poorly ventilated roof spaces.

In determining the ventilation requirements, an assessment of the maximum heat input rate for all appliances needs to be considered. Furthermore, provision for other process equipment and other products of combustion in the heated space, which may include those from motor vehicles, will need to be assessed.

Flueing

10.16   The standard for flueing follows the same general guidelines as for all open or room-sealed appliances as described in Part 6, Flues and Chimneys. Open-flue air heaters shall not be fitted with additional fans in the flue system unless the manufacturer specifically states otherwise. You should always check the CO/CO_2 ratio of the combustion performance, thereby determining if the appliance is burning safely or not.

Commercial Hot Water Storage Heaters

10.17   There are several designs of hot water storage heaters. The basic model consists of a cylinder in which a large volume of water is contained, and at the base is an atmospheric gas burner. The gas products exit up via a central flue, passing right through the water chamber, and are expelled outside. To assist the transference of heat, twisting baffles are positioned to direct the hot flue products onto the flue–water surface. The cylinder itself is well insulated to conserve heat. The gas storage heater may be either open-flued, see Figure 10.11a and b or room-sealed, see Figure 10.12. However, apart from the flueing arrangement, the operating principle is the same for both. One major disadvantage of these units is their tendency to produce large volumes of condensation because the flue passes through the comparatively cold water, cooling the flue products to their dew point. This condensate drops back down the central flue and causes major corrosion problems to the burner and base of the unit.

The heat input to these appliances depends on the design and model chosen. The heat input of larger commercial appliances can have inputs of over 200 kW where blown gas is used. The larger appliances generally have a multi-flue arrangement. Water storage capacities also cover a wide range. High-efficiency models are available with the burner positioned at the top, forcing the products down and up through an extended flueway/heat exchanger, thereby extracting a larger amount of the latent heat from the flue products. Clearly, with this design, an additional condensate pipe would be needed at the lowest point to remove the large volumes of condensation that are generated.

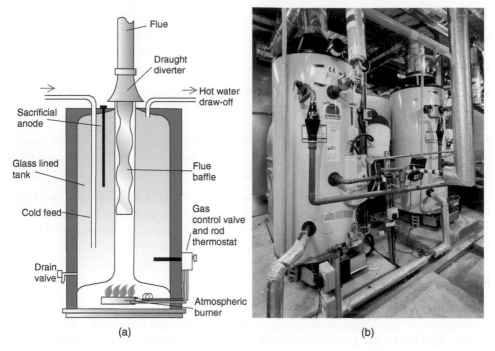

(a) (b)

Figure 10.11 (a) Open-flued hot water storage heater. (b) Installed open-flued hot water storage heaters

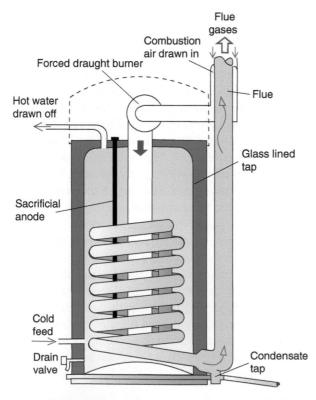

Figure 10.12 Room-sealed high-efficiency hot water storage heater

Water Supplies

10.18   The water supply to these appliances may be taken directly from the water supply main or alternatively a low-pressure system can be installed where the water is fed via a cistern. Mains pressure systems are referred to as an unvented supply, and where these are to be installed, the operative will need to hold the appropriate competency card before attempting to work on the appliance. A description of these systems and the controls used associated with the unvented system is outside the scope of this book, and therefore additional reading/research will be required to understand their operation fully.

Water Temperature

10.19   Some older units have no electrical supply and the temperature control is via a rod-type thermostat connected directly to a multi-functional gas control block. Later models will have an electrical supply and an adjustable electronic thermostat, and an additional energy cut-out, which, if operated, will require a manual reset.

Sacrificial Anode

10.20   The sacrificial anode is simply a rod of magnesium positioned in the top of a storage water heater. It will corrode as the result of electrolysis before any other metal in the system. Electrolysis is the destruction of one metal due to the chemical reaction of another in a damp environment. The anode's condition can be simply checked by withdrawing it from the vessel after isolating the water supply. If it is extensively corroded, it is simple to replace.

Commercial Instantaneous Water Heaters

10.21   Modern instantaneous water heaters typically operate with the use of electronics. The water flow is detected by a control system which activates an electronic ignition system, without a permanent pilot, to ignite the burner. The water temperature is monitored by various sensors that, in conjunction with the control system, modulate the burner accordingly. The hot water temperature can be programmed to a set temperature and can be altered by the user. Some manufacturers have developed system designs for the commercial market that involve linking various water heaters together mechanically and electrically, see Figure 10.13. You will notice that this particular installation is fitted on the outside of a building, saving valuable space inside. The modular set-up increases the potential supply of instantaneous hot water as and when it is required. It must be noted that it is very important to connect pipework to the units using the reverse return method, often referred to as first in and last out, as illustrated in Figure 10.14. Piping the heaters this way ensures that there is a balance in the resistance of water flow through each unit.

Figure 10.13 Commercial instantaneous hot water heaters. Credit: Rinnai UK.

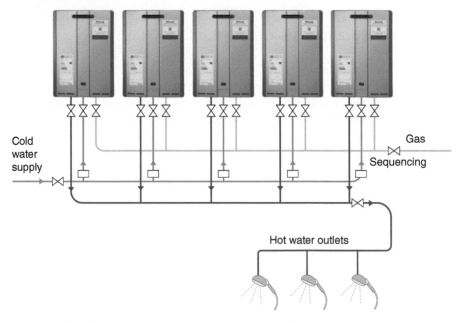

Figure 10.14 Illustration showing piping of hot water using reverse return method. Credit: Rinnai UK.

Overhead Radiant Heaters

Relevant ACS Qualification	Relevant Industry Documents
CORT1	BS EN 416, BSEN 13410 and BS 6896

Radiant Heating

10.22   Before one can begin to grasp the concept of overhead radiant heating, one must understand the principles on which this form of heating works. Radiation is the method of direct heat transference of infrared energy from the source to the point at which the heat rays land, and this is where the heat is absorbed. Anything blocking the path of the heat ray will block the heat transference; see Figure 10.15a. The best analogy is the feeling of the sun's rays directly on your skin on a hot day; the rays cannot reach you when you are in the shade. When heat is emitted, the greatest intensity is on a surface parallel to the heater panel, with the heat ray travelling at right angles to it. As the angle at which the heat strikes the surface falls below 90°, so does the heat intensity falling gradually to the point where no radiation is experienced, see Figure 10.15b. This means that a single heater might not be sufficient to heat an area: several heaters could be required, spaced as evenly as possible

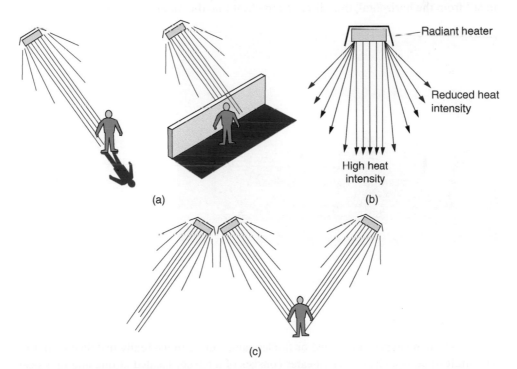

Figure 10.15   (a) Poor heat distribution shadowing occurring. (b) How the angle of radiated heat affects intensity. (c) Good heat distribution with no shadowing

around the area to be heated, as can be seen in Figure 10.15c. Each point in the room should be heated from at least two directions, more if possible, to prevent shadowing.

Heater Design

10.23 There are two basic designs of radiant heaters:

1. luminous heaters – ceramic plaque heaters;
2. non-luminous heaters – radiant tube heaters.

The Ceramic Plaque Heater

10.24 This is a flueless appliance, shown in Figure 10.16, which discharges combustion products directly into the premises. Sometimes the water vapour that is produced condenses onto cold walls and steelwork. The closer these heaters are fitted to the working area, the more the radiant heat intensity and the less the area that is heated. Generally, one should aim to provide a maximum of 240 W/m^2 at head level (2 m). This can be achieved by positioning a typical heater some 4–5 m off the ground, but the manufacturer's instructions will give more information as to the ultimate fixing height based on the heat output. Ceramic heaters may, if permitted by the manufacturer's instructions, be installed at an angle of up to 60° from the horizontal, thus directing the heat into the room.

Figure 10.16 A ceramic plaque heater

Radiant Tube Heater

10.25 These heaters can be flued or flueless and can be individually installed or put in as a multi-tube installation. The heater consists of a burner located at one end of a steel tube, typically 65 mm in diameter. On a forced draught version, the force of the fan blows the combustion products through the tube to the exit. Whereas, on an induced draught version, a fan at the end sucks the combustion products through the tube. The fan suction also draws the air/gas mixture into the burner, see Figure 10.17a and b. The length of the tube and its diameter vary according to the heat output required and the manufacturer's design. Typical individual heaters are approximately 7 m long, with the tube turning back on itself in a U shape, some 3.5 m from the burner. Above the tube is positioned a polished aluminium or stainless steel reflector panel, designed to direct the infrared heat down into

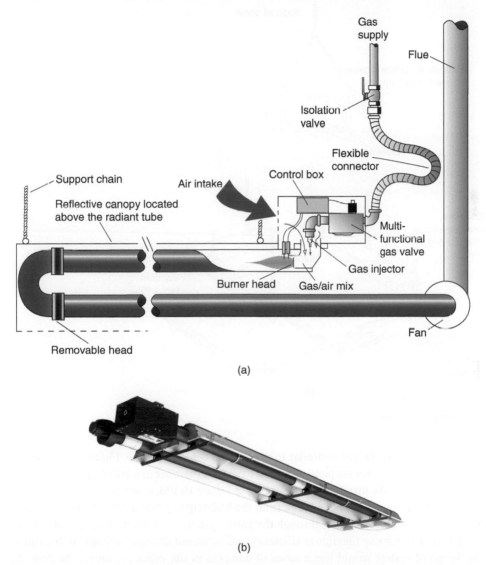

(a)

(b)

Figure 10.17 (a) Induced draught radiant tube heater. (b) Induced draught radiant tube heater. (c) End view of a radiant tube heater. (d) Multi-tube installation. Credit: Powrmatic Ltd.

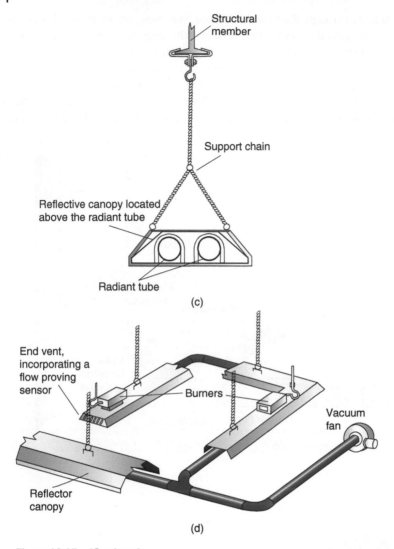

Figure 10.17 (*Continued*)

the room. The fixing height is similar to that for the ceramic plaque. Figure 10.17c shows an end view of the heater suspended from a structural member of a building.

With the multi-tube installation, illustrated in Figure 10.17d, a large centrifugal vacuum fan is located at the point of discharge from the building to provide the necessary suction pressure to pull all the products through the entire system, often with a reduced flue outlet size. In order to achieve maximum efficiency and sufficient draught through all the tubes, this design of system would use a series of dampers in the tubes to control the flow. To maintain a high enough temperature in the tube, several burners are positioned throughout its length.

Figure 10.18   Black bulb sensor. Credit: Titan Products Ltd.

Black Bulb Sensor

10.26   Because radiant heating systems do not heat the air in the room in which they are situated, it would be pointless to use the usual type of room thermostat to control the temperature. Instead, a sensor consisting of a bi-metallic strip or electronic thermostat, enclosed in a bulb-shaped hemisphere coloured matt black, is used, see Figure 10.18. The black finish readily absorbs heat. The heated surface of the bulb warms the trapped air within, which acts on the thermostat making or breaking the electrical supply as necessary, so controlling the operation of the heaters.

Installation Requirements

10.27   The heaters can be fixed high up on a side wall or, alternatively, can be suspended from a bracket or chain, an arrangement that invariably uses a flexible gas connection to an appropriate standard with a quarter-turn isolation valve fitted immediately upstream. The flue pipe, where installed, is generally not subject to high temperatures due to the cooling effect of the long tube.

The ventilation requirements for flued heaters can be found in Part 7: Section 7.27, and for flueless heaters, see Sections 10.29–10.33. Where a flueless heater has been installed, it is essential that the environment is monitored to ensure the discharge of products into the heated space remains acceptable in terms of the CO and CO_2 levels. The maximum exposure limits in any space should not exceed 0.001% CO (10 ppm) and 0.28 % CO_2 (2800 ppm). If the installation is within a hazardous area, it would be appropriate to sufficiently increase the ventilation requirements.

Work Record and Additional Maintenance and Servicing

10.28   When working on overhead heaters, the tasks include the generic gas installation commissioning procedures listed on the checklist in Part 8: Figure 8.1b, plus the following appliance-specific work:

- cleaning and polishing the reflectors;
- removing dust build-up from the face of the ceramic plaques; this may be accomplished with a fine jet of compressed air or by lightly brushing;
- replacing damaged or badly linted plaques;
- brushing through the tubes and removing loose material from the radiant tubes;
- confirming the suitability of the supports and ensuring their sound construction.

Ventilation for Overhead Radiant Heaters (Type A – Flueless)

Relevant ACS Qualification

CORT1

Relevant Industry Documents

BSEN 13410, IGEM/UP/10

10.29   The ventilation requirements for flued overhead heaters have previously been identified in Part 7: Section 7.29. However, where Type A (flueless) heaters are encountered, it must be determined whether the volume of the room meets the minimum size to ensure the combustion products are suitably diluted. This volume is 10 m^3 per kW of the appliance maximum rated net input. For example, where a room has four 35 kW heaters, the minimum room volume would need to be:

$$4 \times 35 \times 10 \, \text{m}^3 = 1400 \, \text{m}^3$$

The size of the room would, therefore, need to be calculated (length × width × height) to ensure this criterion was met.

Removal of the Combustion Products

10.30   Where it has been determined that the air change rate for a building is greater than 1.5 volumes/h and the heater/s used do not exceed 5 W/h total room volume, no additional ventilation would be required; the ventilation is maintained by natural air changes. If, however, additional ventilation is required to safely remove the products of combustion to outside, a low-level ventilation grille located below the appliance would be required. This grille should be equal in size to a high-level extract or exhaust grille located above the appliance. The method of finding this extract/exhaust opening size would be based on one of the following two options:

- thermal evacuation;
- mechanical extraction.

Thermal Evacuation

10.31 This method relies on the hot convection currents rising up through the building space where the combustion products can disperse freely outside into the open air. Therefore, the openings must be above the appliance, ideally at or near any apex or ridge position. Where the opening is located in a side wall, it should not be located further away from the appliance/s than 6 × the height of the vent (e.g. if a wall vent is at 5 m the appliance must be within 30 m). For ceiling vents, this distance is reduced down to 3 × the height of the vent.

To find the exhaust vent opening size the following three stages need to be completed:

1. Determine the total exhaust air volume to be allowed for:
   - 10 m³/h per kW.
2. Find the evacuation velocity:
   - Calculate the temperature difference between the inside temperature and the outside temperature. Once calculated, compare against vent height using the graph found in Figure 10.19. Determining a typical outdoor temperature could be open to variations of opinion. Therefore, Table 10.1 provides guidance on typical outside winter temperatures based on the zone location of the premises, as shown in Figure 10.20.
3. Complete the following calculation:

   total exhaust air volume ÷ (evacuation velocity × 3600) × 10 000

   $$\left[\text{i.e. stage } 1 \div \left(\text{stage } 2 \times 3600\right) \times 10\,000 = \text{air vent size in cm}^2\right]$$

   (3600 is the number of seconds in an hour, and the 10 000 converts m² to cm²).

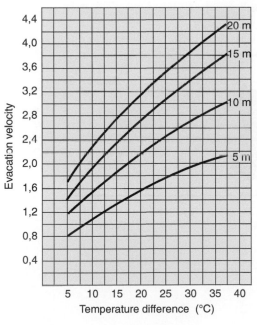

Figure 10.19 Evacuation velocity graph

Table 10.1 Outdoor design temperature zones within the British Isles

Zone shown in Figure 10.20	Typical minimum outside winter temperature (°C)
Zone 1	−1
Zone 2	−2
Zone 3	−3
Zone 4	−4
Zone 5	−5

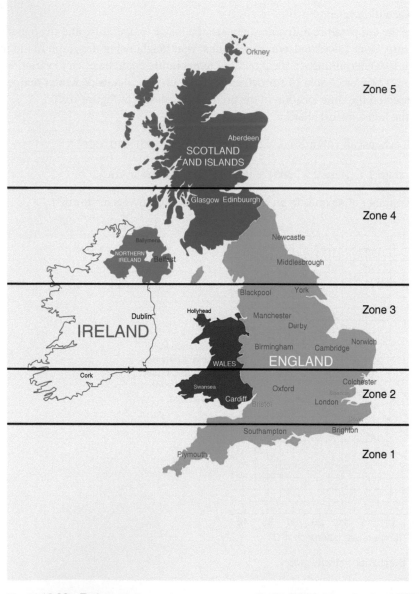

Figure 10.20 Typical minimum temperature zones in the British Isles. Credit: okili77/adobestock.

10.32   Example: A building in Billericay (England) has five 25 kW flueless overhead heaters with the exhaust vent located at 4 m above the inlet vent. The inside design temperature is 19 °C. By referring to Figure 10.20 and Table 10.1, we can determine that Billericay is in Zone 2. Therefore, the temperature difference between the outside temperature at −2 °C and the inside temperature at 19 °C is 21 °C.

The calculation for the minimum size of the ventilation openings is as follows:

1. Determine the total exhaust air volume:

$$10 \, m^3/h \times (5 \times 25 \, kW) = 1250 \, m^3/h$$

2. Find the evacuation velocity:
   Using the graph in Figure 10.19, compare 21 °C to the 5 m graph line = 1.6.
   (Note: There was no 4 m line on the graph, so you go to the next line up.)
3. Calculate the vent size:

$$\text{Total exhaust air volume} \div (\text{evacuation velocity} \times 3600) \times 10\,000$$

$$1250 \div (1.6 \times 3600) \times 10\,000 = 2170 \, cm^2$$

So a high and low-level grille, each totalling 2170 cm^2, would be required.

Mechanical Extraction

10.33   This method of ventilation simply relies on extraction fans to be used to remove the products of combustion. The fans must be located above the appliance, either wall or roof-mounted, and must be interlocked with the gas supply to ensure the appliance cannot operate until an exhaust route has been established.

The fan would need to provide the minimum exhaust flow rate equal to the minimum necessary air volume, which has previously been identified as the total net heat input (kW) × 10 m^3/h.

Thus, in the last example, where mechanical extract was desired, the minimum flow rate would need to be:

$$10 \, m^3/h \times (5 \times 25 \, kW) = 1250 \, m^3/h$$

Gas Boosters

Relevant ACS Qualification	Relevant Industry Documents
BMP1	IGEM/UP/2

10.34   This section deals with the installation and operating requirements of medium-pressure gas boosters with a maximum outlet pressure of 500 mbar (0.5 bar). For outlet pressures exceeding 500 mbar, reference needs to be made to IGEM/UP/6, which covers the installation requirements of compressors with an outlet pressure range greater than 0.5 bar but not exceeding 400 bar.

A booster consists of a centrifugal fan driven by an electric motor. The impellers of the fan are in direct contact with the gas, drawing it in through a central point and forcing it out by centrifugal force with increased pressure. A compressor, on the other hand, increases the pressure by forcing the gas through a rotary screw or reciprocating pump.

Gas boosters are used to increase and maintain gas pressures for plant and equipment that require higher pressures at the appliance burner than can usually be supplied by the gas distribution system. When planning the installation of a booster, the gas transporter needs to be consulted, and full details of the installation submitted 14 days prior to the proposed installation date. The gas transporter will want to confirm that the booster will not cause pressure fluctuations in the upstream gas supply (Figures 10.21–10.23).

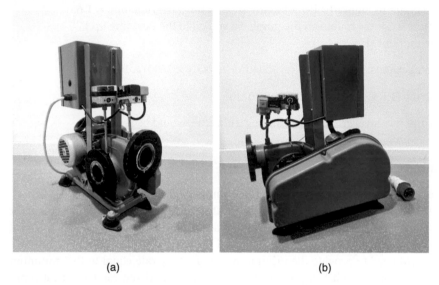

(a)   (b)

Figure 10.21   (a) Gas booster showing the inlet low-pressure cut-off and the low-pressure switch on the outlet. (b) Gas booster showing drive belt guard

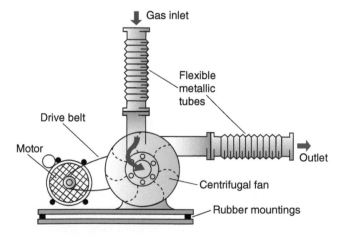

Figure 10.22   Operating principle of a centrifugal gas booster

Figure 10.23 Operating principle of a reciprocating compressor

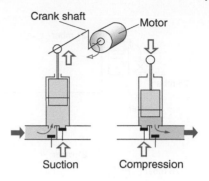

Pressure Increase and the Effect of Boyle's Law

10.35   Although the pressure is increased in the installation, it must be remembered that there is no increase in gas volume. In fact, the opposite will occur, as demonstrated below by applying Boyle's Law (see Part 2: Section 2.75).

Example: If the supply pressure of 1 m³ of gas is increased from 21 to 60 mbar, the resultant volume will be:

> *Pressure* 1 × *Volume* 1 ÷ *Pressure* 2 = *Volume* 2

> ∴ (1013 + 21) mbar × 1 m³ ÷ (1013 + 60) mbar = 0.96 m³

Note: The value 1013 above is atmospheric pressure, which needs to be included in the calculation.

Where the pressure is not to be increased but just maintained, no such loss in volume occurs.

Installation Requirements

10.36   As already stated, the booster does not increase the volume of gas so it is essential that there is sufficient gas supply to deliver the required gas load. Therefore, the correct sizing of the upstream pipework is essential. The booster should ideally be located as close as possible to the point where the elevated pressure is needed in order to reduce the amount of pipework at higher operating pressure, as illustrated in Figure 10.24a. Where the entire installation needs to be boosted for a specific requirement, such as to maintain adequate system pressure during peak demand in a large industrial or commercial installation, the booster can be installed upstream of all equipment as shown in Figure 10.24b. A gas booster should not be regarded as a substitute to correctly sized pipework or to overcome existing installation problems; this practice is not recommended and should only be considered as a last resort.

10.37   The unit needs to be located on a firm, flat horizontal surface. To reduce noise levels transmitting through the pipework, it is desirable to install the unit on anti-vibration mountings, with flexible metallic tubes to the inlet and outlet connections. These also serve the function of reducing the possibility of fatigue fracture caused by excessive movement.

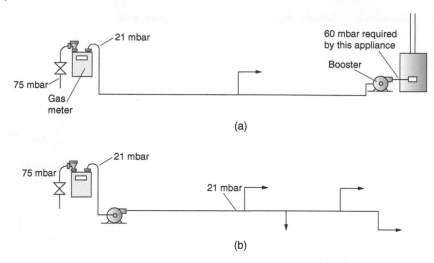

Figure 10.24 Gas booster installation (a) Gas booster used to elevate pressure to a single appliance. (b) Gas booster used to maintain pressure within a large system during peak demand

10.38 A booster should be located in a dry, well-ventilated area. Ventilation is necessary to prevent any minor leak from accumulating in the room and causing the atmosphere to become unsafe, and for cooling purposes to ensure the ambient temperature does not exceed 45°C. Ventilation should be large enough to ensure good air movement and be dispersed equally about the room with any openings at least 0.5 m away from external sources of ignition. Where mechanical ventilation is provided, this needs to be interlocked with the booster to prevent its operation in the event of fan failure. The size of the ventilation will be determined by the manufacturer. However, the amount of ventilation required will depend upon the operating pressure and Table 10.2 gives an indication of the minimum requirements.

Table 10.2 Minimum ventilation requirements for a room housing a gas booster

Operating pressure (mbar)	Airflow (m³/h)	Ventilation	
		Natural – dispersed on more than one wall High and low-level grille in each wall (cm²)	Natural – on one wall High and low-level grille (cm²)
≤50	11.4	290	400
>50–100	18.2	445	620
>100–150	250	15 500	21 700
>150–200	318	17 900	25 100
>200–300	363	21 900	30 700
>300–400	454	25 300	35 400
>400–500	522	28 300	39 600

Source: Data from IGEM/UP/2

10.39 Gas boosters must not be installed in the same room as an air compressor unless the air supply for the air compressor is ducted directly from outside air. This is to prevent the possibility of a gas leak being distributed around the compressed air system.

Pressure Fluctuation in the Pipework

10.40 To prevent pressure fluctuations upstream and downstream of the installation, the booster is sandwiched between a low-pressure cut-off (LPCO) and a non-return valve, as shown in Figure 10.25. In addition to these controls, where pressure instability or surging proves to be a problem, a small controlled bypass may be incorporated and adjusted under a no-flow condition to eliminate the problem. A 25 mm lockshield valve is often used. However, the booster manufacturer should be consulted.

10.41 The LPCO is a pressure switch that senses the pressure in the pipeline preceding the unit. Where the gas pressure falls to a predetermined level, it shuts down the electrical supply to the motor, thereby preventing a dangerous situation from occurring, such as reduced pressure in the upstream pipework or the ingress of air into the system. Where the pressure switch is activated, there should be no automatic re-start, and manual intervention will be required. The response time of the LPCO should not exceed three seconds. The commissioning gas engineer must ensure that the LPCO operates correctly and the setting should be as high as possible to prevent nuisance operation; this will usually be 50% of the gas supply pressure with the appliance running at full load but no lower than 10 mbar. The setting and response time of the LPCO should be checked at least annually. In addition, some boosters will be manufactured with a second low-pressure switch on the outlet of

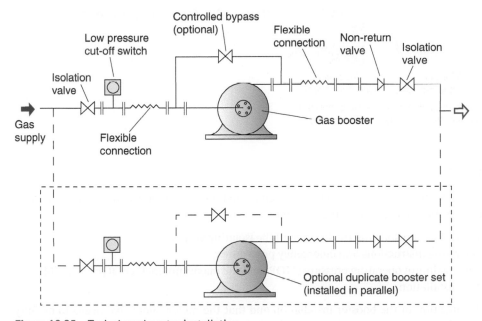

Figure 10.25 Typical gas booster installation

the booster as shown in Figure 10.21a. This switch should be adjusted when the booster is running to ensure that the minimum required boosted pressure is available.

10.42   A non-return valve is required where:

- two boosters are connected in parallel, and gas re-circulation may occur;
- reverse rotation is possible;
- reverse pressure surges occur when the appliance is turned off due to large volume outlet pipework.

The non-return valve prevents air from entering the system in the event of reverse flow and protects against increased back pressure.

10.43   The gas installation pipework design should ensure the safe operation of all plant and equipment under normal working conditions, including during the operation of the booster. The installation should be designed with a maximum working pressure drop across the system of no more than 1 mbar where the incoming supply is regulated at 21 mbar. For systems regulated at operating pressures greater than 21 mbar, the designed pressure drop should be no more than 10% of the design pressure. Where the inlet supply pressure to a booster has a tendency to suffer the continued effects of reduced pressure problems, it is possible to install a volume accumulator or reservoir prior to the unit. This can also be achieved by oversizing the upstream pipework to the booster. This, in effect, stores a supply of gas and thus prevents the supply from drastically fluctuating or dropping as gas is drawn from the pipe. Alternatively, slow start-up and speed inverter control systems may be used.

10.44   Should the gas supply to the booster require a reduction in pipe size, it is necessary to use concentric reducers or tapers in order to prevent unnecessary turbulence, such as those shown in Figure 10.26.

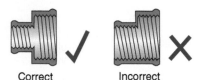

Figure 10.26   Gas booster pipework reduction

Correct
(concentric reducer)   Incorrect
(eccentric reducer)

Where reduction in pipe size is required
concentric reducers should be used

Notices and Instructions

10.45   Where a booster is installed, the following notices and information must be permanently displayed adjacent to the machine:

- a line diagram showing the location of the isolating valves;
- operating instructions and emergency procedures;
- the maximum operating pressure (MOP) and the maximum incidental pressure (MIP) of the installation.

Notification of the booster installation and that the meter inlet valve needs to be fully open while the booster is in operation should be displayed on the meter.

Commissioning, Maintenance and Servicing

10.46 Commissioning of boosters must only be completed by a competent person and follow the procedure set out by the appliance manufacturer. It is recommended that the machine is initially commissioned with air to ensure all safety devices and interlocks are operating correctly. Safety devices must not be removed or bypassed at any time.

The maintenance and service schedule of a booster will be determined by the manufacturer and should be followed at all times. Neglecting regular inspection and maintenance can lead to the failure of components such as bearings, drive elements and fan units. Routine maintenance will typically include the following:

- Inspection of bearing assemblies. These are usually sealed units and lubricated for life. When they reach the end of their working life the whole unit must be replaced. Increased operational noise and vibration may indicate they need replacing. Replacement will likely be required every 3–5 years depending on manufacturer.
- Inspection and cleaning of drive belt and pulley. Ensure the belt guard is in place and that the drive belt is rotating in the correct direction. Drive belts should be replaced at intervals determined by the booster manufacturer (typically annually).
- Inspection of flexible connections (recommended replacement after five years).
- Inspection and testing of safety devices including the LPCO.

The manufacturer's operating, maintenance and servicing instructions must be left with the building owner/responsible person.

11

Commercial Laundry

Commercial Laundry Equipment

Relevant ACS Qualification
CCLNG1 and CLE1

Relevant Industry Document
BS 8446

11.1 Gas tumble dryers, washing machines and rotary gas ironers are available for use in commercial laundries. The appliance manufacturer will specify the correct installation methods and any specific commissioning/servicing checks that should be performed. The correct operation of all safety controls, safety shut-off valves and pressure or flow-sensing valves must also be confirmed.

Location of Appliances

11.2 The location of gas-fired laundry equipment needs careful consideration, particularly when installed in the same location as a dry cleaner. If dry cleaning chemicals were to mix with the air supplied to the appliance, the interaction with the appliance burner flame would produce toxic gasses. To prevent this from occurring, the laundry appliance must be separated by a minimum distance of 5 m from the dry cleaner or installed in a separate room. The make-up air for the laundry equipment must not come from the area of the dry cleaner and must be separate from any make-up air for the dry cleaning equipment. When checking flame picture on the laundry appliance, this must be confirmed with the dry cleaner operating.

All appliances should be positioned on a firm, level surface that is capable of taking the weight of the appliance when fully loaded. There must be adequate room to enable the appliance to be loaded or worked at without causing an obstruction. Appliances should

Gas Installation Technology, Third Edition. Andrew S. Burcham, Stephen J. Denney and Roy D. Treloar.
© 2024 John Wiley & Sons Ltd. Published 2024 by John Wiley & Sons Ltd.

only be stacked where the manufacturer's instructions permit this and be fitted with a suitable restraining device. Care also needs to be taken to avoid combustible materials coming into close proximity with the appliance and any exhaust system.

Gas Supply and Appliance Connection

11.3   By definition, these appliances are non-domestic, and kilowatt ratings will vary depending on the appliance type. However, the pipe size to an individual appliance would not normally exceed 32 mm in diameter. The final connection to the appliance is made by means of a flexible connection in accordance with an appropriate standard, such as BS 699-2, with an individual isolation valve included adjacent to each appliance. A restraint device is also required to prevent excessive movement of the appliance away from the supply connection.

In addition to fitting the appliance isolation valve, a manually operated emergency control valve must be installed on the installation pipework to each laundry area and located as near as practicable to the exit door. Where this manual isolation valve is not in a readily accessible position, then an automatic isolation valve (AIV) conforming to BS EN 161 will need to be installed as close as practicable to the point of entry, downstream of the manual isolation valve and upstream of the appliance isolation valve. The AIV is operated via a stop button, along with an appropriate proving system to ensure down-stream pilots, etc., are turned off before the gas is reinstated. A suitably worded notice explaining what to do in the event of an emergency will need to be affixed at locations where manual isolation valves are fitted, where emergency stop buttons can be operated, or where AIVs can be reset. The notice will advise how to isolate the gas and that all burners and pilot valves on appliances should be turned off before attempting to restore the gas supply.

Tumble Dryer and Rotary Ironer

11.4   Although these appliances do not have what would be considered a 'conventional' flue, both of these appliances move large amounts of air and need to evacuate products of combustion and water vapour. This is achieved by providing 'make-up air' and an exhaust system. Owing to the substantial air movement through the building caused by the exhaust venting, any open-flued appliances may be subjected to increased spillage problems; there-fore, additional spillage checks will be required. These should be carried out with all the appliances running (Figures 11.1 and 11.2).

Make-up Air

11.5   Due to the fact that commercial laundry systems are designed to actively exhaust large amounts of air from the room, air in addition to that required for combustion needs

Figure 11.1 Commercial tumble dryer

Figure 11.2 Rotary ironer

to be provided. This air is termed make-up air. Make-up air is clean, cool air directly from outside, which aids the drying process, replaces the air evacuated by the appliances and includes air for combustion. Air for combustion forms a very small percentage of the overall make-up air, typically around 2–6% of the total volume.

A permanent, non-closable supply of make-up air is vital for correct and safe operation of both tumble dryers and rotary ironers. Manufacturer's instructions will determine the minimum size and volume of the make-up air and should always be consulted. Where these manuals cannot be sourced or found, Tables 11.1 and 11.2 may be referred to for an indication of the amount of air that may be required. Both the make-up air inlet and the exhaust termination should communicate directly to outside air. However, the make-up air inlet must be at least 2 m away from the exhaust termination to ensure that damp, warm air is not reintroduced to the building. If the make-up air is fan-assisted, this must be interlocked to the gas supply to prevent the appliance from operating in the event of fan failure and incorporate a manual reset.

Table 11.1 Make-up air requirements based on exhaust outlet diameter

Appliance exhaust outlet diameter (mm)	Minimum free cross-sectional area of air inlet/per dryer (cm^2)
100	395
150	885
200	1570
250	2455

Note: This table is for guidance only, and manufacturers should always be consulted to determine the correct ventilation requirements.
Source: Data from BS 8446

Table 11.2 Make-up air for rotary ironers based on drying load capacity

Ironer capacity (bed length) (m)	Minimum free cross-sectional area of air inlet/per ironer (cm^2)
0.8–2.1	1000
2.1–3.6	1500

Note: This table is for guidance only, and manufacturers should always be consulted to determine the correct ventilation requirements.
Source: Data from BS 8446

Exhaust and Ducting Systems

11.6   The exhaust system acts as an extractor to remove the vast amount of water vapour, lint and combustion products and move them to the external environment. The exhaust duct will need to be sized to comply with the manufacturer's requirements. Tumble dryers produce combustible lint, which if allowed to accumulate in the exhaust duct, will reduce efficiency and pose a fire risk. Therefore, no grilles/filters should be incorporated in the duct, as they rapidly become blocked. For the same reason, it is essential that adequate access is made to the duct for maintenance purposes and the removal of lint. For maximum efficiency and to minimise the accumulation of lint, the exhaust duct route to the outside air should be as short as possible, and the number of bends should be kept to a minimum to limit the resistance to airflow. The duct should be of a smooth bore metal construction capable of withstanding the high temperature of the exhaust air (80 °C for tumble dryers and 180 °C for rotary irons). If screws or rivets are used, they should be as short as possible so as not to restrict the airflow or allow lint to gather.

Tumble dryers should be connected to the exhaust duct in accordance with manufacturer's instructions. This is usually by means of a flexible connection. The flexible exhaust connection shall be no longer than 500 mm, have no changes in direction or be kinked/deformed in any way and be capable of withstanding the maximum appliance exhaust temperature. Where a single duct is used as a header, serving several appliances, it needs to be sized with an overall cross-sectional area of at least the sum total of all appliances connected to the duct.

Note: The exhaust ducts from tumble dryers and ironing machines must be kept separate.

Figure 11.3 Location of exhaust terminal

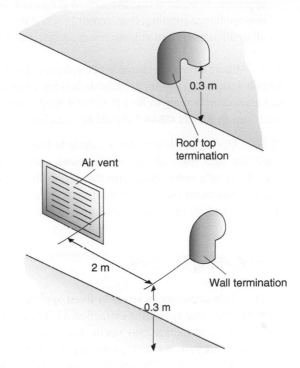

Exhaust System Termination

11.7   An unrestricted, open-ended outlet should terminate facing down towards the ground to prevent rain/wind entering. The outlet should maintain a minimum clearance of 300 mm from ground level, and in order to prevent any re-entry of products into the building, the duct should terminate at least 2 m from any openings such as ventilation air inlets, as shown in Figure 11.3. This is to ensure that damp, warm air is not reintroduced to the building.

Commissioning and Maintenance

11.8   Commissioning should be in accordance with the manufacturer's Instructions and must include all necessary safety checks as required by the Gas Regulations. The exhaust system should be checked in the following sequence:

1. Visual inspection of the ductwork to confirm correct sizing, support, construction, termination, etc.
2. Prove correct airflow by means of a smoke test. With the duct cold, place a lit smoke pellet in the ductway of the appliance and turn it on without the heating being operated. Examine the ductwork and joints for leaks and check all smoke exits the termination point. Repeat the test simulating a worst-case scenario with doors/windows closed and all appliances on.
3. Verify the backflow pressure is at an acceptable level. With the appliance operating at its normal working temperature, the exhaust duct should be tested within 500 mm of the

appliance to confirm that the airflow pressure is less than 1.25 mbar. Again, verify with one appliance running, then repeat in worst-case scenario with doors/windows closed, all appliances on and any other extract systems set to maximum rate.

11.9 Where a carbon monoxide (CO) alarm is fitted, ensure that it conforms to the correct British Standard. Laundry establishments are considered workplaces, and a domestic-type CO alarm conforming to BS EN 50291-1 may not be suitable. Therefore, a CO alarm that conforms to **BS EN 45544-3** should be installed.

11.10 Upon completion, demonstrate to the user the correct and safe operation of the appliance and leave all wiring diagrams, computer program support information and manufacturer's instructions with the responsible person. Advise the responsible person of any manufacturer-specific servicing requirements and of the need to have the ductwork inspected and cleaned annually to remove lint.

Gas Washing Machine

11.11 This appliance is an open-flued appliance and, as such, special care needs to be taken over the vast amount of extraction where a tumble dryer or rotary ironer is used in the same vicinity, as spillage may result. The gas connections are as described in Section 11.3. The size of flue system selected will vary, depending on the heat input of the appliance. This would not be less than 125 mm diameter or 100 mm × 100 mm square section, but sizes up

Figure 11.4 Commercial gas washing machine

Figure 11.5 Commercial gas washing machine

to 250 mm may be necessary. A draught diverter is required. Generally, each appliance will have its own individual flue system; however, should a common flue be used, fan-assisted appliances must not be installed with appliances operating on natural draught. Should a fan-assisted flue be employed, it is essential to incorporate a suitable proving system to isolate the appliance in the event of fan failure (Figures 11.4 and 11.5).

12

Commercial Catering

Commercial Catering

12.1   Commercial catering establishments include a variety of restaurants, cafes, school kitchens, pubs, factories, and bakeries where food is processed on a large scale, producing biscuits, cakes and bread, etc. There are many kinds of appliances in commercial kitchens that cater to cooking, warming and cleaning. The following list provides a guide to the cooking processes that may be involved.

Cooking Processes

Baking
This involves cooking food dry in an oven, with no oil, etc., added to the process.

Roasting
This involves cooking in an oven or on a rotating spit using a little oil or fat to keep the food moist. Often, the food is initially cooked at a high temperature, e.g. at a high oven setting or by boiling prior to roasting at a slower, cooler temperature.

Braising
This is a method of slowly cooking the meat or vegetables in a closed container in an oven. The food is surrounded by a liquid or sauce that provides an accompanying gravy or sauce.

Blanching
This is a process in which the food is usually immersed in boiling water for two to five minutes and then immediately cooled in order to remove strong odours, whiten or assist in the removal of skin from a fruit, e.g. tomatoes. Blanching is also carried out to limit the action of enzymes that would cause frozen food to deteriorate.

Boiling
This is a process in which the food is cooked in water at 100 °C. Food, such as root vegetables, may be placed in cold water and brought to the boil. Other foods, such as green vegetables and meat, are immersed in the boiling water.

Gas Installation Technology, Third Edition. Andrew S. Burcham, Stephen J. Denney and Roy D. Treloar.
© 2024 John Wiley & Sons Ltd. Published 2024 by John Wiley & Sons Ltd.

Simmering
This is a process of cooking in which the liquid is just below the boiling temperature; at this temperature, the liquid shows only an occasional sign of movement. Typically, soups and broths are cooked in this way.

Poaching
Here the temperature is just below boiling point but above the temperature used for simmering, at around 93–95 °C. At this temperature, there is a gentle, occasional bubbling of the liquid. Eggs and fish are typically poached; containers are usually used to contain the food, keeping it separated from the water.

Steaming
This is a process in which the food is cooked in an enclosed container, surrounded by steam from boiling water. Often, to assist in steaming, the pressure of the food container is raised; this has the effect of reducing the cooking time required.

Stewing
This is a kind of simmering process in which meat and vegetables are slowly cooked in a stock mixture that just covers the food. Where the food is cooked in the oven, it is often referred to as a hotpot or casserole.

Deep Frying
This involves submerging the food completely in hot oil or lard/fat at temperatures usually between 180 and 195 °C. Often, the food is coated with a flour and egg mixture (battered) to give a desired taste.

Shallow Frying
Unlike deep frying, in shallow frying, the food is cooked in only a small quantity of oil/fat; the food needs to be turned to ensure that all sides are adequately sealed and cooked. Chips (French fries) and other vegetables cooked this way are known as sautéed.

Dry Frying or Griddling
With this method of cooking a very small dash, if any, of oil is put on a hot plate, and all excess fat from the food is allowed to drain away down the slightly tilted plate or griddle.

Grilling
In this method of cooking, the food is cooked by radiant heat from a heated mesh or grill. Grilling can be achieved in many ways: the heat source may be above, below or horizontal to the food source to give the desired effect. For example, where the grill is located below the food (under-fired or flame-grilled), as the liquid fat drops from the food, it flares up to give the food a distinctive smoky burnt taste and flavour.

Toasting
This is a process, particularly common with bread, in which the food is warmed or browned to give a crisp texture.

Catering Establishments

Relevant Industry Document
BS 6173, IGEM/UP/19, DW/172

Commercial Kitchens

12.2 A typical commercial kitchen is shown in Figure 12.1, and you will note that a wide variety of specialist gas equipment is used in the process of cooking and cleaning. The layout of a commercial kitchen needs to be carefully planned. For example, fryers and fish and chip ranges must not be located next to areas where water is used, which could lead to a hazard unless purpose-made splash shields are in place. This would also include any retractable wash-down hoses from adjacent appliances which could be within reach of the fryer when extended. Neither should narrow appliances or those with hot liquids be located at the end of a row of installed appliances. Adjacent walls and floors will need to be of a non-combustible impervious finish capable of withstanding the maximum temperature permitted by the appliance manufacturer. The floor must also be able to withstand the weight of the appliances when full.

Figure 12.1 A typical commercial kitchen

Gas Supplies and Emergency Isolation

12.3 The gas supply to each catering area requires a manual isolation valve to be located in an accessible position as near to the exit as is practicable for use in an emergency. In addition, an automatic isolation valve (AIV) conforming to BS EN 161 shall be fitted to the installation pipework serving each catering area. The AIV is fitted between the upstream manual isolation valve and the downstream appliance isolation valves and located as near as is practicable to the point of entry to the catering area. The AIV is used as part of the interlock requirements (see Section 12.5) and, in the event of an emergency, is operated by means of emergency stop buttons located near the exit of each catering area. A suitably worded notice explaining what to do in the event of an emergency will need to be affixed at locations where manual isolation valves are fitted, where emergency stop buttons can be operated, or where AIVs can be reset. The notice will advise how to isolate the gas and that all burners and pilot valves on appliances should be turned off before attempting to restore the gas supply.

12.4 Gas supply pipes should be spaced 25 mm away from the wall or floor surface to facilitate cleaning. All pipes, supports and associated fittings shall be of a corrosion-resistant material or will need to be painted to protect them from the effects of grease, water and cleaning fluids. Where the gas pipework enters the catering area by means of a sleeve, the sleeve should extend a minimum of 25 mm through the wall or floor and be sealed with a flexible sealant on the kitchen side to prevent the build-up of any grease, water or corrosive cleaning chemicals inside the sleeve. Each appliance should have its own isolation valve to facilitate cleaning and servicing, and the final connection to mobile appliances should be made using a flexible connector and self-sealing socket in accordance with BS 669-2. Mobile appliances should also be fitted with a suitable restraint that is shorter than the flexible connection to prevent excess movement; wheels need to be fitted with a suitable locking device that is easily accessible to the user. Although fixed appliances would normally be connected with rigid pipework, flexible connections may be used where there is a concern about excessive vibration during operation, provided the appliance is not permanently flued to atmosphere. For fixed appliances connected rigidly to the installation pipework, a suitable means of disconnection, such as a union, will need to be provided between the appliance isolation valve and the appliance to enable withdrawal for cleaning. Whenever an appliance is disconnected from the gas supply, all open ends must be sealed with an appropriate fitting.

Interlocking

12.5 Most catering appliances rely on a mechanical ventilation system in order to create a safe and comfortable work environment by extracting the products of combustion along with cooking odours, heat and vapour safely to outside air. This ventilation system is typically provided by means of a canopy or a ventilated ceiling. The majority of catering appliances fitted in commercial kitchens under a ventilation and extraction system will be Type A (flueless). Other appliances, such as some fryers and convection ovens, which normally require a flue (Type B), may be installed under a canopy or ventilated ceiling provided they are interlocked to the gas supply and extract system; under these circumstances, the canopy or ventilated ceiling performs the same role as a flue. The interlock is designed to shut down the gas supply to appliances dependent on the extract system if the airflow falls below a pre-set minimum level required to ensure safe operation of the appliance. Some commercial kitchen interlock systems also include environmental air quality monitoring systems. These may be designed to measure carbon dioxide, temperature and humidity. Interlock systems must not be capable of being overridden. Figure 12.2 shows an example of a catering kitchen ventilation system interlocked to the gas supply, with Figure 12.3 showing a safety shut-off with an integrated proving system.

The interlock will need to be designed and installed to meet the requirements of appropriate standards such as IGEM/UP/19 and BS 6173. Reference should also be made

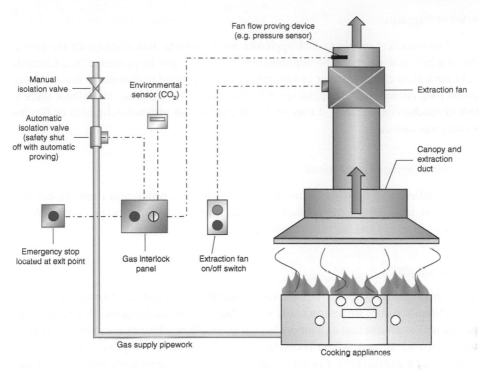

Figure 12.2 Principle of a commercial catering kitchen ventilation system interlocked to the gas supply (not to scale)

Figure 12.3 Safety shut off with integrated proving system

to the Building and Engineering Association's publication DW/172 *Specification for kitchen ventilation systems* for the design and installation of the catering ventilation system. Where an existing system is encountered that does not have an interlock fitted or has an override, the gas operative will need to follow the risk assessment protocol given in IGEM/UP/19. This will determine if the installation constitutes an unacceptable risk to safety, and the responsible person should be advised accordingly in writing.

Electrical Supplies

12.6   The electrical supply to each appliance needs to be in accordance with the manufacturer's instructions, meet the requirements of BS 7671 and be protected by a correctly rated fuse and/or circuit breaker as appropriate. Any flex used to connect to the appliance shall be long enough for the appliance to be withdrawn for cleaning, etc. Where there is a risk of mechanical damage, it may be appropriate to use armoured, braided or flexible abrasion-resistant cable.

Water and Drainage Connections

12.7   Installers will need to ensure that any wholesome water, and drainage connections, are as required by the appliance manufacturer and in line with current Water Regulations/Byelaws and industry standards.

Fire Precautions

12.8   Fire precautions need to be observed in catering establishments to ensure that there is no additional risk of fire, particularly where hot surfaces and liquids are encountered. It is essential to ensure that appliances are not sited where adjacent combustible surfaces could exceed 65 °C, and, where necessary, the surfaces should be protected. In general, manufacturer's instructions give clear guidance as to the separation distances needed; appliances must not obstruct escape routes.

Where ductwork for pipes or ventilation passes through the building's structure, it must be designed so as to prevent the spread of fire or smoke along it.

Ventilation/Extraction in Commercial Kitchens

Relevant Industry Document
BS 6173, DW/172, IGEM/UP/19

Ventilation Requirements

12.9   Every kitchen must have a properly designed and manufactured ventilation system to ensure a safe and comfortable work environment. The ventilation requirements for a catering establishment must provide sufficient clean, cool air for the occupants to breath, remain comfortable and to remove the excess hot air, odours, vapour and steam from the cooking and washing activities. The specification for such systems is of a specialist nature, and the detailed advice in the Building Engineering Services Association publication DW/172 should be followed. However, commercial catering gas operatives will need to understand the principles of the ventilation system design and test its correct operation during commissioning, service and maintenance of the gas appliances installed in the kitchens.

Note: DW/172 uses new unit symbols in its specification text (see examples on below). However, for clarity, this section will continue to use the more familiar notations that have been adopted throughout this book.

Original style	New style (DW/172)
m/s	$m.s^{-1}$
m^3/s	$m^3.s^{-1}$
$m^3/s/m^2$	$m^3.s^{-1}.m^2$

The ventilation system needs to be designed to match the cooking load, the number of staff/customers and the equipment used. This is usually achieved by a combination of hoods and extractor fans and a supply of fresh make-up air to replace that removed. To calculate the amount of air that needs to be evacuated, the 'thermal convection method' is the only design method recommended by DW/172. Using air change rates for designing commercial kitchen ventilation should not be used. However, if an air change rate estimate is required at the early design stage, without precise knowledge of the type and quantity of cooking appliances to be used, it would not be unreasonable to expect a minimum rate of 60 air changes per hour.

Commercial catering ventilation systems and any associated interlocks must be commissioned, tested, and maintained in accordance with the manufacturer's instructions by a competent person. Servicing and testing would be carried out at least annually.

Calculating Extract Requirements – Thermal Convection Method

12.10   The purpose of a canopy (or ventilated ceiling) is to collect any plume created by cooking/washing appliances and effectively discharge this to outside air. For optimum extraction to be achieved, correct flow rates must be calculated: too low and the work environment could become contaminated; too high and excess draughts will be created. You will see from the following that the actual size of the appliances and the cooking demands determine the size of the canopy, but the type of cooking appliance determines the extraction flow rates.

To confirm that the ventilation system is extracting the correct flow rate, each appliance installed under a canopy is given a thermal coefficient value, which is based on $m^3/s/m^2$ of the surface area of the appliance, as shown in Table 12.1. Each appliance value is added together to calculate the total theoretical volume to be extracted. Next, depending on the type of canopy installed and its location, it is necessary to multiply the theoretical extract flow rate figure by the appropriate canopy factor shown in Table 12.3. This determines the actual, or specific, extract flow rate required. A worked example is shown in Section 12.14.

Table 12.1 A selection of appliance coefficient values for canopies/ventilated ceilings

Appliance	Gas m³/s/m²	Electric m³/s/m²	Surface Temp. °C
Atmospheric steamer	0.35	0.20	125
Bain marie	0.20	0.15	57
Bench		0.03	25
Boiling pan	0.35	0.25	146
Boiling table	0.35	0.25	190
Brat pan	0.65	0.50	240
Chargrill	0.95	0.52	350
Chrome griddle	0.45	0.40	290
Combi oven (10 grid)	0.45	0.32	92
Deep fat fryer	0.50	0.45	190
Fan-assisted convection oven	0.40	0.30	86
Mild steel griddle	0.30	0.25	190
Open top range/oven	0.40	0.30	190
Pasta cooker	0.30	0.20	120
Pastry/baking oven - double	0.30	0.20	90
Pizza deck oven - single	0.20	0.15	90
Pressure steamer	0.30	0.20	120
Salamander grill	0.75	0.55	260
Solid top oven range	0.60	0.51	420
Tandoori oven	0.50	0.33	90
Water boiler	0.25	0.20	78

Note: Where the design of the kitchen includes multiple appliances of the same type, such as in food technology kitchens in educational establishments, there exists the risk of exceeding the maximum CO_2 level of 2800 ppm. In these circumstances, it is suggested the coefficient values given in Table 12.2 *are used. This is only applicable to those appliances listed in* Table 12.2.
Source: Data from The Building Engineering Services Association DW172 Specification for Kitchen Ventilation Systems

Table 12.2 Appliance coefficient vales where CO_2 reduction is required due to multiple appliances of the same type

Appliance type	Up to 3 appliances m³/s/m²	Above 3 appliances m³/s/m²
Boiling table/ hob/stock pot stove	0.45	0.55
Deep fat fryer	0.60	0.75
Open top range and oven	0.50	0.60

Source: Data from The Building Engineering Services Association DW172 Specification for Kitchen Ventilation Systems

Table 12.3 Canopy factors

Type	Low level	Passover	Overhead Wall	Overhead wall, Island Mounted	Island
Open both ends	1.15	1.15	1.25	1.60	1.35
Open one end	1.10	1.10	1.20	1.50	1.25
Closed both ends	1.05	1.05	1.15	1.40	1.15

Source: Data from The Building Engineering Services Association DW172 Specification for Kitchen Ventilation Systems

Canopy Design and Extraction Ductwork

12.11 It is advisable to take most of the kitchen extraction from directly above the appliances that generate the heat, odours and vapour; canopies are ideal for handling contaminated air in such concentrated areas. Where vapours, etc., are generated over a wider area, a ventilated ceiling may be considered appropriate. Unless restricted by walls, the plan dimension of the canopy shall have a minimum 250 mm overhang, which is measured to the inside edge of the canopy, as shown in Figure 12.4a. Where a combination oven is installed, the front overhang should be increased to 600 mm to ensure the steam or fumes are effectively extracted when the door is opened. The height of the canopy should be such that it does not form an obstruction, with the lower edge being between 2 m and 2.1 m from the finished floor level. A space of 100 mm at the rear of the appliances should be allowed for services.

It is important to ensure that the bottom edge of any grease separator is at least 450 mm above the cooking surface. Where salamander grills are located at high level, a deflector cowl supplied by the manufacturer should be fitted to the flue opening on top of the grill. This will allow the products of combustion to cool sufficiently before passing through the separator and thereby reduce the risk of igniting grease or oil deposits on the separators; this should also help prevent unsightly discolouration of the stainless steel surfaces.

The canopy should be non-combustible and easy to clean. The material used for fabricated canopies should be ultra-fine-grained stainless steel. The canopy should be designed to enable any grease separators to be easily accessible for cleaning, which should be carried out at regular intervals. The grease extracted by the separators must not be allowed to fall back onto the cooking surface but should be designed to drain off into a collection drawer with a minimum volume of 500 ml. An example of grease separators and a grease collection draw are shown in Figure 12.4b. Some grease vapour will pass through the separators, therefore, the interior of the canopy ductwork must be accessible for inspection and cleaning. Access doors should be provided at 2 m centres. The supplier of each canopy or ventilated ceiling must fit a performance 'rating plate' to the system with the following information:

- supplier contact details;
- date of installation and serial number;
- the design supply and extract flow rates;
- design pressure drop across the canopy.

The rating plate will usually be fitted on the inside left end of the canopy.

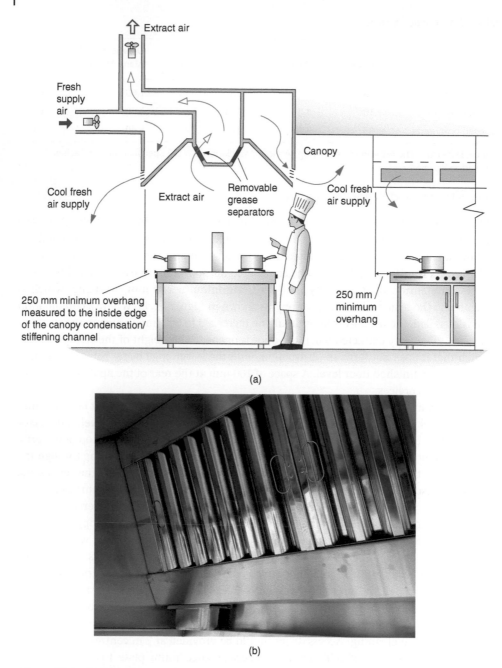

(a)

(b)

Figure 12.4 (a) Typical canopy layout. (b) Removable grease separators and grease collection draw

Commercial kitchen ventilation systems are designed so that the ductwork and kitchen remain under negative pressure; this is to prevent fumes and odours from flowing out of the kitchen and contaminating surrounding areas. Therefore, fan units should be positioned as near as possible to the discharge point to maintain the negative pressure within the ductwork. *Note*: Due to the nature of the negative environment created by a mechanical extract system, open-flued appliances or open fires must not be installed in the same or internally connected space as spillage would result.

In order to minimise vibration and noise transfer all fans need to incorporate anti-vibration mountings. The canopy ductwork shall discharge at a position where it will not cause a nuisance to adjoining properties or allow the extracted air to be drawn back into the supply system. The duct should terminate via a high-velocity terminal at a height determined by the local authority (Figure 12.5). It is important to note that the exhaust system serving wood-burning appliances must be separate from all other systems. It must not be connected to an extract system which serves gas and electrical equipment.

Figure 12.5 Terminal and ductwork with cleaning access panels visible

Make-Up Air

12.12 The ventilation system in commercial kitchens is designed to extract large volumes of excess hot air, odours, vapour and steam from cooking and washing activities. For the safety and comfort of those working in the kitchen, it is essential that an allowance is made for the provision of cool, clean replacement air. This 'make-up air' not only needs to replace the air extracted by the canopy, or ventilated ceiling, but also allow sufficient air for combustion.

Make-up air can be by means of a mechanical inlet supply or natural air infiltration. However, the preferred method is a fan powered mechanical system where the air is often introduced into the kitchen via the canopy, as shown in Figure 12.4a. All make-up air should come from a clean air source, and not from dirty locations such as waste storage areas, and have a minimum inlet temperature of 10 °C for canopies and 16 °C for ventilated ceilings. The system will need to be designed to provide between 75% and 95% of the total extracted air volume, with the remaining air being allowed to infiltrate from adjoining areas. This helps to maintain the necessary negative pressure required in commercial kitchens to ensure that cooking odours do not transfer to areas outside the kitchen.

Air for combustion forms a relatively small percentage of the overall make-up air. Therefore, the incoming air supply can be filtered. In fact, this is necessary in commercial kitchens in order to avoid contaminated air, dust and insects entering the catering environment. Synthetic filters with a minimum filtration efficiency of ePM1 70% to 75% shall be used (fibre glass filters must not be used). Dirty filters will clearly affect the amount of supply air and the balance of the ventilation system. Therefore, some form of visible or audible indication of the filter condition will need to be provided. Screens may be fitted to inlet louvres/grilles to prevent the entry of birds and vermin, but insect screens must not be used as these can easily become blocked. Make-up air must not be drawn through toilet areas or come from adjoining eating areas. Where a natural air duct is used, this should be positioned at high level, be as short as possible and also be filtered.

Carbon Dioxide (CO_2) in Commercial Kitchens

12.13   Carbon dioxide interlock systems aim to ensure a safe working environment where it is possible for combustion products to be emitted into the catering area. Where fitted, these systems should provide a visual and audible warning when the level of CO_2 rises above 2800 ppm (0.28%) and lockout the gas supply should the CO_2 level rise above 5000 ppm (0.50%). Only detectors that are designed for commercial catering equipment are suitable and must be installed and maintained by a competent person in accordance with the manufacturer's instructions.

BS 6173 requires that the level of CO_2 in the working environment of all catering establishments be checked during commissioning. Other occasions when ambient air testing must be carried out include:

- when fitting new gas or ventilation equipment or when carrying out maintenance on the equipment;
- when working on gas equipment in any catering establishment for the very first time;
- when it is suspected that the ventilation system is not working effectively, e.g. there have been complaints regarding high levels of heat and condensation etc. from staff.

The procedure for CO_2 ambient air testing described in Part 8: Section 8.52 can be used as a guide whilst taking note of the following:

- Carryout ambient air testing using test equipment conforming to BS EN 50543 with the ventilation system operating normally and with all appliances and burners operating at full rate without any utensils.
- Allow the atmosphere to settle, this may take at least 10 minutes for a small establishment and longer for a larger kitchen.

- Test the environment at three points within the cooking work area at a height of 2 m above floor level and record the average. Samples should not be taken close to air inlets. The level of CO_2 should not exceed 2800 ppm.
- Where levels are in excess of 2800 ppm endeavour to work with the owner/operatives to remedy the situation, where it is considered that the continued use of the catering equipment is unsafe, advise the responsible person in writing and follow the procedure set out in the Gas Industry Unsafe Situations Procedure (GIUSP).
  - *New installation*
    >2800 ppm CO_2 disconnect the appliance/s and label as uncommissioned.
  - *Existing installation*
    >2800–5000 ppm CO_2 apply GIUSP category **AT RISK (AR)**
    >5000 ppm CO_2 evacuate the premises and make safe the installation. Apply GIUSP category **IMMEDIATELY DANGEROUS (ID)**
- Where remedial work is undertaken, the test should be repeated.

Where carbon monoxide alarms are fitted, these should be suitable for the harsh environment of a commercial catering kitchen. Therefore, domestic models should not be used.

Example of Flow Rate Calculation Based on Thermal Convection Method

12.14   Using the details shown in Figure 12.6, determine the correct extract flow rate and the make-up air requirements for the kitchen illustrated. For the purposes of this exercise, the make-up air to be provided is mechanical and shall be calculated to provide 85% of the total extracted volume

The extract flow rate is calculated by completing the following steps, as shown in Table 12.4:

1. Begin by filling in column 1 with the appliance type.
2. Calculate the plan area in m^2 for each appliance and insert it in column 2.
3. Insert the coefficient for each appliance taken from Table 12.1 into column 3. *Note*: This will be determined by the power source. From Figure 12.6, we see that all appliances are gas-powered except the deep fat fryer, which is electric. A bench always has a coefficient of 0.03.
4. Multiply columns 2 and 3 for each appliance and enter the row total in column 4. This will give you the flow rate in m^3/s for each appliance.
5. Add together all the appliance flow rates in column 4 to give a theoretical extract flow rate.
6. To determine the specific extract air required, multiply the theoretical extract flow rate by the canopy factor, which is found in Table 12.3. In this example, the canopy is overhead wall closed on both ends, which has a factor of 1.15. You now have the specific extract flow rate.
7. Next, calculate the make-up air flow rate. In this example, the required supply flow rate is 85% of the specific extract rate. Therefore, multiply the specific extract rate by 0.85 to determine the supply make-up air rate in m^3/s.
8. Once the supply and extract air flow rates have been calculated, shown in the shaded boxes in Table 12.4, these would be compared with the data on the canopy rating plate

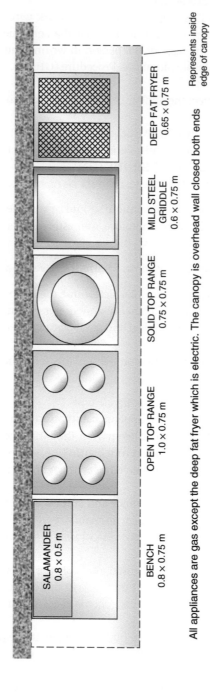

All appliances are gas except the deep fat fryer which is electric. The canopy is overhead wall closed both ends

Appliance	Dimensions
SALAMANDER	0.8 × 0.5 m
BENCH	0.8 × 0.75 m
OPEN TOP RANGE	1.0 × 0.75 m
SOLID TOP RANGE	0.75 × 0.75 m
MILD STEEL GRIDDLE	0.6 × 0.75 m
DEEP FAT FRYER	0.65 × 0.75 m

Represents inside edge of canopy

Figure 12.6   Appliance line up with plan sizes for extract flow rate calculation example

to confirm that the canopy design supply and extract flow rates meet those required by the installed appliances.

9. The performance of the ventilation system should then be confirmed by carrying out an ambient air test, see Section 12.13.
10. Where the correct flow rates cannot be confirmed or where there are other safety concerns, follow the guidance in the Gas Industry Unsafe Situations Procedure.

Table 12.4 Calculations for flow rates based on thermal convection method using appliance line up shown in Figure 12.6

1.	2.	3.	4.
Appliance type	**Appliance area (m²)**	**Coefficient**	**Flow rate (m³/s)**
Salamander	(0.80 × 0.50) **0.400**	× **0.75 =**	0.300
Bench	(0.80 × 0.75) **0.600**	× **0.03 =**	0.018
Open top range	(1.00 × 0.75) **0.750**	× **0.40 =**	0.300
Solid top range	(0.75 × 0.75) **0.563**	× **0.60 =**	0.338
Mild steel griddle	(0.60 × 0.75) **0.450**	× **0.30 =**	0.135
Deep fat fryer	(0.65 × 0.75) **0.488**	× **0.45 =**	0.220
	Theoretical extract flow rate		1.311
	Multiply theoretical rate by the canopy factor		× 1.15
	Specific extract flow rate required		**1.508**
	Supply make-up air @ 85%		**1.282**

Note 1: Although this example refers to a canopy, the extract rate for a ventilated ceiling follows the same procedure as that set out above.
Note 2: The preferred method for ventilation calculations is 'metres per second' (m/s). However, gas engineers will be more familiar with the use of 'metres per hour' (m/h). To convert m/s to m/h simply multiply by 3600.

Commercial Catering Appliances

Relevant ACS Qualification
COMCAT1 – 5

Relevant Industry Document
BS 6173, Gas Safe Technical Bulletin 044

12.15 Commercial catering covers a wide range of appliance types. An overview of some of the appliances that may be encountered is provided in this section. As always, each appliance must be located, installed and commissioned in accordance with the manufacturer's instructions.

Flame Failure Protection

12.16   Before installing either new or previously used appliances, you must first confirm all of the following

- the appliance has a CE or UKCA mark and a readable data plate;
- it is suitable for the gas in use;
- the manufacturer's installation instructions are available;
- that each burner is fitted with a flame supervision device (FSD), sometimes referred to as a flame failure device. *Note*: Where previously used appliances do not have FSDs fitted, upgrading of these controls must only be carried out using the appliance manufacturer's parts and in accordance with their instructions. Where this cannot be completed, the installation must not go ahead.

If appliances are encountered on existing installations that are not fitted with suitable safety devices, e.g. without FSDs, then engineers should advise the equipment owner/operatives to replace or upgrade the equipment. Where this is not practicable, engineers will need to risk assess the appliance in accordance with BS 6173 to determine if it can be used without constituting a danger; a risk assessment should also be carried out where an appliance is installed that does not carry a CE mark. Any unsafe situation identified should be classified in accordance with the current Gas Industry Unsafe Situation Procedure.

Where cooking appliances are installed in food technology classrooms, the installation must comply with IGEM/UP/11; see Part 13: Educational Establishments.

ACS Commercial Catering Appliance Categories

12.17   For the purposes of ACS assessments, commercial catering appliances are grouped in the categories of COMCAT1-5 as shown below. However, in this section, appliances are discussed according to type

COMCAT1 – Boiling Burners, Open and Solid Top (Atmospheric Burners)
   Appliance Range: stockpot stoves, hotplates, warming plates, forced and natural convection ovens, combination ovens, bains marie, hot cupboards, and ancillary equipment (e.g. tandoors, pizza ovens, heated woks etc.). Expansion-type water boilers, boiling pans, bulk liquid and jacketed urns, gas-fired dishwashers, boiling tables, gas-fired rinsing sinks.

COMCAT2 – Pressure Type Water Boilers (Atmospheric Burners)
   Appliance Range: pressure-type water boilers (stills or equivalent), pressure steamers, pressurised steaming ovens, and ancillary equipment.

COMCAT3 – Grills, Deep Fat and Pressure Fryers
   Appliance Range: deep fat fryers, pressure fryers, Bratt pans, griddles, over and under-fired grills, simulated charcoal grills, salamander grills.

COMCAT4 – Fish and Chip Frying Ranges
   These appliances are manufactured to the customer's specification and normally assembled on-site to meet requirements of the shop.

COMCAT 5 – Forced Draught Burners
   COMCAT1-3 appliances with forced draught burners

Appliance Types

General Purpose Boiling Tables, Ovens and Ranges

12.18   There is a large range of ovens and hotplates for commercial kitchens; the choice depends on the kind of cooking that is to be undertaken. The basic models are described here.

Boiling Table

12.19   The boiling table is simply a hob or hotplate mounted at a working height of around 800 mm. It may consist of an open-topped ring burner, or it may have a solid top, with the burner located below the metal surface plate, as shown in Figure 12.7. A storage shelf is often fitted below the cooking surface. A smaller version, between 450 mm and 600 mm high and with one or two burners, is referred to as a stockpot stove and is used to heat a large, heavy container.

Figure 12.7   A solid top oven showing the burner located below the metal surface plate

Ovens and Ranges

12.20   Commercial ovens work by means of natural convection, which provides a stratified heat distribution (i.e. hotter at the top) or forced convection. The forced convection oven uses a fan to provide even heat distribution throughout the oven.

The general-purpose oven, may be an integral or freestanding unit, single or tiered; it may also include an open-top hob and/or hotplate as part of its design, as shown in Figure 12.8.

An example of a freestanding range oven is shown in Figure 12.9. The method of heating the contents of the oven by the appliance burner is referred to as direct, indirect, or semi-indirect, as illustrated in Figure 12.10a–c. The three designs work as follows:

1. **Direct**: With this method of heating, the products of combustion are dispersed directly into the oven chamber with the burner flame usually visible at the back or side of the oven.
2. **Indirect**: With this design a gas burner, located outside the oven chamber, passes the hot combustion products around the outside shell and discharges them out through the flue. In a forced convection oven, a fan located in a section of the oven compartment circulates the warmed air emitted from the oven's inner walls.
3. **Semi-indirect**: With this design, the combustion products pass through the oven itself. The burner is not located inside the oven but passes the flue products through small holes in the side panels of the oven. The combustion products are drawn into the oven by a fan that creates a negative pressure. Owing to its design, with a throat restriction as shown in Figure 12.10c, some of the air is re-circulated via a route at the rear of the oven, while some is discharged through the flue.

Figure 12.8 Oven with open hob and solid top

Figure 12.9 An open-top range

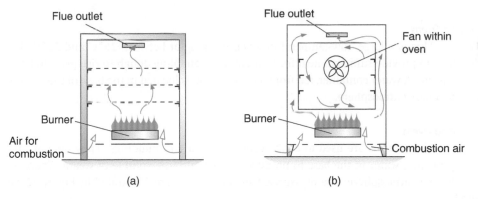

Figure 12.10 Examples of natural convection and forced convection ovens. (a) Natural convection direct (combustion products enter the oven); (b) forced convection indirect (combustion products do not enter the oven)

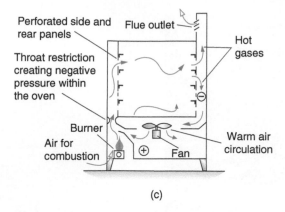

Perforated side and rear panels

Flue outlet

Hot gases

Throat restriction creating negative pressure within the oven

Burner

Air for combustion

Fan

Warm air circulation

(c)

Figure 12.10 (c) forced convection semi-indirect (combustion products indirectly enter the oven)

Specialist Ovens and Steamers

12.21 The ovens described here are found in specialist kitchens, depending on the type of cooking that is to be carried out.

Pastry Ovens

12.22 These are similar in design to the indirect forced convection ovens, previously described. The pastry oven is used for baking cakes and pastries and often has several levels or compartments between 125 mm and 300 mm in height stacked on top of each other, thereby comprising up to four ovens in a group. Each oven may have its own burner, or it may be constructed with a single burner located in the base, however, separate thermostats for each oven are provided. To change the conditions in an individual oven, vents in the doors are incorporated. These create a moist atmosphere when closed and a dryer heat when opened.

Proving Ovens

12.23 This is a large, low-temperature oven working at between 26 °C and 32 °C. It is designed to provide sufficient heat for fermentation, enabling dough to rise prior to bread making, etc. A water container, fed from a feed cistern, is located in the base of this oven to ensure a moist atmosphere.

Steaming Ovens

12.24 These ovens are used to cook vegetables, fish and puddings. They operate simply by heating water in the base to produce a constant large volume of steam. There are two designs: atmospheric and pressurised steaming ovens, as illustrated in Figure 12.11a and b.

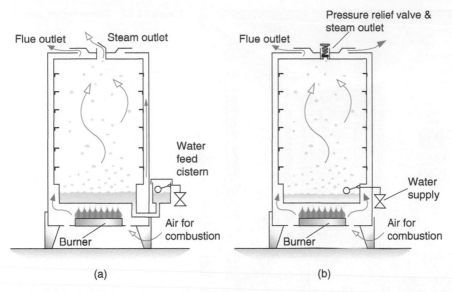

Figure 12.11 Steaming ovens. (a) Atmospheric steaming oven; (b) pressurised steaming oven

The Atmospheric Steaming Oven

12.25   These have a cold-fill cistern located adjacent to the oven. This maintains the water level in the oven for the purpose of heating. The top of the oven is open to allow the steam to pass out slowly and unrestricted.

The Pressurised Steaming Oven

12.26   Pressure steaming ovens incorporate a pressure relief valve located at the top of the oven and pressure locked doors. The door seals require inspection during routine service and damaged seals must be replaced in order to prevent steam escaping during operation, which could cause burns or scalding. Pressurised steaming ovens tend to cook food much faster than the atmospheric types. The water level is maintained by a float-operated valve located in the oven compartment itself.

Combination Ovens

12.27   A combination oven, as the name suggests, has the capability of providing a number of cooking functions all in one unit. This may include proving, steaming, baking, braising, as well as convection cooking. It can be programmed to change humidity levels and temperature to ensure the food is cooked to perfection. An example is shown in Figure 12.12.

Figure 12.12   Combination oven

Specialist Baking Ovens

12.28   There are many types of specialist baking ovens, each designed to produce a different effect on the cooking process. They can be quite large and may include many racks, rotating reels, or use a conveyor system to cook the food for a set period, according to the speed of the gears. These ovens are particularly suitable for large-scale or batch production (Figure 12.13).

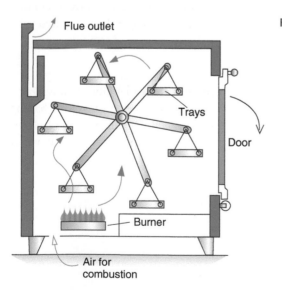

Figure 12.13   Rotating reel oven

Boiling Pans, Hot Cupboards and Bains Marie

Boiling Pan

12.29   These are large-capacity containers that are used to boil up large volumes of water, etc., for bulk cooking, typically holding up to 180 litres in volume. There are several designs, including those that allow for the contents to be directly heated by a burner located beneath a single cooking pot, and those that indirectly heat the pan contents by hot water or steam surrounding the pot; this type is often referred to as a jacketed pan. This indirect method of heating reduces the amount of sticking and burning that can occur when cooking foods such as custards, toffees and jams. Boiling pans often include a large diameter draw-off tap at their base to assist in draining the liquid contents. See Figure 12.14a and b.

Figure 12.14   Boiling pans.
(a) Directly heated boiling pan;
(b) jacketed boiling pan

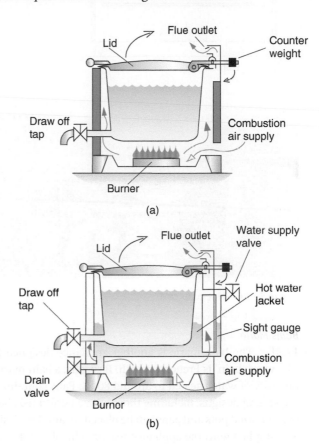

Hot Cupboards

12.30   These are basically insulated cabinets that are designed to keep previously cooked food warm, or they may be used to warm plates. The heat input will depend on their size and the temperature required. 60 °C may be needed to warm plates, whereas a higher temperature of around 80 °C to keep food warm. Two methods are used to warm the cupboard: one uses a burner located inside the base of the cabinet area so that the contents are warmed directly by the hot gases as they pass through to the flue outlet. The other design indirectly warms the cabinet by circulating a flow of hot gases or steam, generated in a water compartment below the base of the cabinet, around through channels inside the unit. Cupboards that are indirectly heated tend to be more humid and have the advantage of ensuring that the food does not dry unduly. Examples are shown in Figures 12.15 and 12.16.

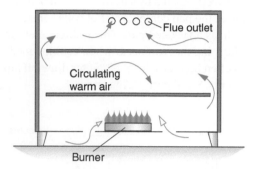

Figure 12.15 Directly heated hot cupboard

Figure 12.16 Hot cupboard

Bains Marie

12.31 The bain marie is another appliance designed to keep food hot prior to serving. The word 'bain' is French for bath. Basically, a bain marie is a trough-like unit that holds a quantity of water heated to around 80 °C. Pans containing the food are placed in it. There are several designs, including those with a perforated shelf, below the water line, on which various sized pots and pans can be placed. In another design, special serving containers that are supplied with the appliance rest on a lip above the water line. There is also a dry heat version in which no water is used, and all the heat is generated from the hot gases of the burner. Various versions are shown in Figures 12.17 and 12.18.

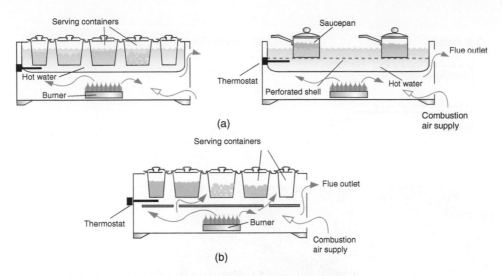

Figure 12.17 Bains Marie. (a) Wet type bain marie; (b) dry type bain marie

Figure 12.18 Wet type bain marie

Grills, Griddles and Fryers

Grills

12.32 *Under-fired grills*, commonly referred to as *char grills* or *flare grills*, give a heat source below the food. The grill consists of a burner firing upwards onto a bed of refractory material such as lava or pumice. The food is placed on a series of bars. As the hot fat from the food drops, it is burnt off in the form of an 'uncontrolled' flame, giving the meat a characteristic appearance and flavour. For examples, see Figures 12.19 and 12.20.

Horizontally fired grills use surface combustion plaques and are often accompanied by a rotisserie to cook foods such as kebabs and chicken. The fat from the food can drop to be collected in a container at the base.

Over-fired grills or *salamanders*, as shown in Figures 12.21 and 12.22, have a heat source above the food. They can also be used for toasting. This type of grill sometimes has a grooved solid aluminium branding plate on which the food is placed. This plate allows the hot fat to drain away, but at the same time, assists in cooking the food on both sides due to the absorbed heat that is held in the aluminium plate.

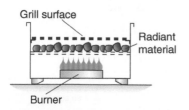

Grill surface

Radiant material

Burner

Figure 12.19   Under-fired or flare grill

Figure 12.20   Underfired-grill, open hob burners and solid top

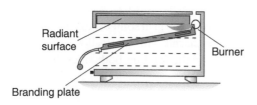

Radiant surface

Burner

Branding plate

Figure 12.21   Over-fired or salamander grill

Figure 12.22 Adjustable salamander grill

Griddles

12.33 The griddle is effectively a hotplate that is heated from below by burners to give a surface temperature of up to 300 °C. It is basically used for dry frying, and as such, it is sometimes called a dry fry plate. A channel is often incorporated to one side to enable the fats from certain foods to be drained away. As with the grill, these appliances are often installed around the counter area of some types of snack bar. Figures 12.23 and 12.24 show examples.

Figure 12.23 Griddle plate

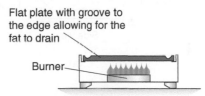

Flat plate with groove to
the edge allowing for the
fat to drain

Burner

Deep Fryers

12.34 These are appliances that hold a quantity of oil or fat in which food, such as chips, fish and doughnuts are cooked. There are two different designs: one that has a flat bottom with the oil heated directly from below, and a second type that has oil that is heated at a location above the base. This second type maintains a cool zone below the flame in which old food particles can sink without carbonising and spoiling the flavour of the food; there are two methods of heating the oil using this design of fryer. In the first method there is what is known as a V-shaped fryer, with the lower V section below the burner. The second method requires the gas flame to be fired directly into tubes that pass through the mass of the oil (for examples, see Figure 12.25a–c). With the flat bottom fryer, it is essential that, before each frying, all old particles are strained from the oil (Figure 12.26).

Figure 12.24 Char grills and griddle plate

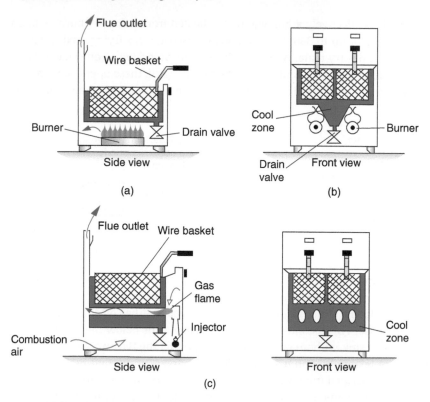

Figure 12.25 A range of deep fryers. (a) Flat bottom fryer. (b) 'V' pan fryer. (c) tube-fired fryer

Figure 12.26   Deep fat fryer

Tilting Fryers and Brat Pans

12.35   These are shallow flat-bottomed appliances used for shallow frying or dry frying. They have a tilting function to enable the surface to be tilted forward by the operation of a lever or hand wheel, see Figure 12.27.

Drinking Water Boilers

12.36   Several different designs of water boilers will be found providing hot water for drinking purposes. These range from the simple urn to the café pressure boiler. Some of these boilers are described below.

The Urn

12.37   The simple urn consists of a container with a lid into which water is poured from a jug. Beneath is a gas burner that simply heats the water. The urn has limited controls and just allows the water to boil away. A more sophisticated version has many extras, including a water fill connection with sight glass and a means of thermostatic control, as illustrated in Figure 12.28. With the urn, water is drawn off from the bottom of the container, and so the water is only hot enough when the whole volume has been sufficiently heated. A variation on this is the jacketed urn, which consists of a container within a container. The inside

Figure 12.27 Brat pan

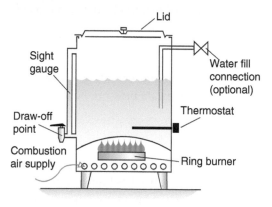

Figure 12.28 Bulk water urn

container contains a liquid that easily burns and sticks when heated, such as milk. The outside container is filled with water. This overcomes the problem of hot spots.

The Expansion Boiler

12.38 Unlike the urn, which requires all the water to be heated before it is ready for use, the expansion boiler draw-off point is located at a high point above the water level of the cold water. As water is heated, it expands, and it is only the boiling water that would reach this draw-off tap. The cold water is fed to the boiler from a cistern with a float-operated valve located at the back of the boiler. The expansion boiler is illustrated in Figure 12.29.

Figure 12.29 Expansion boiler

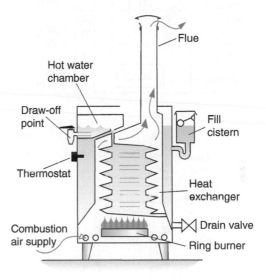

The Pressure Boiler (Café Steam Boiler)

12.39 This is a design of boiler that produces both steam and hot water. The boiler is often located beneath the countertop, and the hot water is forced up to the draw-off point by the steam pressure contained in the boiler. Cold water enters the boiler and stops when the water level reaches a set height, leaving a void at the top of the boiler. As the water is heated and boils, the steam generated cannot escape, as with the two previous boilers, but accumulates in the top void. As the pressure of the steam begins to build, it acts on the pressure stat, which in turn closes down the gas supply. Nothing will happen if the hot draw-off tap or steam outlet is opened prior to the production of steam. However, if there is steam pressure in excess of atmospheric pressure, opening the hot draw-off tap will cause the steam to force the boiling water up the discharge tube and discharge from the outlet. Conversely, if the steam outlet is opened, it too would discharge from its outlet. Both a pressure relief valve and high limit stat, interlocked with a thermocouple interrupter, are incorporated with this type of boiler. Often, these boilers have an adjoining side urn for heating milk and a coffee percolator, named a café set. The control levers of the gas and water supplies to these boilers are connected together, referred to as a safety interlock. This prevents one supply from being open without the

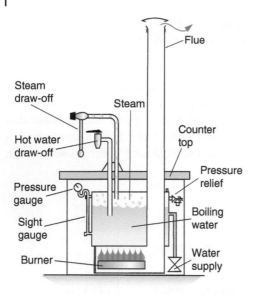

Figure 12.30 Pressure boiler

Flue

Steam draw-off

Steam

Counter top

Hot water draw-off

Pressure relief

Pressure gauge

Sight gauge

Boiling water

Burner

Water supply

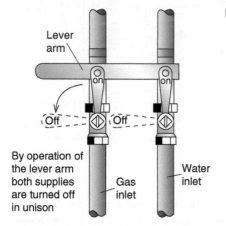

Figure 12.31 Safety interlock

Lever arm

By operation of the lever arm both supplies are turned off in unison

Gas inlet

Water inlet

other. The pressure boiler is depicted in Figure 12.30, and a safety interlock can be seen in Figure 12.31.

Mobile Catering Units

Relevant ACS Qualification
CMC

Relevant Industry Document
UKLPG COP24 Part 3

12.40 The mobile catering unit could be a motor vehicle, trailer or caravan that has been converted or it may be purposely designed and factory-produced. There are many designs, including the burger or hot dog van, fish and chip vehicle and the mobile café (Figure 12.32). These vehicles and the gas appliances contained within them are covered by the Gas Regulations and all work undertaken must be completed by a Gas Safe registered

(a)   (b)

Figure 12.32   (a) Mobile catering unit. (b) Mobile catering vehicle

engineer with the appropriate mobile catering qualification. Although changing the gas cylinders is not considered as gas work, all joints must be checked for leaks using a suitable leak detection fluid prior to use.

Cylinder Carriage and Location

12.41   When transporting gas cylinders to the location in a closed vehicle, there are specific considerations, as described in Part 4: Section 4.56. It is best to site the cylinders on the nearside of the vehicle to reduce the risk of damage in the event of a road accident. The cylinders must be transported in a suitably designed compartment which is separated and gas-tight from the interior of the vehicle, of a fire-resistant construction providing a minimum 30-minute fire resistance and ventilated to the exterior. The size of the ventilation needs to be a minimum of 1% of the compartment floor area and sited at high and low levels with a minimum of 5000 mm^2 (50 cm^2) at each level. The low-level vent should be in the floor or at the bottom of the wall. There must be no fixed sources of ignition within 1 m measured horizontally from the compartment and 0.3 m above the highest vent. The compartment must not be within 1 m of the vehicle exhaust system. Once on site, the cylinders should be located in the outside air at least 1 m from any opening into the vehicle or sources of ignition, and no closer than 2 m when measured horizontally to any opening into a cellar, un-trapped drain or gully. A sign should also be positioned warning of the presence and danger of gas.

Pipework

12.42   Where an automatic changeover device supplies a constant gas supply to the vehicle, an emergency control valve needs to be located next to the vehicle entrance, along with the appropriate warning notice identifying what to do in the event of a gas escape. For single-cylinder installations, the cylinder valve is sufficient. Pipework would generally follow the principles and guidance given in BS EN 1949 (leisure accommodation vehicles), with additional protection provided for pipework fixed below the vehicle to overcome the increased likelihood of damage, such as by stones being 'thrown up' while the vehicle is moving. See also the notes specific to caravans and motor caravans in Part 14.

Appliances

12.43   All appliances must be installed according to manufacturer's instructions, comply with the Gas Regulations and each appliance burner must be fitted with a flame supervision device. Appliances should be secured in position and sited so that they do not obstruct the passageway or where they might create a fire hazard. It is essential that no appliance is ignited while the vehicle is in motion and that the gas supply is also turned off at the cylinder valve when the vehicle is to be moved. Vehicles that carry deep fryers should include a canopy that incorporates a flue to the outside. Any flue needs to be capable of withstanding the vibration of the vehicle movement and the discharge needs to be in a safe place outside the vehicle at a high level and away from any openings into the vehicle. The canopy should extend at least 150 mm beyond the appliance on all sides and have a minimum of 270 mm^2 of flue area for every 10 000 mm^2 of canopy base area.

Ventilation

12.44   The minimum ventilation openings to a mobile unit should be at least 2 500 mm^2 for every 1 kW heat input of all the combined appliances, or 10 000 mm^2, whichever is the greater. This ventilation is then divided equally between high and low levels, with the low-level ventilator grille being positioned in the floor or at the bottom of the wall. This ventilation is in addition to that provided through the serving hatch and windows.

Visual Inspection and Maintenance

12.45   Every time the vehicle is used, the cylinders, pipework, appliances and flues should be inspected for obvious signs of damage. It is recommended that the installation is inspected every six months to ensure the continued safety of the vehicle. Appliances should be maintained in line with the manufacturer's instructions, and all gas work should be carried out by a suitably qualified Gas Safe registered engineer at least every 12 months. Where appliances and flues are frequently used, it is recommended that servicing takes place at six monthly intervals. The tightness of the gas system, which is subjected to the continued movement brought about by travel, should also be checked at regular intervals.

Carbon Monoxide and Carbon Dioxide

12.46   A carbon monoxide (CO) alarm conforming to BS EN 50291 should be fitted in every mobile unit. Operators should also be aware of the symptoms of CO poisoning, and if they suffer from headaches, nausea and dizziness when the appliances are running, the appliances should be isolated and checked by a Gas Safe registered engineer with the appropriate mobile catering qualifications. Carbon dioxide (CO_2) detection systems are also necessary in mobile catering vehicles and must be installed to manufacturer's instructions. It should not be possible to operate the appliances without the serving and ventilation hatches being open. Therefore, serving hatches must be interlocked to the gas supply to prevent appliances from being used with the hatch closed.

13

Educational Establishments

Educational Establishments

Relevant ACS Qualification
COCN1

Relevant Industry Document
IGEM/UP/11

13.1   Educational establishments include schools, colleges, universities and training facilities, both state-run and private. However, training facilities for gas operatives undertaking their competency training and assessment are not within the scope of IGEM/UP/11.

General Considerations

13.2   The design, installation and maintenance of gas systems and appliances in educational establishments will generally follow the requirements of other industry standards and codes of practice outlined in this book. However, there are some additional and specific requirements that need to be met to ensure the safety of young, inexperienced or unqualified persons who will, at times, have access to gas installations and equipment during learning activities. In addition, some areas within educational establishments may well be at a higher risk of unauthorised entry or interference and the risk of such incidents will need to be considered.

Gas Pipework

13.3   Generally, the gas system operating pressures shall not exceed 50 mbar. If there are special circumstances where higher operating pressures are required, such as where experimental equipment requires higher operating pressure or where LPG cartridges are to be used, a full risk assessment must be carried out in line with IGEM/UP/16 and IGEM/G/7. All pipework will need to be suitably protected where there is a foreseeable

Gas Installation Technology, Third Edition. Andrew S. Burcham, Stephen J. Denney and Roy D. Treloar.
© 2024 John Wiley & Sons Ltd. Published 2024 by John Wiley & Sons Ltd.

risk of damage and interference; concealed or steel pipework may be the best option in exposed areas. Where appliances use gas and air/oxygen under pressure, such as brazing hearths, a non-return valve will need to be fitted upstream of the equipment.

13.4 Moveable appliances are required to be connected to the installation pipework with a flexible connection and a self-sealing swivel plug/socket, which conforms to BS EN 15069. For domestic cooking appliances in food technology classrooms, the flexible hose should comply with BS 669-1; for commercial and domestic equipment in a catering setting, BS 669-2 or a flexible hose complying with BS EN 14800 as appropriate. Flexible appliance connections must not exceed 1.4 m in length and must be protected with a stainless steel restraint. The fixing point for restraints on overhead booms will need to be as high as possible to prevent hazards caused by hanging loops.

13.5 Lightweight equipment, such as Bunsen burners, small boiling rings or hand-held gas torches, are often used in teaching areas. These are typically connected to the gas supply via a bench-top laboratory gas tap and a length of flexible rubber tube, as shown in Figure 13.1. In these circumstances, the tubing must be regularly inspected for physical damage and replaced every five years, if not before, and be as short as practicable without

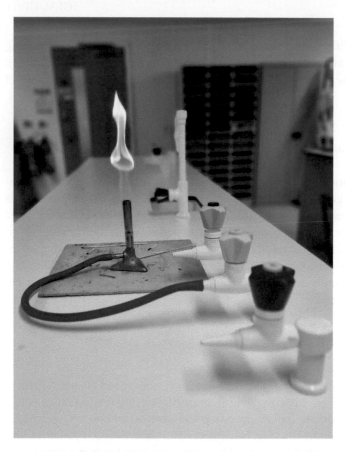

Figure 13.1 Bunsen burner and a science laboratory gas tap

exceeding 1 m in length. Laboratory gas taps should be fitted above the work surface and not in a position where they could be subject to accidental damage or cause injury, such as on the side of a bench. Adequate ventilation must be provided where the gas supply is run inside a bench. For every 1 m^3 of internal bench volume, ventilation of 20 cm^2 at both high and low levels will be required.

13.6 Where a corrugated stainless steel tube (CSST) is used to supply a number of areas within the establishment, each run shall be easily identifiable so that each pipe can be distinguished from another. This can be achieved by permanently marking each 'run' of pipework along its length. CSST must not be used for the final connection on moveable gas appliances.

Isolation Valves and Interlocks

13.7 In addition to any emergency control valve (ECV) provided by the Gas Transporter, an additional emergency control valve (AECV) shall be fitted in a readily accessible position and as near as is practicable to the point of entry to each teaching/preparation area. Where this is not possible, or where the gas needs to be interlocked to other systems for safety reasons, an automatic isolation valve (AIV) complying with BS EN 161 must be fitted. Interlocks are used to ensure that the environment remains at a safe and comfortable level for all occupants when gas appliances are used. Interlocks are typically connected to supply and extract ventilation systems and environmental systems that control the level of CO_2, heat, humidity, etc. When a predetermined level is exceeded, the interlock shuts the gas supply to the appliances. Where a teaching area has burners that are not fitted with flame failure detection, the AIV must incorporate an automatic proving system which will only allow gas to flow if all downstream gas taps/burners are closed. Examples are shown in Figures 13.2 and 13.3. The proving system must not supply more than one teaching area. Figure 13.4 shows an example of a gas safety interlock control panel.

Figure 13.2 Weep bypass proving system

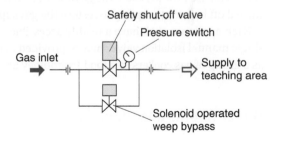

Emergency stop buttons must be fitted near the teacher's desk or board and adjacent to the main light switch for the room or at the point of exit. It is recommended that proving systems are tested for operation by pressing the stop buttons to isolate the gas at the end of each working day. A suitably worded notice explaining what to do in the event of an emergency will need to be affixed at locations where manual isolation valves are fitted, where emergency stop buttons can be operated, or where AIVs can be reset. The notice

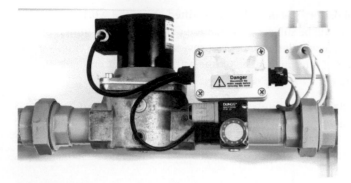

Figure 13.3 Safety shut off with integrated proving system

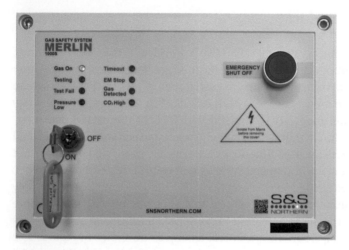

Figure 13.4 Gas safety interlock control panel

should advise how to isolate the gas and that all burners, gas taps and appliances should be turned off before attempting to restore the gas supply after shutting off.

Each appliance shall have a readily accessible manual gas isolation valve. An additional single manual isolation valve shall be provided to isolate each group of gas appliances, such as a group of gas cookers in a food technology area.

Boiler Rooms and Plant Rooms

13.8   It is recommended that boiler/plant rooms in educational establishments be treated as high-risk areas due to the potential of unauthorised entry or interference. Therefore, in addition to the requirements set out in IGEM/UP/10, further safety precautions are required for all new installations or when existing boiler/plant rooms are upgraded or refurbished. An automatic shut-off system should be installed to isolate the supply to the installed equipment in the event of a fire. The system should include emergency stop buttons positioned at the entrance/exit of the plant room, and the location of the ECV should be prominently displayed at the entrance to the boiler/plant room. To further reduce the risk, it would be appropriate to ensure that the boiler/plant room is securely locked and only accessible to approved individuals, pipework is welded steel, and a gas leak detection system is installed.

Where a risk assessment identifies that there is a risk of a gas leak in a boiler/plant room reaching a connected occupied area, a gas detection system must be fitted that will raise an alarm and shut down the gas supply where flammable gas is detected. Similarly, if there is a potential risk that products of combustion could reach the connected occupied area, a carbon monoxide (CO) system complying with BS EN 45544-3 should be installed, which will likewise raise an alarm and shut off the gas supply to the boiler/plant room equipment.

Ventilation and Flues

13.9 Purpose provided and permanent ventilation is necessary to dilute fumes and the vapours in teaching areas where gas appliances are used. The ventilation system must be designed to limit the concentrations of carbon dioxide (CO_2) in the environment to 2800 ppm; in food technology areas, the concentration of CO_2 is measured at a height of 1.5 m above floor level. Carbon dioxide interlock systems must provide a visual and audible warning when the level of CO_2 rises above 2800 ppm (0.28%) and lockout the gas supply should the CO_2 level rise above 5000 ppm (0.50%).

13.10 Canopies and extract ducts should be fitted above kilns to extract excess heat with the ducting exiting the building to outside air by the shortest possible route. Canopies and flue systems are required for Art, Craft, Design and Technology equipment where odours and potentially dangerous gases can be produced (brazing hearths, forges, furnaces, ceramic kilns, etc.). Canopies above domestic cooking appliances in food technology areas should be designed to provide an extraction flow rate in excess of 150 m^3/h per appliance. Guidance on the design and installation of ventilation and canopies for multiple domestic or commercial catering equipment in vocational training areas can be found in BS 6173, IGEM/UP/19, and the British Engineering Services Association publication DW/172 (see Part 12 for more details).

13.11 Ventilation systems in science laboratories and practical workspaces will need to be designed to dilute the combustion products of any appliance burners to an acceptable level. In most circumstances, this will be by means of a mechanical ventilation system, which will also deal with the pollutants that may be generated by science experiments and so forth. Sufficient incoming air will need to be supplied to compensate for the air extracted when using ducted fume cupboards (Figures 13.5–13.7).

13.12 Where a flue passes through an occupied space, the area shall be fitted with a CO detection system that complies with BS EN 45544-3.

Figure 13.5 Fume cupboard

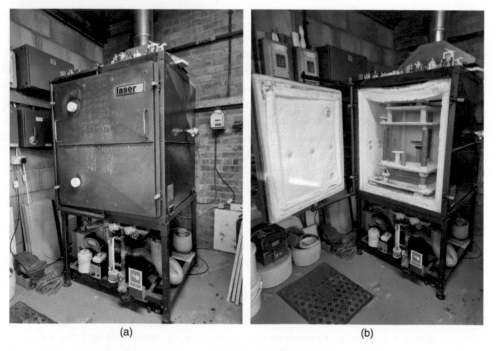

(a)   (b)

Figure 13.6 (a) Gas-fired kiln. (b) Gas-fired kiln (internal view)

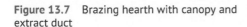

Figure 13.7 Brazing hearth with canopy and extract duct

Appliances

13.13 All appliances must be fitted in accordance with manufacturer's instructions and the relevant industry standards for the appliance type. Type A (flueless) and Type B (open-flued) appliances used for water and space heating must not be installed in teaching areas. Type C (room-sealed) appliances used for water and space heating can only be located in the teaching area if they are suitably protected from damage and unauthorised use. All appliance burners installed in food technology areas must be fitted with a flame supervision device, and all new appliances shall be CE or UKCA marked.

The electrical wiring to gas appliances shall be in accordance with BS 7671, with moveable appliances connected by means of plugs and sockets.

Commissioning, Testing and Maintenance

13.14 Gas work, as defined in the Gas Regulations, can only be performed by an appropriately qualified Gas Safe registered engineer. Maintenance on appliances, interlocks, mechanical ventilation systems, etc. shall be performed as necessary, but at least annually. The level of CO_2 should be checked regularly, and a simple handheld device can be used. However, when it is necessary to check ambient air concentrations during commissioning or maintenance visits, gas engineers should refer to BS 7967-5 for guidance (see also Part 8: Section 8.52).

14

Non-permanent Dwellings

Residential Park Homes

Relevant ACS Qualification
CCLP1 RPH

Relevant Industry Documents
BS 3632, BS 6891

14.1 The residential park home, although used for permanent residence, is regarded as a non-permanent dwelling. The structure is produced off-site, often in two halves and bolted together in situ, see Figure 14.1a and b. Basically, these homes are prefabricated bungalows, often with a pitched and tiled roof. They are manufactured fully fitted to include a kitchen/dining area, sitting room, bathroom and bedrooms. Generally, everything in the home is included, including all fixtures and fittings, plus carpets and soft furnishings. They are set on a concrete base and supplied with water, drainage, electricity and

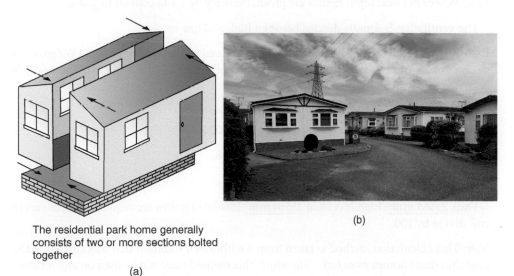

The residential park home generally
consists of two or more sections bolted
together

(a)

(b)

Figure 14.1 (a) The residential park home. (b) Examples of residential park homes

Gas Installation Technology, Third Edition. Andrew S. Burcham, Stephen J. Denney and Roy D. Treloar.
© 2024 John Wiley & Sons Ltd. Published 2024 by John Wiley & Sons Ltd.

gas connections. The gas supply requirements detailed in Section 14.8 for caravan holiday homes apply to many residential park homes. These units may be supplied with natural gas or LPG. The appliances found within this type of building are similar to those of any permanent dwelling. A flue and false chimney breast accommodate a flue box and non-combustible hearth to enable a gas fire to be installed. Alternatively, a caravan holiday home-type fire may be installed; see Section 14.10. Residential park homes may also require the use of drop vents in enclosed housings to prevent the accumulation of gas in low areas; see Section 14.9 where this is described.

Ventilation

14.2   As stated in Section 14.1, because the structure is regarded as non-permanent, it has different ventilation requirements to meet the needs of its structure from those of a permanent dwelling. The ventilation requirement for such buildings depends on the time of construction and is calculated using the following methods:

14.3   Ventilation for habitable areas in residential park homes built prior to 2005, which contain open-flued or flueless gas appliances are calculated using the following formula:

$$(2200G) + (440F) + (650P) + (1000R) = V$$

where

G is the gross kW heat input of all Type A appliances (flueless);
F is the gross kW heat input of all Type B appliances (open-flued);
P is the number of people for whom the home was designed;
R is the gross kW heat input of all open-flued oil-fired appliances and
V is the minimum area of fixed ventilation in mm^2.
Note: Where net heat input figures are given, multiply by 1.1 to convert to gross.

The ventilation is equally divided between high and low levels.

Example: A residential park home has an open-flued radiant gas fire of 5.2 kW (gross), a gas cooker of 12.7 kW (gross) and an open-flued oil boiler of 17 kW (gross). The home is designed for six people.
The ventilation size would need to be:

$$(2200 \times 12.7) + (440 \times 5.2) + (650 \times 6) + (1000 \times 17) = V$$

$$(27940) + (2288) + (3900) + (17000) = 51128$$

$$51128 \div 2 = 25564$$

Thus 25564 mm^2 high-level and 25564 mm^2 low-level grilles are required. To convert to cm^2 divide by 100.

Note: This calculation method is taken from a withdrawn standard that was current at the time that these homes were built. Therefore, this method must not be used on any residential park homes built from 2005 onwards, and the ventilation requirements must follow the manufacturer's instructions.

Background Ventilation Requirements

14.4   Although more a question of design than gas safety, background ventilation is also required to overcome condensation problems, etc., in this type of accommodation. This includes:

- *Bathrooms, shower rooms and WCs* that require 10 cm² at low level and 10 cm² at a high level.
- *Bedrooms and general living rooms* that require 10 cm² at low level and 20 cm² at high level. If the home is designed for more than four occupants, the following formula is to be applied: 6.5 cm²/person apportioned high and low levels with the high-level grille being twice the size of the low-level grille. Alternatively, in the case of a bedroom, a single 40 cm² grille can be positioned at high level in a window frame.

Another method for providing background ventilation utilising trickle vents installed ≥1.75 m above floor level can be adopted as an alternative to the above:

1. 80 cm² to all habitable rooms (e.g. lounge, bedrooms) and 40 cm² to all other rooms (e.g. bathrooms, kitchens, WC).
2. An average ventilation area of 60 cm² to all habitable areas, with bathrooms, kitchens and WCs having a minimum of 40 cm².

14.5   BS 3632 is the standard that manufacturers of new residential park homes will follow to complete their build. The various methods of background ventilation explored in BS 3632 include:

- Background ventilation with intermittent ventilation.
- Continuous mechanical ventilation.
- Mechanical ventilation with heat recovery (MVHR).
- Passive stack ventilation.
- Purge ventilation (openable window).

Leisure Accommodation Vehicles – General Requirements

Relevant ACS Qualification
CCLP1 LAV

Relevant Industry Documents
BS EN 721, BS EN 1949

14.6   Leisure accommodation vehicles are for temporary or seasonal use and fall into one of the following three categories:

- *caravan holiday homes,* delivered to a site by a transporter and manufactured with wheels solely for manoeuvring into place. These are often grouped together, forming a holiday site;
- *touring caravans*, towed by a vehicle;
- *motor caravans*, which are self-propelled.

Examples of these can be seen in Figure 14.2a–c.

Figure 14.2 (a) Caravan holiday home. (b) Touring caravan. (c) Motorcaravan

Leisure accommodation vehicles are supplied fully fitted out and, depending on their size, include kitchen/dining area, sitting area, bedrooms and often a flushing toilet and shower facility. Everything is included, even carpets and soft furnishings. An optional extra in many of the modern units is a central heating system. Connections to this type of caravan form part of the construction and include water, drainage, electricity and gas (LPG), which can be linked temporarily to the services provided by the park operator. If the unit is for hire, then it would be subject to the landlord's gas safety inspections and records, described in Part 8: Section 8.14, as required by the law.

Ventilation Requirements

14.7　Prior to 1999, ventilation sizing in caravans was undertaken using a different formula from that of today. For the static caravan holiday home, the calculation used was the same as that currently used in boats and non-permanent dwellings. If the caravan was built before 1999, this method of ventilation should be applied, and it is described in Section 14.3. Also, where an open-flued appliance, other than a refrigerator, is found to be located within an enclosed space, such as a cupboard, then a minimum ventilation is required. Ventilation openings shall provide 5 cm^2/kW gross heat input at both high and low levels.

However, in order to standardise with Europe, a ventilation method is now used based on BS EN 721, which takes into account the floor plan of the vehicle:

- Where non-room-sealed appliances are installed in *caravans and motor caravans*, safety ventilation is based on the overall floor plan area of the vehicle itself, as shown in Table 14.1a.
- Where non-room-sealed appliances are installed in *caravan holiday homes*, the ventilation size is based on the area of the habitable compartment, as shown in Table 14.1b.
- Even when no gas appliances are installed in a leisure accommodation vehicle, safety ventilation is still necessary to provide renewal air for occupants. Therefore, where a compartment within a leisure accommodation vehicle has no gas appliances fitted or only room-sealed appliances are installed, safety ventilation is as shown in Table 14.2.

Table 14.1a Habitable compartment safety ventilation in caravans and motor caravans containing non-room-sealed appliances

Overall plan area of vehicle	Low-level vent (minimum)	High-level vent (minimum)	
		If in roof	If in walls
≤5 m²	10 cm² (1000 mm²)	75 cm² (7500 mm²)	150 cm² (15000 mm²)
>5 – ≤10 m²	15 cm² (1500 mm²)	100 cm² (10000 mm²)	200 cm² (20000 mm²)
>10 – ≤15 m²	20 cm² (2000 mm²)	125 cm² (12500 mm²)	250 cm² (25000 mm²)
>15 – ≤20 m²	30 cm² (3000 mm²)	150 cm² (15000 mm²)	300 cm² (30000 mm²)
>20 m²	50 cm² (5000 mm²)	200 cm² (20000 mm²)	400 cm² (40000 mm²)

Table 14.1b Habitable compartment safety ventilation in caravan holiday homes containing non-room-sealed appliances

Area of habitable compartment	Low-level vent (minimum)	High-level vent (minimum)	
		If in roof	If in walls
≤5 m²	10 cm² (1000 mm²)	75 cm² (7500 mm²)	150 cm² (15000 mm²)
>5 – ≤10 m²	15 cm² (1500 mm²)	100 cm² (10000 mm²)	200 cm² (20000 mm²)
>10 – ≤15 m²	20 cm² (2000 mm²)	125 cm² (12500 mm²)	250 cm² (25000 mm²)
>15 – ≤20 m²	30 cm² (3000 mm²)	150 cm² (15000 mm²)	300 cm² (30000 mm²)
>20 m²	50 cm² (5000 mm²)	200 cm² (20000 mm²)	400 cm² (40000 mm²)

Table 14.2 Safety ventilation for habitable compartments in leisure accommodation vehicles with no gas appliances fitted or only room-sealed appliances installed

Area of habitable compartment	Low-level vent (minimum)	High-level vent (minimum)	
		If in roof	If in walls
≤10 m²	5 cm² (500 mm²)	15 cm² (1500 mm²)	30 cm² (3000 mm²)
>10 m²	10 cm² (1000 mm²)	75 cm² (7500 mm²)	150 cm² (15 000 mm²)

The ventilation openings should:

- be located more than 300 mm from any terminal;
- be protected by an accessible grille that can be easily cleaned;
- where high-level wall vented, be at least 1750 mm above the floor level and ≥300 mm above the surface of high-level bunk bed mattresses;
- where low-level wall vented, be no more than 100 mm above the floor;
- not be affected by drapes or curtains covering them;
- where ventilation is to pass through a bed box or cupboard, etc., it must not be affected by stored items and, where necessary, be ducted;
- comply with the manufacturer's instructions.

Where a refrigerator has been installed within a cupboard, an additional 13 cm² of effective ventilation with a mesh size of 6–9 mm should be provided on the floor of the vehicle to help stabilise the flame and disperse any gas in the event of a leak, see Section 14.9.

Leisure Accommodation Vehicles – Caravan Holiday Homes

Relevant ACS Qualification
CCLP1 LAV

Relevant Industry Documents
BS EN 721, BS EN 1949, BS 6891

Gas Supply

14.8   LPG is traditionally used to supply a caravan holiday home from a permanently metered supply or from cylinders. The regulator is often mounted on a purpose-built bracket, providing stability, and the supply is taken to the van through a hose, conforming to BS 3212: type 1 or BS EN 16436: Class 1, which has a braided stainless steel outside covering to protect against rodent attack. All LPG hoses should be replaced every ten years or earlier if they show signs of damage or deterioration.

If the caravan is not fully anchored down, and there is a possible risk of movement due to high winds, a quick-release self-sealing coupling or a low-pressure cut-off device should be incorporated. The length of this hose shall be ≥300 mm and kept to the minimum possible length without exceeding 2000 mm. However, in areas at risk of flooding, caravans are having floatation devices installed that anchor the home to the ground and concertina upwards if the water level rises. This enables the home to float on the surface of the water. In this instance, it is necessary to ensure that the length of the hose is sufficient to accommodate the maximum extension of the floatation device.

In static caravans, installation pipework and pressures should follow the same general rules as that for permanent dwellings. Although, it should be noted that the working pressure for road vehicles in Great Britain is 30 mbar; see Part 3: Table 3.2 for regulator set pressures.

Drop Vents (Gas Dispersal Hole)

14.9   To prevent the accumulation of gas in base units and cupboards where unburnt gas may have accidentally been released, holes a minimum size of 13 cm^2, with a mesh size of 6–9 mm, are positioned on the floor to allow it to escape. These may also form part of the low-level ventilation. It is essential that they are designed and positioned so that they do not get covered and that the gas can escape from beneath the caravan itself, see Figure 14.3, where there is a gas dispersal hole beneath the fridge.

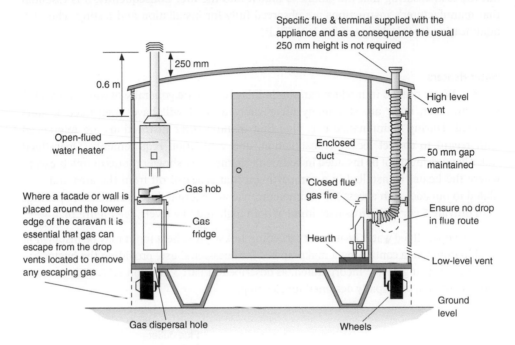

Figure 14.3   Caravan holiday home

Specific Appliances

Closed Flue Gas Fires

14.10   In general, the appliances used in a caravan are of the same design as those found in the permanent home, and for this reason the same ACS assessments and appliance designs that have been previously described apply. However, there is one specific design of gas fire, shown in the illustration of a caravan holiday home in Figure 14.3, that incorporates a continuous 63 mm flexible flue pipe that joins the fire to the terminal. This is located within a concealed part of the van. It is essential that this void is adequately ventilated at the top and bottom with grilles, sized to manufacturer's specifications, to provide air for cooling purposes. Access to the void also needs to be allowed for, and during safety checks, this void must be inspected.

The flue needs to be securely fixed and supported, as per manufacturer's instructions, ensuring that it remains clear of combustible surfaces and passes continually upward to the terminal with no dip in the flue. The combustion ventilation requirements have previously been described in Section 14.7, where the ventilation is calculated in the same way as for an open-flued appliance. Furthermore, it must be noted that the method of testing for spillage differs from conventional gas fires; the test involves closing all doors and windows, lighting the fire for a period of 10 minutes to warm up and then turning off the appliance to carry out the test, ensuring that the smoke is drawn into the fire. Consequently, it is essential that manufacturer's instructions are observed fully for installation and testing. The ACS qualification for this design of fire is HTRLP2.

Water Heaters

14.11   Water heaters in modern caravan holiday homes are generally of the room-sealed type, however, there are still many older caravans that will have open-flued heaters installed. During maintenance it is vital that manufacturer instructions are referenced with regard to correct location, installation, and ventilation requirements. Open-flued water heaters must not be installed in bedrooms, bathrooms, shower rooms or toilets except where the heater is installed in a compartment that is sealed off from the area and ventilated to outside, with access for maintenance via an external door. Ventilation openings shall provide 5 cm^2/kW gross heat input at both high and low levels, see Figure 14.4.

Note: An open-flued water heater not exceeding 14 kW gross heat input may be installed in a habitational area containing a make-up bed for occasional use, providing the heater has a vitiation sensing device incorporated, as described in Part 3: Section 3.57. This does not apply to an area principally designed for sleeping.

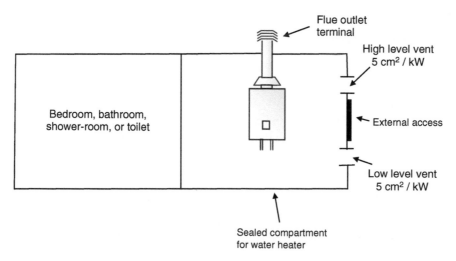

Figure 14.4   Open-flued water heater in a sealed compartment

Cookers

14.12   A leisure accommodation vehicle requires a permanent notice to be fixed adjacent to cooking appliances advising on use; see Figure 14.5. The heading shall have lettering at least 6 mm high in red capitals, and the remaining text shall be black and at least 3 mm in height.

Figure 14.5   A permanent notice to be fixed adjacent to cooking appliances advising on use

⚠️ **WARNING**

When you are cooking, it is essential to provide additional ventilation, such as opening windows near the grill, cooker, and oven. Do not use cooking appliances for space heating.

Leisure Accommodation Vehicles – Caravans and Motor Caravans

Relevant ACS Qualification
CCLP1 LAV

Relevant Industry Documents
BS EN 721, BS EN 1949

Gas Supply

14.13   A supply regulator must be fitted, allowing a fixed operating pressure of 30 mbar with an overpressure shut-off device capable of ensuring that a pressure of 150 mbar is not supplied to any appliance. Installation pipework with capillary joints must employ hard solder. The gas supply to this type of vehicle is from small LPG gas cylinders carried on board. The cylinders are either positioned externally within a frame or contained within an enclosed housing.

The *compartment* must:

- only be accessible from outside the vehicle (with the exception as detailed in Section 14.14);
- be sealed totally from the internal area;
- provide a securing device to hold the cylinders in an upright position;
- allow sufficient access to connections and for replacement of the cylinders;
- ensure hot surfaces, e.g. exhaust pipes, are at least 250 mm from the side wall and 300 mm beneath unless a thermal shield is included, as shown in Figure 14.6, with a 25 mm air gap;
- housings are to have ventilation direct to the outside. This may be at low level only, to a minimum size of 2% of the floor area of the enclosure, with a minimum of 100 cm^2 *or* high- and low-level ventilation can be provided of not less than 1% of the floor area with a minimum of 50 cm^2.

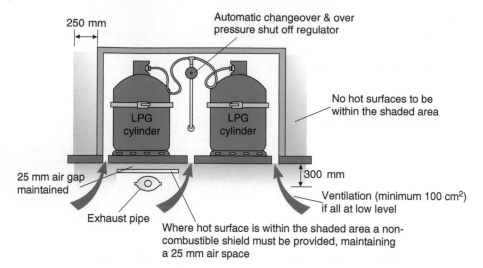

250 mm

Automatic changeover & over
pressure shut off regulator

LPG
cylinder

LPG
cylinder

No hot surfaces to be
within the shaded area

25 mm air gap
maintained

300 mm

Ventilation (minimum 100 cm²)
if all at low level

Exhaust pipe

Where hot surface is within the shaded area a non-
combustible shield must be provided, maintaining
a 25 mm air space

Figure 14.6   Gas storage compartment

14.14   Some road vehicles have type-approved base vehicle bodywork, which would require penetration to provide external access for a cylinder compartment. Therefore, internal access is permitted on these occasions, provided the following conditions are met:

- Must have a sealed access, with its bottom edge at least 50 mm above the floor.
- Can contain no more than two 11 kg sized cylinders, and the ventilation provided must comply with that stated in Section 14.13

Note: If a duct is attached to the ventilation opening, the maximum length of the duct can be no more than five times the diameter of the duct, it must fall continuously to outside. The maximum length of the duct can be increased to ten times the duct diameter in order to avoid flue outlets, if necessary.

Specific Appliances

14.15   The appliances used within a motorised or touring caravan are quite specialist and different from the appliances found in all other types of non-permanent and permanent dwellings. Some smaller boats make use of some of these appliances. Space heaters, water heaters and gas refrigerators must be of a room-sealed design. Cookers and gas lights will be flueless. Many of the larger caravan holiday homes are no longer fitted with gas lights, however, for touring vehicles this may not be the case, as many vans have this method of lighting in addition to electrical lamps. The installation of any appliance must carefully follow the manufacturer's instructions. Where flueing below the floor of a vehicle, care needs to be taken to ensure that the outlet is not within an area, such as within the chassis, where air is drawn and the products could be pulled back into the vehicle. Also, terminals should not be positioned within 500 mm of the refuelling point. The warm air heaters shown in Figure 14.7 are typical. A concentric flue/ventilation duct to the outside is made to the connection at the top. The unit shown is also suitable for installation in a boat and can be mounted in almost any position, as the flue does not necessarily need to come from the top. A warm air and hot water combination unit that may also be installed in this type of caravan is shown in Figure 14.12.

If a heating system or other appliances are intended for use while the vehicle is moving, a means to prevent the uncontrolled release of LPG shall be incorporated into the design of the gas system. Currently, there are two alternative methods available, these are:

- an excess flow valve which closes when there is an excess flow of LPG;
- a crash sensor which closes the LPG supply when there is an abnormal acceleration or deceleration of the vehicle, such as when an accident occurs.

The ventilation requirements have previously been described in Section 14.7.

Figure 14.7   Warm air heater. Source: Truma Ltd

Gas Supply on Boats

Relevant ACS Qualification
CCLP1 B

Relevant Industry Documents
IGEM/G/6, BS EN ISO 10239 and PD 54823

Shore Fed Gas Supply

14.16   This type of supply would only be considered for a permanently moored boat. The boat would need to be adequately secured to prevent any movement apart from that expected due to tidal variations, flooding conditions and possible drainage of canals or due to drought. The method of gas connection can be achieved by one of the following two methods:

1. by using a loop of adequate length, allowing for the maximum movement, between the fixed rigid pipes of the boat deck and bank; precautions need to be taken to ensure that the loop cannot be fouled up due to the actions of the movement, see Figure 14.8;
2. by using two flexible connectors and a fixed, rigid section secured to a gangplank that is hinged at one end with a sliding way at the other, see Figure 14.9.

For an LPG installation, the type of hose that can be used for the flexible connections is PVC-coated, braided, armoured hose with flexible tubing manufactured to BS 3212 Type 2 or BS EN 16436 Class 3. If the supply is natural gas, the flexible connectors need to be PVC coated, braided, armoured hose with manufactured heavy-duty stainless steel to BS 6501.

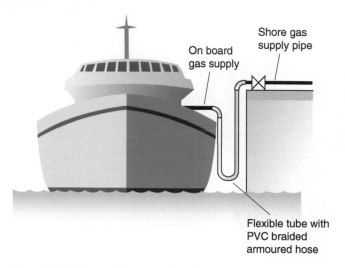

Figure 14.8 Shore-fed gas supply using a loop of adequate length

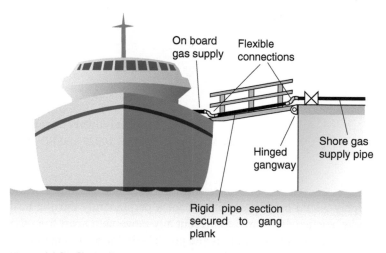

Figure 14.9 Shore-fed gas supply using flexible connectors and fixed rigid section secured to a gangplank

If there is a possibility that the boat could break from its mooring, allowing the potential of a gas escape, then a quick-release, self-sealing coupling should be incorporated. For LPG supplies this safety arrangement may be achieved by the use of a low pressure cut-off device.

Where a gas meter is to be used, this needs to be located on the shore within a securely fixed meter box, above normal flood level. In addition, at the point of supply there should be a warning notice advising of the need to ensure safe isolation.

On-board LPG Supply

14.17 Ideally, the gas cylinders should be stored on an open deck or within a housing that is suitably ventilated so that any leakage of LPG will fall and drain overboard. Drain and housing ventilation openings need to be sited at least 500 mm from any openings/hatches below deck and from possible sources of ignition. It is possible to store the cylinders in a compartment below deck or within a recess in the deck. From the base of this compartment there should be a low-level drain vent made of a material suitable for the passage of LPG and of no less than 19 mm internal diameter that discharges, with a fall to outside, at least 75 mm above the water line, even when the boat is fully laden, see Figure 14.10.

On-board Cylinder Housing

14.18 Points to observe where on-board cylinder storage within a housing is used:

- Provide the housing with an equivalent fire resistance to the surrounding structure of the vessel.
- The compartment should not be used for general storage.
- The compartment needs to be sealed (vapour-tight) from the hull.
- Access should be gained via a lid or cover at the top.
- In addition to low-level ventilation, high-level ventilation should be provided above the level of the cylinder.

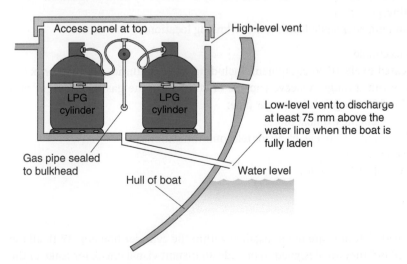

Figure 14.10 Onboard LPG supply

- The high-pressure stage regulators should be installed in the same compartment and above the cylinder in a position that protects them from mechanical damage. The high-pressure hoses should be a maximum of 1 *m* in length but kept to the minimum practical.
- An overpressure device should be included and non-return valves incorporated in the high-pressure stage connections where two cylinders are connected together.

Pipework Installation on Boats

Pipework

14.19   Special precautions need to be observed when running pipework on boats as the pipework will be subjected to very harsh conditions in terms of the vast amount of water and vapour that can be expected, which may contain salt, oil and possibly petrol. The materials to be used should be restricted to the following:

- seamless copper tube to BS EN 1057;
- rigid stainless steel, suitable for use with LPG in a marine environment, conforming to BS 6362;
- corrugated stainless steel tube to BS7838 or BS EN 15266;
- for existing systems only; copper-nickel alloy, suitable for use with LPG in a marine environment.

All joints to the pipework should either be hard soldered (using a solder with a melting point $\geq 450\,^{\circ}\text{C}$) or have compression or screwed fittings. Compression fittings should use copper or stainless steel olives as appropriate. Soft soldered joints should not be used. The method of pipe sizing given in Part 2: Section 2.62 can be used for boat installations.

All pipework should be supported at intervals no greater than 500 mm and where it penetrates bulkheads, the pipe needs to be adequately protected by a suitable sealed sleeve or bulkhead fitting/grommet.

The pipe should not be installed within the following locations:

- where it is inaccessible;
- in areas dedicated to electrical equipment, including where batteries are used or petrol engines, unless run through a sleeve capable of containing the gas in the event of a leak – there should be no joints in the gas pipe;
- within 100 mm of engine exhaust pipes;
- in ducts used for ventilation, electricity or telecommunications;
- below the bilge water level;
- within 30 mm of electrical cables.

Leak Detectors

14.20   Special leak detectors are often installed within the cylinder housing. With all the appliances turned off, they are designed to provide an instant visual check for leaks in the low-pressure pipework. There are two types of detectors.

Figure 14.11 Bubble leak tester.
Source: Alde International (UK) Ltd

The Bubble Leak Tester

14.21 When the top is fully depressed, if there is any leak on the supply pipework, it will be indicated by bubbles, which can be seen in the liquid through the viewing window, see Figure 14.11. When installed, a pressure gauge should also be fitted in the high-pressure side of the system.

The 'gasflow' Indicator

14.22 This is fitted to the regulator. If there are no leaks, the indicator stays green, but if there is a leak, the indicator changes to red.

Tightness Testing on Boats

Tightness Testing with Air

14.23 BS EN 10239 is primarily concerned with the manufacture of new craft and therefore restricts its scope from identifying the requirements of testing with gas. It suggests before gas is put into the pipework for the first time, the pipework should be subjected to a test of three times its operating pressure, but no greater than 150 mbar. This pressure should, after a five-minute stabilisation period, hold constant during a following five-minute period. To complete a tightness test with gas, one should follow the procedure in Section 14.24:

Tightness Testing with Gas

14.24 To tightness test with gas, carry out the following procedure:

(1) Turn off the supply control valve and burner control taps.
(2) Ensure appliance isolation valves are open and cooker lids are raised.

(3) Burn off the gas in the system using an appliance burner.

(4) Connect a manometer at a suitable test point.

(5) Slowly open the supply control valve until the regulator locks up. Record the lock-up pressure.

(6) Turn off the supply control valve.

    I. If there are no high-pressure hoses in the section to be tested, light a burner on one appliance and reduce the pressure in the system by about 5 mbar. Turn off the appliance. This is to release lock-up.

    II. If there are high-pressure hoses on the system to be tested, reduce the pressure until the dropping rate speeds up and immediately turn off the appliance. The pressure at which this happens varies, typically, it will be about 30 mbar for propane or 20 mbar for butane.

(7) Record the gauge reading.

(8) Wait five minutes for temperature stabilisation

    I. If pressure has stabilised or dropped during stabilisation period, record the reading, and move on to step (9)

    II. If the pressure rises without stabilising or has risen to lock-up pressure, this could be due to a rise in temperature. If this is suspected, repeat the procedure from (6) I.

    III. A pressure rise could also indicate that the supply valve is letting by. If this is the case, it will need to be rectified before repeating the whole procedure.

    IV. Alternatively, if the test includes high-pressure hoses between the supply control valve and the regulator, the rise in pressure may be due to hose relaxation. If this is suspected, install a further supply control valve downstream of the high-pressure hoses and repeat the whole procedure using this additional valve as the control, thus eliminating the hoses from the test.

(9) Wait for two minutes and record any pressure loss. If there is no discernible drop, proceed to (11). If there is a discernible drop, proceed to (10).
Note: A pressure loss of ≤0.25 mbar on a water gauge is considered not discernible and therefore, any gauge movement would not be acceptable. Although electronic gauges may be able to register movement more accurately than a water gauge, they have the same pass criteria. 'No discernible movement' for an electronic gauge would be a maximum rise or fall of 0.25 mbar during the test. For gauges that can only register to 1 decimal place, this figure is rounded down to a maximum of 0.2 mbar.

(10) If there is a discernible drop or a smell of gas:

    I. Close the isolation valves to the appliances and retest the supply line pipework only. If there is no discernible drop proceed to stage (10) II. If a discernible drop is present, it must be located and repaired before proceeding to stage (10) II.

    II. Once the supply pipework has been confirmed with no discernible drop, re-open the isolation valves to the appliances. A pressure drop can be deemed acceptable provided the drop is within the limits stated in Table 14.3 and there is no smell of gas. If the pressure drop, with appliances connected, exceeds the limit stated in Table 14.3, repeat the test with only one appliance's isolation valve open at a time to identify the appliance or appliances that are leaking. Carry out repairs as necessary and re-test. Once there is no discernible drop, you can proceed to Stage (11).

(11) Light a burner at an appliance to drop the pressure on the gauge to approximately half the pressure recorded in Stage (7), and immediately close the burner control tap.

Operate the UPSO if there is one downstream of the supply control valve and allow it to reclose. This may result in a small sudden rise in the gauge reading. Record the reading.

(12) Wait for two minutes, operate UPSO again if present and record the gauge reading.

    I. If the is no discernible rise, the supply control valve can be recorded as satisfactory. If the pressure rises, return to stage (8) II, III or IV.

    II. If there is no discernible drop with all appliance isolation valves open, proceed to stage (13). If there is a discernible drop, return to Stage (10).

(13) Following an acceptable result, burn off the gas, remove the gauge, reseal test point, turn the gas back on and spray the test point with LDF.

Table 14.3 Permissible pressure drop, with appliances connected and LPG supplied by cylinders

		Initial test pressure	
		37 mbar	28–30 mbar
Installation volume m³	Number of appliances installed	Pressure drop over two minutes mbar	
≤0.001 m³	1	1.5	1.0
	2	3.0	2.0
	≥3	4.0	3.0
>0.001 m³	1	1.0	1.0
	2	1.5	1.5
	≥3	2.0	2.0

Source: Data from PD 54823

Appliances and Ventilation for Boats

Relevant ACS Qualification
CCLP1 B

Relevant Industry Documents
PD 54823 and BS EN 10239

Appliances

14.25   On larger boats, there are all the typical gas appliances associated with domestic dwellings, including boilers, gas fires, cookers and refrigerators. Older installations may include flueless and open-flued appliances (Types A and B). However, for all new installations and replacement of existing appliances, apart from gas cookers, room-sealed appliances (Type C) must be installed. They need to be recommended by the manufacturer for use in a marine environment. All appliances need to incorporate a flame supervision device and be installed with particular care if they are near to any combustible materials. A typical warm air/hot water combination unit is shown in Figure 14.12. A second design that is also suitable for installation within a boat can be found in Figure 14.7.

Figure 14.12 Warm air and hot water combination unit. Source: Truma Ltd

Ventilation Requirements

14.26 Ventilation on board a boat is found by applying the following formula:

$$(2200U) + (440F) + (650P) = V$$

where

U is the gross kW heat input of all Type A appliances (flueless);
F is the gross kW heat input of all Type B appliances (open-flued);
P is the number of people for whom the vessel was designed;
V is the minimum area of fixed ventilation in mm^2.
Note: Where net heat input figures are given, multiply by 1.1 to convert to gross.

The ventilation is divided equally between high and low level, low-level ventilation being achieved by the use of suitable ductwork.

The size of the ventilation opening depends on the age of the craft. Since 1999, as already stated, all gas appliances installed, apart from the cooker, need to be room-sealed. However, open-flued appliances may be found in older boats. Where there are no open-flued appliances on board, the 440F is simply dropped from the calculation.

Example: A four-berth boat has the following gas appliances installed: a cooker of 12.6 kW (gross), a room-sealed water heater of 14.8 kW (gross) and room-sealed warm air heater of 10 kW (gross). The ventilation required would be:

$$(2200U) + (650P) = V$$

$$\therefore \ (2200 \times 12.6) + (650 \times 4) = \underline{30320 \text{ mm}^2 \ (303.2 \text{ cm}^2)}.$$

If the water heater and warm air heater had been open-flued, as was sometimes fitted in older craft, then the ventilation needed would be much larger:

$$(2200U) + (440F) + (650P) = V$$

$$\therefore \ (2200 \times 12.6) + (440 \times 24.8) + (650 \times 4) = \underline{41232 \text{ mm}^2 \ (412.3 \text{ cm}^2)}.$$

Compartment Ventilation

14.27 Appliances in compartments should be ventilated at high and low levels in accordance with the instructions from the relevant manufacturer. In their absence ventilation size from Table 14.4 should be provided. Ventilation for room-sealed appliances requires air for cooling purposes. Open-flued appliances also require air for cooling as well as for combustion purposes.

Table 14.4 Minimum free area compartment ventilation (cm^2/kW of appliance maximum net input

Appliance flue type	Compartment ventilated to internal space (cm^2/kW)	Compartment ventilated to outside (cm^2/kW)
Type C	High level 10 cm^2 Low level 10 cm^2	High level 5 cm^2 Low level 5 cm^2
Type B	High level 10 cm^2 Low level 20 cm^2	High level 5 cm^2 Low level 10 cm^2

Source: Data from PD 54823

15

Electrical Work

Electricity

15.1   Before reading this section, it must be understood that *all* electrical work must be performed by a 'competent person'. It is evident that this short section will not make the reader competent in the skills of an electrician. The purpose of this section is designed to enable gas operatives to identify the basic needs associated with electrical installation work and the risks involved. With most gas appliances having an electrical connection, this section will assist gas operatives to identify these needs and the hazards associated with electricity. The scope and content are limited to essential knowledge and the basic safety aspects that a gas engineer should be able to competently perform. To go beyond this, further training would be required to ensure safety.

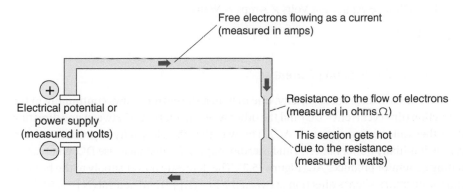

Figure 15.1   Electricity flowing through a conductor in a circuit

What is Electricity?

15.2   Electricity is basically the flow of infinitely small particles, called free electrons, along a conductor in a circuit (Figure 15.1). If there is no circuit, there can be no flow of electricity. As with water in a stream, if it is flowing, it is referred to as a current. Electrical current is measured in amperes (amps). To make the current flow, a force is

Gas Installation Technology, Third Edition. Andrew S. Burcham, Stephen J. Denney and Roy D. Treloar.
© 2024 John Wiley & Sons Ltd. Published 2024 by John Wiley & Sons Ltd.

required, referred to as an electromotive force (EMF). This can be obtained from several sources, including batteries, thermocouples and generators. The EMF is measured in volts. Basically, where two conductors have different voltages, a current will flow between them. The current needs to flow through/along a material that has free electrons within its structure, such as a metal, and the resistance to the flow, due to the conductors size and length, will decrease or increase the speed of the current, causing the conductor to warm up due to friction. The resistance to flow is measured in ohms (Ω), and the warming effect, being the power generated, is measured in watts.

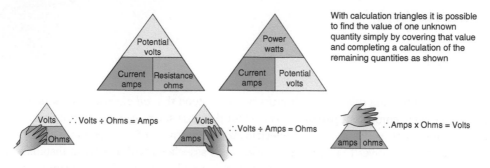

Figure 15.2 Ohm's law illustrated as a calculation triangle

The current, EMF, resistance and power relate to each other, and their relationship is given by Ohm's law:

Volts ÷ Ohms = Amps and Volts × Amps = Watts

(See Figure 15.2)

Direct Current and Alternating Current

15.3   There are two forms of electricity: one in which the current flows continually in the same direction (direct current – DC) and the other where it alternates between one direction and the other (alternating current – AC). The way that the electricity is produced determines how it initially flows. Batteries and thermocouples, for instance, use DC, whereas an alternating generator produces AC (Figure 15.3). The domestic power supply to the home is AC, and the current switches direction between the line (L [+]) and neutral (N [−]) 50 times per second (50 Hz).

Magnetism

15.4   One final point in this quick introduction to electricity is that a force is generated by the flow of electricity. As the current flows, it creates a small flux or flow of energy from the conductor. Where a length of conductor wire is wound to form a coil, this flux is then magnified to produce a magnetic field. This phenomenon is used in many electrical systems, operating motors and solenoids (magnetic valves) and is also used to induce a flow of electricity in another conductor, such as occurs in a transformer (Figure 15.4).

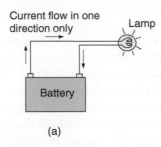

Current flow in one direction only

Lamp

Battery

(a)

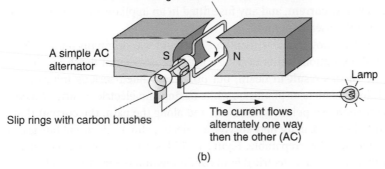

As the loop rotates it cuts through the magnetic flux generated between the two poles of the permanent magnet and in so doing induces the electrons to flow

A simple AC alternator

S   N

Lamp

Slip rings with carbon brushes

The current flows alternately one way then the other (AC)

(b)

Figure 15.3   Electrical current. (a) Direct current (DC); (b) alternating current (AC)

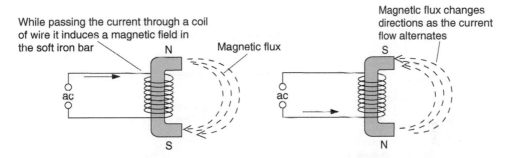

While passing the current through a coil of wire it induces a magnetic field in the soft iron bar

N   Magnetic flux

ac

S

Magnetic flux changes directions as the current flow alternates

S

ac

N

Figure 15.4   Electricity and magnetism

Electrical Safety

Relevant Industry Document
Electricity at Work Regulations, BS 7671, Gas Safe Technical Bulletin 118a and
HSE Guidance Note GS38

15.5   Using gas can involve hazards. Death can be caused instantly by a gas explosion or by the slower, hidden, but equally deadly effect of carbon monoxide poisoning. Electricity can also kill by electric shock or cause a fire as a result of some poor design. Users and installers must be aware of these potential hazards.

Electrical Shock

15.6   No one is immune to electric shock, and every time a gas installer works on an appliance that requires an electrical supply, they are working on something that has the potential to kill. Therefore, electricity should always be treated with the utmost care and respect. Many people, at some point in their lives, will have experienced an electric shock. Thankfully, for most of them, the current was small, and the body's reflex action broke contact with the supply before too much current passed through the body to cause lasting harm.

The size of electric current that can kill a human is generally accepted to be as small as 50 mA (0.05 amps). At ≥50 mA, the muscles may contract so that the individual will not be able to release themselves and break contact. If a 50 mA current was to cross a person's chest, this is more than enough to cause a serious cardiac disturbance which could result in death. 50 mA is a very low current, and any fuse fitted in an appliance or supply would be insufficient to protect the user from harm.

Death by electric shock is known as electrocution. The word was originally derived from the words 'electro' and 'execution' but is now also used for accidental death by electric shock. Not all electric shocks result in death, but other associated serious injuries can result, including falls from ladders or platforms while testing and electrical burns. Electrical burns occur at entry and exit points of the body, and along the path, the current flows through the body. Although a casualty may survive an electric shock, electrical burns can be severe and may sometimes be irreparable. Figure 15.5 illustrates the danger posed to a person touching an appliance under electrical fault conditions and the passage of electrical current through the body either directly to the general mass of earth or through earthed pipework.

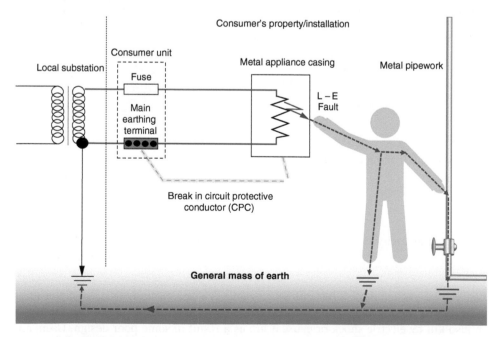

Figure 15.5   Electric shock and the passage of current through the body

The effect of electrical current on the human body is well documented. The resistance of the human body varies from person to person but, under certain conditions, can be as low as 1000 Ω. Therefore, if an individual was to come in contact with a 230 *V* live cable, the amount of current passing through the body could be over 4.5 times the level required to cause a serious cardiac disturbance. This can be demonstrated by applying Ohm's law:

Volts ÷ Resistance = Amps

$230V \div 1000R = 0.23\,A$ or 230 mA

Because of this, all electrical installations must provide some form of protection against electrical shock. One common form of additional protection is a residual current device (RCD). A 30 mA RCD with a disconnection time of not more than 0.2 s may be provided to give protection against contact with a live supply, as it should disconnect the circuit at a level well below the 50 mA often considered enough to cause electrocution. However, an RCD must not be used as the primary protection and is no substitute for basic protection such as electrical insulation, enclosures and barriers, and, of course, safe isolation (see Section 15.11 and Figure 15.15). Therefore, it is imperative that gas operatives understand the inherent risks associated with electricity and do not remove any covers to electrical components unless they are fully competent in the work that they are about to undertake. In fact, the Electricity at Work Regulations, like the Gas Safety Regulations, state that no one should undertake work unless *competent* to do so.

Electrical Fires

15.7   Electrical shock is not the only way in which electricity can kill; electricity can also cause fires. A fire caused by an electrical fault is often the result of a poor earthing arrangement or high resistant joints. In a correctly designed installation, if a live wire was to touch an earth conductor, the current should flow so rapidly that it exceeds the ampere rating of the fuse, causing the fuse to trip and break the circuit. Therefore, it is essential to confirm that all electrical work is inspected and tested in accordance with approved standards to eliminate the risk of electrical fires.

When a gas installer completes the installation of a new boiler or cooker, etc., the work often involves the final connection to the electrical supply. The gas installer often approaches this with an 'I can do that' attitude and proceeds to make the final connection, possibly from a local spur outlet point. However, it must be understood that such electrical work should still be designed, installed, tested and certified in accordance with BS 7671 and be carried out by a competent person. By making these types of connections the gas operative is taking responsibility for the electrical connections and has a legal duty to ensure that the installation is left safe.

Bonding

15.8   The term bonding here relates to the joining of all metalwork within a building by means of cables, if necessary, called the main protective bonding conductor so that everything is at the same potential. At specific points, such as where the gas or water supply pipes enter the building, a cable, known as the equipotential bonding wire, is taken back to the main earthing conductor. Thus, should any live conductor touch any piece of metalwork within the building, it allows a current to flow to earth, causing the fuse to blow or the circuit breaker to trip. If the metal was not bonded to earth, this would not happen, and it would become live causing a risk of electrocution should anyone touch it.

Stray Currents and the Temporary Bonding Wire

15.9   For the fuse to trip, a current larger than the fuse rating needs to flow through it, and if a live conductor touches a piece of bonded metalwork, this will certainly occur (see Figure 15.6). However, if the neutral conductor touches a section of metalwork, although it is live, it will not necessarily cause the fuse to blow. This is because in the modern electrical supply system [called the protective multiple earthing system (PME or TNCS system)], the neutral conductor is used as the protective earthed neutral (PEN) conductor. In Figure 15.7, the neutral wire is broken and is touching the earth, and therefore the metalwork, but the light is still working. This is because the current can still flow back and forth through the series of conductors between the local supply transformer and the light without incurring a current fault; thus, the fuse will not blow. When the unsuspecting gas operative removes a section of pipe, this might be what is occurring. As the two sections of pipe are pulled apart, it is like breaking the switch, so a spark may jump between the two sections as they are separated. It may also be that the operative is holding one section of the pipe in each hand, in which case the current would continue to flow through the operative, resulting in electrocution.

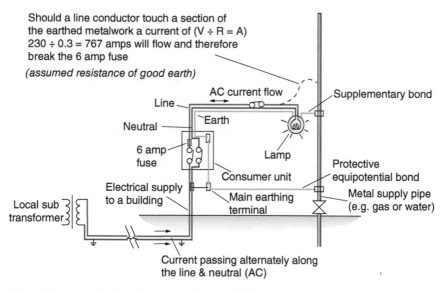

Figure 15.6 Installation showing no fault condition

In recent years, due to a number of electrocutions, it has become good practice for gas operatives to check for stray currents before touching any pipework or appliances. One way

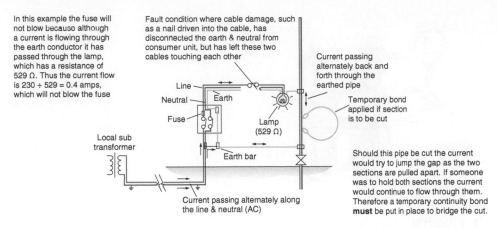

In this example the fuse will not blow because although a current is flowing through the earth conductor it has passed through the lamp, which has a resistance of 529 Ω. Thus the current flow is 230 ÷ 529 = 0.4 amps, which will not blow the fuse

Fault condition where cable damage, such as a nail driven into the cable, has disconnected the earth & neutral from consumer unit, but has left these two cables touching each other

Current passing alternately back and forth through the earthed pipe

Line

Neutral — Earth

Fuse

Lamp (529 Ω)

Temporary bond applied if section is to be cut

Local sub transformer

Earth bar

Should this pipe be cut the current would try to jump the gap as the two sections are pulled apart. If someone was to hold both sections the current would continue to flow through them. Therefore a temporary continuity bond **must** be put in place to bridge the cut.

Current passing alternately along the line & neutral (AC)

Figure 15.7 Installation showing neutral fault condition and current flowing through the pipework

Figure 15.8 Examples of single pole non-contact voltage indicator or 'voltage stick'

to achieve this is by the use of a single pole non-contact voltage indicator, such as a 'voltage stick' shown in Figure 15.8. Anyone working in the industry should develop the habit of using such a voltage indicator on all exposed metal surfaces, including pipework and appliance casings, before touching and beginning work to assess whether a dangerous voltage is present due to fault conditions. It is important to note that such a voltage stick should only be used to identify live equipment and **not** for *proving* that it is dead. Further tests will be required to prove the installation/equipment is electrically dead before any work can be undertaken (see Section 15.11 and Figure 15.15).

Since a voltage stick does not prove that no voltage is present, one must assume there is. So, as a precaution, when cutting into a run of pipework or when removing a meter, one should place a temporary continuity bond of not less than 10 mm² across the section of pipe or the meter installation as shown in Figure 15.9. Apart from reasons of safety, as just described, it is also a legal requirement to do this. Any temporary bond must be left in place until a positive bond has been remade (Figure 15.10). If in any doubt about the electrical safety of an installation, a competent electrician should be consulted.

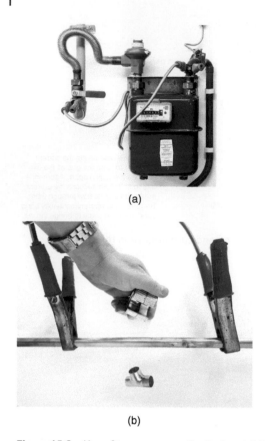

(a)

(b)

Figure 15.9 Use of temporary continuity bond. (a) Temporary removal of meter; (b) section of pipework to be cut for a tee piece

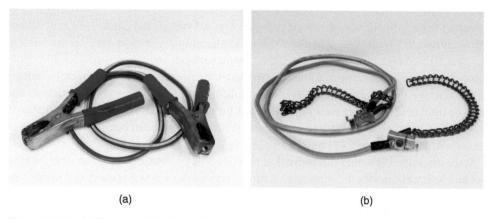

(a)   (b)

Figure 15.10 (a) Temporary bonding wires for domestic-size pipework. (b) Temporary bonding strap for larger diameter pipework

Absence of Bonding

15.10   Where a gas operative installs new pipework or discovers that there is no bonding to an installation, it is not their responsibility to correct this, and they should not unless they are competent to do so. However, if a gas operative cannot confirm the presence of adequate equipotential bonding, they have a legal obligation to notify the responsible person for the property that bonding may be required. This advice should be given in writing and advise the responsible person to have the installation checked by an electrically competent person.

Safe Isolation

15.11   Under the Electricity at Work Regulations, it is illegal to work on live electrical circuits unless the operative is undertaking essential testing work that can only be completed with the supply on. Working on live systems should be regarded as highly dangerous and only to be performed by a fully trained individual who clearly knows all the dangers.

Most electrical accidents occur when someone is working on or near equipment that is thought to be dead but which is in fact live. All exposed electrical work must be treated as live and you must never assume otherwise. Therefore, it is important to know how to safely isolate an electrical supply and confirm that the equipment to be worked on is, in fact, dead.

Before an appliance or installation is isolated, thought must be given to the effect that this will have on building equipment and services and the building occupiers. Firstly, determine what needs to be isolated. In some instances, this will be a single appliance. In others, a single circuit or sometimes the whole installation. Isolating an electrical supply can have serious implications for users of the electrical equipment, so before commencing work, speak to the client or responsible person to discuss their requirements during the duration of your work to ensure that any disruption is kept to a minimum. In some premises, particularly when working in a non-domestic setting, a 'permit to work' may be required.

Once permission to isolate the circuit/appliance has been granted by the person responsible for the electrical installation, the following procedure can be used to confirm that the supply is dead:

1. **Select an Approved Voltage Indicator (AVI)**

Many accidents are caused by unsuitable test equipment. Therefore, all test equipment should conform to the recommendations set out in the HSE guidance note GS38. For example, test probes, leads and clips should be manufactured to the appropriate British Standards: BS EN 61010, BS EN 61557 or BS EN 61243-3.

Probes and clips should have finger barriers to prevent accidental contact with live conductors during the test and be insulated to leave a maximum exposed tip of 4 mm. It is strongly recommended that spring-loaded retractable tip covers are used or the exposed tip is reduced to 2 mm or less. Voltage detector leads should also provide protection against excess current and have the correct category rating for the installation, e.g. CAT IV. An example is shown in Figure 15.11.

Note: The use of a multimeter is not recommended due to the risk of the device being set to the wrong function, and electrical screwdrivers that light up do not comply and should be regarded as unsafe.

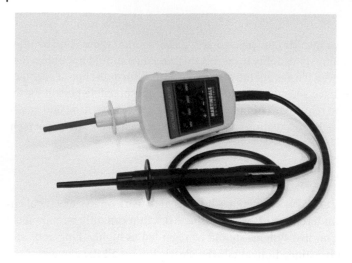

Figure 15.11   Example of an approved voltage indicator to BS EN 61243-3 with finger guards and retractable tip covers

2. **Examination of test equipment**

Ensure the test equipment is in good condition: no cracks, cuts or abrasions and loose terminals or damaged retractable covers or finger guards.

3. **Check that the AVI is working correctly**

Any device that is used to prove an installation is dead can fail to operate correctly. Therefore, the AVI should be checked for correct operation before and after use, preferably on a voltage-proving unit (see Figure 15.12), or a known supply of a similar voltage to the test circuit if this can be done safely.

4. **Locate the means of isolation**

Turn off the supply and affix a locking device to the switch to ensure that the power cannot be reinstated without your knowledge. Where fuses are used, remove the fuse, retain it in your possession and lock the fuse holder. Attach a warning label to warn others that work is being carried out on the installation and that the supply should not be reinstated. Typical lock-off devices available for this purpose are shown in Figures 15.13a and b.

5. **Confirm safe to touch equipment**

Use a single pole non-contact voltage stick to check for stray currents before touching the appliance casing. Touch the voltage stick in a number of places on exposed metal surfaces, including any connecting pipework, before touching and beginning work to assess whether a dangerous voltage is present due to fault conditions. If voltage is detected, do not proceed until investigated by a suitably qualified electrician.

6. **Expose the appliance conductors**

Use appropriately insulated tools, taking great care not to touch any exposed conductive parts.

Figure 15.12 Example of a voltage-proving unit

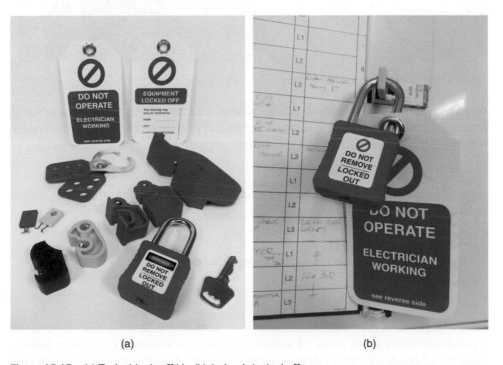

(a)   (b)

Figure 15.13 (a) Typical lock-off kit. (b) A circuit locked off

7. **Verify that the circuit is dead**

Using the AVI test between:

- Line to Line (3 phase only: L1, L2, and L3)
- Earth to Line
- Earth to Neutral
- Neutral to Line.

No voltage should be detected (see Figure 15.14)

Figure 15.14   Verifying the circuit is electrically dead

Important note: When checking voltage between an earth terminal and live (including neutral), always make the earth connection first and remove it last. This is to reduce the risk of a dangerous voltage appearing on the remaining probe tip if the terminal is still live.

8. **Confirm that AVI is still working**

Re-check that the test instrument is still working correctly by testing on a proving unit or known supply.

9. **Safe to begin work**

An easy-to-follow flowchart for safe electrical isolation is provided in Figure 15.15.

Capacitors and their Ability to Hold a Charge

15.12   Even when everything has been done to check that a circuit is dead, there is still a possibility that a shock could be administered from an electrical component called a capacitor or condenser. These devices, found in several shapes and sizes, although typically cylindrical, like a small cotton reel, are designed to store a charge of electricity that

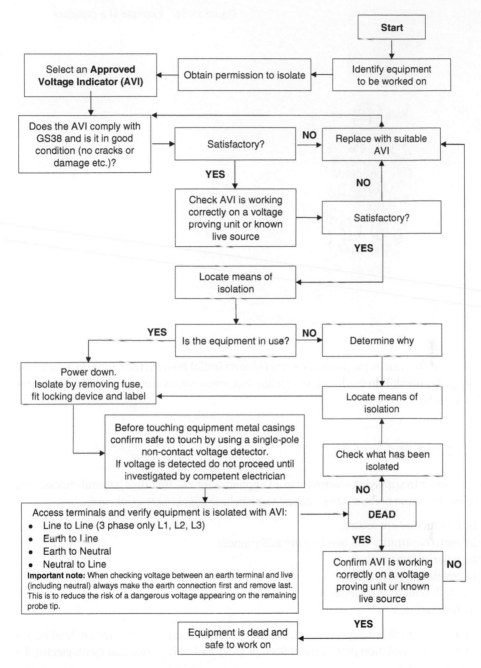

Figure 15.15 Safe isolation flow chart

Figure 15.16   Example of a capacitor

can be used for various purposes, such as to give an initial boost to the supply to get a motor turning. It is possible to discharge a capacitor, but generally, awareness is all that is required (Figure 15.16).

Inspection and Testing

15.13   Prior to supplying electricity to an appliance for the first time, several checks need to be made. These preliminary electrical tests include checking the following:

- the conductors are secure;
- the earth continuity and bonding are maintained;
- the polarity is correct;
- the insulation resistance;
- the earth fault loop impedance;
- the fuse rating.

Prior to undertaking any of these tests, the installation must be confirmed dead by following the safe isolation procedure in Section 15.11 and only performed by an electrically competent person.

Fuse Rating

15.14   The fuse is designed to protect the cable and appliance. It will not protect you against electric shock. The fuse rating will be determined by the appliance manufacturer.

For most gas appliances this is 3 amps, but lower fuse ratings may be required, and these should not be exceeded. Failure to observe this simple rule may lead to appliances becoming live, damaged or cause a fire.

Fault Diagnosis of Basic Electrical Controls

15.15 With the supply established, it may be that an appliance fails to operate. All the electrical connections need to be inspected to ensure that there are good sound connections and that the PVC insulation has been sufficiently stripped back at the ends, thus exposing the conductor and making a suitable contact. These tests should be made with the power isolated. In fact, the Electricity at Work Regulations stipulate that the normal policy should be to only work on electrically dead equipment. Working on live equipment is permitted only in very limited circumstances. Any potential live working needs to be carefully planned, and three conditions must be met before this is permitted. If just one of the following conditions cannot be met, then only dead working is permitted:

- it is unreasonable in all circumstances for the equipment to be dead;
- it is reasonable in all circumstances for the person to be near the equipment while it is live and
- suitable precautions are used to prevent injury.

Fault finding is one of the limited areas where live working may be necessary as it could be impossible to trace a fault when the equipment is dead. If it is decided that dead working is unreasonable, then a suitable and comprehensive risk assessment should be carried out and recorded. If, after the risk assessment, it is decided to continue with live working, suitable precautions should be taken to prevent injury. These precautions might include:

- erecting temporary barriers with warning notices to exclude unauthorised access to the danger area;
- informing others in the dwelling or workplace of the dangers and that they should stay away from the area and ask for children and pets to be kept away;
- allow plenty of space to work and keep the area clear of obstructions. If you receive a shock in an enclosed space, it may be difficult to break free;
- use robust and properly insulated tools and personal protective equipment;
- never leave an appliance unattended with live conductors exposed and
- avoid lone working.

It must be stressed that working with live electrical systems poses a substantial hazard and work must not proceed unless the operative is fully competent.

Flow charts showing the logical sequence that needs to be undertaken for fault diagnosis are commonly found in manufacturer's instructions. These diagrams are a useful tool in order to determine a particular fault. They will take you step by step through the sequential operation of an appliance and often include more than just the electrical controls that may be at fault; examples are shown in Figures 15.17 and 15.18.

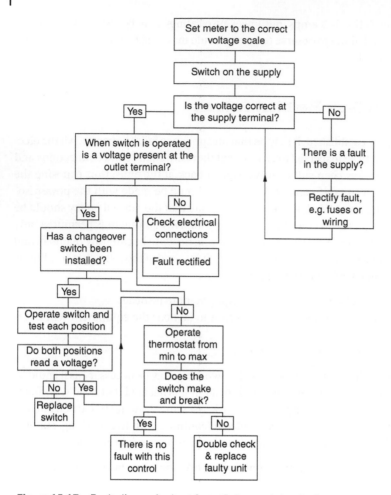

Figure 15.17   Fault diagnosis chart for switches and simple thermostats

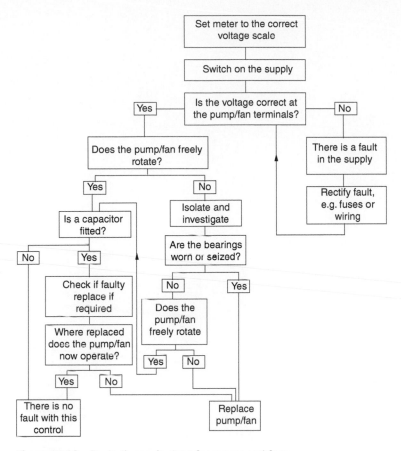

Figure 15.18   Fault diagnosis chart for pumps and fans

16

The ACS Assessment Process

The ACS Assessment Process

16.1 Since 1991, anyone working within the scope of the Gas Safety (Installation and Use) Regulations must be a member of a class of persons recognised by the Health and Safety Executive (HSE). Since 2009, the only authorised gas safety registration body in Great Britain and Northern Ireland, the Isle of Man, Jersey and Guernsey is the Gas Safe Register.

For an application to be accepted by Gas Safe, valid evidence of competence must be provided. The Gas Safe Register will only accept limited evidence, such as:

- Nationally Accredited Certification Scheme for Individual Gas Fitting Operatives (ACS).
- ACS aligned Gas Service National Vocational Qualifications (NVQ) or Scottish Vocational Qualifications (SNQ).

Note: There may be other suitable accredited vocational qualifications for gas services. A list of suitable evidence of gas safety competence recognised by the Gas Safe Register can be found on the Energy and Utility Skills (EU Skills) website – http://euskills.co.uk/about/our-industries/gas.

Initial ACS assessments and other suitable gas qualifications, are valid for a period of five years. Upon expiry, if the individual wishes to continue registration with Gas Safe and work in the gas industry, a reassessment ACS must be undertaken; this, too, is valid for a five year period.

Initial Assessments

16.2 Anyone taking an ACS category for the first time will take an initial assessment. These assessments are usually in more depth than reassessment. There is often a range of theory papers and practical assessments that must be completed, and at the time of writing, the pass mark for all ACS assessments is 100%, although most assessments are currently open book.

Renewal/Reassessment

16.3 Renewal, or reassessment, of competence is required every five years. However, reassessments are usually shorter than initial assessments as they are designed for

Gas Installation Technology, Third Edition. Andrew S. Burcham, Stephen J. Denney and Roy D. Treloar.
© 2024 John Wiley & Sons Ltd. Published 2024 by John Wiley & Sons Ltd.

experienced gas operatives and simply cover key knowledge and skills along with any changes that have been introduced to the industry.

To benefit from taking this shorter *reassessment* ACS the individuals current certificate must be in date or no more than 12 months past the expiry date. If the certificate has lapsed by more than 12 months, an *initial* assessment must be taken.

It must be noted that both Gas Safe registration and current ACS competencies are required in order to work legally on gas installations, fittings or appliances. Once an ACS certificate has expired, the holder can no longer legally work in the areas covered by that certificate; that aspect of their Gas Safe registration will be suspended until they re-sit the relevant ACS assessment. ACS is a certificate of competence and is essentially a permit to work, along with Gas Safe registration.

Applicants must provide the original certificate as evidence to the assessment centre if they wish to renew an ACS competency. As with initial assessments, the pass mark is 100% during both theory and practical assessments.

MOT-Style Certification

16.4   MOT-style certification is now available for ACS assessments. This means that a renewal can take place up to six months prior to the certificate expiry date without losing any time on original qualification. The new expiry will be five years plus any remaining time on the original certificate. For example, if the original certificate expires on 1 December 2025, the assessment can be taken any time from 1 June 2025 to 1 December 2025 without the loss of any remaining time on the certificate; the new expiry will be 1 December 2030. However, if the period is greater than six months, the new expiry date will be five years from the date of assessment. If new categories are added, these too will expire five years from the date of assessment whenever it is taken.

Extending Scope or Range

16.5   If an individual already holds ACS categories, for example, CCN1 and CENWAT, and they are due for renewal, it is not compulsory to have any training prior to assessment, although most training and assessment centres would recommend it, and the vast majority of gas operatives opt for the training. However, if any individual wants to extend the range of categories held, or the scope of work undertaken by adding, say, CKR1 or a changeover, then a number of conditions need to be met prior to assessment:

1. The applicant must already hold the core competency, in this case, CCN1.
2. Training must be undertaken at a recognised ACS training centre.
3. The applicant must be Gas Safe registered, and, in most cases, a minimum period of 12 months must have passed since they gained their initial ACS core competency or qualification.
4. Where it has not been possible to gain supervised on-site experience in the new category, documented experience must be gained in a realistic work environment at the training centre.

Note: Where an engineer wishes to extend the scope or range of gas work within 12 months of completing their initial core qualification or ACS, it is possible to undertake an MLP – Bridge with a recognised training provider (for more details, see Part 1: Section 1.2).

Matters of Gas Safety Criteria

16.6   Undertaking any assessment or examination can be a stressful experience, although having an insight into the process before you begin can often lessen anxiety. So, what is known about the ACS scheme and the range of assessments?

For details of ACS assessment categories, see Part 1: Section 1.5.

The purpose of the ACS scheme is to ensure individual gas operatives are assessed against a set of national and industry standards, best practices and technology. EU Skills provides the standard-setting body services for ACS, and the gas industry Strategic Management Board is deemed to be the scheme owner. To ensure that individuals are assessed against this national standard of assessments, Certification Bodies measure competence against a single set of approved criteria for matters of gas safety. The criteria for each ACS category are published by EU Skills on their website – http://euskills.co.uk/about/our-industries/gas/standards-setting-body.

Each ACS assessment is based on these **Matters of Gas Safety Criteria,** and Certification Bodies will base their assessments on the information given under the title of each assessment listed. This information is publicly available on EU Skills website, so nothing should come as a surprise when being assessed, and individuals can only be assessed against the published criteria.

For example, by looking at the Matters of Gas Safety Criteria for **CCN1** (2024), it can be seen that the assessment will be designed to test the knowledge and understanding of gas safety and competence in the following subjects:

1. Gas safety legislation and standards.
2. Gas emergency actions and procedures.
3. Products and characteristics of combustion.
4. Ventilation.
5. Installation of pipework and fittings.
6. Tightness testing and purging.
7. Checking operating pressure at outlet of meter.
8. Unsafe situations, emergency notices and warning labels.
9. Operation and positioning of emergency isolation controls and valves.
10. Checking and setting appliance burner pressures and gas rates.
11. Operation and checking of appliance gas safety devices and controls.
12. Chimney Standards.
13. Chimney inspection and testing.
14. Installation of open, balanced and fan-assisted chimneys.
15. Re-establish existing gas supply and re-light appliances.

There are in fact 16 pages of detailed criteria that individuals will be assessed against. The information contained in the Matters of Gas Safety for CCN1 also includes details of how to carry out certain checks and tests which must be performed during the assessment.

Assessment centres are also obligated to provide the **legislative, normative** and **informative** documents relevant to each assessment for reference by the candidate during the assessment process – see Part 1: Section 1.11.

Although many aspects of the assessment are designed to be unpredictable, the information contained in the Matters of Gas Safety on http://euskills.co.uk will be beneficial to review during preparations for an assessment.

Your ACS Assessment (CCN1)

16.7   In this section, you will find some details of what to expect when you take the CCN1 assessment and may prove useful to you during your preparations for your ACS assessment. Information has been taken from various parts of this book and compiled in one place. At the conclusion of each section, reference is also made to the parts of this book where you can study the material in greater detail if so desired.

CCN1 Assessment

16.8   The CCN1 ACS assessment is designed to test your knowledge and understanding of core domestic natural gas safety. *Knowledge and understanding* (K&U) assessments are theory-based questions, often by means of multiple-choice question papers, but in some cases, practically assessed worksheet (PAWS) questions will be used, which require a written answer. You will also be required to prove your competence against a defined *performance criteria,* which will involve completing a number of practical tasks and interpreting your results. Please note that not all K&U and performance criteria apply to reassessment; for full criteria details, visit – http://euskills.co.uk.

You will be assessed on the following subjects:

1. **Gas safety legislation and standards**

16.9   This part of the assessment is to test your K&U of the Gas Regulations, the HSE's Approved Code of Practice (ACOP) L56 and Building Regulations. The assessment will usually be by means of a multiple-choice question paper and possibly by some PAWS questions.

From the Gas Regulations, one key regulation that all gas operatives should know by rote is **Regulation 26 Gas appliances – safety precautions** and, in particular, 26 (9). This regulation details the minimum essential tasks that every gas operative must undertake whenever *working* on a gas appliance. These checks include:

a) the effectiveness of any flue;
b) the combustion air supply;
c) the operating pressure or heat input, or where necessary both *(if it is not reasonably practicable to test the operating pressure or heat input, or where necessary both, then the combustion performance must be checked) and*
d) the safe functioning of the appliance.

You should try to commit these essential checks to memory. The checks include:

Flue   *including visual inspection, flue flow, spillage and general flue integrity.*
Air   *visual inspection of combustion/compartment air bricks/vents and size.*
Gas   *operating pressure/heat input and/or combustion performance.*
Safety *this would include checking safety devices, case seals, fail safes, etc.*

For more details, see Parts 1 and 8.

2. Gas emergency actions and procedures

16.10   This section is also a theory-based K&U assessment. For example, you will need to know the specific gravity of gas and the limits of its flammability. In addition, you should be familiar with ACOP L56, Regulation 37 (1)–(4) and know what to do in the event of an escape of gas and what action to take if gas continues to escape after turning off the supply.

For more details, see Parts 2 and 8.

3. Products and characteristics of combustion

16.11   This section includes both K&U and practical tasks and forms quite a large part of the assessment. It includes the ACS competency CPA1 (combustion performance analysis). You will be required to carry out combustion performance analysis on three appliances with different flue types: Type A (flueless), Type B (open-flued) and Type C (room-sealed), and interpret your results against manufacturers' instructions or relevant industry standards.

You will also be required to demonstrate an accurate knowledge of carbon monoxide (CO), be able to recognise the symptoms of CO poisoning and explain the correct advice to give someone describing these symptoms. You will also need an awareness of ambient levels of carbon dioxide (CO_2) in the atmosphere and the critical levels of CO_2 that could cause vitiation affecting the combustion process. In addition, you will need to identify complete and incomplete combustion at appliance burners, explain the causes of incomplete combustion and correctly identify suitable/unsuitable CO detectors and indicator cards.

For more details, see Parts 2 and 8.

4. Ventilation

16.12   This is predominantly a K&U assessment with a small practical section where you will be required to inspect, measure and calculate the area of a number of air vents. You will need to identify correct and incorrect installations and understand what could affect the performance of air vents, such as fly screens, closable vents and ducting. The K&U section of this assessment is quite extensive and is largely based on BS 5440-2.

With many gas engineers now only installing and working on modern room-sealed condensing appliances, they may seldomly come across this area of K&U. However, you will need to demonstrate that you could identify faults with all aspects of ventilation if you were to encounter them, including ventilation for open-flued and flueless appliances; location of ventilation for single and multiple DFE fires; the effect of oil and solid fuel appliances, double glazing, extraction and draught proofing on ventilation; ventilation for internal kitchens; passive stack ventilation; intumescent air vents; ventilation for multi-appliance installations; vents in series; compartment ventilation and suitable labels and notices.

For more details, see Part 7.

5. Installation of pipework and fittings ≤ 35 mm

16.13   The practical tasks for this section involve identifying pipework defects and the installation of copper pipework and mild steel pipework using appropriate methods and jointing agents. You will be required to test the installation for gas tightness, safely remove a meter, use temporary continuity bonds, capping the meter and all open ends before starting work. The installation will include capillary, compression and threaded fittings. Upon completion, you will reconnect the meter, remove temporary bonding, re-test and purge the installation.

You will need an understanding of the information contained in BS 6891 and the Gas Regulations to answer the theory questions in the K&U unit. Some less familiar areas that will be covered include a pipe sizing exercise; pliable corrugated stainless-steel tubing (CSST) and fittings jointing requirements; pipework within timber/light steel frame walls and passing through a timber/light steel frame/masonry wall; ventilation required for pipework in ducts; pipework in multi-occupancy buildings; installing pipework in fire escapes and protected areas, and the suitability and purpose of a single pole non-contact voltage tester.

For more details, see Parts 4, 5 and 15.

6. **Tightness testing and purging**

Tightness Testing

16.14   Tightness testing is one area of gas work that should be second nature to all gas operatives, but do not be complacent. In the Matters of Gas Safety Criteria, there are 20 steps that are required to carry out a successful tightness test. It is surprising how easy it can be to miss one of these steps while being observed during an assessment.

You will need to know the meaning of 'no perceptible movement' and how to confirm let-by if the pressure rises by more than perceptible movement during the let-by test. You should be able to explain why for existing installations with appliances connected and no smell of gas, a permissible pressure loss is allowed depending on pipe size and meter type. Finally, you must confirm the installation is left safe by spraying the sealed pressure test point, ECV outlet connection and all connections between the ECV and the meter regulator with an approved leak detection fluid (LDF) to complete the test.

You will also need to perform a successful tightness test on a medium-pressure installation without a meter inlet valve (MIV) fitted. Your K&U of where to find guidance on testing pipework of diameter > 35 mm or total installation volume (IV) > 0.035 m^3 will also be confirmed.

For more details, see Part 5.

Purging

16.15   Along with performing a practical purge, and taking all precautions necessary to work safely, you will also be required to undertake a purging calculation exercise to determine the installation volume (IV) and purge volume (PV) of various meters connected to ≤ 35 mm pipework. You will also need the K&U of the different procedures for purging installations of IV ≤ 0.02 m^3 and those of IV > 0.02 m^3.

For more details, see Part 5.

7. **Checking and/or setting meter regulators**

16.16   During this section, you will be required to correctly check and confirm the operating pressure of an allocated meter regulator using the appropriate appliance/s.

For the K&U unit, you will need to identify both low and medium-pressure installations and how they operate. You will also know the correct procedure for notifying the gas transporter for pressures outside the acceptable range and the effects of pressure absorption across a primary meter installation.

For more details, see Part 2.

8. Unsafe situations, emergency notices and warning labels

16.17   For this task, you will need K&U of IGEM/G/11 (the Gas Industry Unsafe Situations Procedure). This document will help you to correctly categorise unsafe situations as:

- Immediately Dangerous.
- At risk (where turning off will remove the risk).
- At risk (where turning off does not remove the risk).
- The correct procedure to follow in each instance.

You will need to correctly identify which labels and notices to issue in each situation and demonstrate what action to take when you encounter installations that do not meet current standards but are not unsafe.

Advisory notices and situations reportable under RIDDOR are also assessed in this K&U unit.

For more details, see Part 8.

9. Operation and positioning of emergency isolation controls and valves

16.18   You will be assessed on your K&U of the correct positioning, labelling and identification of ECV/AECV/MIV. Including those positioned inside buildings, outside buildings and in multi-occupancy buildings.

For more details, see Part 4.

10. Checking and setting appliance burner pressures and gas rates

16.19   During this assessment, you will be required to safely check and record the operating pressure (burner pressure) of an appliance and then check and record its gas rate using the gas meter test dial or index. You must be able to correctly interpret your results against the appliance manufacturer's instructions. Within this section is the need to explain the requirements of a range-rated appliance. You will also need to know how to check and record the gas rate where a smart meter is installed.

For more details, see Part 2.

11. Operation and checking of appliance gas safety devices and controls

16.20   Many assessment centres will use a control rig for this assessment with a number of controls fitted to a single gas supply. You will likely be more familiar with some controls than others, so take your time and try to visualise the type of appliance you would find each control connected to and its principle of operation.

You will be required to identify and check each control for correct operation, demonstrating diagnosis of any faults that you encounter. Where necessary, be sure to isolate the gas and electricity supply before attempting to make any repairs, ensuring that the installation is gas-tight before retesting the controls.

Finally, you will need to explain the safe operation of each control/device to your assessor. Manufacturers' instructions should be available.

For more details, see Part 3.

12. Chimney Standards

16.21   Your K&U of a wide range of chimney/flue standards will be thoroughly assessed in this section of CCN1. The content will be largely based on relevant British Standards, Gas Regulations and manufacturers' instructions and is theory-based.

Your K&U of the following chimney/flue systems will be assessed:

- existing solid fuel chimneys;
- pre-cast flues;
- chimneys for single open-flue natural draught appliances;
- shared open-flue chimneys for natural draught appliances;
- fan draught chimneys for open-flue appliances;
- balanced compartments for open-flue appliances;
- room-sealed natural draught chimney configurations;
- room-sealed fanned draught chimney configurations;
- room-sealed appliances for shared chimneys (SE-ducts, U-ducts and CFS);
- condensing flues;
- chimneys for vertex appliances; and
- exchange information and planning requirements for chimneys.

An industry document that some domestic gas operatives may be less familiar with is IGEM/UP/17 *Shared chimney and flue systems for domestic gas appliances*. Not all gas operatives will encounter these types of flue systems, but you will be required to recognise types of shared or communal flue systems and explain their operation.

Some operatives may find the chimney/flue sections of the CCN1 assessment challenging as they seldom, if ever, encounter all types of chimney/flues included in the assessment. But do remember that you may, and it could be critical that you can identify any unsafe situations on all types of domestic chimney/flue systems. You may not encounter them regularly now, but your ACS certificate is valid for five years, and who knows where your career will take you during that time period.

For more details, see Part 6.

13. Chimney inspection and testing

16.22   After demonstrating your K&U of chimney standards, you will be required to perform a number of practical tasks. This will include a thorough visual inspection of a number of both correct and incorrect installations to verify that they meet current standards and industry best practices. The flue survey detailed in Part 8: Figure 8.2 will be a useful reference for this part of the CCN1 assessment.

The practical tasks in this section will also include a flue flow test on an open-flue/chimney, a spillage test with appliance connected and in operation, and case seal checks on a room-sealed fan-assisted positive pressure appliance.

For the K&U section of this unit, you will need to explain alternative methods of compliance when inspection hatches are not available for existing flues in voids (Gas Safe TB 008). You will also need to explain what actions to take where fumes, smells or spillage have been reported/encountered (BS 7967).

For more details, see Parts 6 and 8.

14. Installation of open, balanced and fan-assisted chimneys

16.23   You may find that the practical tasks for this unit are of a similar format to *13. Chimney inspection and testing* as you will be required to identify correct and incorrect installations of open-flue chimney installations, balanced and fan-assisted chimney systems. Be prepared to refer to relevant British Standards, manufacturers' instructions and the gas industry's unsafe situations procedure (IGEM/G/11).

For more details, see Parts 6 and 8.

15. Re-establish existing gas supply and re-light appliances

16.24   The criteria for this unit are based on Regulation 33 of the Gas Regulations – *Testing of appliances*. You will be required to confirm the installation is gas-tight, visually inspect the appliance and re-light, including – satisfactory operation of user controls and checking for any unsafe situations. You should also be in a position to recognise the correct cooker hose for use in a multi-occupancy building and describe the action to take when an un-commissioned appliance is identified or if pipework and appliances are not commissioned when the gas supply is re-established.

For more details, see Parts 8 and 9: Section 9.51.

Glossary

ADVENTITIOUS AIR This is the air that enters a room through natural means such as through the cracks around window frames and door openings.

AIR/GAS RATIO The proportion of gas to air required in order to achieve a combustible mixture.

ANACONDA The name given to the flexible stainless steel pipe between the emergency control valve (ECV) and the gas meter regulator.

ATMOSPHERIC SENSING DEVICE (ASD) A device designed to sense possible spillage from an appliance or the lack of oxygen within a room that is available to burn safely in an appliance.

AUTOMATIC CHANGEOVER VALVE A special LPG regulator that automatically takes gas from another gas cylinder or bank of LPG cylinders when the supply has run out.

BACK BOILER A boiler sometimes found located behind a gas fire within a domestic dwelling.

BADGE RATING The manufacturer's specification ratings applicable to a specific appliance to include the gas type, heat input and pressure with which the appliance is designed to operate.

BALANCED COMPARTMENT A situation where an open-flued appliance has been installed within a small room/cupboard, with suitable door seals, etc. with the intention to convert the appliance to be room-sealed to habitable areas.

BALANCED FLUE The name given to a design of gas appliance that takes its air supply at a point adjacent to the position where the flue gas is discharged, this being outside the building.

BAYONET FITTING A special valve designed to automatically close when a hose connected to it has been withdrawn.

BI-METALLIC STRIP A special strip of two metals that have different expansion rates bonded together. Its intention is to bend as it warms up and in so doing makes/breaks electrical contacts or opens/closes a gas line.

BLACK BULB SENSOR A special thermostat that is used to more accurately sense the building temperature when an overhead radiant heating system is used. When the desired temperature is reached, the sensor reacts to turn the appliance off.

BLACK DUST Copper sulphide in the form of a fine black film that sometimes accumulates inside gas pipework.

Gas Installation Technology, Third Edition. Andrew S. Burcham, Stephen J. Denney and Roy D. Treloar.
© 2024 John Wiley & Sons Ltd. Published 2024 by John Wiley & Sons Ltd.

BOILER The name given to the appliance where water is heated, although rarely to boiling temperature.

BONDING The green and yellow cables connected to pipework with the intention of ensuring all the pipes are at the same electrical potential.

BOOSTER A device that passes gas through a set of rotating fins with the intention to increase its pressure.

BOURDON GAUGE A flattened tube, in the shape of a horseshoe, with a central cog that turns as gas or water enters the tube as it tries to straighten it out, and in so doing indicates the pressure within on a dial face.

BRANCH FLUE SYSTEM The name given to an open-flue system found in blocks of flats that takes the flue products from several appliances located over several floors.

BULK TANK The name given to an enclosed tank used to store a supply of LPG.

BURNER PRESSURE The gas pressure, as recommended by the appliance manufacturer, on the burner side of the appliance regulator.

CALIBRATION CERTIFICATE A certificate issued to confirm that a test instrument has been checked for accuracy.

CALORIFIC VALUE The amount of energy that is contained within a known quantity of fuel.

CANOPY A capture hood installed above gas appliances that is designed to receive rising vapour, fumes, etc. and safely remove to outside air by means of an extraction system.

CARBON DIOXIDE (CO_2) A gas consisting of one carbon atom double bonded to two oxygen atoms. It is one of the most important heat-trapping, or greenhouse, gases. It is abundant in the atmosphere and is produced by burning carbon-based fossil fuels such as gas, coal and oil. It is also produced through natural processes such as wildfires and volcanic eruptions.

CARBON MONOXIDE (CO) A gas consisting of one carbon and one oxygen atom. It is highly toxic and is produced as the result of incomplete combustion of a fuel.

CATALYTIC CONVERTER A device that converts poisonous gases such as CARBON MONOXIDE and aldehydes, produced as the result of incomplete combustion, into CARBON DIOXIDE.

CATCHMENT SPACE (DEBRIS COLLECTION SPACE) The location or void at the base of a chimney with the intention to collect any items that may fall down the flue system, possibly causing a blockage where the appliance connection has been made.

CHIMNEY The name given to the structure or wall of the material enclosing the flue, to include masonry, metal or plastic pipe materials.

CLOSED-FLUE SYSTEM An OPEN-FLUED APPLIANCE that does not have a draught diverter. These are sometimes given the flue/chimney classification of B2.

CLOSURE PLATE A piece of sheet metal secured to a fireplace opening, with appropriate holes in, through which to pass the flue of a gas fire. It is designed to ensure the correct flue draught, assist in preventing products passing back into the room and to provide access to inspect and clean the catchment space.

CO/CO_2 RATIO A comparison between the CARBON DIOXIDE and CARBON MONOXIDE content in a sample of flue gases. Maximum combustion performance levels for different appliances can be found in Part 8: Table 8.2.

COMBINATION BOILER A design of a boiler that heats both central heating and domestic hot water on demand, unlike a regular boiler, which uses a separate domestic hot water storage cylinder.

COMMISSIONING The term used to mean the appliance has been tested and set up as intended by the manufacturer.

COMMON FLUE The name of a flue system that receives the pipes from several appliances.

COMMUNAL FLUE SYSTEM (CFS) A modern shared flue system that allows multiple boilers to be connected to the same common flue. They provide a common air supply and flue outlet to suitable fan-assisted boilers. Typically found in multiple-storey apartment buildings.

CONDENSATE The water that has condensed from a vapour to its liquid form. Water vapour within flue gases typically condenses when it touches a surface at around 55°C in temperature.

CONDENSATE PIPE A pipe designed to remove the condensate water to a suitable drainage disposal point from condensing boilers and large open exposed flue systems.

CONDENSING BOILER A highly efficient boiler that removes as much of the latent heat from the flue gases as possible. The appliance removes so much heat from the products of combustion that the water produced as a result of the combustion process condenses within the appliance itself.

DEBRIS COLLECTION SPACE See CATCHMENT SPACE.

DECORATIVE FUEL EFFECT (DFE) FIRE A gas fire designed to burn and look like a natural solid fuel-burning appliance.

DILUTION AIR Air that is drawn into a flue system with the intention of reducing the CO_2 content. Dilution air is drawn naturally into draught diverters by the negative pressure caused as the gas products pass up the flue. This is achieved mechanically in FAN DILUTION systems.

DIRECT FIRED APPLIANCE A flueless warm air heating system that directly warms the air as it is drawn through the combustion chamber and then blown into and around a building, unlike a system that employs a heat exchanger and flue system to indirectly warm the airflow.

DIRECT PURGE Putting natural gas, or air as the case may be when decommissioning, directly into a system of installation pipework. See also; INDIRECT PURGE.

DISC VALVE A valve that operates by simply bringing a disc, usually with a washer attached, onto a seating through which the gas flows, such as a SOLENOID VALVE.

DISPLACEMENT GAS METER A gas meter that measures the quantity of gas used as it passes through a series of chambers of a known size, unlike inferential gas meters, where gas speeds and pressures are recorded.

DOUBLE-WALLED FLUE PIPE See TWIN- (DOUBLE-) WALLED FLUE PIPE.

DOWN DRAUGHT A situation where the flow of air or combustion products is forced to blow down through an open-flued chimney. The down draught is caused by the terminal outlet being affected by positive wind conditions outside the building.

DRAUGHT BREAK An opening into an open-flue chimney system, such as a DRAUGHT DIVERTER or stabiliser, installed to alleviate pressure fluctuations within the flue system.

DRAUGHT DIVERTER A DRAUGHT BREAK through which excessive up or down draughts can be eliminated from affecting the flames within the combustion chamber. During a DOWN DRAUGHT situation products are dispelled into the room, whereas during an excessive updraught additional air is pulled into the opening in the flue system.

DRAUGHT STABILISATION DOOR A DRAUGHT BREAK in which a weighted hinged door is used instead of the opening as found with a DRAUGHT DIVERTER, designed to open or close where there are positive or negative pressure fluctuations within the chimney system.

DROP-FAN VALVE A quarter-turn gas control valve in which the handle, once in the closed position, can be laid down to lock it in the off position.

DROP VENT The name is sometimes used to identify the gas dispersal hole found in the base of a cupboard in a leisure accommodation vehicle designed to prevent an escape of LPG accumulating.

EFFLUX VELOCITY The speed of the products of combustion passing up through a chimney system.

ELECTRO-FUSION WELDING A method of joining a plastic pipe that relies on heat being used to melt the surface of the plastic to form a solid bond. The heat is generated by passing a small current through an element located within the plastic of the fitting.

ELECTRO-HYDRAULIC VALVE An electrically operated gas control valve that opens by oil being pumped from one chamber to another.

EMERGENCY CONTROL VALVE (ECV) The first gas control valve to be found as gas enters the building and used to close off the gas supply to the property.

EXCLUSION ZONE A perimeter distance that is set up, preventing entry to everyone, when a commercial gas installation is being strength tested in excess of 1 bar pressure.

EXPLOSIVE LIMITS See FLAMMABILITY LIMITS.

FAMILY OF GASES The characteristics of a gas identified by its Wobbe number. Natural gas falls into the second family of gas, whereas LPG falls into the third family. The old manufactured town gas was a first family gas.

FAN DILUTION A chimney system that draws a quantity of fresh air into the flue system for the purpose of diluting the products of combustion, in particular the CARBON DIOXIDE.

FAN-FLUED APPLIANCE A gas appliance that incorporates a fan within its flue system to assist the removal of the combustion products. The system may be open-flued or room-sealed.

FILAMENT IGNITOR A coil of high-resistance metal that glows red hot when a current is passed through it for the purpose of igniting a gas appliance.

FIRE STOP The sealing of the void through which a pipe or similar object passes through a wall or ceiling. The space is sealed with the intention to slow down and prevent the spread of flames, heat and smoke within a building.

FIREBOX A name sometimes used to represent the location where the combustion takes place within a gas appliance, in particular a gas fire.

FIRST-STAGE REGULATOR An LPG regulator designed to receive the gas directly from a bulk tank or cylinder and reduce it down to an intermediate pressure of 0.75 – 1.5

bar prior to its being further reduced in pressure for use. The purpose of the first-stage regulator is to avoid a pressure drop in the pipework from the bulk tank to the building.

FINAL-STAGE REGULATOR See SECOND-STAGE REGULATOR.

FLAMBEAUX A decorative gas-burning torch.

FLAME CHILLING The cooling of a gas flame below its ignition temperature, causing it to be extinguished. One cause of flame chilling is where a flame is allowed to touch the surface of a heat exchanger.

FLAME FRONT The point where the unburnt gases finish, and combustion begins.

FLAME LIFT A term used to indicate that a flame has lifted off from the surface of a burner.

FLAME RECTIFICATION DEVICE A FLAME SUPERVISION DEVICE that works by detecting if a flame is present or not by passing a small current of electricity into the area where the combustion is to take place. If a flame is present, the electricity will flow; if there is no flame present, no electrical current will be sensed, and the gas flow will be closed down.

FLAME SPEED The speed at which the flame will pass through a gas/air mixture.

FLAME SUPERVISION DEVICE (FSD) A gas control valve incorporated in the gas supply to a burner installed with the intention to close off the supply of gas if no flame is detected. Also referred to as a flame failure device (FFD).

FLAMMABILITY LIMITS (EXPLOSIVE LIMITS) The minimum and maximum percentage of gas that needs to be present in a quantity of air for combustion to occur.

FLARE STACK A pipe that terminates with a burner and has an inline flame arrestor. It is used where purging of commercial gas installations is required.

FLOW CUP A device that is used to determine the rate of water flow from a tap.

FLOW SWITCH A paddle-type device located in a gas or water stream to sense whether or not the medium is flowing. As the paddle moves with the flow, it makes the electrical switch contacts, which in turn supplies electricity to a component.

FLUE The passage through which the combustion products pass.

FLUE BLOCK A collection of pre-cast concrete blocks designed for the removal of combustion products from domestic gas appliances. A flue block system is designed to be constructed within the wall of a building, therefore saving space.

FLUE BOX A metal box for the location of a gas fire, for use where there is no existing brick chimney or the existing chimney is unsuitable.

FLUE FLOW TEST (PULL TEST) The testing of an open-flue system for a suitable updraught.

FLUE GAS ANALYSIS The sampling of the products of combustion from an appliance with the intention to determine the appliance efficiency and its safe operation.

FLUE LINER A material used to line a flue. Brick chimneys are constructed using either concrete or clay liners, whereas existing brick chimneys are lined with materials such as flexible stainless steel.

FLUE PIPE The pipe used to transfer the flue products to the external environment.

FLUELESS APPLIANCE A gas appliance that discharges its combustion products into the room where it is installed.

FORCED DRAUGHT BURNER A gas burner that draws the air to be used for combustion into the appliance by the use of a fan and in turn blows the gas/air mixture into the combustion chamber.

GAUGE READABLE MOVEMENT (GRM) The lowest change in pressure that it is deemed possible to read on a gauge. This will vary depending on the accuracy of the gauge in use.

GOVERNOR See REGULATOR.

GROSS HEAT INPUT The gross heat input refers to all the gas consumption/energy that has been put into an appliance ignoring any energy/heat that has been lost to the surrounding areas or that discharged out through the flue.

HEARTH The non-combustible area located below and to the front of a gas-burning appliance, such as a fire.

HEARTH PLATE A notice plate located in a building, at one of several possible locations, such as at the opening into a flue system/fireplace or possibly at the gas meter, to identify the type of chimney system that has been installed and what it is suitable for.

HEAT EXCHANGER The location within an appliance where the heat is transferred from one source to another. For example, from the hot burning gases to the water or from the hot primary water from a boiler to the water inside a domestic hot water cylinder.

HEAT INPUT See GROSS HEAT INPUT or NET HEAT INPUT.

HIGHER FLAMMABLE LIMIT (HIGHER EXPLOSIVE LIMIT) The upper end of the percentage limits for combustion to occur. See FLAMMABILITY LIMITS.

HOB The open burner often located on top of an oven, used for boiling or frying, etc.

HYDROCARBON The term hydrocarbon refers to a mixture of carbon and hydrogen atoms. Paper and wood are hydrocarbons, as is gas. For example, methane is CH_4, consisting of one carbon atom and four hydrogen atoms.

IMPINGEMENT A term often used to indicate that the flame has touched the heat exchanger within an appliance. This invariably causes FLAME CHILLING and the production of soot.

INCOMPLETE COMBUSTION A term used to indicate that the products of combustion have not been completely consumed. Incomplete combustion can be caused by several means to include FLAME CHILLING and the starvation of sufficient oxygen to completely consume the fuel.

INDIRECT PURGE Filling the system with nitrogen to remove gas or air. See also DIRECT PURGE.

INDUCED DRAUGHT The through draught that occurs within a gas-burning appliance in order to assist in the combustion process and to expel the combustion products to the outside environment. The induced draught is created by the fan located on the outlet side of the appliance.

INDUSTRY UNSAFE SITUATIONS PROCEDURE This relates to the categorising of a situation where a gas appliance or installation may be At Risk (AR) of danger or be Immediately Dangerous (ID), for which, in both cases, action needs to be taken to make the installation safe.

INFERENTIAL GAS METER See details under DISPLACEMENT GAS METER.

INJECTOR The small hole through which the gas exits the pipe as it enters the burner.

INSET FIRE A gas fire that is set into the builder's opening.

INSTANTANEOUS WATER HEATER A water heater that heats the water only as and when it is required.

INTERMEDIATE PRESSURE The pressure within an LPG installation between the first- and second-stage regulator where the pressure has been reduced down to 0.75 – 1.5 bar.

INTUMESCENT VENT A special ventilation grille that is designed to close in the event of a fire, thereby preventing the spread of smoke or the supply of air to feed a fire.

J GAUGE A special manometer in which the liquid is enclosed within an enclosed tube, so as to prevent the liquid falling out if the manometer is laid down. The J gauge reads the pressure in one leg only as opposed to the two legs found in a U GAUGE.

LEISURE ACCOMMODATION VEHICLE (LAV) A touring caravan, self-propelled motor caravan, or a caravan holiday home which is designed for holiday accommodation. The caravan holiday home can come in a variety of designs from the traditional caravan to a log-cabin style and will have a chassis and wheels to manoeuvre them into place.

LINT The term given to represent fine fibres from fabric, dust particles, fluff and animal hairs, etc.

LINT ARRESTOR The small filter found on the air inlet supply to an atmospheric burner.

LOCK-UP PRESSURE The pressure at which a regulator will close to prevent over-pressurisation of a system.

LOW-PRESSURE CUT-OFF (LPCO) A device fitted in the gas supply to detect low pressure on its outlet side. Pressure cannot progress beyond the valve until a set minimum pressure has been achieved.

LOWER FLAMMABLE LIMIT (LOWER EXPLOSIVE LIMIT) The lower end of the percentage limit for combustion to occur. See FLAMMABILITY LIMITS.

LPG Abbreviation for liquefied petroleum gas.

LUMINOUS FLAME See POST AERATED FLAME.

MAGNETIC VALVE A SOLENOID VALVE.

MAKE-UP AIR Air that is introduced to a space to replace air that has been lost due to extraction.

MANOMETER A gauge for detecting the small pressures (in millibars) found inside gas installations. These can be fluid-filled as with a J GAUGE and U GAUGE, or electronic devices.

MAXIMUM INCIDENTAL PRESSURE (MIP) The potential pressure that could enter a gas installation under fault conditions, such as where a regulator fails to close off the supply.

MEDIUM PRESSURE The pressures found within a natural gas installation ranging from >75 mbar to ≤2 bar.

MENISCUS The curvature on the surface of the water as seen inside the tube of a U GAUGE.

MICROPOINT A bayonet-type gas connection used on leisure equipment, such as a barbecue.

MODULAR BOILER Two to six boilers installed together and connected by a series of headers to provide a total heat output, depending on the number of boilers in use.

MODULAR FLUE A range of boilers that share the same flue system.

MULTI-FUNCTIONAL GAS VALVE A special gas control valve that has many functions to include a filter, FSD, regulator, solenoid valve/s and test points, etc. It was designed to save room where so many gas controls are needed in the gas line to an appliance.

NATURAL GAS The naturally occurring gas found trapped below ground. It consists primarily of methane gas.

NEEDLE VALVE A small valve that opens and closes by allowing a pointed tapered head to meet against the seating of a similar shaped orifice.

NET HEAT INPUT The net heat input refers to the gas consumption/energy that has been put into an appliance with any energy/heat that has been lost to the surrounding areas or that discharged out through the flue subtracted from the total. The net heat input is basically what you get, ignoring any losses.

NET ZERO The term used to describe the balance between the amount of greenhouse gases (carbon dioxide, methane, etc.) emitted into the atmosphere and the amount that is removed. Net zero is achieved when the amount of gases emitted is no more than the amount removed. It can be achieved through reducing emissions and/or removing emissions from the atmosphere.

NON-RETURN VALVE A valve that will allow gas or liquid to flow in one direction only.

NOTICE PLATE See HEARTH PLATE.

OPEN-FLUED APPLIANCE A fuel-burning appliance that takes the air, as required for combustion, from the room in which the appliance is situated and discharges the products of combustion to the outside environment by the use of a chimney system.

OPERATING PRESSURE The pressure recorded when gas is flowing through the system.

OVERPRESSURE SHUT-OFF (OPSO) A valve designed to automatically close off the gas supply where a pressure greater than that designed for the system to operate safely enters the valve.

OXYGEN DEPLETION SYSTEM A device designed to close down an appliance where it senses that there is insufficient oxygen in the room to support complete combustion of the fuel.

PERCEPTIBLE MOVEMENT The lowest possible movement that it is deemed possible for the human eye to perceive on a water gauge.

PHOTO-ELECTRIC DEVICE An ultraviolet light-sensitive control that closes off the gas supply to a burner where no flame is detected. Flames emit ultraviolet light.

PIGTAIL The name given to the LPG hoses that connect the gas cylinders to the house/accommodation supply pipework.

PILOT FLAME A small, established flame sometimes used within a gas-burning appliance, designed to ignite the gas flow through the main burners.

PIPE SLEEVE A pipe, located within a wall, through which a smaller pipe passes, thereby allowing for movement and preventing corrosion between the smaller pipe and the wall.

PLAQUE HEATER A radiant room heater that emits heat from a hot glowing surface.

PLUG VALVE A quarter-turn valve consisting of a tapered plug with a hole in. As the hole in the plug is turned through 90° the liquid or gas will flow.

PLUMING The appearance of mist or clouds coming from a flue terminal. It is a situation where the water contents resulting from the combustion process are saturated within the flue gas products and are close to their dew point. Pluming is typically found where there is a low flue gas temperature.

POSITIVE DISPLACEMENT METER The typical 'U' type gas meter consisting of four chambers through which the gas flows and in so doing causes a dial to turn to record the total volume of gas used. See also DISPLACEMENT GAS METER.

POST AERATED FLAME (LUMINOUS FLAME) A flame that has no 'pre-mix' air added to the gas prior to the combustion of the fuel. These flames are very luminous and yellow in appearance.

PRE-MIX BURNER A gas burner that utilises a fan to pull in air where it is pre-mixed with the fuel gas prior to being blown into the combustion chamber, where the mixture is to be consumed.

PRESSURE ABSORPTION (PRESSURE DROP) A term used to indicate that a loss of pressure has been experienced due to frictional losses as the gas flows through the system, meter, pipe and fittings.

PRESSURE RELIEF VALVE A special valve designed to open and release any excess pressure that has built up within a system or vessel. The excess pressure is generally expelled safely to the outside environment.

PRESSURE STAT A thermostat found on pressure/steam boilers. It is designed to close down the supply of gas as necessary when a predetermined pressure has been achieved.

PRESSURE SWITCH A control device consisting of a diaphragm washer with a small electrical contact switch on one side, the switch contacts being made to close as the diaphragm flexes. This diaphragm moves in response to a negative or positive pressure detected on one side of the diaphragm.

PRIMARY AIR The air supply introduced to mix with gas prior to the combustion of the fuel.

PRIMARY FLUE The section of an open flue that connects to the appliance prior to any DRAUGHT BREAK.

PROVING DEVICE A pressure-sensing device that detects the flow of air through an appliance or flue duct system prior to allowing gas to flow to a burner.

PTFE The abbreviation for 'polytetrafluoroethylene'; it is a white plastic tape or thread used in making joints to threaded pipework.

PULL TEST See FLUE FLOW TEST.

PURGE POINT The 'capped off' gas connection point used to connect a hose and PURGE STACK, if necessary, to enable gas to be discharged from a gas installation when purging the gas into or out from a system of pipework and fittings.

PURGE STACK An arrangement of pipework, hose and fittings used in the process of purging industrial and commercial gas installations.

PURGING Purging basically refers to putting gas into pipes when commissioning a system or putting air into the pipes when de-commissioning.

RADIANT TUBE HEATER A tube, mounted at a high level, that has hot gases passed through it in order to emit the heat generated down into the room, thereby warming the occupants.

RADON GAS A gas that sometimes seeps from the ground in areas where uranium or radium is present. The gas is colourless, odourless and radioactive.

REGISTER PLATE A piece of thin sheet metal through which an open-flue pipe passes as it enters a brick chimney.

REGULAR BOILER A boiler that heats water that is not drawn off through taps. Typically it is a boiler used in conjunction with a hot water storage cylinder, i.e. it is not a COMBINATION BOILER.

REGULATOR (GOVERNOR) A device designed to reduce the pressure within a gas pipe to that required.

RELATIVE DENSITY See SPECIFIC GRAVITY.

RELAY VALVE In electricity, a relay valve is an electrical switch that operates another switch or series of switches remotely. As a gas control, a relay valve is a valve designed to control the flow of gas to a burner. It is used in conjunction with a weep pipe and by-pass. If the weep line is open gas will flow to the burner, if it is closed the burner will go out or drop to a predetermined simmer flame. These types of gas valve are now very unlikely to be encountered.

RESIDENTIAL PARK HOME (RPH) A detached bungalow-style home that is manufactured off site, often in two halves, and delivered to site where it is assembled.

RESPONSIBLE PERSON This can be the occupier/owner of the premises or any person with authority at a point in time who can take appropriate action in relation to the property or a gas fitting within.

RETENTION FLAME A small stable gas flame located next to the main burner port, it is designed to overcome the problem of the main flame lifting off from the base of the burner.

RIBBON BURNER A small burner consisting of a series of flat corrugated strips of metal fixed together through which the gas flows.

RIDGE TERMINAL An open-flue terminal that is located at the apex or ridge of a roof.

ROOM-SEALED APPLIANCE A gas appliance that has its combustion system sealed from the room in which it is located. The air supply for combustion is either taken directly from outside air or an unoccupied area of the building and the combustion products are discharged directly to open air outside the building.

SAFETY CUT-OFF DEVICE A device designed to cut off the flow of gas to an appliance. One such device is that found located on the gas supply to a gas cooker with a drop-down lid.

SAFETY INTERLOCK A valve that requires the operation of another such valve to be used in conjunction with it. For example, you would need to turn on one control valve before a second valve is permitted to turn or as one valve turns, it automatically turns a second control valve.

SECONDARY FLUE That part of an open-flued system that takes the flue pipe to the terminal.

SECONDARY METER An additional gas meter installed for the purpose of separate billing or used to check the amount of gas used for a specific purpose.

SECOND-STAGE REGULATOR An LPG regulator that reduces the gas pressure to the design operating pressure of the building, usually 37 mbar.

SECTIONAL BOILER A boiler that is made up on site consisting of several sections.

SE-DUCT An older design of open-flue duct used in high-rise buildings for the purpose of connecting several domestic room-sealed gas appliances to a common flue.

SEMI-RIGID COUPLING A flexible disconnecting joint used on larger gas pipelines.

SHELL BOILER A commercial gas boiler consisting of a series of fire tubes that are surrounded by water. It is through these tubes that the flue products pass and in so doing heat the water.

SINGLE POINT A water heater located at the point of use, heating the water for washing purposes when required.

SINGLE-STAGE REGULATOR An LPG regulator that reduces the high gas pressure contained within the storage cylinder (typically between 2 and 7 bar) down to that required for use (28 – 37 mbar) in one single operation.

SLOW IGNITION DEVICE A device found in older instantaneous water heaters that slows down the speed at which the water diaphragm lifts and in so doing prevents the burner lighting with explosive force.

SMOKE MATCH A small match that produces a quantity of smoke which is held at the draught diverter of an open-flue appliance during a spillage test.

SMOKE PELLET A pellet designed to produce a large volume of smoke. Its purpose is to determine the effectiveness and integrity of an open-flue system.

SOLENOID VALVE A valve that opens by the passing of an electrical current through a coil of wire. When a current flows through a coil of wire, it becomes magnetised, so as the current flows the iron armature of the valve is pulled into the coil.

SOUNDNESS TEST The name sometimes used to refer to the test undertaken to check the integrity of a pipe or flue system.

SPACE HEATER A name used to describe a heater that is designed to warm up a room, i.e. the space. This would typically include a gas fire, but may also in fact refer to a boiler.

SPECIFIC GRAVITY (RELATIVE DENSITY) A number that is given to a solid or gas to identify its weight when compared to water or air. Those with a number smaller than 1 will sink; conversely, those with a number greater than 1 will float. The number given to a particular gas is compared to air at 1 and for all solids the number given is compared to water at 1. Natural gas, at 0.58, will rise upwards.

SPILLAGE The unwanted flow of combustion products into a room discharging from the appliance draught diverter or flue system.

SPILLAGE TEST A test undertaken using a SMOKE MATCH at the DRAUGHT DIVERTER on an OPEN-FLUED APPLIANCE to check that the combustion products are being conveyed up into the flue system and not discharging into the room out from the draught diverter.

STABILITY BRACKET The bracket that is located low down at the back of a freestanding cooker with the intention to ensure the cooker will not tilt forward should someone lean onto an open door. Sometimes a chain is used for this purpose.

STANDING PRESSURE The pressure within a gas installation when no gas is flowing.

SYSTEM BOILER A central heating boiler designed for a sealed heating system, including all the necessary components pre-installed, from the sealed expansion vessel to the pressure relief valve. System boilers utilise a separate hot water cylinder if it is required for domestic hot water purposes.

SYSTEM PRESSURE DROP The pressure absorption between two ends of a section of pipework due to frictional loss when the gas is flowing.

TEMPORARY BONDING WIRE A 10 mm^2 green and yellow wire that can be fixed to either side of a section of pipework or meter that has to be removed. It is designed to prevent sparks and electric currents flowing through unsuspecting operatives as the two sections are pulled apart, which is possible in a faulty electrical installation.

TERMINAL The point of discharge from a flued gas appliance.

THERMISTOR A special heat resistor, designed to sense temperature change, making/breaking the electrical circuit.

THERMOCOUPLE A component consisting of two different metals bonded together at one end, which when heated produces a small voltage to energise an electromagnet in a THERMOELECTRIC VALVE.

THERMOCOUPLE INTERRUPTER A special THERMOCOUPLE that has a break within the circuitry of wire to enable a branch to be taken off to a switch that could be operated to break the circuit and therefore close down the supply in the event of a fault occurring.

THERMO-ELECTRIC VALVE A FLAME SUPERVISION DEVICE that operates by the use of heat playing onto a THERMOCOUPLE.

TIGHTNESS TEST The name given to the test used on a gas installation to test its integrity, thereby ensuring that it is gas tight.

TRANSITIONAL FITTING A pipe fitting that converts one material to another, e.g. plastic to steel pipe.

TWIN- (DOUBLE-) WALLED FLUE PIPE A flue pipe consisting of one pipe within another pipe. The purpose of a twin-wall flue pipe is to assist keeping the internal flue pipe warm, this being the pipe through which the combustion products pass.

U GAUGE A manometer in which the liquid is enclosed within a U-shaped tube. The gauge reads the pressure in both legs as opposed to the one leg found in a J GAUGE.

ULLAGE SPACE A space above the liquid level within an LPG cylinder. The ullage space is where the gas is stored.

UNDER-PRESSURE SHUT-OFF (UPSO) A design of LOW-PRESSURE CUT-OFF.

VENTURI A device that creates a difference in pressure in a pipeline as water or air flows through it. This pressure differential is used to open and close gas valves or make electrical contacts make/break via a PRESSURE SWITCH.

VERTEX FLUE (SOLVER FLUE) The trade names given to two manufacturers' designs of room-sealed appliances in which the air required for combustion is taken from the roof void.

VITIATION This relates to the quantity of air within a room, the amount of oxygen that it contains and whether there is sufficient oxygen for complete combustion of the fuel. Vitiated air will lack sufficient oxygen for complete combustion.

VITIATION-SENSING DEVICE A device that senses a lack of oxygen at a burner to ensure complete combustion takes place, closing down the appliance if necessary. One such vitiation device includes a THERMISTOR located within a DRAUGHT DIVERTER.

WEEP PIPE A small pipe running from the top of a relay valve and passing into the combustion chamber. See RELAY VALVE.

WOBBE NUMBER A number determined by undertaking the following calculation:

Calorific value of a fuel gas $\div \sqrt{\text{specific gravity of a fuel gas}}$

It is used to place a gas into a specific FAMILY OF GASES.

ZERO GOVERNOR A gas control that opens as a result of a negative pressure acting below its diaphragm, sucking the gas from the pipe.

Appendix A

Abbreviations, Acronyms and Units

±	plus or minus
<	less than
≤	less than or equal to
>	greater than
≥	greater than or equal to
∴	therefore
Σ	sum total
Ω	ohm
∝	proportional
cm	centimetre
cm^2	square centimetre
ft^3	cubic foot
mbar	millibar
m	metre
m^2	square metre
m^3	cubic metre
mm	millimetre
mm^2	square millimetre
m/s	metre per second
m^3	cubic metre
m^3/h	cubic metres per hour
m^3/s	cubic metres per second
$m^3/s/m^2$	cubic metres per second per square metre
m^3/kW	cubic metres per kilowatt
AC	alternating current
ACOP	Approved Code of Practice
ACS	Nationally Accredited Certification Scheme for Individual Gas Fitting Operatives

Gas Installation Technology, Third Edition. Andrew S. Burcham, Stephen J. Denney and Roy D. Treloar.
© 2024 John Wiley & Sons Ltd. Published 2024 by John Wiley & Sons Ltd.

AECV	additional emergency control valve
AR	at risk
ASD	atmospheric sensing device
BS	British Standard
BS EN	British Standard Europaische Norm (European Standard)
CE	Conformité Européenne (European conformity)
CO	carbon monoxide
CO_2	carbon dioxide
CORGI	Council of Registered Gas Installers
COSHH	Control of Substances Hazardous to Health
COSSVM	carbon monoxide safety shut-off void monitoring system
CV	calorific value
DC	direct current
DFE	decorative fuel effect fire
dm	decimetres
ECV	emergency control valve
ESP	emergency service provider
EMF	electromotive force
FFD	flame failure device
FSD	flame supervision device
GIUSP	gas industry unsafe situations procedure
GM	gauge movement
GRM	gauge readable movement
GRP	glass reinforced plastic
GT	gas transporter
H_2O	water
HSE	Health and Safety Executive
Hz	Hertz
ID	immediately dangerous
IGEM	Institution of Gas Engineers and Managers
ILFE	inset live fuel effect fire
IV	installation volume
IV_f	installation volume (fittings)
IV_m	installation volume (meter)
IV_p	installation volume (pipe)
IV_t	installation volume (total)
LAV	leisure accommodation vehicle
LCD	liquid crystal display
LDF	leak detection fluid
LFL	lower flammable limit
LPCO	low-pressure cut-off

LPG	liquefied petroleum gas
MAM	meter asset manager (gas supplier)
MIP	maximum incidental pressure
MIV	meter inlet valve
MOP	maximum operating pressure
MPLR	maximum permitted leak rate
N_2	nitrogen
NO_x	nitrogen dioxide and nitric oxide mix (NO_2 NO)
O_2	oxygen
ODS	oxygen depletion system
OP	operating pressure
OPSO	overpressure shut-off
PCB	printed circuit board
PE	polyethylene
PME	protective multiple earthing (another name for TNCS)
ppm	parts per million
PTFE	polytetrafluoroethylene
PV	purge volume
PVC	polyvinyl chloride
RIDDOR	Reporting of Injuries, Diseases and Dangerous Occurrence Regulations
RPH	residential park home
SAP	standard assessment procedure
SEDBUK	Seasonal Efficiency of Domestic Gas Boilers in the UK
SG	specific gravity
STP	strength test pressure
TNCS	terra neutral combined separated (basically earth and neutral combined at supply but separated at the building); see also PME
TTD	tightness test duration
TTP	tightness test pressure
UFL	upper flammable limit
UKCA	United Kingdom Conformity Assessed
UPSO	under-pressure shut-off
UV	ultraviolet

Index

Page numbers in **bold** type refer to tables and page numbers in *italics* refer to illustrations.